Foundations of Data Science with Python

Foundations of Data Science with Python introduces readers to the fundamentals of data science, including data manipulation and visualization, probability, statistics, and dimensionality reduction. This book is targeted toward engineers and scientists, but it should be readily understandable to anyone who knows basic calculus and the essentials of computer programming. It uses a computational-first approach to data science: the reader will learn how to use Python and the associated data-science libraries to visualize, transform, and model data, as well as how to conduct statistical tests using real data sets. Rather than relying on obscure formulas that only apply to very specific statistical tests, this book teaches readers how to perform statistical tests via resampling; this is a simple and general approach to conducting statistical tests using simulations that draw samples from the data being analyzed. The statistical techniques and tools are explained and demonstrated using a diverse collection of data sets to conduct statistical tests related to contemporary topics, from the effects of socioeconomic factors on the spread of the COVID-19 virus to the impact of state laws on firearms mortality.

This book can be used as an undergraduate textbook for an Introduction to Data Science course or to provide a more contemporary approach in courses like Engineering Statistics. However, it is also intended to be accessible to practicing engineers and scientists who need to gain foundational knowledge of data science.

Key Features:

- Applies a modern, computational approach to working with data
- Uses real data sets to conduct statistical tests that address a diverse set of contemporary issues
- Teaches the fundamentals of some of the most important tools in the Python data-science stack
- Provides a basic, but rigorous, introduction to Probability and its application to Statistics
- Offers an accompanying website that provides a unique set of online, interactive tools to help the reader learn the material

John M. Shea, PhD is a Professor in the Department of Electrical and Computer Engineering at the University of Florida, where he has taught classes on stochastic methods, data science, and wireless communications for over 20 years. He earned his PhD in Electrical Engineering from Clemson University in 1998 and later received the Outstanding Young Alumni award from the Clemson College of Engineering and Science. Dr. Shea was co-leader of Team GatorWings, which won the Defense Advanced Research Project Agency's (DARPA's) Spectrum Collaboration Challenge (DARPA's fifth Grand Challenge) in 2019. He received the Lifetime Achievement Award for Technical Achievement from the IEEE Military Communications Conference (MILCOM) and is a two-time winner of the Ellersick Award from the IEEE Communications Society for the Best Paper in the Unclassified Program of MILCOM. He has been an editor for IEEE Transactions on Wireless Communications, IEEE Wireless Communications magazine, and IEEE Transactions on Vehicular Technology.

Chapman & Hall/CRC

The Python Series

About the Series

Python has been ranked as the most popular programming language, and it is widely used in education and industry. This book series will offer a wide range of books on Python for students and professionals. Titles in the series will help users learn the language at an introductory and advanced level, and explore its many applications in data science, AI, and machine learning. Series titles can also be supplemented with Jupyter notebooks.

Image Processing and Acquisition using Python, Second Edition
Ravishankar Chityala, Sridevi Pudipeddi

Python Packages
Tomas Beuzen and Tiffany-Anne Timbers

Statistics and Data Visualisation with Python
Jesús Rogel-Salazar

Introduction to Python for Humanists
William J.B. Mattingly

Python for Scientific Computation and Artificial Intelligence
Stephen Lynch

Learning Professional Python Volume 1: The Basics
Usharani Bhimavarapu and Jude D. Hemanth

Learning Professional Python Volume 2: Advanced
Usharani Bhimavarapu and Jude D. Hemanth

Learning Advanced Python from Open Source Projects
Rongpeng Li

Foundations of Data Science with Python
John Mark Shea

For more information about this series please visit: https://www.crcpress.com/Chapman--HallCRC/book-series/PYTH

Foundations of Data Science with Python

John M. Shea

CRC Press is an imprint of the
Taylor & Francis Group, an **informa** business
A CHAPMAN & HALL BOOK

Designed cover image: © Agnes Shea

First edition published 2024
by CRC Press
2385 NW Executive Center Drive, Suite 320, Boca Raton FL 33431

and by CRC Press
4 Park Square, Milton Park, Abingdon, Oxon, OX14 4RN

CRC Press is an imprint of Taylor & Francis Group, LLC

ISBN: 978-1-032-34674-8 (hbk)
ISBN: 978-1-032-35042-4 (pbk)
ISBN: 978-1-003-32499-7 (ebk)

DOI: 10.1201/9781003324997

Typeset in Latin Modern font
by KnowledgeWorks Global Ltd.

Publisher's note: This book has been prepared from camera-ready copy provided by the authors.

For Tucker, Charlotte, and Amelia—

proof that events with zero probability
(having three children that are this wonderful)
happen!

Contents

Acknowledgments xi

Preface xiii

1 Introduction 1
1.1 Who is this book for? 1
1.2 Why learn data science from this book? 1
1.3 What is data science? 2
1.4 What data science topics does this book cover? 5
1.5 What data science topics does this book **not** cover? 6
1.6 Extremely Brief Introduction to Jupyter and Python 6
1.7 Chapter Summary 16

2 First Simulations, Visualizations, and Statistical Tests 17
2.1 Motivating Problem: Is This Coin Fair? 17
2.2 First Computer Simulations 18
2.3 First Visualizations: Scatter Plots and Histograms 20
2.4 First Statistical Tests 27
2.5 Chapter Summary 31

3 First Visualizations and Statistical Tests with Real Data 33
3.1 Introduction to Pandas 34
3.2 Visualizing Multiple Data Sets – Part 1: Scatter Plots 39
3.3 Partitions 47
3.4 Summary Statistics 48
3.5 Visualizing Multiple Data Sets – Part 2: Histograms for Partitioned Data . 56
3.6 Null Hypothesis Testing with Real Data 61
3.7 A Quick Preview of Two-Dimensional Statistical Methods 73
3.8 Chapter Summary 76

4 Introduction to Probability 77
4.1 Outcomes, Sample Spaces, and Events 77
4.2 Relative Frequencies and Probabilities 78
4.3 Fair Experiments 83
4.4 Axiomatic Probability 86
4.5 Corollaries to the Axioms of Probability 94
4.6 Combinatorics 99
4.7 Chapter Summary 112

5 Null Hypothesis Tests **113**
5.1 Statistical Studies . . . 113
5.2 General Resampling Approaches for Null Hypothesis Significance Testing . 120
5.3 Calculating p-Values . . . 127
5.4 How to Sample from the Pooled Data . . . 132
5.5 Example Null Hypothesis Significance Tests . . . 138
5.6 Bootstrap Distribution and Confidence Intervals . . . 141
5.7 Types of Errors and Statistical Power . . . 147
5.8 Chapter Summary . . . 150

6 Conditional Probability, Dependence, and Independence **151**
6.1 Simulating and Counting Conditional Probabilities . . . 152
6.2 Conditional Probability: Notation and Intuition . . . 157
6.3 Formally Defining Conditional Probability . . . 159
6.4 Relating Conditional and Unconditional Probabilities . . . 162
6.5 More on Simulating Conditional Probabilities . . . 164
6.6 Statistical Independence . . . 166
6.7 Conditional Probabilities and Independence in Fair Experiments . . . 172
6.8 Conditioning and (In)dependence . . . 176
6.9 Chain Rules and Total Probability . . . 178
6.10 Chapter Summary . . . 186

7 Introduction to Bayesian Methods **187**
7.1 Bayes' Rule . . . 187
7.2 Bayes' Rule in Systems with Hidden State . . . 195
7.3 Optimal Decisions for Discrete Stochastic Systems . . . 197
7.4 Bayesian Hypothesis Testing . . . 206
7.5 Chapter Summary . . . 215

8 Random Variables **217**
8.1 Definition of a Real Random Variable . . . 217
8.2 Discrete Random Variables . . . 229
8.3 Cumulative Distribution Functions . . . 237
8.4 Important Discrete RVs . . . 247
8.5 Continuous Random Variables . . . 265
8.6 Important Continuous Random Variables . . . 279
8.7 Histograms of Continuous Random Variables and Kernel Density Estimation 299
8.8 Conditioning with Random Variables . . . 301
8.9 Chapter Summary . . . 305

9 Expected Value, Parameter Estimation, and Hypothesis Tests on Sample Means **306**
9.1 Expected Value . . . 306
9.2 Expected Value of a Continuous Random Variable with SymPy . . . 312
9.3 Moments . . . 315
9.4 Parameter Estimation . . . 326
9.5 Confidence Intervals for Estimates . . . 333
9.6 Testing a Difference of Means . . . 342
9.7 Sampling and Bootstrap Distributions of Parameters . . . 356
9.8 Effect Size, Power, and Sample Size Selection . . . 360
9.9 Chapter Summary . . . 362

10 Decision-Making with Observations from Continuous Distributions 364
10.1 Binary Decisions from Continuous Data: Non-Bayesian Approaches 364
10.2 Point Conditioning 375
10.3 Optimal Bayesian Decision-Making with Continuous Random Variables . . 380
10.4 Chapter Summary 386

11 Categorical Data, Tests for Dependence, and Goodness of Fit for Discrete Distributions 387
11.1 Tabulating Categorical Data and Creating a Test Statistic 388
11.2 Null Hypothesis Significance Testing for Dependence in Contingency Tables 394
11.3 Chi-Square Goodness-of-Fit Test 400
11.4 Chapter Summary 408

12 Multidimensional Data: Vector Moments and Linear Regression 409
12.1 Summary Statistics for Vector Data 410
12.2 Linear Regression 422
12.3 Null Hypothesis Tests for Correlation 435
12.4 Nonlinear Regression Tests 437
12.5 Chapter Summary 445

13 Working with Dependent Data in Multiple Dimensions 446
13.1 Jointly Distributed Pairs of Random Variables 446
13.2 Standardization and Linear Transformations 456
13.3 Decorrelating Random Vectors and Multi-Dimensional Data 467
13.4 Principal Components Analysis 475
13.5 Chapter Summary 484

Index 485

Acknowledgments

I want to thank the many people who helped make this book possible.

My love and thanks go to my wife, Jill, for twenty-five years of love and friendship; this book would not be possible without the wonderful family and home we have created together. I love that we can enjoy quiet times at home or adventures together all over the world. This book is better because of your careful editing.

To my children, Tucker, Charlotte, and Amelia: I treasure every moment with you, whether it is trading funny memes, solving the New York Times crossword puzzle, battling at Ping Pong, running our family 5Ks, trying to guess the winner of Survivor, reading together, or just working side-by-side on the couch.

I thank my parents, Larry and Agnes Shea, for their unwavering love and support. Mom, you are missed every day.

I have been inspired by too many dedicated engineering educators to list them all. I have to especially thank my Ph.D. advisor, Dr. Michael Pursley, for mentoring me on how to conduct research, how to teach, and how to write technically. Thank you to Dr. John Harris, who convinced me to develop the data-science course on which this book is based. Thank you to Dr. Catia Silva, who was my co-instructor for one of the semesters teaching the data-science course.

Several students provided feedback on the book, including Caleb Bowyer, Walter Acosta, Cortland Bailey, Brennan Borchert, Alexander Braun, Patrick Craig, Justin Nagovskiy, Allison Neil, Michael Russo, Dieter Steinhauser, Phillip Thompson, and Marisa Younger.

Preface

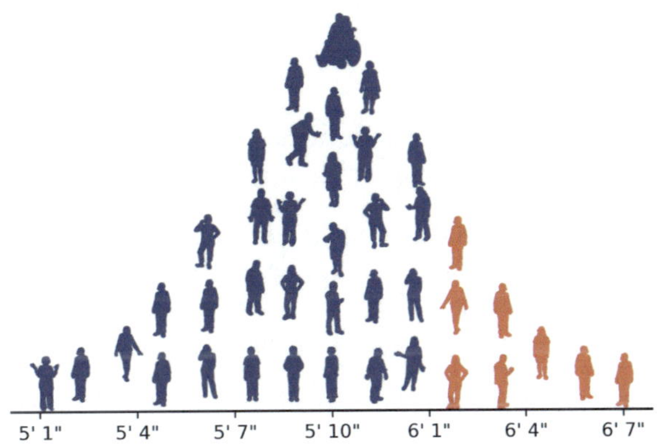

This book is an introduction to the foundations of data science, including loading and manipulating data, data visualization, statistics, probability, and dimensionality reduction. This book is targeted toward engineers and scientists, but it should be easily accessible to anyone who knows basic calculus and the essentials of computer programming. By leveraging this background knowledge, this book fits a unique niche in the books on data science and statistics:

- **This book applies a modern, computational approach to work with data.** In particular, it uses simulations (an approach called *resampling*) to answer statistical questions.
  - Many books on statistics (especially those for engineers) teach students how to answer statistical questions using only analytical approaches that many learners find difficult to understand. Most learners can easily understand how resampling works, in contrast to some arcane formula.
- **This book uses real data sets and addresses contemporary issues.**
  - Many statistics books use contrived examples that are small enough to print in a book and work with using a calculator, but this results in data sets that are unrealistic and uninteresting. The computational approach used in this book allows the use of data sets from across the web to conduct statistical tests on topics from the effects of socioeconomic factors on the spread of the COVID-19 virus to the impact of state firearm laws on firearm mortality.
- **This book provides a basic, but rigorous, introduction to probability and its application to statistics.**
  - Some of the other books that use the resampling approach to statistics omit the mathematical foundations because they are targeted toward a broader audience who may not have the rigorous mathematical background of engineers and scientists.

- **This book shows how to work with some of the most important tools in the Python data-science stack, including:**
  - NumPy for working with vectors and matrices, as well as many types of numerical functions,
  - SciPy for working with random distributions and conducting statistical tests,
  - Pandas for loading, manipulating, and summarizing data,
  - Matplotlib for plotting data, and
  - `scikit-learn` for accessing standard data sets and for advanced statistical processing.
- **This book was co-written with a book that covers linear algebra and its application to data science** using Python and NumPy.
  - Techniques like dimensionality reduction require linear algebra. Although linear algebra is not covered in this book, the companion book *Introduction to Linear Algebra for Data Science with Python* provides the necessary background knowledge with the same mix of analysis and Python implementation.
- **This book provides a unique set of online, interactive tools to help students learn the material**, including:
  - interactive self-assessment quizzes,
  - interactive flashcards to aid in learning terminology,
  - interactive Python widgets and animated plots.

Interactive elements are available on the book's web site: fdsp.net.

View the quiz and flaschards for this preface at

fdsp.net/intro,

which can also be accessed using this QR code:

Credits: The image at the top of this page is made using the Wee People font made by ProPublica: https://github.com/propublica/weepeople. Inspired by a Tweet by Matthew Kay: https://twitter.com/mjskay/status/1519156106588790786.

1

Introduction

Welcome to *Foundations of Data Science with Python*! This chapter provides an introduction to the book and its place in the field of data science. It then provides a brief introduction to some of the tools that are used throughout the book. By the end of this book, you will learn how to analyze and interpret data, formulate hypotheses about the data, perform statistical tests, and communicate your findings accurately and effectively.

1.1 Who is this book for?

This book is targeted toward engineers and scientists, whether working or still in school. Given this target audience, I assume that the reader has a basic working knowledge of:

- computer programming (knowing Python is helpful, but not required), and
- one-dimensional differential and integral calculus.

This book is written by an engineer with degrees in both electrical and computer engineering. This book and its companion, *Introduction to Linear Algebra for Data Science with Python*, were written to provide the main textbooks for a 4-credit, semester-long course for engineers, taught in the Department of Electrical and Computer Engineering at the University of Florida. These books are intended to be a broad introduction to data science, but they are also designed to replace courses in *Engineering Statistics* and *Computational Linear Algebra*.

1.2 Why learn data science from this book?

This book uses a **computational first** approach to data science. You will learn how to leverage the power of modern computers and scientific software to visualize, transform, and simulate data. For instance, one of the main approaches used in this book is to conduct statistical tests by carrying out simulations that draw samples from the data being analyzed. This approach has the following benefits:

- **We start working with real data sets quickly** because this approach does not require a lot of mathematical background and the computer does all the mathematical manipulation and plotting.
- **Simulation models are easy to create and understand.** The results do not rely on any arcane formulas but only the ability to build very simple simulations that draw from the experimental data.

DOI: 10.1201/9781003324997-1

- **This approach is more general than the traditional approach.** It does not rely on the data coming from specific probability distributions, and the same simulation can be used to generate frequentist or Bayesian statistics.

Interactive flashcards and self-assessment quizzes are provided throughout the book to help learners master the material and check their understanding. The entire set of interactive materials can be accessed on the book's website at fdsp.net.

The interactive materials use spaced repetition to help readers retain knowledge as they progress through the book. Starting with Chapter 2, the interactive chapter reviews also give a random subset of review problems from earlier chapters. Research shows that spaced repetition improves the retention of material.

Each chapter of the book ends with "take-aways" that help summarize the important points from the chapter and address issues that may be topics of questions in data science interviews.

1.3 What is data science?

Our world is filled with information. In fact, the amount of information we have access to can often be overwhelming. We start by considering when information becomes *data*:

> Definition
>
> **data**
>
> Collections of measurements, characteristics, or facts about a group.

Then a simple definition of data science is:

> Definition
>
> **data science**
>
> The process of extracting meaning from data.

Data consists of *data points*:

> Definition
>
> **data points**
>
> A collection of one or more pieces of information collected about a single individual or entity.

Each data point may contain *variables* and *features*:

Definitions

variables

Particular characteristics, measurements, or facts that make up a data point.

features

Individual pieces of information in a data set. While variables typically represent unprocessed or raw data, features can include both variables and processed versions of the variables.

In the machine-learning (ML) literature, the term *feature* is often used for both raw and processed data, especially if the data are used as the input for some ML process. Until Chapter 12, we will primarily work with existing data sets and refer to the pieces of information that make up the data points as variables. In Chapter 13, we will consider how to transform data to create new features.

Variables and features may be either *quantitative* or *qualitative*:

Definitions

quantitative data

Numeric data. Quantitative data may be either discrete (such as the number of people in a family) or continuous (such as grade point average).

qualitative data

Non-numeric data. Qualitative variables are generally non-numeric categories that data may belong to (such as hair color). Some categories may have an order associated with them, but the order does not imply a numeric nature to the categories. For example, a survey question may have responses from Strongly Disagree to Strongly Agree.

Examples of quantitative variables:

- height
- weight
- yearly income
- college GPA
- miles driven commuting to work
- temperature
- wind speed
- population

Examples of qualitative variables:

- hair color (blond, brown, black, red, gray, ...)

- current precipitation status (no precipitation, raining, snowing, sleeting, ...)
- car type (sedan, coupe, SUV, minivan, ...)
- categories of hurricanes or earthquakes
- model of smartphone used to access a mobile app
- blood type

As engineers and scientists, our goal is to make sense of the world and to use what we learn to take action. Data science applies computational tools and mathematical methods to process and transform data for the purpose of better understanding what the data can (and cannot!) tell us about the world.

Data scientists often start with a *research question*:

> Definition
>
> **research question**
>
> A question that can be answered using research, including data collection and analysis.

For instance:

- Does more education translate to more wealth?
- Do state gun laws affect firearms mortality?
- Is the climate changing?
- How fast was the COVID-19 coronavirus spreading when it first became prevalent in the United States in the Spring of 2020?

One of the goals of a data scientist is to take broad research questions and translate them into questions that can be answered using data. One set of criteria (from *Designing Clinical Research* by Hulley, Cummings, Browner, Grady, and Newman) has the acronym **FINER**, which stands for **Feasible**, **Impactful**, **Novel**, **Ethical**, and **Relevant**.

In this book, I have tried to address these issues by working with existing data sets to answer questions that are important and timely. For instance, here are some ways that the questions above may be reformulated so that they can be answered with data:

- Instead of "Does more education translate to more wealth?", a data scientist may ask a more specific question, such as "For people in the United States, does post-baccalaureate education increase median net family wealth?".
- Instead of "Do gun laws affect firearm mortality?", a data scientist may ask "Are state permitless-carry laws associated with a difference in average firearm mortality rates?".
- Instead of "Is the climate changing?", a data scientist may ask "Has the average annual temperature in Miami, FL increased over the past 40 years?".
- In assessing the rate of spread of the COVID-19 virus, the data scientist may ask "Was the number of cases growing exponentially in March of 2020? If so, what was the exponential growth rate?".

Terminology review

Interactive flashcards to review the terminology introduced in this section are available at fdsp.net/1-3, which can also be accessed using this QR code:

1.4 What data science topics does this book cover?

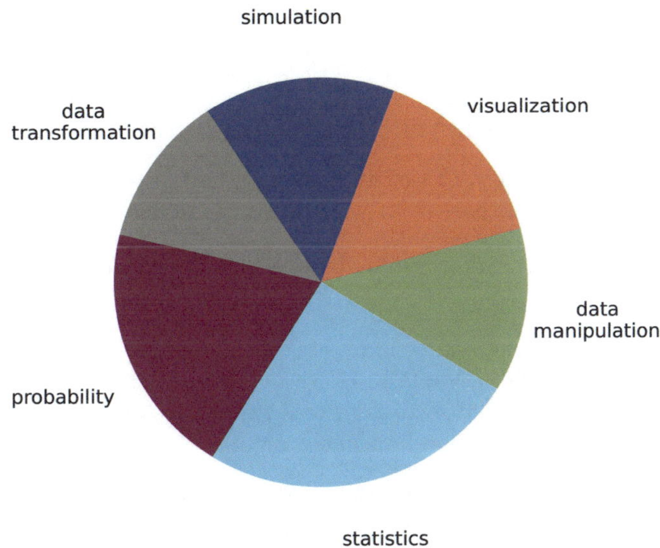

FIGURE 1.1
Data science topics covered in this book.

Data science is a very broad topic. Data scientists use many different tools to make sense of data, from databases to advanced ML algorithms. This book focuses on some of the fundamental tools used to create meaning from data, and I have tried to introduce skills and tools that will be useful to engineers and scientists in other contexts. The main topics covered in this book are shown in Fig. 1.1 and described below:

- **Simulation** is used to emulate random phenomena and to carry out statistical tests.
- **Visualization** is used to transform data into graphical forms that help reveal trends in the data or tell a story about the data.
- **Data manipulation** is the process by which data is loaded and prepared for analysis.
- **Statistics** is used to determine whether observations made from the data are meaningful or could just be attributed to randomness in the data.
- **Probability** is used to create mathematical models for random phenomena; these models can be used to develop optimal estimators and make optimal decisions.

- **Data transformation** consists of mathematical processing to achieve goals such as creating new features or reducing the size of the data.

Throughout this book, I use two key tools to enable this computational approach:

- **Jupyter** is a web-based notebook environment that combines features of an integrated development environment (IDE) with those of a word-processing or web-development application. Jupyter notebooks can include text, mathematics, graphics, executable program code, interactive widgets, and more. Specifically, this book uses JupyterLab.
- **Python** is a versatile programming language that has a rich set of libraries that support data science activities.

1.5 What data science topics does this book not cover?

This book focuses on the foundations of data science, and there are many important topics that could not be included. In particular, a prospective reader should know that:

- This is not a machine learning book. In particular, this book does not cover neural networks at all.
- This book assumes that the reader knows the basics of linear algebra and how to work with vectors and matrices in NumPy. However, I wrote this book simultaneously with another book called *Introduction to Linear Algebra for Data Science with Python* that provides the necessary background on linear algebra.
- This book does not cover many practical aspects of working with data, such as using databases to retrieve and store data. Nor does it provide much coverage of different approaches to "cleaning" data, such as dealing with missing or mislabeled data.
- This book does not provide comprehensive coverage of the libraries that are used, such as NumPy, Pandas, and `scikit-learn`; rather, it focuses on showing how to use these libraries for some foundational data science techniques.
- This book does not address many important issues related to the ethics of data science.

On the website for this book (fdsp.net), I provide a list of suggested "Next Steps" that include books and online materials that address these important topics.

1.6 Extremely Brief Introduction to Jupyter and Python

The purpose of this and the following two sections is to briefly introduce users to Jupyter and Python. The content here should be treated as an introduction to explore further and is not meant to be comprehensive. There are a broad variety of tutorials on the web for both of these topics, and links are provided for users who need additional instruction.

If you are already familiar with Jupyter and/or Python 3, feel free to skip ahead.

1.6.1 Why Jupyter notebooks?

According to the Project Jupyter web page (https://jupyter.org), "The Jupyter Notebook is an open-source web application that allows you to create and share documents that contain live code, equations, visualizations and narrative text. Uses include: data cleaning and transformation, numerical simulation, statistical modeling, data visualization, machine learning, and much more".

The reasons that Jupyter notebook was chosen for this book include:

- Jupyter notebooks can integrate text, mathematics, code, and visualization in a single document, which is very helpful when conveying information about data. In fact, this book was written in a series of over 140 Jupyter notebooks.
- Jupyter notebooks allow for an *evolutionary approach* to code development. Programs can start as small blocks of code that can then be modified and evolved to create more complex functions.
- Jupyter notebooks are commonly used in the data science field.

1.6.2 Why Python?

Python is a general-purpose programming language that was originally created by Guido van Rossum and maintained and developed by the Python Software Foundation. Python was chosen for this book for many reasons:

- **Python is very easy to learn.** Python has a simple syntax that is very similar to C, which many engineers and scientists will be familiar with. It is also easy to transition to Python from MATLAB scripting, which many engineers will be familiar with.
- **Python is an interpreted language,** which means that programmers can run the code directly without having to go through extra steps of compiling their programs.
- **Python interpreters are freely available and easy to install.** In addition, Python and Jupyter are available on all major operating systems, including Windows, MacOS, and Linux.
- **Python is popular for data science and machine learning.** Python is widely used for data science and machine learning in both industry and universities.
- **Python has rich libraries for data science.** Python has many powerful libraries for data science and machine learning. In addition, Python has powerful libraries for a broad array of tasks beyond the field of data science, which makes learning Python have additional benefits.

1.6.3 How to get started with Jupyter and Python

Python and Jupyter are often packaged together in a *software distribution*, which is a collection of related software packages. The creators of several Python software distributions include additional Python software libraries for scientific computing. This book assumes the use of the Anaconda distribution, which its creators bill as "The World's Most Popular Data Science Platform"[1].

Anaconda's *Individual Edition* is freely available to download from the Anaconda website at https://www.anaconda.com/products/individual. Choose the proper download based on

---

[1] https://www.anaconda.com/, retrieved May 30, 2023.

your computer's operating system. You may also have to select a version of Python. This book is based on **Python 3**, which means that any version of Python that starts with the number 3 should work with the code included in this book. For instance, as of May 2023, the Anaconda distribution included Python version 3.10.

WARNING

Python version 2 or Python versions after 3 may have syntax changes that cause the programs in this book to not run without modification.

After downloading, install Anaconda however you usually install software(for instance, by double-clicking on the downloaded file). Anaconda will install Python and many useful modules for data science, as well as Jupyter notebook and JupyterLab.

Note:

The term "Jupyter notebook" refers to a file format (with *.ipynb* extension), while "Jupyter Notebook" (with a capital N) refers to an application with a web interface to work with those files. To help avoid confusion, I will write *Jupyter notebook file* or simply *notebook* whenever referring to such a file, and we will use JupyterLab as the web application for opening and working with such files.

As of January 2023, JupyterLab "is the next-generation web-based user interface for Project Jupyter". (from https://jupyterlab.readthedocs.io/en/stable/). The Jupyter Notebook application offers a simple interface for working with notebooks and a limited number of other file types. JupyterLab has a more sophisticated interface and can include many different components, such as consoles, terminals, and various editors. The interface for working with notebooks is similar in both, and most users will be able to use either one interchangeably.

1.6.4 Getting organized

We are almost ready to start using Jupyter and Python. Before you do that, I recommend you take a minute to think about how you will organize your files. Learning data science requires actually working with data and performing analyses. This will result in you generating a lot of Jupyter notebook files, as well as a lot of data files. I suggest that you create a folder for this data-science book (or for the course if you are using this as a course textbook). This folder should be easily accessible from your home directory because that is the location where JupyterLab will open by default. You may wish to add additional structure underneath that folder. For instance, you may want to create one folder for each chapter or each project. If you create separate folders for the data, I suggest you make them subfolders of the one containing the notebooks that access that data.

An example layout is shown in Fig. 1.2.

1.6.5 Getting started in Jupyter

Let's begin exploring JupyterLab using an existing notebook:

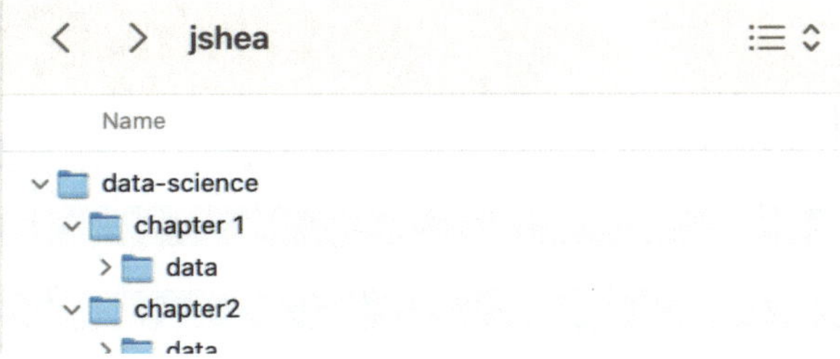

FIGURE 1.2
Example directory structure for organizing files for working through the examples and exercises in this book.

1) Download a Jupyter notebook file.

We will use the file "jupyter-intro.ipynb", which is available on this book's website at:

https://www.fdsp.net/notebooks/jupyter-intro.ipynb

If your browser displays the notebook as text, you will need to tell it to save it as a file. You can usually do this by right-clicking or control-clicking in the browser window and choosing to save the page as a file. For instance, in Safari 14, choose the "Save Page As..." menu item. Be sure to name your file with a `.ipynb` ending.

Hint: If your file was saved to your default Downloads folder, be sure to move it to an appropriate folder in your `data-science` folder to keep things organized!

2) Start JupyterLab.

JupyterLab can be started from the Anaconda-Navigator program that is installed with the Anaconda distribution. Start Anaconda-Navigator, scroll to find JupyterLab, and then click the *Launch* button under JupyterLab. JupyterLab should start up in your browser.

Alternative for command-line users: From the command prompt, you can start JupyterLab by typing `jupyter lab` (provided the Anaconda bin directory is on the command line search path). Because setting this up is specialized to each operating system and command shell, the details are omitted. However, details of how to set up the path for Anaconda can be found at many sites online.

Your JupyterLab should open to a view that looks something like the one in Fig. 1.3.

WARNING

If you have used JupyterLab before, it may not look like this – it will pick up where you left off!

The JupyterLab interface has many different parts:

1. The **menu bar** is across the very top of the JupyterLab app. I will introduce the use of menus later in this lesson.
2. The **left sidebar** occupies the left side below the menu bar. It includes several different tabs, which you can switch between by clicking the various icons on the very far left of the left sidebar. In Fig. 1.3, the folder icon is highlighted, which

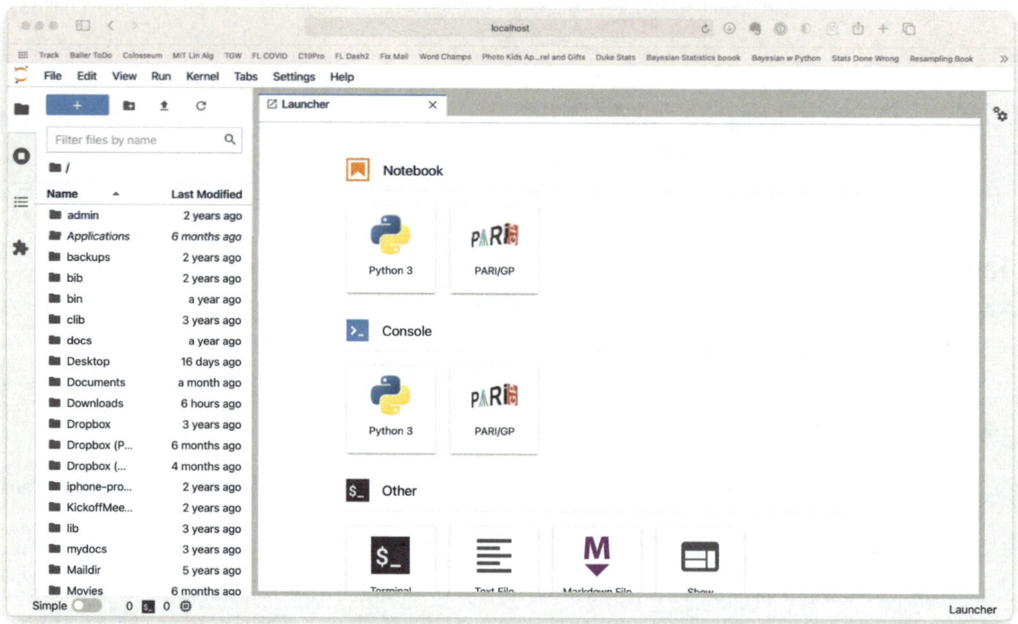

FIGURE 1.3
The JupyterLab interface.

indicates that the file browser is selected. For this book, we will use the left sidebar only to access the file browser.

3. The **main work area** is to the right of the left sidebar. The main work area will usually show whatever document you are working on. However, if you have not opened any document yet, it will show you different types of notebooks that you can open and other tools that you can access. To start a completely new Jupyter notebook file that can run Python 3 code, you could click on the Python 3 icon under `Notebooks`. For now, you do not need to do that.

Detailed documentation for JupyterLab is available at

https://jupyterlab.readthedocs.io/.

3) Navigate to the downloaded notebook.

Use the file browser in the left sidebar of JupyterLab to navigate to the downloaded file. *If the file browser is not already showing your files, click on the folder icon (on the very left-hand side of the window) to switch to it.*

Navigation using the file browser should be similar to navigating in most file selection boxes:

- Single click on items to select them.
- Double click on a folder to navigate into it.
- Double click on a file to open it.

- As you navigate into folders, the current path (relative to your starting path) is shown above the file list. You can navigate back out of a folder by clicking on the parent folder's *name* in the current path.

If you downloaded the file `jupyter-intro.ipynb` to the `chapter1` subdirectory of the `data-science` directory, which lies in your home directory, then you would:

- Double click on the `data-science` folder.
- Double click on the `chapter1` folder.
- Double click on the file `jupyter-intro.ipynb`.

The file `jupyter-intro.ipynb` should open in the main work area.

1.6.6 Learn the basics of JupyterLab

After opening the `jupyter-intro.ipynb` notebook, take a minute to scroll through the notebook before interacting with it. Note that the notebook includes formatted text, graphics, mathematics, and Python programming code. You will learn to use all of these features as you work through this book.

Notebook structure

Jupyter notebooks are subdivided into parts called **cells**. Each cell can be used for different purposes; we will use them for either Python code or for Markdown. Markdown is a simple markup language that allows the creation of formatted text with math and graphics. Code cells are subdivided into Input and Output parts. Single click on any part of the `intro.ipynb` notebook to select a cell. The selected cell will be indicated by a color bar along the entire left side of the cell.

JupyterLab interface modes

The JupyterLab user interface can be in one of two modes, and these modes affect what you can do with a cell:

- In **Edit Mode**, the focus is on one cell, which will be outlined in color (blue on my computer with the default theme), and the cursor will contain a blinking cursor indicating where typed text will appear.
- In **Command Mode**, you cannot edit or enter text into a cell. Instead, you can navigate among cells and use keyboard shortcuts to act on them, including running cells, selecting groups of cells, and copying/cutting/pasting or deleting cells.

There are several ways to switch between modes:

In **Command Mode**, here are two ways to switch to **Edit Mode** and begin editing a cell:

- Double-click on a cell.
- Select a cell using the cursor keys and then press Enter.

In **Edit Mode**, here are two ways to switch to **Command Mode**:

- Press Esc. The current cell is *not evaluated*, but it will be selected in **Command Mode**.

- If editing a cell that is *not* the last cell in the notebook, press Shift+Enter to *evaluate* the current cell and return to **Command Mode**. (If you are in the last cell of the notebook, Shift-Enter will evaluate the current cell, create a new cell below it, and remain in **Edit Mode** in the newly created cell.)

More on cells

In **Edit Mode**, code or Markdown can be typed into a cell. Remember that each cell has a *cell type* associated with it. The cell type does not limit what can be entered into a cell. The cell type **determines how a cell is evaluated**. When a cell is evaluated, the contents are parsed by either a Markdown renderer (for a Markdown cell) or the Python kernel (for a Code cell). A *kernel* is a process that can run code that has been entered in the notebook. JupyterLab supports different kernels, but we will only use a Python kernel. Cells may be evaluated in many different ways. Here are a few of the typical ways that we will use:

- Most commonly, we will evaluate the current cell by pressing Shift+Enter or Shift+Return on the keyboard. This will always evaluate the current cell. If this is the last cell in the notebook, it will also insert a new cell below the current cell, making it easy to continue building the notebook.
- It is also possible to evaluate a cell using the toolbar at the top of the notebook. Use the triangular "play" button (pointed to by the red arrow in the image in Fig. 1.4) to execute the currently selected cell or cells.

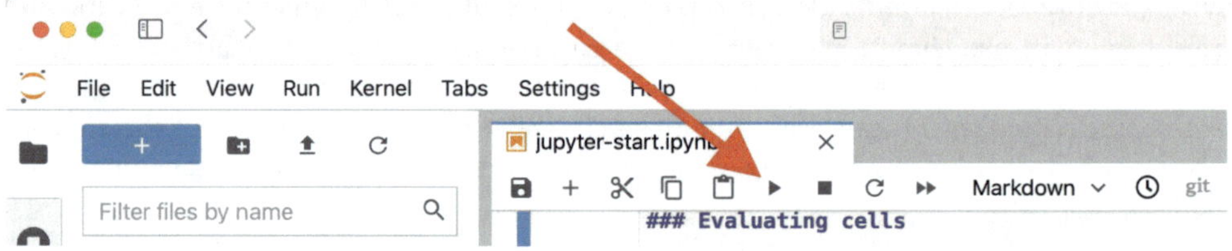

FIGURE 1.4
Image of Jupyter interface indicating the location of "play" button for executing cells.

- Sometimes we wish to make changes in the middle of an existing notebook. To evaluate the current cell and insert a new cell below it, press Alt+Enter on the computer keyboard.
- Cells can also be run by some of the commands in the `Kernel` menu in the JupyterLab menu. For example, it is always best to reset the Python kernel and run all the cells in a notebook from top to bottom before sharing a Jupyter notebook with someone else (for example, before submitting an assignment). To do this, click on the `Kernel` menu and choose the `Restart Kernel and Run All Cells...` menu item.

If you enter Markdown into a Code cell or Python into a Markdown cell, the results will not be what you intend. For instance, most Markdown is not valid Python, and so if Markdown is entered into a Code cell, a syntax error will be displayed when the cell is evaluated. Fortunately, you can change the cell type afterward to make it evaluate properly.

Important!

! New cells, including the starting cell of a new notebook, start as *Code* cells.

Cells start as *Code* cells, but we often want to enter Markdown instead. We may also wish to switch a *Markdown* cell back to a *Code* cell. There are three easy ways to change the cell type:

- As seen in Fig. 1.5, you can use the drop-down menu at the top of the notebook to set the cell type to *Code*, *Markdown*, or *Raw*.

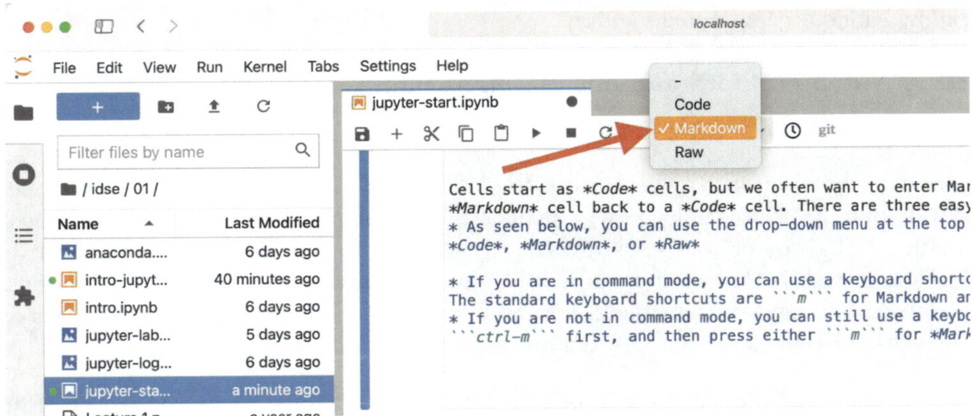

FIGURE 1.5
Picture of JupyterLab interface showing the cell type drop-down menu.

- If you are in command mode, you can use a keyboard shortcut to change the type of a selected cell. The standard keyboard shortcuts are `m` for *Markdown* and `y` for *Code*.
- If you are not in command mode, you can still use a keyboard shortcut, but you will need to press `Control`+`m` first, and then press either `m` for *Markdown* or `y` for *Code*.

Intro to Markdown in Jupyter

In this book, you will use Jupyter both to perform data analysis and to create documents that describe the results of your analysis. Although we will use code to process the data and generate graphs, that is usually not enough to tell the story to a reader of your document. Markdown is used to add text, heading, mathematics, and other graphics.

The example notebook `jupyter-intro.ipynb` demonstrates the main features of Markdown that we will need in this book. Recall that you can double click on any cell in the notebook to see the Markdown source. The `jupyter-intro.ipynb` notebook illustrates the features of Markdown listed below. A tutorial on how to create each of these in Markdown is given online at fdsp.net/1-6.

1. Headings are written like `# Heading`, where more # can be added for subheadings.
2. Text and paragraphs. Paragraphs are indicated by blank lines.
3. Emphasis can be added to text using asterisks, with single asterisks indicating `*italics*` and double asterisks indicating `**bold**`.
4. Bulleted lists can be created by putting items after an asterisk followed by a space: `* my list item`.
5. Numbered lists can be created by putting items after a number, a period, and a space: `1. my numbered item`.

6. Links can be created by putting the link text in square brackets, followed by the link URL in parentheses, like `[Example link](http://google.com)`
7. Images are similar way to URLs, except have an exclamation point (!) before the square brackets: `![Image example]{my_image.jpg}`.
8. Math can be entered using LaTeX notation.

A good reference for Markdown syntax is Markdown Guide: https://www.markdownguide.org/extended-syntax/.

Getting Notebooks into and out of JupyterLab

There are several ways to get notebooks into JupyterLab:

- As previously mentioned, you can use the left-hand files pane to navigate to the current location of a file. Note that you will be constrained to only navigating to files in the directory in which Jupyter was started or in any subdirectory below that. One disadvantage of this approach is that your work will be saved wherever that file currently resides. For instance, if you have downloaded a notebook from the internet into your `Downloads` folder, your work on that notebook will remain in the `Downloads` folder.
- You can use drag-and-drop to copy any file into a directory that you are currently browsing using JupyterLab's files pane. To do this:
  - Open the files panel in Jupyter and navigate to the directory where you want to work.
  - In your operating system's file manager (e.g., Windows Explorer or Mac Finder), open the folder containing the file you want to copy.
  - Position and resize the folder and your web browser's window so you can see both simultaneously.
  - Click and hold on the icon for the Jupyter notebook file that you want to move. Then drag it onto the files panel.
  - When the Jupyter notebook is over JupyterLab's files panel, the outline of the files panel will change to indicate that it is ready for you to drop the file. Release the mouse button or trackpad to copy the file into the selected directory.
  - **Note that this makes a copy of the file from its original location.**
- As an alternative to drag-and-drop, you can click on the upload icon (an arrow with a line under it) at the very top of the files panel. This will bring up a file selector that you can use to copy a file from anywhere on your computer.

You can save your work by choosing `Save Notebook` in JupyterLab's `File` menu or by pressing the keyboard sequence listed next to that item in the menu. When you manually save your work in this way, Jupyter actually saves two copies of your work: it updates the `.ipynb` file that you see in the file list, and it also updates a *hidden* checkpoint file. When you are editing or running your notebook file, Jupyter will also autosave your work periodically – the default is every 120 s. When Jupyter autosaves, it only updates the `.ipynb` file. If Jupyter crashes or you quit it without saving your notebooks, your last autosaved work will be what you see in the `.ipynb` files. However, you can always revert to the version you purposefully saved by using the `Revert Notebook to Checkpoint` item in the `File` menu.

When starting new Jupyter notebooks, their initial name will be `"Untitled.ipynb"`. You can easily rename your notebook in a couple of ways. First, you can choose the `Rename`

`Notebook...` option from the file menu. As an alternative, you can right-click on the notebook in the left-hand Files panel and choose `Rename`. In both cases, be sure to change only the part of the notebook name that is in front of the `.ipynb` extension. Jupyter uses that file extension to recognize Jupyter notebook files.

Important!

!

When you are finished working with a Jupyter notebook, I recommend you perform the following steps:

1. First, from the `Kernel` menu, choose `Restart Kernel and Run All Cells...` This will clear the previous output from your work and rerun every cell from the top down.
2. Check over your notebook carefully to make sure you have not introduced any errors or produced any unexpected results from having executed cells out of order or from deleting cells or their contents. By performing these first two steps, you help make sure that someone else loading your notebook file will be able to reproduce your work.
3. Check the notebook file name and update it if necessary.
4. Save the notebook.
5. Choose `Close and Shutdown Notebook` from Jupyter's `File` menu.
6. If you are finished working in JupyterLab, then choose `Shut Down` from JupyterLab's `File` menu.

Another common workflow in JupyterLab is to use an existing notebook as a starting point for a new notebook. Again, there are several ways to do this:

- If you already have the existing notebook open, then you can save it under a new name by choosing `Save Notebook As...` from Jupyter's `File` menu and giving the notebook a new name. **Note that after you use this option, the notebook that is open in the main work will be the notebook with the new name. You will no longer be working on the original notebook.**
- You can also duplicate a notebook by right-clicking on the notebook's name in the File panel on the left-hand side and choosing `Duplicate`. A copy of the notebook will be created with the name of the existing notebook appended with a suffix like `-Copy1` before the `.ipynb`.

Jupyter magics

Code cells can also contain special instructions intended for JupyterLab itself, rather than the Python kernel. These are called *magics*, and a brief introduction to Jupyter magics is available on this book's website at fdsp.net/1-6.

1.6.7 Getting started in Python

Python is an interpreted language, which means that when any Code cell in a Jupyter notebook is evaluated, the Python code will be executed. Any output or error messages will appear in a new output portion of the cell that will appear just after the input portion of the cell (that contains the Python code). At the bottom of the `jupyter-intro.ipynb` notebook, there is an empty cell where you can start entering Python code. If there is not already an empty cell there, click on the last cell and press `Alt-Enter`.

A detailed introduction to Python is available on this book's website at fdsp.net/1-6. Although there are many good Python tutorials online, the one at fdsp.net/1-6 is especially tailored to the features of Python that are most often used in this book. For users who want to learn more about Python, the following resources are recommended:

- *A Whirlwind Tour of Python* (https://jakevdp.github.io/WhirlwindTourOfPython/) by Jake VanderPlas is a free eBook that covers all the major syntax and features of Python.
- Learn Python for Free (https://scrimba.com/learn/python) is a free 5-hour online introduction to Python (signup required).
- The Python documentation includes a Python Tutorial: https://docs.python.org/3/tutorial/.

Terminology review and self-assessment questions

Interactive flashcards to review the terminology introduced in this section and self-assessment questions are available at fdsp.net/1-6, which can also be accessed using this QR code:

1.7 Chapter Summary

This chapter introduced the topics that will be covered in this book, as well as two of the main tools used throughout the book. JupyterLab is used to provide a computational notebook environment. These notebooks can combine programming code, text, graphics, and mathematics. We will use these to conduct simulations and data analysis and to present results. Python is used because it is widely adopted by the data science and machine learning communities, as well as being a general-purpose programming language. Python has well-developed libraries for data science and many other applications.

Access a list of key take-aways for this chapter, along with interactive flashcards and quizzes at fdsp.net/1-7, which can also be accessed using this QR code:

2

First Simulations, Visualizations, and Statistical Tests

We want to start working with real data as quickly as possible, but using real data introduces a lot of complexity. This chapter starts with a simple example that does not use real data but allows us to quickly start writing computer simulations, visualizing data, and performing statistical tests. In Chapter 3, we will use these skills to analyze a real data set on per-state COVID-19 case rates and socioeconomic factors.

2.1 Motivating Problem: Is This Coin Fair?

You find a strange coin. You assume it is equally likely to come up heads and tails – we say the coin is *fair*. You would like to conduct a statistical test to determine if it is fair. So, you flip the coin 20 times and count how many times it comes up heads. This is our first example of a *random experiment*:

> Definition
>
> **random experiment**
>
> An experiment for which the outcome cannot be predicted with certainty.

Note that we generally can provide some predictions about the output of a random experiment. For instance, if the coin is fair, then heads and tails are equally likely. You probably have the intuition that on 20 flips, the number of heads observed will most likely be close to 10.

Suppose the number of heads observed is significantly different from 10. Under what conditions would we decide that the coin might not be fair?

If you observe only 6 heads on the 20 flips, should you reject the idea that the coin is fair?

What if you observe only 4 heads?

To answer this question, we first need to understand some basics of what a probability is. The following is a very general definition that should match most people's intuition for the meaning:

DOI: 10.1201/9781003324997-2

Definition

probability

A number that we assign to an event that is proportional to how likely that event is to occur and that is between 0 and 1.

If the probability of something occurring is very close to 0, then that thing is very unlikely to occur. If the probability is very close to 1, then that thing is very likely to occur.

To determine whether the coin could be fair, we can try to find the probability that a fair coin would produce a result that is as extreme (or more extreme) than what we observed. In other words, if we observe 6 heads, we can determine the probability that a fair coin would produce 6 or fewer heads. If the probability is small, then we can reject the possibility that the coin is fair. How we define "small" is up to the experimenter and anyone who reviews their result. For our work, let's say that we will reject the possibility that the coin is fair if the probability that we would see 6 or fewer heads on 20 flips of a fair coin is less than 0.05 (i.e., 5%).

So, now we just need some way to calculate the required probability for 20 flips of a fair coin. There are two common ways to do this:

1. Analyze the probability mathematically.
2. Estimate the probability experimentally.

We are not ready to explain the mathematical analysis yet, so we will estimate the probability experimentally. So get ready to start flipping coins!

Not really – to estimate the probability we are interested in will require flipping a coin 20 times, recording the outcome, and then repeating those steps thousands of times! Rather than actually flipping the coin, we will **simulate** flipping the coin.

Terminology review and self-assessment questions

Interactive flashcards to review the terminology introduced in this section and self-assessment questions are available at fdsp.net/2-1, which can also be accessed using this QR code:

2.2 First Computer Simulations

A **computer simulation** is a computer program that models reality and allows us to conduct experiments that:

- would require a lot of time to carry out in real life
- would require a lot of resources to carry out in real life
- would not be possible to repeat in real life (for instance, simulation of the next day's weather or stock market performance)

We will simulate the fair coin experiment using Python. We can simulate a fair coin by randomly choosing a result from a list that contains values representing heads and tails, where each element in the list is equally likely to be chosen. Let's use the string `'H'` to denote heads and the string `'T'` to denote tails:

```
faces = ['H', 'T']
```

To randomly choose one of the faces, we will utilize the *random* module, which is one of the standard modules included with Python. We will import it in the usual way:

```
import random
```

To choose one face at random, we can use the `random.choice()` function, as below. Note that if you are running this instruction, the result may be different than what is shown below because it is random.[1]

```
random.choice(faces)
```

```
'T'
```

Now let's simulate the scenario described in Section 2.1. We could repeatedly choose random faces using a loop, but the `random` module offers a more efficient way to choose all 20 faces at the same time, using the `choices()` function:

```
coins = random.choices(faces, k=20)
print(coins)
```

```
['T', 'T', 'T', 'H', 'H', 'T', 'T', 'T', 'T', 'H', 'T', 'H', 'H', 'T', 'T', 'H
↪', 'H', 'H', 'H', 'T']
```

To count the number of 'H' results in the list, we can use the `count()` method of the list object:

```
coins.count('H')
```

```
9
```

Every time a new list of coin faces is generated, the result will be different. To estimate the probability, we will have to run this statement many times and keep track of the results; however, that would be very time-consuming and error-prone. Fortunately, the computer can automatically run it for us. In the simulation below, I start introducing some *best practices*:

- The number of times to simulate the experiment is defined at the very top of the simulation as `num_sims`.
- Any parameters of the experiment are defined near the top of the simulation. In this simulation, we only need to define the number of times to flip the coin, `num_flips`.

Note: In the code below, I use the `end` keyword parameter of the `print` function to cause each print statement to output a space at the end instead of a new line. This is only done to make the output more concise.

[1] The methods used in this book actually create *pseudorandom* outputs, which are outputs that behave as if they were random but are repeatable if the random source is initialized in the same way (represented by a number called a *random seed*).

```
num_sims = 20
num_flips = 20
for sim in range(num_sims):  # The simulation loop

    # Simulate all coin flips for one experiment
    coins = random.choices(faces, k=num_flips)

    print(coins.count('H'), end=' ')
```

```
8 10 8 9 12 8 8 11 6 9 12 8 7 9 12 12 11 6 12 11
```

2.3 First Visualizations: Scatter Plots and Histograms

Before we try to answer the question about whether the coin is fair, we take a minute to consider how we can visualize the experimental results generated in the simulation. The most straightforward plot to create from a sequence of numerical data is a *scatter plot*:

Definition

scatter plot

A (two-dimensional) scatter plot takes two sequences $\mathbf{x} = (x_0, x_1, \ldots)$ and $\mathbf{y} = (y_0, y_1, \ldots)$ and plots symbols (called *markers*) that represent the locations of the points $(x_0, y_0), (x_1, y_1), \ldots$ in two dimensions.

For our purposes, we will plot the observed data versus the experiment number. Let's collect the data we need for a scatter plot. We will store the number of `'H'` seen in each iteration of the simulation. A list is a good container for this purpose:

```
num_sims = 100
flips = 20

results = []
for sim in range(num_sims):  # The simulation loop

    # Simulate all coin flips for one experiment
    coins = random.choices(faces, k=flips)

    # Add the number of 'H's in this experiment to our results list
    results += [ coins.count('H') ]
```

There are many different libraries for generating plots in Python. The most common and popular is Matplotlib, which is based on MATLAB's plotting commands. We usually import the `pyplot` module from Matplotlib to the `plt` namespace:

```
import matplotlib.pyplot as plt
```

You may also want to use a Jupyter magic to make sure plots show up inline with your code and writing in your Jupyter notebook. We will typically use `%matplotlib inline`, but an alternative is `%matplotlib %notebook`, which will make plots interactive.

```
%matplotlib inline
```

Then we can generate a scatter plot by calling `plt.scatter()` and passing it two arguments: a list of x-coordinates and a list of corresponding y-coordinates. Since we are plotting against the simulation number, we can pass it a `range` object for the x-coordinates:

```
plt.scatter(range(num_sims), results);
```

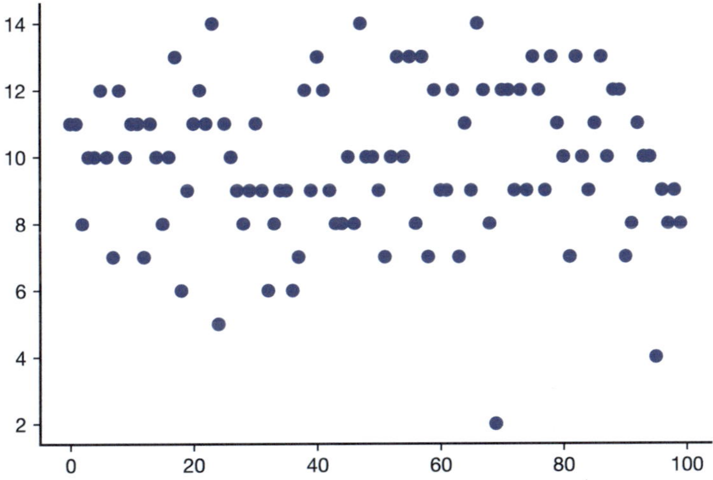

FIGURE 2.1
Output of `plt.scatter()` showing simulation results for number of heads observed on 20 flips of fair coin versus simulation iteration.

The output of this command is shown in Fig. 2.1. Two comments about the command above and the plot:

- I have placed a semi-colon (;) at the end of the plot command. This will suppress the output of the command, which is different than the graph that is shown. The command `plt.scatter()` returns a Matplotlib PathCollection object that represents the plotted data. I will use semi-colons regularly for this purpose with Matplotlib commands.
- If you have run this command and your plot has different colors and additional border lines, that is to be expected. I have a custom matplotlibrc file that styles the plots in this book. These differences only relate to style and not to the plot's substance.

The following code shows how to add axis labels and a title. The resulting plot is shown in Fig. 2.2.

```
plt.scatter(range(num_sims), results)
plt.xlabel('Experiment number')
plt.ylabel('Number of heads observed')
plt.title('Simulation of no. of heads observed on 20 coin flips');
```

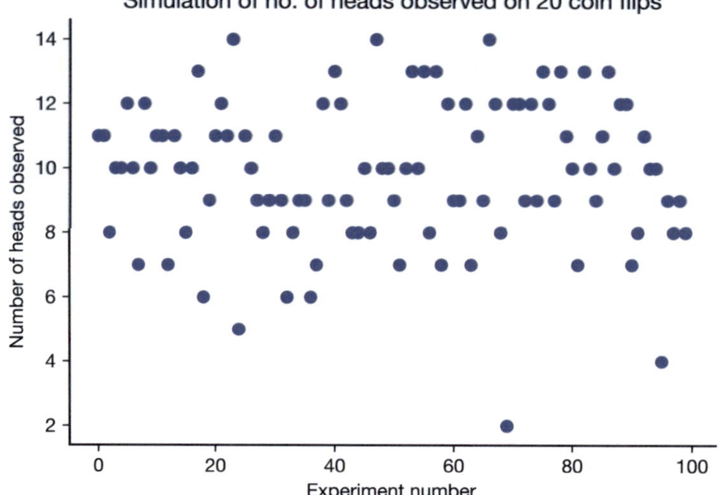

FIGURE 2.2
Scatter plot for coin-flip simulation with axis labels and title.

It should be clear that most of the results are around 8 to 12. It might be easier to visualize this if we plot the number of experiments in which each value for the number of heads is observed. We can understand how to create this from a physical perspective: imagine if we took the graph in Fig. 2.2 and rotated it 90° to the left and that the circular markers were turned into balls that were allowed to drop down until they fell onto the axis or another ball. Let's build exactly this type of visualization one ball at a time. An animation that illustrates exactly this is available at fdsp.net/2-3, which can also be accessed using the QR code shown.

When the balls have finished dropping, the resulting plot shows the number of occurrences of each outcome. For reference, a static version of the final figure is shown in Fig. 2.3. The code to generate this is in the notebook www.fdsp.net/notebooks/ball-drop-histogram.ipynb.

This plot is a type of *histogram*:

Definition

histogram

A type of bar graph in which the heights of the bars are proportional to the number of occurrences of the outcomes spanned by the width of the bars.

Histograms are often used for continuous data that may take on any real value within some range. The width of the bars is determined by the spacing of the *bins*, and outcomes anywhere within those bins are counted toward that bar.

Matplotlib offers the `plt.hist()` function to create histograms, but it is designed for continuous data, and the bins that `plt.hist()` chooses will often result in confusing visualizations for discrete data, where the data can only take on specific values. For instance, the following code includes the data used to create the visualization above and code to generate a histogram using the default bins. The resulting plot is shown in the left plot of Fig. 2.4.

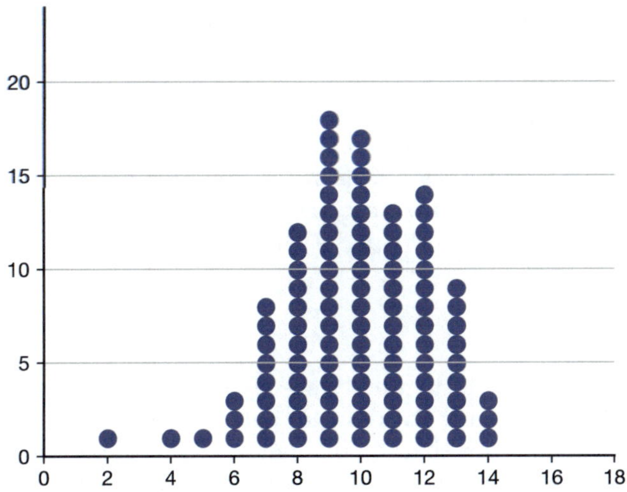

FIGURE 2.3
Final output from dropping balls onto numbers representing counted number of heads on 20 flips of a fair coin.

```
plt.hist(results)
```

```
(array([ 1.,  1.,  1.,  3.,  8., 30., 17., 13., 14., 12.]),
 array([ 2. ,  3.2,  4.4,  5.6,  6.8,  8. ,  9.2, 10.4, 11.6, 12.8, 14. ]),
 <BarContainer object of 10 artists>)
```

The first two outputs of the `plt.hist()` function are the counts and bin edges, respectively. The third output is a Matplotlib object that we will not cover in this book.

The histogram created by Matplotlib with the default bins looks distinctly different than the one we created by “dropping balls”. This is because the default is to create 10 bins that are evenly spaced between the minimum and maximum values. In this case, it results in

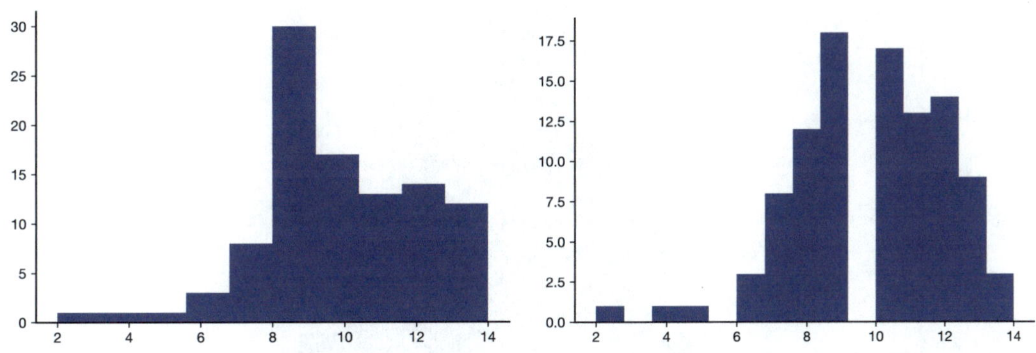

FIGURE 2.4
Two Matplotlib histograms for the coin flipping data with 10 bins (right) and 15 bins (left).

bins that are 1.2 units wide, so some bins contain multiple different numbers of heads. For instance, there is a bin from 8 to 9.2 that includes both the values of 8 heads and 9 heads, resulting in a large spike in the middle. There is also a bin from 12.8 to 14, which includes both the values of 13 and 14 heads, resulting in the last bin on the right side being much taller than in the ball-drop histogram.

In some cases, applying Matplotlib's `hist()` function to discrete data will result in bins with the expected heights mixed with empty bins. For instance, we can specify `bins=15` to get 15 uniformly spaced bins, as shown in the following code. The resulting plot is on the right side of Fig. 2.4 and shows that the result still does not accurately reflect the distribution of the data because now it has a large empty space in the middle.

```
plt.hist(results, bins=15);
```

We can make a much better histogram by specifying the bins. We will do so by specifying a list of bin edges. When working with integer data, it is typically best to make bins that are one unit wide with bin edges that are 0.5 units to each side of the bin centers. To span the observed values in this dataset, we would need bin edges at $1.5, 2.5, \ldots, 14.5$.

We could make a list of these bin edges by typing them individually or by writing a **`for`** loop to add them to a list. However, let's take the opportunity to introduce the NumPy function `np.arange()` that can create a uniformly spaced array of points between two edges. It takes a starting value, a stopping value (which will not be included in the output), and an increment. We want points spaced 1 unit apart from 1.5 to 14.5, so we can generate them as follows:

```
import numpy as np

edges =  np.arange(1.5, 15.5, 1)
print(edges)
```

```
[ 1.5  2.5  3.5  4.5  5.5  6.5  7.5  8.5  9.5 10.5 11.5 12.5 13.5 14.5]
```

Now we are ready to regenerate the histogram with the correct bins:

```
plt.hist(results, bins=edges);
```

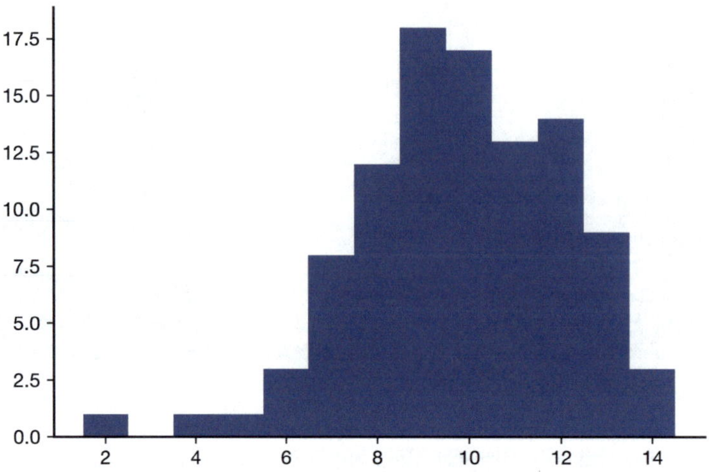

This graph is not ideal for a variety of reasons. First, all the bars run together because they are the same color. Second, the y-axis labels are not appropriate because we have quantities that are integers, not real numbers. Third, it is hard to read off the heights of the bars if they are far from the labeled y-axes. Finally, the axes are not labeled, so someone looking at the graph will have no idea what it is trying to convey. Fortunately, these are all easily remedied:

1. We can pass the `edgecolor` keyword parameter to the `plt.hist` function with value `black`.
2. We can specify the locations of the labeled "ticks" on the y-axis using `plt.yticks()`.
3. We can add grid lines for the y-axis (at each tick) using `plt.grid()`.
4. We can label our axes using `plt.xlabel()` and `plt.ylabel()`.
5. If the figure were displayed without a caption, we could add a title using `plt.title()`.

The following code implements these, and the resulting figure is shown in Fig. 2.5.

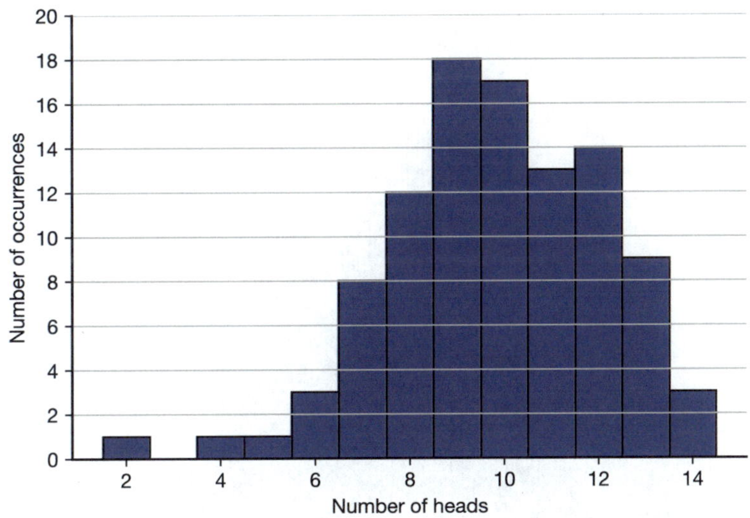

FIGURE 2.5
Improved histogram for number of heads on twenty flips of a fair coin.

```
plt.hist(results, bins=edges, edgecolor='black')
plt.yticks(np.arange(0, 22, 2))
plt.grid(axis='y')
plt.xlabel('Number of Heads')
plt.ylabel('Number of occurrences')
```

The **counts** of the outcomes can be used to estimate the **probabilities** of the outcomes if we turn them into *relative frequencies*:

Definition

relative frequency

The proportion of times that we observe a result matching our criteria during repeated experiments (including simulation); i.e., the number of times an event occurs divided by the number of times the experiment is conducted.

For many experiments, the relative frequencies (results measured from experimentation) will converge to the true probabilities (mathematical descriptions fundamental to the experiments) when the number of experiments is large. (At this point in the book, we do not have any formal framework for calculating probabilities, but we will refine these statements in later chapters.)

Let's consider two different approaches to determining the relative frequencies. The first is via `plt.hist()`. If the bin widths are one unit wide, then the relative frequencies can be found by setting the keyword parameter `density=True`:

```
plt.hist(results, bins=edges, edgecolor='black', density=True)
```

```
(array([0.01, 0.  , 0.01, 0.01, 0.03, 0.08, 0.12, 0.18, 0.17, 0.13, 0.14,
        0.09, 0.03]),
 array([ 1.5,  2.5,  3.5,  4.5,  5.5,  6.5,  7.5,  8.5,  9.5, 10.5, 11.5,
        12.5, 13.5, 14.5]), <BarContainer object of 13 artists>)
```

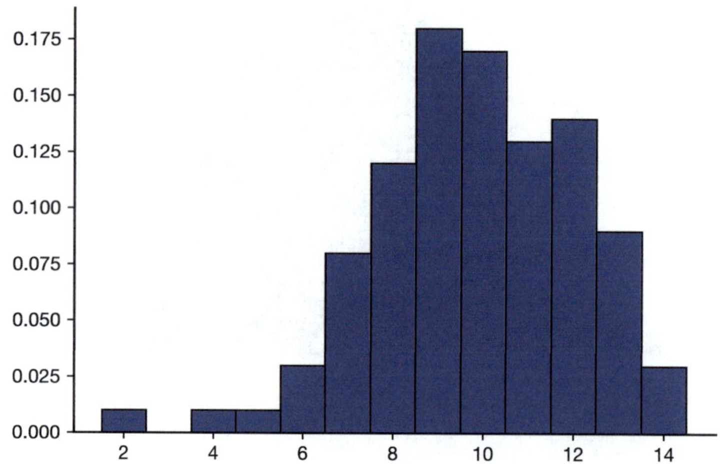

Note that when `plt.hist()` is called with `density=True` and bins that are one unit wide, the first output will be the relative frequencies (the `density` parameter is useful in other situations when the bins are not one unit wide, but we defer the discussion of this to Section 8.7. We can verify the relative frequencies from `plt.hist` by directly counting and then normalizing. We could do this with a `for` loop, but NumPy makes this much easier using the `np.unique()` function, which will return a list of the unique values in a list or array. If `return_counts=True` is also passed to that function, it will also return an array that contains a count of each unique item:

```
vals, counts = np.unique(results, return_counts=True)
print(vals, counts)
```

```
[ 2  4  5  6  7  8  9 10 11 12 13 14] [ 1  1  1  3  8 12 18 17 13 14  9  3]
```

Then we can calculate the relative frequencies by dividing the counts by the sum of the counts:

```
counts / counts.sum()
```

```
array([0.01, 0.01, 0.01, 0.03, 0.08, 0.12, 0.18, 0.17, 0.13, 0.14, 0.09,
0.03])
```

The relative frequencies calculated using NumPy are identical to those returned via `plt.hist()`. As we anticipated, the relative frequencies are highest around 10 heads and lowest for very few or very high numbers of heads. For instance, we see that the relative frequency of 6 heads is 0.03, which gives a good indication that seeing 6 heads is rare. However, determining whether it is rare enough for us to conclude that the coin must be biased requires some additional work, which we will do in the next section.

Terminology review and self-assessment questions

Interactive flashcards to review the terminology introduced in this section and self-assessment questions are available at fdsp.net/2-3, which can also be accessed using this QR code:

2.4 First Statistical Tests

Now let's focus on finding how often 6 or fewer heads occur when we flip a fair coin 20 times. Let's start by printing information about those extreme results. To do this, we will use an `if` statement inside our simulation loop to check if the number of heads observed is less than or equal to 6 and call a `print` statement if that condition is satisfied.

```
import random

random.seed(1324)

faces = ['H', 'T']

num_sims = 100
flips = 20

# best practice: put any thresholds outside the simulation loop
threshold = 6

for sim in range(num_sims):
    coins = random.choices(faces, k=flips)
    num_heads = coins.count('H')
    if num_heads <= threshold:
        print(f'{sim:2} : {num_heads} heads')
```

```
 5 : 6 heads
17 : 4 heads
28 : 6 heads
67 : 6 heads
79 : 6 heads
```

We will use variations on this basic function with different parameters later. Whenever *you anticipate doing this, it is a good idea to turn the code block into a function.* This is very easy in Jupyter:

1. Copy the last version of the simulation and paste it into a new code cell.
2. Select the entire code block in the new Input cell, either by using the mouse or Command + A (Mac) or Control + A (Windows/Linux).
3. Indent the selected block by pressing Command +] (Mac) or Control +] (Windows/Linux).
4. Click at the very left edge of the first line of the cell and type `def coinsim_print():` and press Enter.
5. Now finish populating the function by cutting the parameters from the code block and pasting them into the function signature as parameters with default values. Put the parameters in the order shown.
6. Finally, add a docstring that explains what the function does.

```
def coinsim_print(num_sims=100, threshold=6, flips=20):
    '''
    Simulate multiple experiments, where each experiment involves flipping a coin
    a specified number of times and printing the results of the experiment if the
    number of heads observed is <= a threshold

    Inputs:
    threshold: will print if the number of heads observed is <= this value
    num_sims: the total number of experiments to simulate
    flips: the number of coin flips in one experiment
    '''

    for sim in range(num_sims):
        coins = random.choices(faces, k=flips)
        num_heads = coins.count('H')
        if num_heads <= threshold:
            print(f'{sim:2} : {num_heads} heads')
```

```
coinsim_print()
```

```
37 : 6 heads
44 : 5 heads
55 : 6 heads
79 : 6 heads
86 : 5 heads
```

We do not really care about the particular experiment in which those extreme results occur. Instead, our goal is to estimate the probability of seeing a result that satisfies our threshold condition, and this probability can be estimated using the relative frequency of those results.

Let's create a new function that calculates the relative frequency of getting 6 or fewer heads on 20 flips of a fair coin. The primary changes are to:

1. Add an "event counter" to count how many times we see an experimental result that matches our criterion.
2. Instead of printing information about the experiments that meet the criterion, increment the event counter.
3. After the simulation loop is completed, calculate and print the relative frequency.

WARNING

Note that I am giving the new function a different name. It is possible to reuse the same function name, but this will produce ambiguities in Jupyter. When you call the function, the function definition that is used is the last one to be run. You can go back and rerun cells in Jupyter, which can then result in you not knowing which version of a function you are running.

Best practice: Do not reuse function names unless you are completely deleting the previous function definition and re-running all of the notebook cells.

First, we prototype the new functionality:

```
num_sims = 1000
flips = 20

threshold = 6

# count how many experiments satisfy the given criteria
event_count = 0

for sim in range(num_sims):
    coins = random.choices(faces, k=flips)
    num_heads = coins.count("H")
    if num_heads <= threshold:
        event_count += 1

print(f'Relative frequency of {threshold} or fewer heads',
      f'is {event_count / num_sims}')
```

```
Relative frequency of 6 or fewer heads is 0.054
```

Once it works successfully, we can convert it into a function:

```python
def coinsim(num_sims=1000, threshold=6, flips=20):
    '''
    Simulate multiple experiments, where each experiment involves flipping a coin
    a specified number of times and printing the results of the experiment if the
    number of heads observed is <= a threshold

    Inputs:
    threshold: will print if the number of heads observed is <= this value
    num_sims: the total number of experiments to simulate
    flips: the number of coin flips in one experiment
    '''

    # count how many experiments satisfy the given criteria
    event_count = 0
    for sim in range(num_sims):
        coins = random.choices(faces, k=flips)
        num_heads = coins.count('H')
        if num_heads <= threshold:
            event_count += 1

    print(f'Relative frequency of {threshold} or fewer heads',
          f'is {event_count / num_sims}')
```

```python
coinsim()
```

```
Relative frequency of 6 or fewer heads is 0.059
```

Note that the relative frequency can change when the simulation is rerun. How much it changes depends on the experiment, the criterion that defines the result we are looking for, and the number of experiments simulated. To provide an accurate estimate of the probability, the number of experiments simulated should be sufficiently large that the relative frequency does not change significantly when the simulation is run again. (For this experiment with a threshold of 6 heads, one million simulation experiments is sufficient.)

```python
coinsim(1_000_000)
```

```
Relative frequency of 6 or fewer heads is 0.05807
```

So, could an observation of 6 heads on 20 flips be reasonable with a fair coin?

With 1,000,000 experiments in the simulation, the relative frequency is approximately 0.058. That means that with a fair coin, we will see 6 or fewer heads about 6% of the time. Since we previously decided to use a criterion that the probability must be less than 0.05, we cannot reject the possibility that the coin is fair.

If we observed 4 heads on 20 flips, could that be reasonable with a fair coin?

Let's rerun the simulation and estimate the relative frequency of seeing 4 or fewer heads:

```python
coinsim(1_000_000, threshold=4)
```

```
Relative frequency of 4 or fewer heads is 0.00592
```

In this case, the estimate of the probability is about 6×10^{-3}, which falls below the 0.05 threshold we selected. So, we would reject the possibility that the coin is fair.

We would believe that the coin is biased toward heads but have no estimate of the size of that bias.

2.4.1 Basic simulation to estimate a probability

Let's use E to denote some result for which we are trying to estimate the probability via computer simulation. As before, we will actually calculate the relative frequency of E and use that as an estimate of the probability of E.

Then the basic simulation structure is as follows:

0. Initialize two counters to zero:
    - an event counter, `event_counter = 0`
    - a loop counter, `i=0`; in Python, the loop counter can be implicitly initialized and tracked using a `for ... in range()` statement

Then loop over the specified number of simulation iterations. For each iteration:

1. Simulate the outcome of the experiment.
2. If E occurred, increment the event counter: `event_counter += 1`
3. Increment the loop counter: `i += 1`. Usually, this is done automatically using a `for` loop.
4. If `i` matches the target number of iterations, then calculate and print the relative frequency; otherwise go to step 1.

We will use variations on this basic computer simulation structure throughout this book.

Terminology review and self-assessment questions

Interactive flashcards to review the terminology introduced in this section and self-assessment questions are available at fdsp.net/2-4, which can also be accessed using this QR code:

2.5 Chapter Summary

This chapter uses a simple experiment to introduce key concepts that will be used throughout this book. I introduced simulation as a way to carry out an experiment many times without requiring a lot of time or effort. Scatter plots were introduced as a way to view the raw data from a simulation, and histograms were introduced as a way of visualizing the distribution of one-dimensional data. The concept of relative frequency was introduced, and relative frequencies were measured using simulation. Finally, I showed how to use simulation to test a statistical hypothesis.

Access a list of key take-aways for this chapter, along with interactive flash-cards and quizzes at fdsp.net/2-5, which can also be accessed using this QR code:

3

First Visualizations and Statistical Tests with Real Data

In Chapter 2, we considered a simple statistical question that we could study through simulation, data visualization, and estimating probabilities. However, the type of problem considered in Chapter 2 is different than many that we encounter in data science because we had an exact model for the random experiment (repeated flips of a fair coin). This model allowed us to build a very simple simulation of the experiment.

In this chapter, we consider how to apply the ideas from Chapter 2 to a real data set, where we do not know the underlying model for the experiment.

Example: Early Numbers of COVID-19 Cases Across States

The novel coronavirus called COVID-19 spread quickly in the United States (US) during the spring of 2020 and soon became a pandemic, affecting countries throughout the world. However, the disease spread at different rates in different areas, and we would like to investigate that to see what types of socioeconomic factors might be related to the spread of COVID-19 in the US.

By early summer 2020, states were implementing a variety of mitigation measures, from mandatory masking to stay-at-home orders. Here I will focus on the spread of COVID-19 prior to most of these measures being put in place. I have chosen to perform our statistical analyzes using the cumulative number of COVID-19 cases through the end of April 2020.

We will analyze data on the number of COVID-19 cases for each state (from a dataset maintained by the New York Times) with data about the states and their populations that come from the US Census Bureau and the US Department of Commerce's Bureau of Economic Analysis. The goal is to investigate whether factors such as economic productivity and how urban a state is were statistically significant indicators of the number of COVID-19 cases.

All of this data is available in two common data formats:

- **Comma Separated Values (CSV)** files have data points separated by commas. The CSV format has the benefit that it is a pure text file and is not tied to any particular company's software. On the other hand, CSV files are not ideal if any of the entries contain commas. In this case, the entry can be enclosed in quotation marks, but then special treatment is necessary to deal with entries that contain quotation marks. There is no single standard for dealing with all these cases.

- **Excel File (.xls or .xlsx)** files are spreadsheet files for Microsoft Excel. These have become so common that many non-Microsoft products and tools can read and import data from these files.

In the next section, we begin by introducing the Pandas library, which is one of the most common Python libraries for importing and working with tabular data.

DOI: 10.1201/9781003324997-3

3.1 Introduction to Pandas

NOT THAT TYPE OF PANDAS

This section introduces the Pandas (or `pandas`) library. The description from the Pandas website (https://pandas.pydata.org) says: "pandas is an open source, BSD-licensed library providing high-performance, easy-to-use data structures and data analysis tools for the Python programming language".

We will import the Pandas library into the `pd` namespace:

```
import pandas as pd
```

Pandas is for working with tabular data, i.e., data that can be tabulated into rows and columns. We will generally store such data in a Pandas *dataframe*. A dataframe is similar to a spreadsheet or a database table. It is a two-dimensional structure in which each row corresponds to a single data point. Each column stores one variable or feature. Like tables in spreadsheets or databases, the columns can be labeled, and the rows can be indexed by consecutive integers or by the data in one of the columns.

We will work with the following data:

- **Cases** and **deaths** from COVID-19 were retrieved from The New York Times "Coronavirus (Covid-19) Data in the United States" GitHub repository at https://github.com/nytimes/covid-19-data/blob/master/us-states.csv. We are going to restrict our analysis to COVID cases in the early part of the pandemic; the data below is for cumulative COVID-19 cases up to April 30, 2020. (Retrieved April 26, 2021)

- **Population** data is from "Annual Estimates of the Resident Population for the United States, Regions, States, and Puerto Rico: April 1, 2010 to July 1, 2019" 2010-2019 Population Estimates. United States Census Bureau, Population Division. December 30, 2019. Link: https://www2.census.gov/programs-surveys/popest/tables/2010-2019/state/totals/nst-est2019-01.xlsx. We will use the data for the population estimates for 2019. (Retrieved April 26, 2021)

- **Gross Domestic Product (GDP)** data is from the US Department of Commerce Bureau of Economic Analysis, Regional Data: GDP and Personal Income, Current-dollar Gross Domestic Product (GDP) (Millions of current dollars) 2019:Q4. GDP is a measure of the total economic output of a state. Link: https://apps.bea.gov/itable/iTable.cfm?ReqID=70&step=1&acrdn=1. (Retrieved April 27, 2021)

- Data on **the percentage of the population living in urban areas** is from the United States Census Bureau, List of Population, Land Area, and Percent Urban and Rural in 2010 and Changes from 2000 to 2010, "Percent Urban and Rural in 2010 by State". We will refer to the percentage of a state's population living in urban areas as the *urban index*. Link: https://www2.census.gov/geo/docs/reference/ua/PctUrbanRural_State.xls. (Retrieved April 27, 2021)

To make it easier to start working with these data sets, I have merged the data into a single CSV file that is available at https://www.fdsp.net/data/covid-merged.csv.

Pandas provides a method to read CSV files, and it can read them directly from a URL:

```
df = pd.read_csv( 'https://www.fdsp.net/data/covid-merged.csv' )
```

This CSV file can be reconstructed from the raw data by running the commands in the notebook available on the book's website at https://www.fdsp.net/data/Download-Covid-Data.ipynb.

3.1.1 Working with dataframes

Let's start by looking at the dataframe. When run inside JupyterLab, Pandas will pretty-print a dataframe if the dataframe variable is evaluated as the last line of a code cell. Note that the column labels are imported from the first line of the CSV, and the unlabeled column on the left-hand side is the index. The index acts like labels for the rows and defaults to consecutive integers, starting at row 0. In the example below, I have truncated the output after the first ten:

```
df
```

	state	cases	population	gdp	urban
0	Alabama	7068	4903185	230750.1	59.04
1	Alaska	353	731545	54674.7	66.02
2	Arizona	7648	7278717	379018.8	89.81
3	Arkansas	3281	3017804	132596.4	56.16
4	California	50470	39512223	3205000.1	94.95
5	Colorado	15207	5758736	400863.4	86.15
6	Connecticut	27700	3565287	290703.0	87.99
7	Delaware	4734	973764	77879.4	83.30
8	Florida	33683	21477737	1126510.3	91.16
9	Georgia	25431	10617423	634137.5	75.07
10	Hawaii	609	1415872	97001.1	91.93

. . .

Note:

Because dataframes often occupy many rows, in the remainder of the book, I will instead only print the top of most dataframes or parts of dataframes. This is done easily in Pandas by applying the `head()` method to whatever is being returned. It will default to returning the first five results.

In Pandas, a particular column can be retrieved by putting the column's name in square brackets after the name of the dataframe. For instance, here is how to retrieve the content of the 'state' column:

```
df['state'].head()
```

```
0       Alabama
1        Alaska
2       Arizona
3      Arkansas
4    California
Name: state, dtype: object
```

A column of a dataframe is returned as a Pandas Series object, which is a one-dimensional object (like a list) that has a row index.

```
type(df['state'])
```

```
pandas.core.series.Series
```

We can retrieve multiple columns simultaneously by including them in a list within the square brackets. This means that there will be double square brackets – one set to tell Pandas we are selecting columns and one to form the list of columns to be selected. For example, we can retrieve both the `state` and `cases` columns as follows:

```
df[['state', 'cases']].head()
```

```
        state  cases
0     Alabama   7068
1      Alaska    353
2     Arizona   7648
3    Arkansas   3281
4  California  50470
```

When we select multiple columns, the data is no longer one-dimensional, and thus the returned data is a dataframe, not a series:

```
type(df[['state', 'cases']])
```

```
pandas.core.frame.DataFrame
```

Note:

When we index into a dataframe like this, the dataframe that is returned is not a new dataframe. Instead, it is a *view* into the original dataframe. If we make changes to the view, those changes also affect the original dataframe.

We can also retrieve the contents of a Pandas row. The general approach is the same as retrieving a column, but we must index into the `.loc` member of the dataframe to retrieve a row:

```
df.loc[1]
```

```
state           Alaska
cases              353
population      731545
gdp            54674.7
urban            66.02
Name: 1, dtype: object
```

We can retrieve multiple rows by providing a list of indices:

```
df.loc[1:5]
```

	state	cases	population	gdp	urban
1	Alaska	353	731545	54674.7	66.02
2	Arizona	7648	7278717	379018.8	89.81
3	Arkansas	3281	3017804	132596.4	56.16
4	California	50470	39512223	3205000.1	94.95
5	Colorado	15207	5758736	400863.4	86.15

We can combine both row and column selection by using `.loc` and specifying the desired columns after a comma inside the square brackets:

```
df.loc[45:50, ['state', 'cases']]
```

	state	cases
45	Virginia	15848
46	Washington	14814
47	West Virginia	1126
48	Wisconsin	6973
49	Wyoming	559

It is often convenient to use the values in one of the columns to provide a more meaningful index for the rows. Pandas does not require that the entries in the index be unique, but using non-unique entries can significantly impact performance and limit the ability to retrieve data by the index values. To set the index, use Pandas' `set_index()` method:

```
df.set_index('state').head()
```

	cases	population	gdp	urban
state				
Alabama	7068	4903185	230750.1	59.04
Alaska	353	731545	54674.7	66.02
Arizona	7648	7278717	379018.8	89.81
Arkansas	3281	3017804	132596.4	56.16
California	50470	39512223	3205000.1	94.95

Note that by default, the `set_index()` method returns a new dataframe; the original dataframe is unchanged:

```
df.head()
```

```
state   cases  population        gdp  urban
0     Alabama   7068     4903185   230750.1  59.04
1      Alaska    353      731545    54674.7  66.02
2     Arizona   7648     7278717   379018.8  89.81
3    Arkansas   3281     3017804   132596.4  56.16
4  California  50470    39512223  3205000.1  94.95
```

If we wish to work with the original one, we can tell Pandas to change the index *in place*:

```
df.set_index('state', inplace=True)
```

This makes finding the data by state much easier:

```
df.loc['Florida']
```

```
cases              33683.00
population      21477737.00
gdp              1126510.30
urban                 91.16
Name: Florida, dtype: float64
```

Note that the index (i.e., these row labels) carries over to the Pandas series that is returned by indexing a particular column of the dataframe:

```
df['cases'].head()
```

```
state
Alabama        7068
Alaska          353
Arizona        7648
Arkansas       3281
California    50470
Name: cases, dtype: int64
```

If all we want is the numerical values in the data series, we can convert it to a list or a NumPy array. Although there are multiple ways to achieve this, the best practice is to use the conversion methods of the dataframe object. To convert to a list, use the `to_list()` method, and to convert to a NumPy array, use the `to_numpy()` method:

```
print(df['cases'].to_list())
```

```
[7068, 353, 7648, 3281, 50470, 15207, 27700, 4734, 33683, 25431, 609, 2016,␣
↪52918, 18099, 7145, 4305, 4708, 28044, 1095, 21825, 62205, 41348, 5136,␣
↪6815, 7563, 452, 4332, 5053, 2146, 118652, 3411, 309696, 10507, 1067, 18027,
↪ 3618, 2510, 48224, 8621, 6095, 2450, 10506, 29072, 4672, 866, 15848, 14814,
↪ 1126, 6973, 559]
```

```
df['cases'].to_numpy()
```

```
array([ 7068,    353,   7648,   3281,  50470,  15207,  27700,   4734,  33683,
        25431,    609,   2016,  52918,  18099,   7145,   4305,   4708,  28044,
         1095,  21825,  62205,  41348,   5136,   6815,   7563,    452,   4332,
         5053,   2146, 118652,   3411, 309696,  10507,   1067,  18027,   3618,
         2510,  48224,   8621,   6095,   2450,  10506,  29072,   4672,    866,
        15848,  14814,   1126,   6973,    559])
```

Now you know the basics of accessing data in Pandas dataframes. In the next section, we will look at different ways to visualize this data.

Terminology review and self-assessment questions

Interactive flashcards to review the terminology introduced in this section and self-assessment questions are available at fdsp.net/3-1, which can also be accessed using this QR code:

3.2 Visualizing Multiple Data Sets – Part 1: Scatter Plots

Our first step in working with data sets is typically to gain some basic understanding of the data and relationships among the different variables. As in Section 2.3, we will create visualizations to help us understand the data. I will introduce several important concepts in statistics, including partitioning data and summarizing data using *summary statistics*. Finally, we will conduct our first statistical test using real data.

Since we will be creating several plots of these data sets, it is convenient to store the values in the appropriate columns of the dataframe as lists or NumPy arrays. Here we store the values as NumPy arrays to allow us to use some of NumPy's built-in methods later in this chapter:

```
cases = df['cases'].to_numpy()
pop = df['population'].to_numpy()
gdp = df['gdp'].to_numpy()
urban = df['urban'].to_numpy()
```

As previously mentioned, we are interested in how socioeconomic factors may have impacted the spread of COVID-19, which we measure through the number of cases. Thus, every statistical test we will conduct will be based on the number of cases.

Let's begin with a scatter plot of the number of cases by state (indexed from 0 to 49):

```
N = len(cases)
plt.scatter(range(N), cases)
plt.xlabel('State number')
plt.ylabel('No. of cases')
plt.title('COVID-19 cases through April 30, 2020');
```

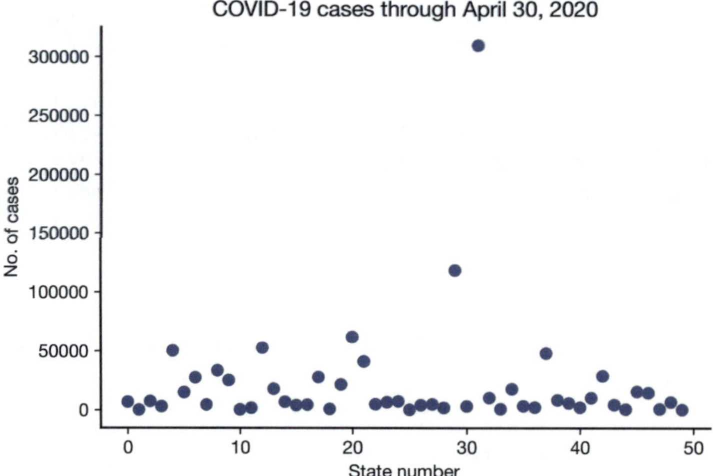

There are one or two values that seem to be unusually large in comparison to others, but this could be caused by the fact that states have different populations. Let's try to get some understanding of how these two factors interact.

3.2.1 Scatter plots for multiple datasets

The easiest way to create a scatter plot with multiple datasets is to simply call `plt.scatter()` multiple times from within a single cell. Let's try that with our cases and population data:

```
plt.scatter(range(N), cases)
plt.scatter(range(N), pop);
```

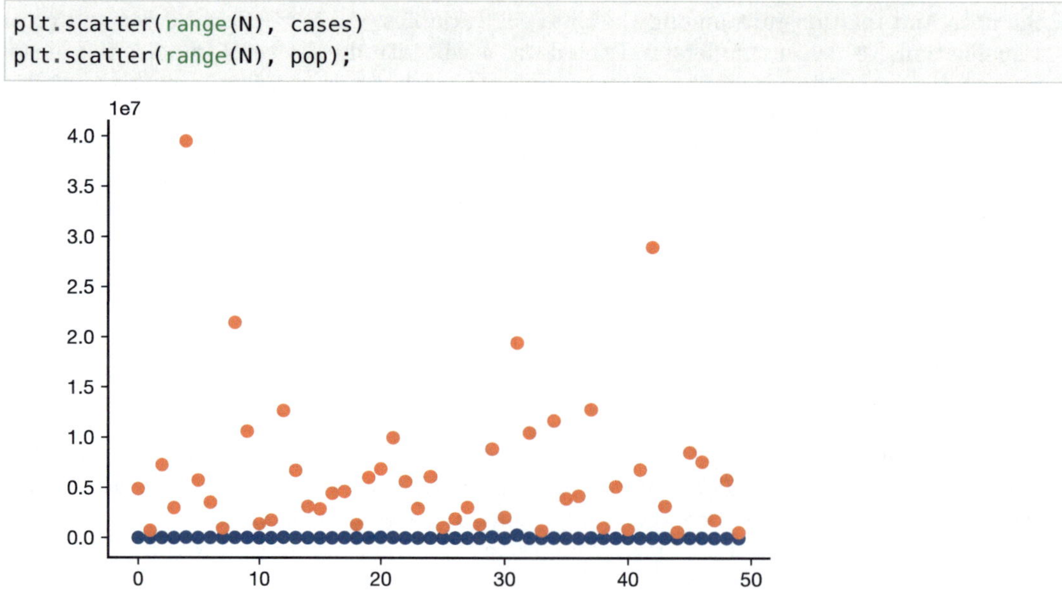

The good news is that Matplotlib knows how to plot multiple data sets on the same plot and automatically picks different colors for the scatter plot markers. We can improve the plot in several ways. First, we can choose a different marker for the population data. Markers can be specified using the `marker` keyword of the scatter function. The value passed is a string that indicates which marker to use.

Some commonly used marker types include:

```
"."            point
"o"            circle
"s"            square
"+"            plus
"x"            x
```

For a complete list of marker types, consult the Matplotlib documentation on markers:

https://matplotlib.org/stable/api/markers_api.html

Let's test this on our plot:

```
plt.scatter(range(N), cases)
plt.scatter(range(N), pop, marker='x');
```

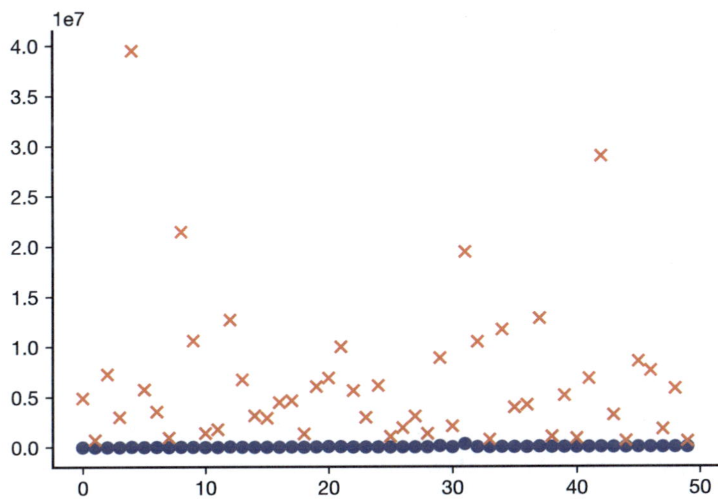

Now we can be sure that the reddish x markers are for the population data – but someone coming across the graph will not! We can add a *legend* to convey this information to the viewer:

> **Definition**
>
> **legend**
>
> When multiple data variables are shown on a single plot, different visual indicators (such as marker styles, line styles, and colors) are used to indicate which plotted items correspond to which variables. A *plot legend* is a key that shows which visual indicators correspond to which variables.

This is most easily done in two steps:

1. When calling `plt.scatter()` for each data set, pass the text that you would like to appear in the legend for that data set using the `label` keyword argument.
2. After all `plt.scatter()` calls are complete, call `plt.legend()` to draw the legend.

```
plt.scatter(range(N), cases, label='# of COVID cases')
plt.scatter(range(N), pop, marker='x', label='Population')

plt.legend();
```

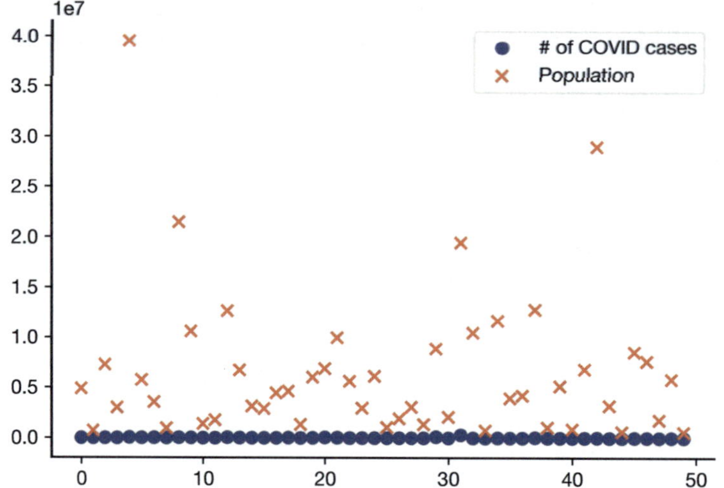

Adding some axis labels finishes the graph:

```
plt.scatter(range(N), cases, label='# of COVID cases')
plt.scatter(range(N), pop, marker='x', label='Population')

plt.legend()
plt.xlabel('State index')
plt.ylabel('Number of residents or cases');
```

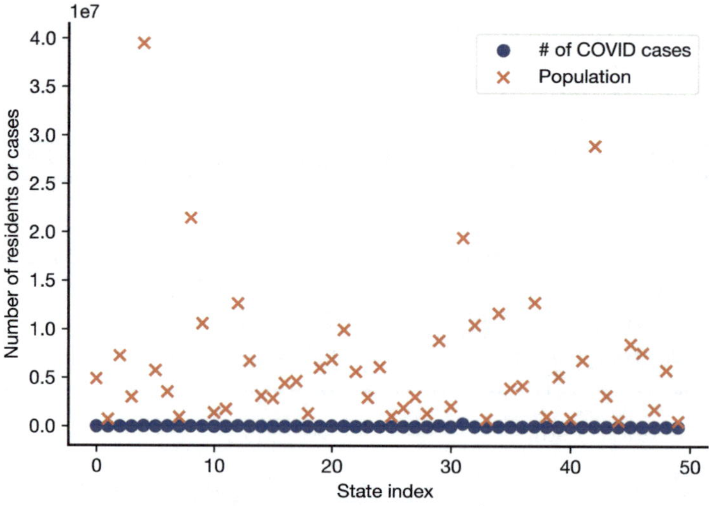

The bad news is that this graph is not very useful for the number of COVID cases. The differences in the numbers of cases cannot be evaluated from the graph because the range of that data is so much smaller than the range of the populations. The NumPy array variables have a max method that we can use to check this:

```
cases.max(), pop.max()
```

```
(309696, 39512223)
```

We can make a better plot showing both variables if we use a different y-axis for each variable. To do this, we can use Matplotlib's `plt.subplots()` function to create a figure and a first axis, and then the `plt.twinx()` function to share the **same** x-axis but allow a new y-axis to be created for the second data set. Because the style of this book is to **not** draw the axis on the right side of the figure, we have to add a command to make that axis visible:

```
fig, ax = plt.subplots()
ax.scatter(range(N), cases)

ax2 = ax.twinx()
ax2.scatter(range(N), pop)
# Now make right axis visible:
ax.spines['right'].set_visible(True);
```

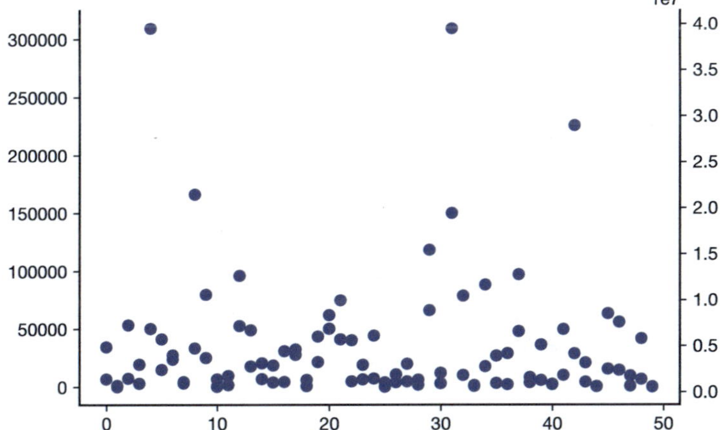

However, when we create the second set of axes, we lose the differentiation of markers. We can specify the marker color and type of marker symbol using the `color` and `marker` keywords, respectively. Matplotlib has a cycle of colors it uses to differentiate multiple variables plotted on the same plot. In the following code, I pass the keyword argument `color='C1'` for the second variable to tell Matplotlib to use color 1 in the cycle of colors (where the first variable is plotted with color 0 in the cycle). I have also added labels, separate legends, and improved the y axes by expressing the data in terms of millions or thousands:

```
fig, ax = plt.subplots()
ax.scatter(range(N), cases / 1000, label='No. of Cases')
ax.set_ylabel('Number of cases (thousands)')
ax.set_xlabel('State index')

ax2 = ax.twinx()
ax2.scatter(range(N), pop / 1e6, color='C1', marker='x', label='Population')
```

(continues on next page)

(continued from previous page)

```
# Now make right axis visible:
ax.spines['right'].set_visible(True);
ax2.set_ylabel('Population (millions)')

ax.legend(loc=2)
ax2.legend()

ax.set_ylim(0, 4e2)
ax2.set_ylim(0, 48);
```

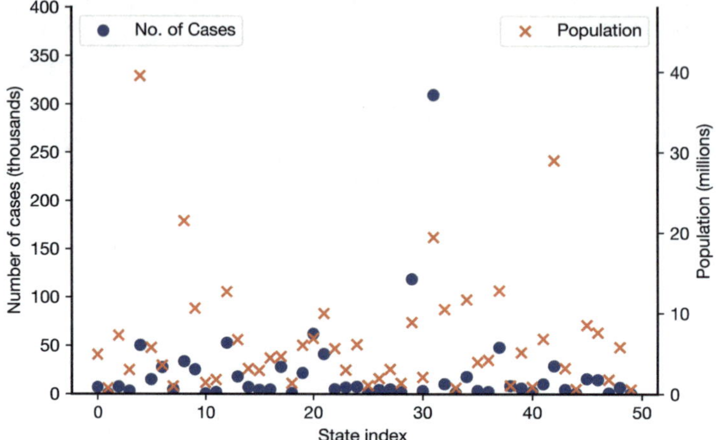

The relation among these data might be more clear if the data were sorted by population. We can sort the rows of the dataframe by population using the dataframe's `sort()` method:

```
df.sort_values('population', inplace=True)
```

```
popS = df['population'].to_numpy()
casesS = df['cases'].to_numpy()
```

```
fig, ax = plt.subplots()
ax.scatter(range(N), casesS/1000, label='No. of cases')
ax.set_ylabel('Number of cases (thousands)')
ax.set_xlabel('States sorted by increasing population')

ax2 = ax.twinx()
ax2.scatter(range(N), popS/1e6, color='C1', marker='x', label='Population')
# Now make right axis visible:
ax.spines['right'].set_visible(True)
ax2.set_ylabel('Population (millions)')

ax.legend(loc=2)
ax2.legend()

ax.set_ylim(0, 4e2)
ax2.set_ylim(0, 48);
```

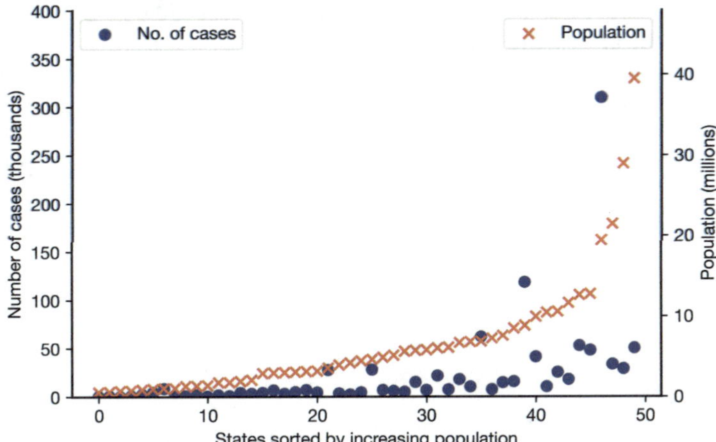

The number of cases generally increases with the population of the states. We might make some guesses about how these two variables might be related. Such a guess is one type of *statistical hypothesis*, which is formally defined in Section 3.6. Two possible hypotheses about how the population of a state might affect the number of cases are:

- Each person in the United States has the same probability of getting COVID-19. Thus, the number of cases per state should increase in proportion to the state's population.
- The probability of getting COVID-19 increases with the population of a state because of what are known as network effects: there is an increasing amount of mixing among the people within the state. Then the number of cases per state should increase with population even faster than under the previous assumption.

Of course, there may be many other hypotheses that explain this relationship. In the remainder of this chapter, we will try to remove some of the underlying effects of different state populations by normalizing the number of COVID-19 cases and state GDP on a per capita basis. Let's look at how we would do this within our Pandas dataframe. We can create a new column of our dataframe by simply assigning to it. Here, we just use functions of our other dataframe columns:

```
df['cases_pp'] = df['cases'] / df['population']
df.head()
```

	state	cases	population	gdp	urban	cases_pp
49	Wyoming	559	578759	40764.3	64.76	0.000966
44	Vermont	866	623989	34320.2	38.90	0.001388
1	Alaska	353	731545	54674.7	66.02	0.000483
33	North Dakota	1067	762062	57471.9	59.90	0.001400
40	South Dakota	2450	884659	56051.9	56.65	0.002769

```
df['gdp_pp'] = df['gdp'] / df['population']
```

As before, we create NumPy arrays of these variables for convenience:

```
cases_pp = df['cases_pp'].to_numpy()
gdp_pp = df['gdp_pp'].to_numpy()
```

The plot below shows the COVID rates (number of cases per population), plotted as a function of the state's order in increasing population:

```
plt.scatter(range(N), cases_pp);
```

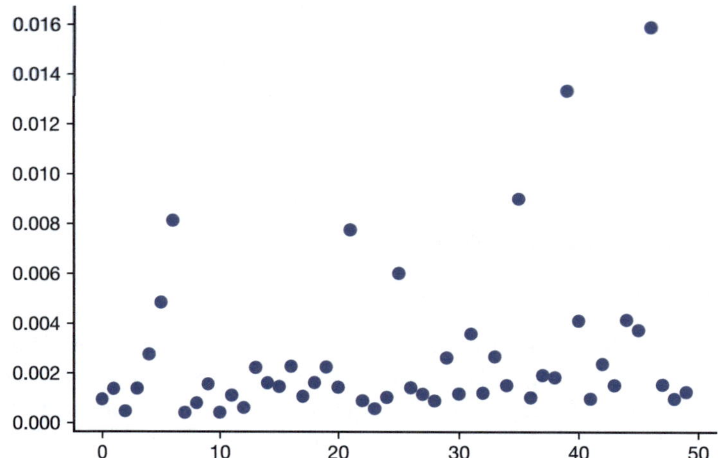

Even with this adjustment, we see that there are some states that have COVID rates that are very different than most of the other states. One reason for using scatter plots early in data analysis is to help identify any *outliers*:

> **Definition**
>
> **outlier**
>
> A value in a data set that takes on a value that is not reasonable, based on the other values in the dataset or other domain knowledge. Outliers are often due to data-entry, measurement, or unit-conversion errors.

We often start data visualization with scatter plots because, unlike histograms, scatter plots show the whole data and are helpful in spotting outliers.

Let's re-sort the dataframe based on our new `cases_pp` column, and we will pass it the keyword argument `ascending=False` to put the highest cases per population first:

```
df.sort_values('cases_pp', inplace=True, ascending=False)
df.head()
```

	state	cases	population	gdp	urban	cases_pp	gdp_pp
31	New York	309696	19453561	1791566.8	87.87	0.015920	0.092095
29	New Jersey	118652	8882190	642967.7	94.68	0.013358	0.072388
20	Massachusetts	62205	6892503	603209.6	91.97	0.009025	0.087517
38	Rhode Island	8621	1059361	62335.4	90.73	0.008138	0.058842
6	Connecticut	27700	3565287	290703.0	87.99	0.007769	0.081537

The top two states in terms of both raw cases and cases per person are New York and New Jersey, with COVID-19 rates over 0.013, whereas other states' rates were all below 0.01. In this case, we know that these states had a very rapid early spread of COVID-19 in comparison to other states, so these data points are probably the true values and would *not* be considered outliers.

Terminology review and self-assessment questions

Interactive flashcards to review the terminology introduced in this section and self-assessment questions are available at fdsp.net/3-2, which can also be accessed using this QR code:

3.3 Partitions

In this chapter, we will use a relatively simple approach to assessing whether a particular socioeconomic factor is associated with different rates of COVID-19 cases. We will use the values of a socioeconomic factor to *partition* the COVID case data into two groups. Below, we provide some definitions to clarify exactly what we mean by a partition. We start by introducing the concept of *disjoint* collections of data:

> Definition
>
> **disjoint**
>
> Groups of data are *disjoint* if no data point belongs to more than one of the groups.

If the data points are unique, we can use the mathematics of sets to formalize these ideas. If you are unfamiliar with sets and their operations, please review the online Appendix for this book at fdsp.net/appendix for a brief introduction. For our purposes, we can always consider a collection of data points to be unique if we include a unique index value with each data point.

A collection of sets $A_0, A_1, \ldots, A_{n-1}$ are disjoint if $A_i \cap A_j = \emptyset$ for every $i \in \{0, 1, \ldots, n-1\}$, $j \in \{0, 1, \ldots, n-1\}$, where $i \neq j$.

We can use the disjoint sets to define a *partition* of a set:

> Definition
>
> **partition**
>
> A partition of a data set is a group of **disjoint** collections of data, such that every data point in the original data set belongs to exactly one of the collections.

If S is a set of data, then the collections $A_0, A_1, \ldots A_{n-1}$ partition S if:

$$\bigcup_{i=0}^{n-1} A_i = S,$$

and $A_0, A_1, \ldots, A_{n-1}$ are disjoint. A *binary partition* of a set S is a pair of sets A_0 and A_1 such that $A_0 \cup A_1 = S$ and $A_0 \cap A_1 = \emptyset$.

To use a particular socioeconomic metric to partition our data, we will use a binary partition: one with higher values of that metric and one with lower values of that metric. To determine what constitutes a “higher value” or a “lower value”, we compute a threshold from the data, and that threshold is used to partition the data. Any such value that is computed from the data is called a *summary statistic*. Summary statistics are discussed in the next section.

Terminology review and self-assessment questions

Interactive flashcards to review the terminology introduced in this section and self-assessment questions are available at fdsp.net/3-3, which can also be accessed using this QR code:

3.4 Summary Statistics

We often characterize data using *summary statistics*:

> Definition
>
> **summary statistic**
>
> A numerical value calculated from data that measures some characteristic of the data.

Most people use the **average** of the data as the standard summary statistic; for instance, when receiving their exam scores, students will usually ask what the class average was for the exam. The average is also the most commonly used summary statistic by data scientists. Most statisticians use the term **sample mean** for this statistic and often refer to it as simply the **mean**. However, using the term *mean* to refer to the average of a set of numbers can create confusion with another type of mean for random data: the **ensemble mean**, which is also usually just called the **mean**. The ensemble mean is introduced in Chapter 9.

Because of this ambiguity in the **mean**ing of the word **mean**, I will use the terms *average* or *sample mean* to refer to the value that is computed from data.

When teaching a class on this subject, I usually ask the class: "What does the **average** or **sample mean** mean?". Here are some of the common answers:

1. The value that most of the data values are centered around
2. The value that is most likely to occur
3. The value that has minimum distance from every value
4. The value that divides the data into two sets of equal size

Only one of these descriptions of the average is always accurate, and even then, the description is ambiguous.

To understand average (and also some other summary statistics), we first have to understand that representing data by a summary statistic results in errors, and we can use these errors to help choose a "good" summary statistic. Let $d_0, d_1, \ldots, d_{N-1}$ be a set of numerical data values, and let ν be a summary statistic. The *error* e_i between data value d_i and ν is simply $e_i = d_i - \nu$. Note that the error may be positive or negative. Intuitively, we should try to choose ν to minimize the errors to the data. However, how to do this is not entirely clear because we have many different errors, and if we adjust the value of ν to decrease the error to one data value, that may increase the error to another data value.

The typical way to overcome this problem is to combine the errors in some way to create a single numerical value. Below I show some possible choices of functions for combining the errors. Since the errors depend on the choice of summary statistic, ν, we write these combinations as functions of ν:

1. **Sum of errors:**

$$E_s(\nu) = \sum_{i=0}^{N-1} e_i$$

2. **Number of nonzero error values:**

$$E_0(\nu) = \sum_{i=0}^{N-1} \mathbf{1}_{\mathbb{R}-\{0\}}(e_i)$$

Here, the indicator function $\mathbf{1}_{\mathbb{R}-\{0\}}(e_i)$ will return 1 for any nonzero errors.

3. **Sum of absolute errors:**

$$E_1(\nu) = \sum_{i=0}^{N-1} |e_i|$$

4. **Sum of squared errors:**

$$E_2(\nu) = \sum_{i=0}^{N-1} (e_i)^2$$

The code below calculates each of these error functions as the value of ν is varied, using a simple data set: $D = \{-1, -1, 0, 2, 5\}$:

```
# The data:
D = [-1, -1, 0, 2, 5]

# Sweep the value of nu from -2 to 6
nus = np.arange(-2, 6.01, 0.01)

# For clarity, store the different error metrics in different variables.
# Initialize them here
sum_errors = 0
num_nonzero_errors = 0
sum_abs_errors = 0
sum_square_errors = 0

# Calculate the error metrics
for d in D:
    sum_errors += d - nus
    num_nonzero_errors += (d - nus) != 0
    sum_abs_errors += np.abs(d - nus)
    sum_square_errors += (d - nus) ** 2
```

Fig. 3.1 shows all four error functions as the value of the summary statistic ν is varied. The data is also shown on this plot using stars, and the repeated data values have been vertically offset so the we can see both repeated values. Code to generate this plot is

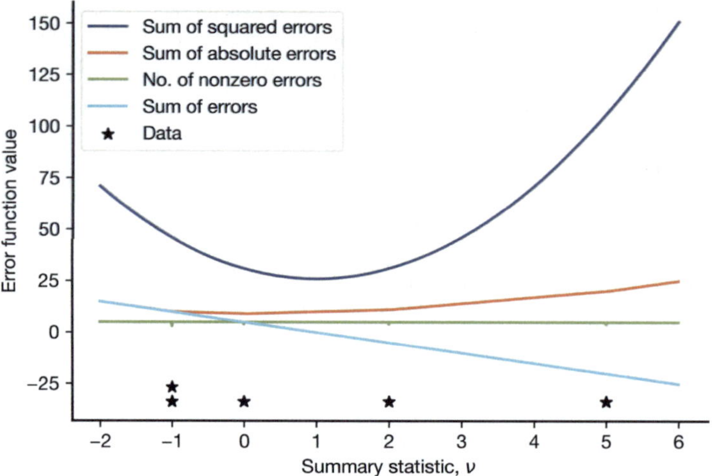

FIGURE 3.1
Values of four different error functions comparing a summary statistic ν to a data set. Data are shown as stars.

available online at fdsp.net/3-4. Since E_0 and E_1 have a much smaller range than the other two error functions, I have shown a zoomed in plot of these in Fig. 3.2.

Before putting much thought into what this means, let's just find the minimums for these different error metrics. To do this, we will find the **index** of the minimum value of each metric using NumPy's `np.argmin()` function. Then we will use that index value as the index of the `nus` array to determine the minimizing value of ν. (The code to print the results is omitted but available at fdsp.net/3-4.)

```
nu_e = nus[np.argmin(sum_errors)]
nu_0 = nus[np.argmin(num_nonzero_errors)]
nu_1 = nus[np.argmin(sum_abs_errors)]
nu_2 = nus[np.argmin(sum_square_errors)]
```

Metric	Minimizing value of nu
Sum of errors	6.0
No. nonzero errors	-1.0
Sum of abs errors	0.0
Sum of squared errors	1.0

Interestingly, even for this very small data set, each of these error metrics results in a different minimizing value of ν on the range $\nu \in [-2, 6]$. None of these metrics is inherently "correct" – although we explain below why one of them is not useful.

From inspecting the graph and further thought, we can make the following observations:

1. The sum of errors decreases without bound as a function of ν. Thus, there is no minimizing value of ν if ν can be any real value. Because of this, the sum of errors is not useful in determining a summary statistic.
2. The number of nonzero errors is equal to the total number of data values, *except at each of the data values.* The number of nonzero errors at a data value is equal

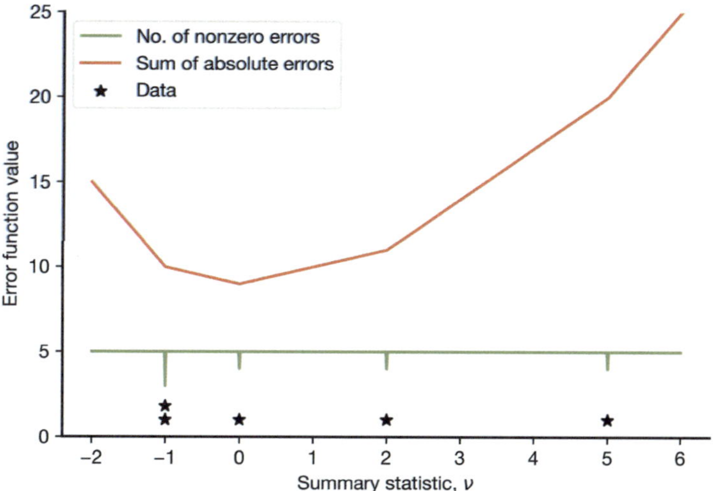

FIGURE 3.2
Zoomed in view of two error functions comparing a summary statistic ν to a data set. Data are shown as stars.

to the number of other data values. This metric will be minimized by setting ν equal to one of the data values that appears the most in the data set.

3. The sum of absolute errors is a continuous, piecewise-linear function. Furthermore, the linear segments connect between data values, between the minimum data value and $-\infty$, and between the maximum data value and $+\infty$. It is not hard to see that the absolute error will increase linearly as ν is decreased from the minimum data value or increased from the maximum data value. For such a function, the minimizing value will be at one of the values where the linear segments intersect, which is at one of the data values.
4. The sum of squared errors is a parabola. For this example, the minimum is not at one of the data values.

In the subsections below, we use the metrics E_0, E_1, and E_2 to define three common summary statistics.

3.4.1 Minimizing the number of nonzero errors

Consider first E_0, the number of nonzero errors. As previously mentioned, E_0 is minimized at the data value that appears most often in the data set. This value is called the **mode**:

> Definition
>
> **mode(s) (data)**
>
> A value (or values) in the data set that appears (or appear) most often. As there may be more than one most common value, the mode of a data set is not unique.

Unfortunately, NumPy does not offer a function to calculate the mode directly. There are NumPy functions that can find the mode indirectly, but here we use two straightforward approaches. Since we will often have our data in a Pandas dataframe, we can use the `mode()` method of a Pandas dataframe. To convert our example data into a simple dataframe, just pass it to the Pandas' `pd.DataFrame()` constructor:

```
import pandas as pd
```

```
df = pd.DataFrame(D)
df
```

```
   0
0 -1
1 -1
2  0
3  2
4  5
```

```
df.mode()
```

```
   0
0 -1
```

The dataframe's `mode()` method returns an indexed set of modes. Since this data set only has one mode, the index 0 value is that mode: -1.

If we are not working with Pandas already, an easier alternative is to use the `mode()` function from the `stats` submodule of the SciPy (pronounced "Sigh Pie") library. The keyword argument `keepdims=False` prevents the mode from being wrapped in an array.

```
import scipy.stats as stats

stats.mode(D, keepdims=False)
```

```
ModeResult(mode=-1, count=2)
```

As with Pandas, SciPy's `mode()` function can return all of the modes. If we just want one value, we can just get the first element of the first return:

```
stats.mode(D, keepdims=False)[0]
```

```
-1
```

Exercise

Consider the data set

$$D_2 = [-1, -1, 2, 2].$$

Copy and modify the code from above to calculate and plot the number of nonzero errors for this data set as a function of ν.

1. What values can the mode take on?
2. What values do SciPy.stats and Pandas return for the mode?

3.4.2 Minimizing the sum of absolute errors

Now we turn to metric E_1, which is the sum of the absolute errors. Because E_1 is a continuous function of ν, we can take its derivative to find its extreme values. Recall that

$$E_1(\nu) = \sum_{i=0}^{N-1} |e_i|,$$

so its derivative is

$$\begin{aligned} E_1'(\nu) &= \sum_{i=0}^{N-1} \operatorname{sgn}(e_i) \frac{d}{d\nu} e_i \\ &= \sum_{i=0}^{N-1} \operatorname{sgn}(d_i - \nu)(-1) \\ &= \sum_{i=0}^{N-1} \operatorname{sgn}(\nu - d_i), \end{aligned}$$

where $\operatorname{sgn}(x)$ is the *signum* function, which is defined as

$$\operatorname{sgn}(x) = \begin{cases} -1, & x < 0 \\ 1, & x > 0 \\ \text{undefined}, & x = 0. \end{cases}$$

Let $L(\nu)$ be the number of data values d_i such that $d_i < \nu$, and let $M(\nu)$ be the number of data values that are greater than ν. Then the derivative of E_1 can be written as

$$E_1'(\nu) = L(\nu) - M(\nu).$$

When ν is less than every data value, then $E_1'(\nu) = 0 - N = -N$; i.e., the slope of $E_1(\nu) = -N$. When ν is greater than every data value, then the slope of $E_1(\nu)$ is $E_1'(\nu) = N - 0 = N$. At each data value in between, the slope of $E_1(\nu)$ changes by the number of data values in the data set that have that same value. The minimum value of $E_1(\nu)$ is achieved wherever the slope changes from negative to positive, which must occur wherever there are an equal number of data values on each side of ν. We call this value the **median**:

Definition

median (data)

A value m such that the number of data values that are less than m is equal to the number of data values that are greater than m. For some data sets, multiple values may satisfy this criterion, so the median is not always unique.

When a data set has an even number of values, any value between $d_{\frac{N}{2}-1}$ and $d_{\frac{N}{2}}$ will minimize the sum of absolute errors and is thus a median. In practice, the value that is halfway between these two data values is often returned as the median:

$$m = \frac{d_{\frac{N}{2}-1} + d_{\frac{N}{2}}}{2}.$$

NumPy and Pandas provide `median()` functions and/or methods; the SciPy stats library does not have a `median()` function as of version 1.10.1.

```
np.median(D)
```

```
0.0
```

```
df.median()
```

```
0    0.0
dtype: float64
```

Exercise

Consider again the data set

$$D_2 = [-1, -1, 2, 2].$$

Copy and modify the code from above to calculate and plot the sum of absolute errors for this data set as a function of ν.

1. What values can be the median?
2. What values do NumPy and Pandas return for the median?

3.4.3 Minimizing the sum of squared errors

Now consider the sum of square errors, E_2. As noted, this function defines a parabola in ν, and we can find the extrema by taking the derivative of $E_2(v)$ with respect to ν and setting it equal to zero:

$$\begin{aligned}
E_2'(\nu) &= \frac{d}{d\nu} \sum_{i=0}^{N-1} (e_i)^2 \\
&= \sum_{i=0}^{N-1} \frac{d}{d\nu} (e_i)^2 \\
&= \sum_{i=0}^{N-1} 2\,(e_i)\, \frac{d}{d\nu} e_i \\
&= 2 \sum_{i=0}^{N-1} (d_i - \nu)\,(-1) \\
&= 2N\nu - 2 \left[\sum_{i=0}^{N-1} d_i \right].
\end{aligned}$$

Setting $E_2'(\nu)$ equal to zero yields:

$$\begin{aligned}
2N\nu - 2 \left[\sum_{i=0}^{N-1} d_i \right] &= 0 \\
N\nu &= \sum_{i=0}^{N-1} d_i.
\end{aligned}$$

Solving for ν gives the minimizing value:

$$\nu = \frac{1}{N} \sum_{i=0}^{N-1} d_i.$$

Most people are familiar with this value: it is the **average** of the data:

Definition

average (data)
sample mean (data)

The *average*, or *sample mean*, of a data set $D = d_0, d_1, \ldots, d_{N-1}$ is a value μ defined by

$$\mu = \frac{1}{N} \sum_{i=0}^{N-1} d_i.$$

Thus, we see that **the average is the value that minimizes the total squared error to the data**. Unlike the median or the mode, the average is unique.

NumPy and Pandas provide `mean()` functions and methods to calculate the average (sample mean); again, SciPy.stats does **not** provide a `mean()` method, since it is already available in NumPy:

```
np.mean(D)
```

```
1.0
```

```
df.mean()
```

```
0    1.0
dtype: float64
```

3.4.4 Some remarks on mode, median, and average

As our example illustrates, a data set's mode, median, and average can all be different values. The mode is usually not meaningful if the data take on values from a continuous set: usually, there are not any repeats in the data. For this reason, the mode is the summary statistic that we will use least for data sets.

As previously mentioned, the average is the most commonly used summary statistic. It is widely accepted for this purpose by the scientific community. Many people consider the median to be a more robust statistic because it is less sensitive to extreme values (such as outliers). For example, suppose we inserted the value 100 into the example data that we have been using:

```
D3 = D + [100]
```

Compare how the average and median are affected:

```
np.mean(D3)
```

    17.5

```
np.median(D3)
```

    1.0

The median is slightly larger because the new value causes it to shift one data value to the right. However, the average is drastically larger, completely outside the original set of values. On the other hand, if the value of 100 is not an outlier, the change in the median may be considered to not represent the new data set as well.

In general, the choice of summary statistic is up to the data scientist or statistician who is analyzing the data. If the data set may contain outliers, an alternative to removing the suspected outliers from the analysis is to use the median instead of the average.

Terminology review and self-assessment questions

Interactive flashcards to review the terminology introduced in this section and self-assessment questions are available at fdsp.net/3-4, which can also be accessed using this QR code:

3.5 Visualizing Multiple Data Sets – Part 2: Histograms for Partitioned Data

Now we return to our COVID-19 data set and apply what we have learned about partitions and summary statistics to set up binary comparisons: these are comparisons between two groups. As before, we begin by loading the data from the CSV file:

```
df = pd.read_csv( 'https://www.fdsp.net/data/covid-merged.csv' )
df.set_index('state', inplace=True)
df['cases_norm'] = df['cases'] / df['population'] * 1000
```

In the following subsections, we will partition the data based on normalized GDP and how urban the state's population is. In each case, we will partition the data into two sets by comparing the metric being studied to the median value of that metric. Note that the median is used instead of the mean because the median will result in the data set being partitioned into subsets that are approximately equal in size.

3.5.1 Partitioning based on GDP

The gross domestic product (GDP) provides a measure of how affluent a state is. As GDP can be expected to increase along with population, we first add normalized GDP (in millions) per capita (in thousands) to our dataframe[1].

[1]Since both COVID cases and GDP are being normalized per 1000 residents, the normalization could actually be dropped. However, we keep it here to be consistent with the analysis of the effect of the urban index.

```
df['gdp_norm'] = df['gdp'] / df['population'] * 1000
```

Since the combined effect of expressing GDP in millions and population in thousands is that this variable is equal to the GDP per capita, expressed in thousands of dollars ($K). For convenience, we will refer to it simply as the "GDP per capita" or "normalized GDP".

We will partition the data around the median of the normalized GDP, which is

```
m_gdp = df['gdp_norm'].median()
m_gdp
```

```
61.04737129841516
```

To partition the data, we introduce a new dataframe method called `query()`. The argument of the `query()` method is a string that contains one or more comparisons of column names to numbers, other column names, or variables. If a variable is used, then it must be prefixed with an @ symbol. This will be more clear with an example. To select the rows of the dataframe for which the `gdp_norm` column is greater than the variable `m_gdp`, we can use the following call to `df.query()`:

```
higher_gdp = df.query('gdp_norm > @m_gdp')
higher_gdp.head(3)
```

	cases	population	gdp	urban	cases_norm	gdp_norm
state						
Alaska	353	731545	54674.7	66.02	0.482540	74.738670
California	50470	39512223	3205000.1	94.95	1.277326	81.114143
Colorado	15207	5758736	400863.4	86.15	2.640684	69.609616

By changing the inequality to $\leq$, we can create a dataframe with states at or below the median GDP per 1000 residents:

```
lower_gdp = df.query('gdp_norm <= @m_gdp')
lower_gdp.head(3)
```

	cases	population	gdp	urban	cases_norm	gdp_norm
state						
Alabama	7068	4903185	230750.1	59.04	1.441512	47.061267
Arizona	7648	7278717	379018.8	89.81	1.050735	52.072199
Arkansas	3281	3017804	132596.4	56.16	1.087214	43.938042

As expected, the two dataframes have the same size:

```
len(higher_gdp), len(lower_gdp)
```

```
(25, 25)
```

Let's generate histograms for case data for both the higher GDP and lower GDP data sets. We plot them on the same axes to make it easier to compare them:

```
plt.hist(higher_gdp['cases_norm'], alpha=0.7, label='GDP/pop > median')
plt.hist(lower_gdp['cases_norm'], alpha=0.7, label='GDP/pop <= median')

plt.legend()
plt.xlabel('Cases per thousand residents')
plt.ylabel('Count');
```

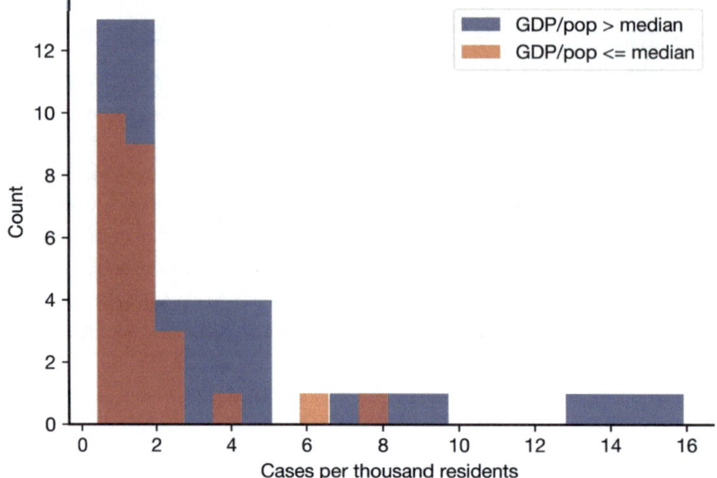

An issue that arises when plotting histograms on the same axes is that it becomes hard to compare values when the histograms use different bins. For example, from about 0.5 to 1.8, the histogram for the higher GDP group has a single bar with 13 entries. On the other hand, the histogram for the lower GDP group has two bars, but together those two bars sum to 19.

To resolve this issue, let's specify the same set of bins for use in each histogram. The first step is to use `np.arange()` to create an array of bin edges that span the range of the data:

```
import numpy as np

mybins = np.arange(0, 18, 2)
mybins
```

```
array([ 0,  2,  4,  6,  8, 10, 12, 14, 16])
```

```
plt.hist(higher_gdp['cases_norm'], alpha=0.7, label='GDP/pop > median',
         bins=mybins)
plt.hist(lower_gdp['cases_norm'], alpha=0.7, label='GDP/pop <= median',
         bins=mybins)

plt.legend()
plt.xlabel('Cases per thousand residents')
plt.ylabel('Count');
```

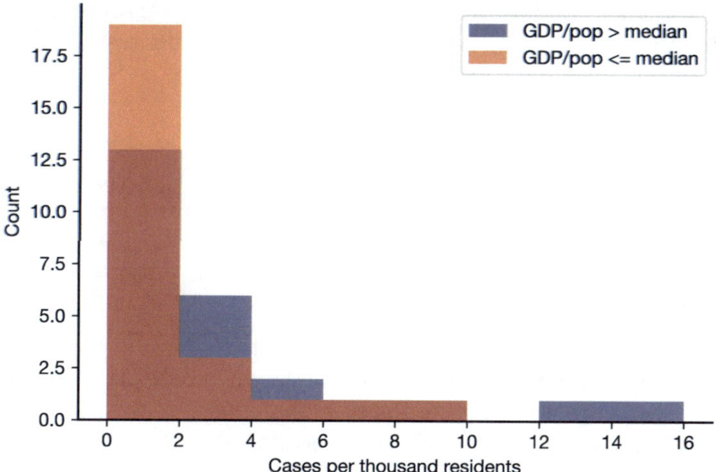

The histogram shows that states with GDPs per capita over the median have higher case counts in general (for instance, the larger bars in the range 2-5), as well as larger maximum values (above 8). Unfortunately, the data does not provide any insight into why that might be the case. One reason might be that states with higher GDPs may have larger network effects, where more of the population interacts with a larger number of people. For instance, such states may have more of their population employed in office jobs and more likely to travel for their work.

When an effect is observed visually, we usually want to quantify that effect in some way and then test whether any observed difference is "real" (rather than just caused by randomness). Let's compare the average of the normalized number of cases for these two different groups:

```
hgdp_mu = higher_gdp['cases_norm'].mean()
lgdp_mu = lower_gdp['cases_norm'].mean()
hgdp_mu, lgdp_mu
```

```
(3.511241636831491, 1.9258062634390531)
```

There is a substantial difference between the two average values. However, the average is more sensitive to extreme values, such as the large normalized case rates of New York and New Jersey. One way to deal with this is to use a statistic that is not as sensitive to extrema; for example, the median is such a statistic:

```
higher_gdp['cases_norm'].median(), lower_gdp['cases_norm'].median()
```

```
(1.9453983135416348, 1.3878449780364717)
```

Exercise

A new dataframe can be created that excludes New York and New Jersey using the following command:

```
df2 = df.drop(['New York', 'New Jersey'])
```

Using this reduced data set, partition the data around the median normalized GDP, plot the histograms for the COVID rates of the two partitions, and find the average COVID rates for the two partitions.

Exercise

Partition the data around the **mean** GDP. Plot the histograms and find the averages for this new grouping.

3.5.2 Partitioning based on urban index

Some of the networking effects mentioned in the previous section may be associated with populations living in large urban areas. In this section, we partition the data based on the urban index, which is the percentage of the population that lives in an urban area. As before, we start by finding the median urban index for the states:

```
m_urban = df['urban'].median()
m_urban
```

```
73.735
```

Now we use that median to partition the data. Histograms of the data in these partitions are shown in Fig. 3.3.

```
higher_urban = df.query('urban > @m_urban')
lower_urban  = df.query('urban <= @m_urban')
len(higher_urban), len(lower_urban)
```

```
(25, 25)
```

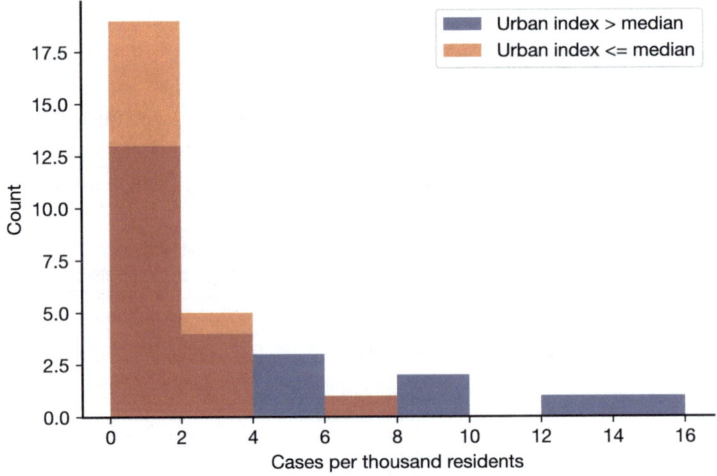

FIGURE 3.3
Histograms for states with lower urban index versus higher urban index.

As with per capita GDP, a substantial difference in the distribution of the COVID case rates is seen for states that are more urban versus less urban. The average normalized case rates for these partitions are:

```
higher_urban["cases_norm"].mean(), lower_urban["cases_norm"].mean()
```

```
(3.8908749904320756, 1.5461729098384682)
```

Again, a substantial difference in averages is observed. Some additional tests are left as exercises for the reader.

Exercise

Compare the medians for these two groups.

Exercise

Remove New York and New Jersey from the data set and recreate the histograms and averages for the partitioned data set without those two states.

3.5.3 Recommended reading

Histograms are commonly used to understand the distribution of data, and we will explore them more throughout this book. However, I recommend that you also review the following website that uses interactive visualizations to explore histograms in detail:

Exploring Histograms: https://tinlizzie.org/histograms/

Terminology review and self-assessment questions

Interactive flashcards to review the terminology introduced in this section and self-assessment questions are available at fdsp.net/3-5, which can also be accessed using this QR code:

3.6 Null Hypothesis Testing with Real Data

When dealing with data from different groups, we will generally observe differences. For instance, we may measure the means or medians of the data sets, and we will usually find that these summary statistics differ between the groups. At this point, we wish to understand whether the observed difference is "significant": is this difference likely to be a property of the underlying groups, or is it just caused by random variations in the data?

As we saw in Chapter 2, the number of available samples has a significant impact on how accurately we can estimate random phenomena. When it comes to real data, the problem is amplified because we do not know the ground-truth characterization of the random phenomena producing the data.

Let's start by restricting our analysis to a small subset of our data to see how effects observed with small data may be caused simply by random sampling. By restricting our data, we can show examples of the technique that we will be using, and the examples will be small enough to understand easily.

3.6.1 Small data example using COVID rates

Start by loading the data and calculating the COVID cases per 1000 residents and GDP per capita ($K):

```
import pandas as pd

df = pd.read_csv( 'https://www.fdsp.net/data/covid-merged.csv' )
df.set_index('state', inplace=True)
df['cases_norm'] = df['cases'] / df['population'] * 1000
df['gdp_norm'] = df['gdp'] / df['population'] * 1000;
```

Now consider the normalized COVID case rates from the first six states and partition them into two sets using alphabetical order. Since Arizona is the third state alphabetically, we can get the first three as follows:

```
first3 = df.loc[:'Arizona']
first3
```

state	cases	population	gdp	urban	cases_norm	gdp_norm
Alabama	7068	4903185	230750.1	59.04	1.441512	47.061267
Alaska	353	731545	54674.7	66.02	0.482540	74.738670
Arizona	7648	7278717	379018.8	89.81	1.050735	52.072199

Similarly, we can get the next three using the following:

```
second3 = df.loc['Arkansas':'Colorado']
second3
```

state	cases	population	gdp	urban	cases_norm	gdp_norm
Arkansas	3281	3017804	132596.4	56.16	1.087214	43.938042
California	50470	39512223	3205000.1	94.95	1.277326	81.114143
Colorado	15207	5758736	400863.4	86.15	2.640684	69.609616

Because the data is so small, histograms do not make sense. Instead, we directly turn to using summary statistics. The means for the normalized COVID rates for these two groups are:

```
first3['cases_norm'].mean()
```

```
0.9915956672540028
```

```
second3['cases_norm'].mean()
```

```
1.6684081069437913
```

Note that the mean COVID case rate for states in `second3` is much larger than for states in `first3`. To carry out a statistical test, we usually want to reduce the observed summary statistics to a single statistic, called the *test statistic*:

Definition

test statistic

A single numerical value that can be used in a statistical test.

When we are comparing two groups and observe different values for some summary statistic of the two groups, the usual test statistic is the difference between the values of the summary statistic. For our example, this is the difference of the means. The terms in the difference are ordered so that the difference will be a positive value. Thus, for the difference of means, the test statistic is:

```
diff = second3['cases_norm'].mean() - first3['cases_norm'].mean()
diff
```

```
0.6768124396897884
```

Seeing such a large difference, we should consider possible hypotheses (i.e., possible explanations) for the difference. Some possible hypotheses for the observed difference between `first3` and `second3` are:

1. States that come later alphabetically might have higher COVID rates. It seems hard to justify why the name of a state would affect its COVID rate.
2. The states in `second3` might differ from the states in group `first3` in some significant way. This could be true, but we should consider another possibility first.
3. The differences could be caused by random sampling. That is, the COVID rates all come from the same underlying random phenomena, but because of randomness, some states end up with higher rates in the observed data than others. In `first3` and `second3`, it just happened by randomness that `second3` ended up with more states with higher COVID rates, and `first3` ended up with more states with lower rates.

In statistics, we want to create hypotheses that can be *tested*, which means that we can assess the likelihood that the hypothesis is true or false. In this case, we use the term *statistical hypothesis*:

Definition

statistical hypothesis

An explanation for phenomena observed in a data set that can be formally tested using the data.

For conciseness, I will typically just write *hypothesis*.

Hypothesis 3 is a very common type of hypothesis when working with two groups of data that have different values for some summary statistic of interest. This type of hypothesis is called the *null hypothesis*:

Definition

null hypothesis

The hypothesis that the feature(s) being measured comes from the same random phenomena; any observed differences come from issues related to random sampling, such as having sets that are too small to accurately measure the true value of some summary statistic.

The concept of the null hypothesis was introduced in *The Design of Experiments* by Sir Ronald A. Fisher, Hafner Press, New York, 1935.

It is helpful to introduce mathematical notation to refer to hypotheses. The null hypothesis is usually denoted by H_0, which is typically read as "H naught". Intuitively, the zero can be read as implying zero difference between the populations (in the feature being compared).

When there is a null hypothesis, we typically compare it with the hypothesis that the feature being measured actually comes from different random distributions that depend on which group it came from. We call this the *alternative hypothesis* and denote it by H_a. Thus, we will typically conduct a form of *binary hypothesis test*:

Definition

binary hypothesis test

A statistical test that decides between two competing statistical hypotheses.

Usually, H_a cannot be tested directly because we do not know ahead of time *how* the underlying phenomena differ. Thus, we will usually conduct tests by assuming the null hypothesis is true. Then, we apply an approach similar to falsification in the natural sciences: we try to determine the probability that the observed value of the test statistic could have come from the null hypothesis. If there is a very low probability of observing such a value under the null hypothesis, then some alternative hypothesis is likely true. (Randomness in the data will usually prevent us from deciding that any hypothesis is absolutely true; i.e., we will only be able to estimate the probabilities of hypotheses.)

When we estimate the probability of some observed statistic occurring under the assumption that H_0 is true, it is called a *null hypothesis significance test*:

Definition

null hypothesis significance test (NHST)

A type of binary hypothesis test that estimates the probability of observing such an extreme value of the test statistic under the condition that some null hypothesis, H_0, is true.

There are several methods to test the null hypothesis. Before introducing the technique we will use, we note that statistical methods can generally be put into one of two categories, *model-based* or *model-free*:

Definitions

model-based methods

In *model-based methods*, the data or test statistic is assumed to come from some known statistical distribution. Such methods often allow the use of analytical methods.

model-free methods

In *model-free methods*, no assumption is made about the data or test statistic fitting to some statistical model. Thus, analytical methods are usually not possible. Instead, techniques must be applied to the data itself to answer questions about the data.

It is important to note that either approach will rely on assumptions. We will primarily use the model-free method because it has several important advantages:

1. Model-based methods typically require the data reach a certain size before the model is a reasonable approximation for the test statistic. This is not an issue with model-free methods (although data size is important for other reasons).
2. It does not require analytical formulas to be available for every test statistic that may be considered.
3. It does not rely on mathematical formulas that may seem obscure or that require a significant amount of foundational work in probability to understand.
4. The model-free approach we will use is easier to get started with, especially for engineers who are experienced in programming and simulating phenomena.

The approach we will use is called *resampling*:

Definition

resampling

A type of statistical simulation in which new samples are repeatedly drawn from the existing data and the statistical measures being used are evaluated for each set of new samples.

When using two-group null hypothesis testing via resampling, the null hypothesis is that the data in the two groups come from the same underlying random phenomena. Thus, we can pool the data from the two groups, and then we draw samples from the pooled data to represent samples from the underlying random phenomena under H_0.

We can put all the data into one long NumPy vector by horizontally stacking the data. To do this, we use NumPy's `np.hstack()` function:

```
pooled = np.hstack((first3['cases_norm'], second3['cases_norm']))
pooled
```

```
array([1.441512  , 0.48254038, 1.05073463, 1.08721441, 1.27732626,
       2.64068365])
```

Since we have numerical data, we leverage the NumPy.random submodule to draw samples. We will import this library as `npr`:

```
import numpy.random as npr
```

The NumPy.random module produces pseudorandom numbers, and its internal state can be set by passing a number (called the seed) to the `npr.seed()` function. To allow readers to obtain the same results as those shown in this book, I set the seed of the random number generator as shown below (where the seed value was chosen arbitrarily). However, be aware that if you re-run cells or run cells out of order, you will get different results because the seed changes every time random values are generated.

```
npr.seed(21341)
```

Now we can randomly draw data from `pooled` using `npr.choice()`, which takes two arguments. The first argument is the variable from which to draw random samples. The second argument is the number of samples to draw (if not given, the default is one).

```
npr.choice(pooled, 3)
```

```
array([1.05073463, 1.05073463, 2.64068365])
```

Note that the resulting array has a repeated value, even though there are no repeated values in the variable `pooled`. There are two ways to sample from data:

1. **Sampling with replacement:** Items drawn are placed back into the array from which data is being sampled. Any number of items may be drawn.
2. **Sampling without replacement:** Items drawn are removed from the array from which data is being sampled. The maximum number of items that can be drawn is the size of the original array.

`npr.choice()` defaults to sampling with replacement but can perform sampling without replacement if passed the keyword argument `replace=False`. For example, here are two draws of 6 items using each approach:

```
npr.seed(21398475)
print(npr.choice(pooled, 6))
print(npr.choice(pooled, 6, replace=False))
```

```
[1.27732626 0.48254038 1.05073463 1.27732626 1.27732626 0.48254038]
[1.05073463 1.27732626 1.08721441 0.48254038 1.441512   2.64068365]
```

To perform resampling under H_0, we simulate how `first3` and `second3` **could have been created** if they came from the same random phenomena. Thus, we create new vectors `first3_sample` and `second3_sample` that contain the same number of samples as `first3` and `second3`, respectively, but where the samples that make up each vector are drawn randomly from the pooled data.

In this chapter, we use the most popular method, which is drawing the data **with replacement**. This is commonly called **bootstrap sampling**, and bootstrap sampling is often used to simulate random values of the test statistic under H_0.

When creating the new groups, it is important to use the same sizes as the original groups. The cell below will perform one such random draw of `first3_sample` and `second3_sample`. Try running the following code cell a few times:

```
first3_sample = npr.choice(pooled, len(first3))
second3_sample = npr.choice(pooled, len(second3))
print(f'new group first3_sample: {first3_sample}')
print(f'new group second3_sample: {second3_sample}')
```

```
new group first3_sample: [0.48254038 2.64068365 1.05073463]
new group second3_sample: [0.48254038 0.48254038 1.05073463]
```

The goal of an NHST is to determine whether the observed value of the test statistic could be attributed just to randomness. To do this, we use resampling to determine how often we see such a large test statistic (difference in means) under H_0. Copy the previous code cell and add the line shown at the end of the following cell to print out the difference in means. Run the cell a few times to see what types of values are observed in the sample test statistic:

```
first3_sample = npr.choice(pooled, len(first3))
second3_sample = npr.choice(pooled, len(second3))
print(f'new group first3_sample: {first3_sample}')
print(f'new group second3_sample: {second3_sample}')

print(f'original value of test statistic: {diff}')
print(f'sample value of test statistic: ',
      f'{second3_sample.mean() - first3_sample.mean()}')
```

```
new group first3_sample: [1.27732626 1.05073463 0.48254038]
new group second3_sample: [1.08721441 1.27732626 0.48254038]
original value of test statistic: 0.6768124396897884
sample value of test statistic:  0.012159927814834104
```

One thing to note is that the sample test statistic can be either positive or negative in a particular simulation iteration. We previously mentioned that we want to determine "how often we see such a large difference", but this turns out to be ambiguous because we could use the magnitude (absolute value) of the sample test statistic, or we could use the signed sample test statistic. For now, let's use the magnitude, which is called a two-tail or two-sided test. I will have more to say about one-tail tests vs. two-tail tests in Section 5.3.

Now, we will build a simulation to estimate the probability of seeing such a large value of the test statistic under H_0. This probability is called the **p-value**. For now, we will say that the difference is *statistically significant* if the observed p-value is smaller than a threshold; i.e., if it is very unlikely to observe such an extreme difference in summary statistics between the groups under the null hypothesis.

An important part of statistical testing is to declare what is being tested and the criterion for statistical significance **before** performing the statistical tests. Thus, we need to determine the p-value threshold **before** running the simulation. For instance, if H_0 is true, should we consider the observed difference significant if we see differences as large as we observed in our data 10% of the time? What if it were only 5% of the time? How about only 1% of the time? There is no right answer to that question.

The threshold for rejecting the null hypothesis will determine the maximum probability that we will allow for making the error of rejecting the null hypothesis when it is actually true. For instance, if the threshold is 10%, then if there is no actual difference between the groups for the random phenomena being measured, a difference in summary statistics as large as the one observed could still occur as often as 10% of the time. In many scientific fields, a *p*-value threshold of 5% is used, and that is the threshold used in this book.

Note that if the estimated *p*-value is above the threshold, that does not mean that the null hypothesis is true or that the alternative hypothesis is false. We say that we "fail to reject the null hypothesis". In many such cases, the null hypothesis will actually be false, but we just do not have enough data to support rejecting the null hypothesis.

With these details out of the way, we can build our simulation. As in Chapter 2, the simulation is a `for` loop in which each iteration represents one random draw. In each iteration, the simulation creates two samples, `first3_sample` and `second3_sample`, by drawing randomly from the pooled data. For each sample, we calculate the averages, and we calculate the sample test statistic as the difference of the sample averages. A counter is used to track how often the magnitude of the sample test statistic is as large as the observed value of the test statistic in the original data.

```
# These are common to most simulation:
# 1) Set up the number of iterations (i.e., number of samples from the pool)
# 2) Initialize our counter to zero
num_sims = 10_000
count = 0

# Put these outside the loop to save execution time since they don't change
# Even though we know these, it is good to get in the habit of setting
# them dynamically from the data
first3_len = len(first3)
second3_len = len(second3)

for sim in range(num_sims):
    # Bootstrap sampling
    first3_sample = npr.choice(pooled, first3_len)
    second3_sample = npr.choice(pooled, second3_len)

    # Calculate the absolute value of the difference of means
    newdiff = abs(first3_sample.mean() - second3_sample.mean())

    # Update the counter if it exceeds the difference from the original groups
    if newdiff >= diff:
        count += 1

print('Prob. of seeing a result this extreme =~', count / num_sims)
```

```
Prob. of seeing a result this extreme =~ 0.2038
```

Run the cell above a few times. The results will vary but should be close to 0.2. Since this *p*-value is much larger than our threshold of 5% (i.e., 0.05), we **fail to reject the null hypothesis**. The observed difference will occur approximately 20% of the time, even if there is no difference in normalized COVID rates among these states.

3.6.2 Testing the observed differences in COVID rates

Now let's apply bootstrap resampling to test the previously observed differences in COVID rates based on GDP per capita using the full set of states:

```
m_gdp = df['gdp_norm'].median()
m_gdp
higher_gdp = df[df['gdp_norm'] > m_gdp]
lower_gdp = df[df['gdp_norm'] <= m_gdp]
hgdp_mu = higher_gdp['cases_norm'].mean()
lgdp_mu = lower_gdp['cases_norm'].mean()
```

As a reminder, we previously found the average normalized COVID rates for higher GDP and lower GDP states to be:

```
hgdp_mu, lgdp_mu
```

```
(3.511241636831491, 1.9258062634390531)
```

We will use the difference in sample means as the test statistic for an NHST, so one of our first steps is to calculate the value of this test statistic:

```
diff_gdp = hgdp_mu - lgdp_mu
diff_gdp
```

```
1.585435373392438
```

For this case, the pooled data is simply all of the normalized COVID data:

```
pooled_covid = df['cases_norm']
```

Copy the simulation from the previous section and modify it to draw data to represent new `higher_gdp` and `lower_gdp` groups:

```
# These are common to most simulation:
# 1) Set up the number of iterations (draws from the pool)
# 2) Initialize our counter to zero
num_sims = 10_000
count = 0

# Put these outside the loop to save execution time, since they don't change
higher_gdp_len = len(higher_gdp)
lower_gdp_len = len(lower_gdp)


for sim in range(num_sims):
    # Bootstrap sampling
    higher_gdp_sample = npr.choice(pooled_covid, higher_gdp_len)
    lower_gdp_sample = npr.choice(pooled_covid, lower_gdp_len)

    # Calculate the absolute value of the difference of means
    newdiff = abs(higher_gdp_sample.mean() - lower_gdp_sample.mean())

    # Update the counter if it exceeds the difference from the original groups
    if newdiff >= diff_gdp:
```

(continues on next page)

(continued from previous page)

```
        count += 1

print(f'Prob. of seeing a result this extreme =~ {count / num_sims: .3f}')
```

```
Prob. of seeing a result this extreme =~  0.072
```

Run the simulation a few times. The results should vary but will generally be around 0.07 (7%). Since this is larger than our p-value threshold of 5%, we fail to reject the null hypothesis. The data does not provide enough evidence for us to be sure that the states with higher GDP per capita are associated with higher COVID rates.

The simulation code above can easily be turned into a function that can be applied to any data when we are using resampling to perform a two-sided test for an observed difference of means. To make this into a function:

1. Click in the cell with the simulation above and choose "Copy Cells" from Jupyter's Edit menu. (Alternatively, you can use the Copy command from your browser's Edit menu or the corresponding keyboard shortcut.)
2. Paste it into a new cell block using "Paste Cells Below" from the Edit menu. (If you used the Copy command from your browser, then click in an empty cell and use your browser's Paste command.)
3. The simulation code will be the body of the new function, and so it will need to be indented. Select the entire contents of the simulation code in the new cell using your mouse, the Select All command from your browser's Edit menu, or the keyboard shortcut. Then press the keyboard shortcut to indent the block (Control+] or Command+]).
4. Add a function signature at the top to name your function and determine the arguments. Provide arguments for the pooled data, the observed mean difference, and the number of observations in each group. It is also helpful to make the number of simulation iterations be an argument of the function in case we need to run more iterations to accurately estimate the p-value.
5. I have revised the names of the variables inside the function to make them more generic, and I recommend you do the same.

The final function should look as follows:

```
def resample_mean(pooled_data, diff, len1, len2, num_sims=10_000):
    '''Resample from pooled data and conduct a two-tailed NHST
    on the mean-difference

    Inputs
    ------
    pooled_data: NumPy array of all data in the original 2 groups
    diff: observed difference in sample means in the original groups
    len1, len2: the lengths of the original groups
    num_sims: the number of simulation iterations

    Output
    ------
```

(continues on next page)

(continued from previous page)

```
    prints resulting p-value
    '''

    count = 0

    for sim in range(num_sims):
        # Bootstrap sampling
        group1 = npr.choice(pooled_data, len1)
        group2 = npr.choice(pooled_data, len2)

        # Calculate the absolute value of the difference of means
        newdiff = abs(group1.mean() - group2.mean())

        # Update the counter if observed difference as large as original
        if newdiff >= diff:
            count += 1

    print(f'Prob. of seeing a result this extreme =~ {count / num_sims}')
```

We can use this function to test the p-value when the states are grouped according to GDP per 1000 residents as follows:

```
resample_mean(pooled_covid, diff_gdp, len(higher_gdp), len(lower_gdp))
```

```
Prob. of seeing a result this extreme =~ 0.0746
```

Now let's apply bootstrap resampling when the states are grouped using the urban index:

```
m_urban = df["urban"].median()
higher_urban = df[df["urban"] > m_urban]
lower_urban = df[df["urban"] <= m_urban]
```

```
higher_urban["cases_norm"].mean(), lower_urban["cases_norm"].mean()
```

```
(3.8908749904320756, 1.5461729098384682)
```

```
diff_urban = higher_urban["cases_norm"].mean() - lower_urban["cases_norm"].mean()
diff_urban
```

```
2.344702080593607
```

We can use our `resample_mean()` function to perform the two-sided test via bootstrap resampling:

```
resample_mean(pooled_covid, diff_urban, len(higher_urban), len(lower_urban))
```

```
Prob. of seeing a result this extreme =~ 0.0104
```

The observed p-value is .01, which is below our 5% threshold, so we conclude that the 25 states with a higher urban index have a statistically significant difference in COVID rates in comparison to the 25 states with a lower urban index.

3.6.3 Distribution of the bootstrap mean-difference

In each iteration of the bootstrap simulation, random sampling produces a new difference of sample means, which is a numerical random value. These random values can be characterized by the set of values and the mapping of probability to those values. (I.e., how likely different values are.) We call this the *distribution* of the random values.

Let's see how the bootstrap means are distributed by modifying our simulation to also produce a histogram of those values. Copy the simulation from above, update the function name, and add the marked lines to store every mean difference and plot the histogram:

```
import matplotlib.pyplot as plt

def resample_mean_hist(pooled_data, diff, len1, len2, num_sims=10_000):
    '''(Docstring omitted to conserve space)'''
    count = 0
    sample_means = []  # New variable to store sample means

    for sim in range(num_sims):
        # Bootstrap sampling
        group1 = npr.choice(pooled_data, len1)
        group2 = npr.choice(pooled_data, len2)

        # Calculate the absolute value of the difference of means
        newdiff = abs(group1.mean() - group2.mean())

        sample_means += [group1.mean() - group2.mean()]  # Store sample mean

        # Update the counter if it exceeds the difference from the original groups
        if newdiff >= diff:
            count += 1

    print(f'Prob. of seeing a result this extreme =~ {count / num_sims}')
    plt.hist(sample_means, 40)  # Plot the histogram with 40 bins
```

Note that the histogram does not depend on the mean difference observed in the data. Since the pooled data and the group sizes are the same for both the GDP grouping and the urban index grouping, both will produce the same histograms (up to variations from the random sampling). Let's call the new function with the arguments used for the urban index data. Because the focus of this simulation is the histogram and 10,000 points will already produce a very detailed histogram, we leave `num_sims` set to the default.

```
resample_mean_hist(pooled_covid, diff_urban, len(higher_urban), len(lower_urban))
```

```
Prob. of seeing a result this extreme =~ 0.0078
```

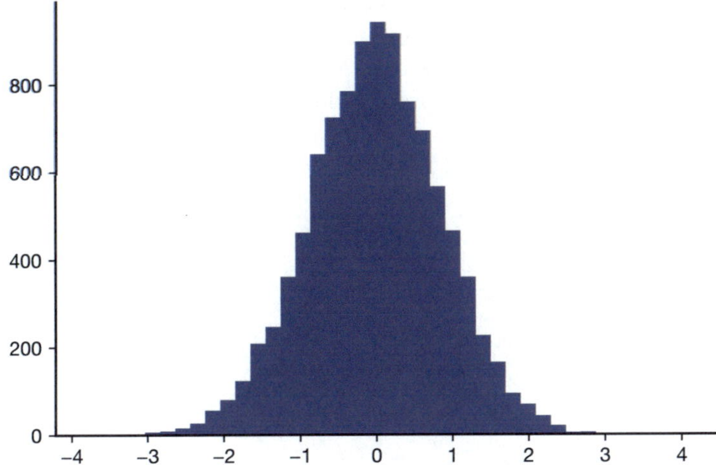

A few observations:

1. The difference of means has a bell shape – we saw this shape before. Why do you think that is?
2. The majority of the values fall between -2 and +2. Thus, it is not surprising that getting a mean difference as large as 2.34 is rare.

Later in the book, we will investigate how we can use the histogram of the bootstrap distribution of the test statistic to provide confidence intervals on the mean difference under the null hypothesis. This is a way to provide more information about how an observed difference in means compares to what is expected under the null hypothesis.

We will also use a model for the distribution of the difference of means and show how to use the model to give an equation for the p-value of the NHST.

Terminology review and self-assessment questions

Interactive flashcards to review the terminology introduced in this section and self-assessment questions are available at fdsp.net/3-6, which can also be accessed using this QR code:

3.7 A Quick Preview of Two-Dimensional Statistical Methods

Until now, we have evaluated the interplay between multiple variables of the COVID data set only when we partition the data using one of the variables and then look at the effect on the COVID rates across the two sets in the partition. The summary statistic is applied to only one variable, the COVID rates. We refer to the number of variables that are used in a statistical method as the *dimension* of the method. So, apart from partitioning the data, we have limited ourselves to one-dimensional methods until now.

Here, I briefly introduce two-dimensional methods as a preview of techniques explored in Chapter 12. To extend our work to two dimensions, we start by creating a scatter plot with on variable on each axis. Usually, the variable that is used as the x coordinate is the one that we intuitively think of as causative of the other variable. For example, it makes sense to think that the rate of COVID cases may increase or decrease with the urban index at the time that COVID became prevalent, but it makes little sense to think that the urban index at that time is affected by the rate of COVID cases.

Fig. 3.4 shows a plot of COVID-19 rates versus urban index for the 50 US states. The plot shows that COVID rates tend to increase with urban index. Thus, it is not surprising that partitioning the data based on the urban index produced statistically significant differences for the mean COVID rates.

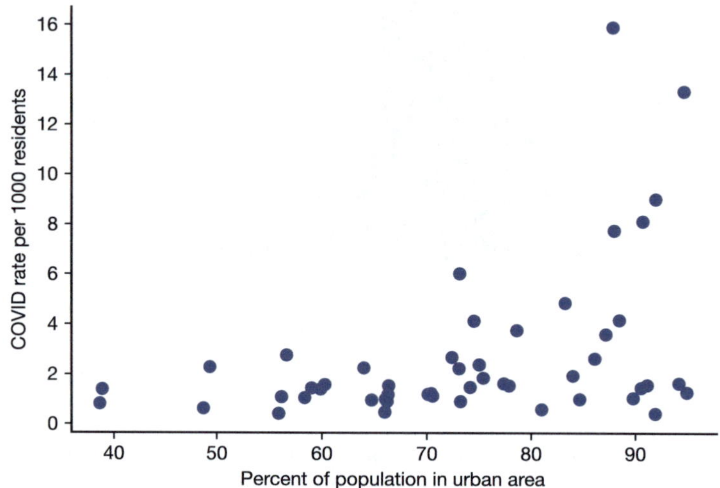

FIGURE 3.4
Covid rates versus urban index for the 50 US states.

When we have data with such an evident relation, it makes sense to fit the data with a curve. Let's try fitting this data with a line, which we will write as

$$y = mx + b.$$

To determine what line makes sense, we will again need some measure of error, so let's use the sum of squared errors again. Thus for each x value (urban index), we will calculate a y value using the formula for the line, and we will calculate the squared error to the actual COVID rate in the data.

Suppose we first fix the y-intercept to zero ($b = 0$) and find the slope m. By inspection of Fig. 3.4, the slope should be somewhere between 0 and 15/100=0.15. Fig. 3.5 shows the squared error for several values in this range. (Code to produce this figure is online at fdsp.net/3.7.) We can use it to determine which gives the best result. The best value of m is approximately 0.039, with total squared error of 427.69, and the resulting fit is shown in Fig. 3.6.

To find the optimal linear fit to minimize the total squared error, we would have to optimize m and b simultaneously. This is called *simple linear regression.* We can use the `linregress()` function from the SciPy.stats library to find the optimal slope and y-intercept:

```
from scipy.stats import linregress

lr = linregress(df['urban'], df['cases_norm'])
lr
```

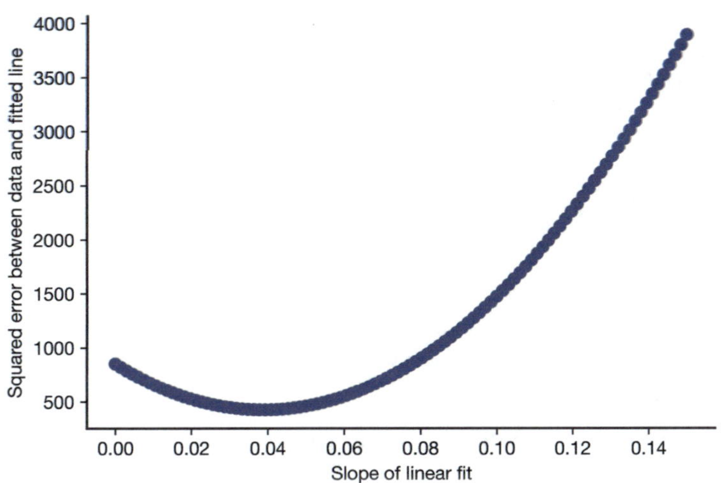

FIGURE 3.5
Sum of squared errors for linear fit between percent urban and COVID rate as a function of m when $b = 0$.

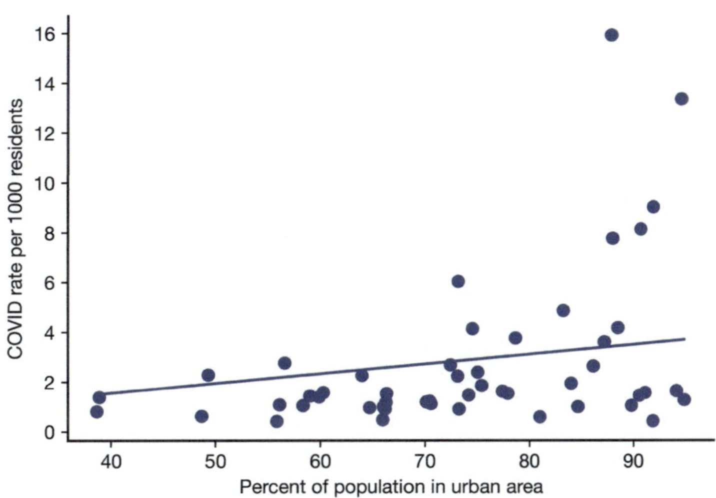

FIGURE 3.6
Best linear fit (for y-intercept of 0) for covid rate as function of urban index.

```
LinregressResult(slope=0.0917027723921469, intercept=-4.029131107469203,␣
↪rvalue=0.42451161644162405, pvalue=0.0021221565036770016, stderr=0.
↪02823082347614785, intercept_stderr=2.116778144832277)
```

The optimal fit has an `intercept` value of approximately -4 and a much steeper `slope` of approximately 0.092 (compared to the best slope of 0.039 when the intercept was fixed to 0). The resulting total squared error is:

```
ylr = lr.slope * x + lr.intercept
np.sum( (true_y -ylr)**2 )
```

```
397.62981896613957
```

We will investigate linear and nonlinear regression techniques in Chapter 12.

Terminology review and self-assessment questions

Interactive flashcards to review the terminology introduced in this section and self-assessment questions are available at fdsp.net/3-7, which can also be accessed using this QR code:

3.8 Chapter Summary

In this chapter, we started working with real data. I introduced Pandas dataframes, including how to load data from a CSV file and how to select and operate on data in dataframes. I then introduced partitions as a way to separate data into distinct groups. Various summary statistics were introduced and explained in terms of the error metrics that they minimize. Next, I introduced the concept of null hypothesis significance testing and showed how to conduct NHSTs using the state data on COVID rates and socioeconomic factors. Finally, some two-dimensional statistics approaches were briefly introduced.

Access a list of key take-aways for this chapter, along with interactive flashcards and quizzes at fdsp.net/3-8, which can also be accessed using this QR code:

4

Introduction to Probability

In this chapter, I provide an introduction to the field of Probability, which provides an essential mathematical foundation for data science. Until now, some of the wording has been intentionally vague because we do not yet have the correct terminology. This chapter will provide a rigorous mathematical foundation for dealing with random phenomena, as well as carefully defined terminology for describing these phenomena and how they interact.

4.1 Outcomes, Sample Spaces, and Events

We begin with building some intuition about probability by further investigating relative frequency. We start by defining what can be an *outcome* of a random experiment:

> Definition
>
> **outcome**
>
> A result of a random experiment that cannot be further decomposed.

Consider rolling a standard six-sided die and observing the top face. The possible outcomes are 1, 2, 3, 4, 5, and 6. We place all possible outcomes into a set called the *sample space*:

> Definition
>
> **sample space**
>
> The set of all possible outcomes of a random experiment.

We will denote the sample space by S. Another common notation is Ω (capital omega). When referring to multiple experiments, we often use subscripts to distinguish the different sample spaces. So, for our six-sided die, the sample space is

$$S_1 = \{1, 2, 3, 4, 5, 6\}.$$

If you need a brief review of sets, see the online Appendix at fdsp.net/appendix.

Note that outcomes are not the only types of experimental results that we can ask about. For the six-sided die, we could ask if the result is even or if the result is less than 3. These are not outcomes. Instead, they can be written as sets of outcomes. We call these *events*:

DOI: 10.1201/9781003324997-4

Definition

event

An event is a set of outcomes. The event occurs if the result of the experiment is any of the outcomes in the set.

If we roll a six-sided die, the event that the outcome is even can be written as

$$E = \{2, 4, 6\}.$$

Note that we have assigned the notation E to refer to this event. We will often create notations to refer to events to formalize and simplify our mathematical representations.

If we roll a six-sided die, the event that the outcome is less than 3 can be written as

$$G = \{1, 2\}.$$

Sample spaces can contain any type of item. For instance, we can refer to the sample space for flipping a coin as

$$S_2 = \{\text{heads, tails}\}.$$

More typically, we would introduce mathematical notation H = outcome is heads, T = outcome is tails. Then we can use the more concise notation $S_2 = \{H, T\}$.

Terminology review and self-assessment questions

Interactive flashcards to review the terminology introduced in this section and self-assessment questions are available at fdsp.net/4-1, which can also be accessed using this QR code:

4.2 Relative Frequencies and Probabilities

It is useful to use the relative frequencies of the outcomes of an experiment to give some insight into the probabilities of outcomes. In this section, we give a mathematical definition for the relative frequency of an outcome and then analyze its properties.

Consider a random experiment with a finite sample space, S. Let $|S| = K$; i.e., there are K possible outcomes. Consider repeating the experiment a fixed number of times and observing the sequence of outcomes. This is a *repeated experiment*, which is a special type of *compound experiment*:

Definition

compound experiment

A compound experiment is any experiment that consists of multiple sub-experiments.

Compound experiments are made up of *trials*:

Definition

trial (compound experiment)

One of the sub-experiments that make up the compound experiment.

We determine relative frequency using a special type of compound experiment called a *repeated experiment*:

Definition

repeated experiment

A compound experiment in which the trials are identical and independent of each other.

By *independent*, we mean that what happens in one experiment cannot affect the probabilities of what happens in another experiment. We will define this concept of independence more carefully later.

Let N denote the number of trials in the repeated experiment, and consider a particular outcome of the experiment. If we indexed the sample space S, then we can assign an index k to this outcome, and we will refer to it as outcome k for convenience. Note that we use the lowercase letter k to refer to one of the outcomes, whereas the capital letter K refers to the total number of outcomes.

WARNING

Using uppercase and lowercase forms of the same letter to refer to related phenomena is a common convention in probability. **It is important to use fonts or write letters so that uppercase and lowercase are distinguishable to readers**. When writing letters, the convention is to use curvier forms and loops on lowercase versions and straighter, blockier forms for the uppercase versions. This helps keep the two cases distinct. Here are some examples in my handwriting:

Lowercase Letters	Uppercase Letters
k	K
m	M
n	N
p	P

The relative frequency of outcome k is the proportion of times that outcome occurred in the N trials. Let $n_k(N)$ denote the number of times outcome k is observed. We can immediately see that

$$0 \leq n_k(N) \leq N, \tag{4.1}$$

since the result is a count of occurrences and the maximum number of times that outcome k can occur is if it occurs on all N trials.

Since each trial results in one outcome, then the sum of the number of times each outcome is observed must equal the number of trials:

$$\sum_{k=1}^{K} n_k(N) = N. \tag{4.2}$$

Let $r_k(N)$ denote the relative frequency of outcome k:

$$r_k(N) = \frac{n_k(N)}{N}.$$

We can immediately derive a few properties of relative frequencies. Dividing (4.1) by N yields

$$0 \leq r_k(N) \leq 1,$$

so relative frequencies are values between 0 and 1 (inclusive of both endpoints). Dividing (4.2) by N yields

$$\sum_{k=1}^{K} r_k(N) = 1,$$

so the sum of the relative frequencies is equal to 1.

Many random experiments possess properties that are sometimes referred to as *statistical regularity*:

Definition

statistical regularity

An experiment has statistical regularity if, under repeated experiments, the relative frequencies converge (in some sense) to some fixed values.

To see this, let's simulate rolling a fair six-sided die and plot the relative frequencies for several increasing values of the number of trials, N. We can generate random numbers between 1 and 6 (inclusive) using NumPy.random's `randint()` function. Since `randint()` uses the Pythonic convention that the upper endpoint is **not included**, we have to pass an upper endpoint of 7 instead of 6. Here is how to generate 20 dice values:

```
import numpy as np
import numpy.random as npr

npr.seed(9823467)
outcomes = npr.randint(1, 7, size=20)
outcomes
```

```
array([6, 3, 6, 4, 2, 2, 4, 3, 3, 6, 5, 2, 5, 6, 1, 5, 5, 6, 1, 4])
```

Now we need to count the number of occurrences of each value. We can achieve this by passing the outcomes to NumPy's `np.unique()` function and using the keyword argument `return_counts=True`. The function returns two arrays. The first array contains the unique values in the `outcomes` array, and the second array returns the corresponding count for each observed value.

```
vals, counts = np.unique(outcomes, return_counts=True)
print(vals, counts)
```

```
[1 2 3 4 5 6] [2 3 3 3 4 5]
```

Then the relative frequencies of the values are equal to the returned counts divided by the total number of outcomes:

```
rel_freqs = counts / len(outcomes)

print("Relative frequencies:")
for i, val in enumerate(vals):
    print(val, rel_freqs[i])
```

```
Relative frequencies:
1 0.1
2 0.15
3 0.15
4 0.15
5 0.2
6 0.25
```

When we visualize relative frequencies, we typically use a stem plot, which shows the value as a circle at the top of a "stem" that connects to the x-axis:

```
import matplotlib.pyplot as plt

plt.stem(vals, rel_freqs)
plt.xlabel("Dice values")
plt.ylabel("Relative frequencies")
plt.title("20 trials");
```

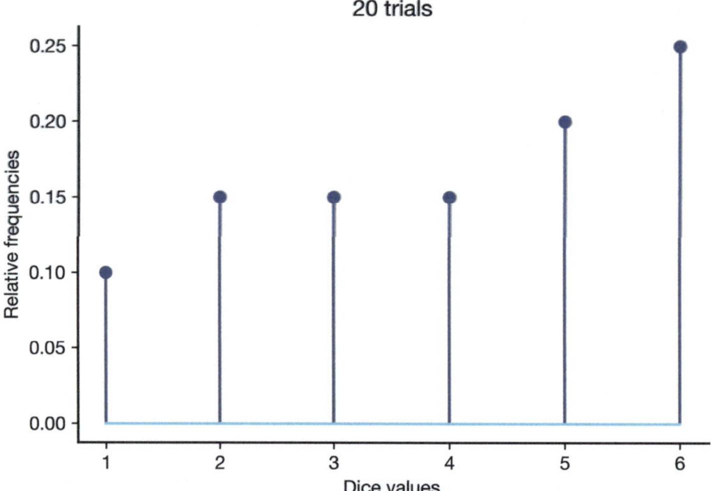

The values show some significant differences in the relative frequencies (from 0.1 to 0.25). Let's see how the relative frequencies change as we vary N. We will generate all 100,000 values at once and then subscript the array when we want to limit the analysis to a particular part. For instance, we can get the first 10 values as follows:

```
outcomes100k = npr.randint(1, 7, size=100_000)
print(outcomes100k[:10])
```

```
[3 4 6 4 2 3 5 3 5 3]
```

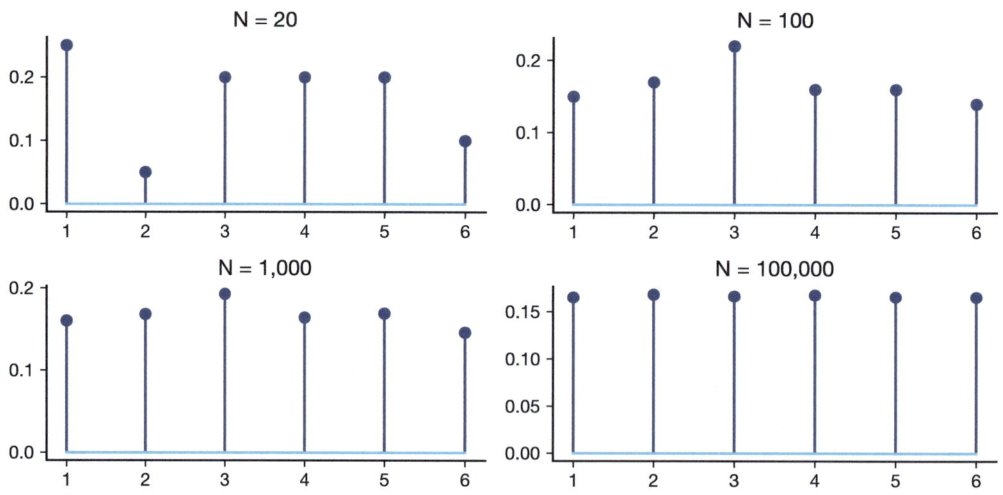

FIGURE 4.1
Relative frequencies for rolling a fair six-sided die for different numbers of trials (N).

Fig. 4.1 shows the relative frequencies as the number of trials varies from 20 to 100,000. As the value of N increases, the relative frequencies converge to a common value:

```
print(rel_freqs)
```

```
[0.16541 0.16857 0.1666  0.16792 0.16585 0.16565]
```

The relative frequencies converge to a value around 0.166, which is close to 1/6. If we further increase the number of trials, the relative frequencies will get closer and closer to 1/6. For an experiment with statistical regularity like this, we can say that the values that the relative frequencies converge to are the *probabilities* of the outcomes. The probability of outcome k can be denoted by p_k.

Note that probabilities then inherit the properties of relative frequencies:

$$0 \leq p_k \leq 1, \text{ and}$$

$$\sum_k p_k = 1.$$

We will require these properties to hold for all probabilities.

Although using relative frequency to define probabilities is simple, it has problems for more general use:

- Not all experiments possess statistical regularity. In particular, some experiments can never be repeated, such as the change in the S&P 500 Index of the U.S. stock market on a given day.
- Even if an experiment has statistical regularity, some experiments cannot be repeated without great difficulty or expense, such as characterizing the amount of fuel required to send a SpaceX rocket to Mars.
- If an experiment has statistical regularity and can be repeated, it is not clear how to determine how many trials are needed to evaluate the probabilities with some prescribed precision.
- Even the meaning of "converge" is not clear in the context of experiments because there is no way to relate precision with a required number of trials.

Because of these limitations, we usually try to provide more mathematical definitions of probabilities for an experiment that are informed by experimental results or other knowledge of the system. The simplest such approach works for a class of experiments that are said to be "fair", which is the topic of the next section.

Terminology review and self-assessment questions

Interactive flashcards to review the terminology introduced in this section and self-assessment questions are available at fdsp.net/4-2, which can also be accessed using this QR code:

4.3 Fair Experiments

Let's start by defining what we mean by a *fair experiment*:

Definition

fair experiment

An experiment is said to be fair if every outcome is equally likely.

Fair experiments form the basis for the oldest ways of calculating probabilities. Much of the early work on probability was motivated by gambling, and games involving cards, dice, and coins are generally based on the assumption that the underlying experiment is fair.

Consider a fair experiment with $|S| = N$ outcomes, and let p_i denote the probability of outcome i, then

$$\begin{aligned}\sum_{i=0}^{N-1} p_i &= 1 \\ \sum_{i=0}^{N-1} p_0 &= 1 \quad \text{(since all the probabilities are equal)} \\ Np_0 &= 1 \\ p_0 &= \frac{1}{N}.\end{aligned}$$

Since the probabilities of the outcomes are equal,

$$p_i = \frac{1}{N} = \frac{1}{|S|}.$$

So, for instance, the probability of getting any number on a fair six-sided die is $1/6$. For a coin, let p_H and p_T denote the probabilities of heads and tails, respectively. If the coin is fair, then we have $p_H = p_T$, and

$$\begin{aligned}p_H + p_T &= 1 \\ p_H + p_H &= 1 \\ 2p_H &= 1 \\ p_H &= \frac{1}{2}.\end{aligned}$$

So $p_H = p_T = 1/2$, where $|S| = 2$.

Although this is an accurate and easy way to calculate probabilities, it is extremely limited in its application because most probabilities we will encounter are not from fair experiments. Even very simple compound experiments that are based on fair experiments are not necessarily fair.

Example 4.1: Monopoly Dice

Monopoly uses two fair six-sided dice. The player moves the sum of the amounts shown on the two dice. Simulate rolling the pair of dice one million times and plot the relative frequencies:

```
import matplotlib.pyplot as plt
import numpy as np
import numpy.random as npr

die1 = npr.randint(1, 7, size=1_000_000)
die2 = npr.randint(1, 7, size=1_000_000)
dice = die1 + die2

vals, counts = np.unique(dice, return_counts=True)
rel_freqs = counts / len(dice)

plt.stem(vals, rel_freqs)
plt.xlabel("Sum of top faces of two dice")
plt.ylabel("Relative frequencies");
```

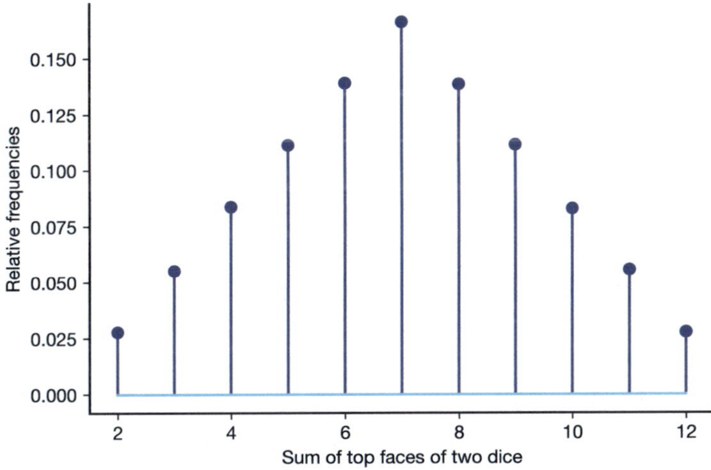

Example 4.2: Flip A Fair Coin Until Heads Occurs

Another simple example is the number of coin flips needed to observe heads on repeated flips of a fair coin:

```
num_sims = 10000
outcomes = []

for sim in range(num_sims):
    count = 0
    while 1:
        coin = npr.randint(2)
        count = count + 1
        if coin == 1:  # Use 1 to represent heads
            break
    outcomes += [count]
```

(continues on next page)

(continued from previous page)

```
vals, counts = np.unique(outcomes, return_counts=True)
rel_freqs = counts / len(outcomes)

plt.stem(vals, rel_freqs)
plt.xlabel("No. of flips to first heads")
plt.ylabel("Relative frequencies");
```

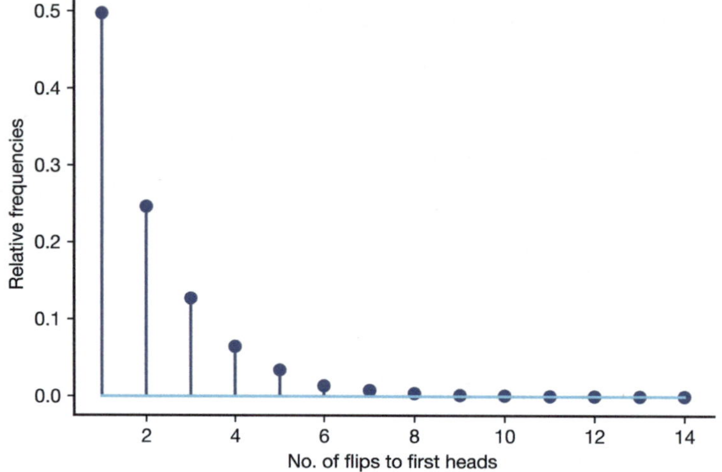

These limitations lead us to our ultimate formulation for probabilities, which is the topic of the next section.

Terminology review and self-assessment questions

Interactive flashcards to review the terminology introduced in this section and self-assessment questions are available at fdsp.net/4-3, which can also be accessed using this QR code:

4.4 Axiomatic Probability

The methods of the previous two sections that define probabilities using relative frequencies or based on properties of fair experiments are helpful to develop some intuition about probability. However, these methods have limitations that restrict their usefulness to many real-world problems. These problems were recognized by mathematicians working on probability and motivated these mathematicians to develop an approach to probability that:

- is not based on a particular application or interpretation,
- agrees with models based on relative frequency and fair probabilities,
- agrees with our intuition (where appropriate), and
- is useful for solving real-world problems.

The approach they developed is called *Axiomatic Probability*. Axiomatic means that there is a set of assumptions or rules (called axioms) for probability but that the set of rules is made as small as possible. This approach may initially seem unnecessarily mathematical, but I believe you will soon see that this approach will help develop a fundamentally sound understanding of probability.

4.4.1 Probability spaces

The first step in developing Axiomatic Probability is to define the core objects to which the axioms apply. These objects are collected in a *probability space*:

> Definition
>
> **probability space**
>
> An ordered collection (tuple) of three objects, denoted by $(S, \mathcal{F}, P)$. These objects are the *sample space*, the *event class*, and the *probability measure*, respectively.

Since there are three objects in a probability space, it is sometimes said that **a probability space is a triple**.

1. The Sample Space

We have already introduced the sample space in Section 4.1. It is a **set** containing all possible outcomes for an experiment.

2. The Event Class

The second object, denoted by a calligraphic F ($\mathcal{F}$), is called the *event class*:

> Definition
>
> **event class**
>
> For a sample space S and a probability measure P, the event class, denoted by $\mathcal{F}$, is a collection of all subsets of S to which we will assign probability (i.e., for which P will be defined). The sets in $\mathcal{F}$ are called events.

We require that the event class be a σ-algebra (read "sigma algebra") of S, which is a concise and mathematically precise way to say that combinations of outcomes and events using the usual set operations will still be events in $\mathcal{F}$.[1]

For many readers of this book, the above explanation will be sufficient to understand what events are in $\mathcal{F}$. If you feel satisfied with this explanation, you may skip ahead to the heading **Event class for finite sample spaces**. If you want more mathematical depth and rigor, here are the properties that $\mathcal{F}$ must satisfy to be a σ-algebra on S:

1. $\mathcal{F}$ contains the sample space: $S \in \mathcal{F}$.
2. $\mathcal{F}$ is **closed under complements**: If $A \in \mathcal{F}$, then $\overline{A} \in \mathcal{F}$.
3. $\mathcal{F}$ is **closed under countable unions**: If $A_1, A_2, \ldots$ are a finite or countably infinite number of sets in $\mathcal{F}$, then

[1] Technically, only countably infinite combinations are guaranteed to be in the event class.

$$\bigcup_i A_i \in \mathcal{F}.$$

Note that DeMorgan's Laws (see the online Appendix at fdsp.net/appendix) immediately imply a few other properties:

- The null set $\emptyset$ is in $\mathcal{F}$ by combining properties 1 and 2. $S \in \mathcal{F}$, and so $\overline{S} = \emptyset \in \mathcal{F}$.
- $\mathcal{F}$ is **closed under countable intersections**. If A_1, $A_2, \ldots$ are a finite or countably infinite number of sets in $\mathcal{F}$, then by property 2, $\overline{A_1}, \overline{A_2} \ldots$ are in $\mathcal{F}$. By property 3,

$$\bigcup_i \overline{A_i} \in \mathcal{F}.$$

If we apply DeMorgan's Laws to this expression, we have

$$\overline{\bigcap_i A_i} \in \mathcal{F}.$$

Then by applying property 2 again, we have that

$$\bigcap_i A_i \in \mathcal{F}.$$

Event class for finite sample spaces

When S is finite, we almost always take $\mathcal{F}$ to be the *power set* of S, i.e., the set of all subsets of S.

Example 4.3: Event Class for Flipping a Coin

Consider flipping a coin and observing the top face. Then $S = \{H, T\}$, and

$$\mathcal{F} = \{\emptyset, H, T, \{H, T\} = S\}.$$

Note that $|S| = 2$, and $|2^S| = 4 = 2^{|S|}$.

Exercise

Consider rolling a standard six-sided die. Give the sample space, S, and the power set of the sample space, 2^S. What is the cardinality of 2^S?

When $|S| = \infty$, weird things can happen if we try to assign probabilities to every subset of S. For typical data science applications, we can assume that any event we want to ask about will be in the event class, and we do not need to explicitly enumerate the event class.

3. The Probability Measure

Until now, we have discussed the probabilities of outcomes. However, this is not the approach taken in probability spaces. Instead, we define a function that assigns probabilities to events, and this function is called the *probability measure*:

Definition

probability measure

The probability measure, P, is a real-valued set-function that maps every element of the event class to the real line.

Note that in defining the probability measure, we do not specify the range of values for P. This is because, at this point, we are only defining the structure of the probability space through the types of elements that make it up.

Although P assigns outcomes to events (as opposed to outcomes), every outcome in S is typically an event in the event class. Thus, P is more general in its operation than we have considered in our previous examples. As explained in Section 4.1, an event occurs if the experiment's outcome is one of the outcomes in that event's set.

4.4.2 Axioms of probability

As previously mentioned, axioms are a minimal set of rules. There are three Axioms of Probability, and they specify the properties of the probability measure:

The Axioms of Probability

I. For every event E in the event class $\mathcal{F}$, $P(E) \geq 0$ *(the event probabilities are non-negative).*

II. $P(S) = 1$ *(the probability that some outcome occurs is 1).*

III. For all pairs of events E and F in the event class that are disjoint ($E \cap F = \emptyset$), $P(E \cup F) = P(E) + P(F)$ *(if two events are disjoint, then the probability that either one of the events occurs is equal to the sum of the event probabilities).*

When dealing with infinite sample spaces, an alternative version of Axiom III should be used:

III'. If $A_0, A_1, \ldots$ is a sequence of events that are all disjoint ($A_i \cap A_j = \emptyset \; \forall i \neq j$), then

$$P\left[\bigcup_{k=0}^{\infty} A_k\right] = \sum_{k=0}^{\infty} P\left[A_k\right].$$

Many students of probability wonder why Axiom I does not specify that $0 \leq P(E) \leq 1$. The answer is that the second part of that inequality is unnecessary because it can be proven from the other axioms. Anything that is not required is removed to ensure that the axioms are a minimal set of rules.

Axiom III is a powerful tool for calculating probabilities. However, it must be used carefully.

Example 4.4: Applying Axiom III, Example 1

A fair six-sided die is rolled twice. What is the probability that the top face on the first roll is less than 3? What is the probability that the top face on the second roll is less than 3?

First, let's define some notation for the events of interest:

Let E_i denote the event that the top face on roll i is less than 3. Then

$$E_1 = \{1_1, 2_1\},$$

where k_l denotes the **outcome** that the top face is k on roll l. Similarly,

$$E_2 = \{1_2, 2_2\}.$$

Note that we can rewrite

$$E_i = \{1_i\} \cup \{2_i\}.$$

Because outcomes are always disjoint, Axiom III can be applied to yield

$$\begin{aligned} P(E_i) &= P\left(\{1_i\} \cup \{2_i\}\right) \\ &= P\left(\{1_i\}\right) + P\left(\{2_i\}\right) \\ &= \frac{1}{6} + \frac{1}{6}, \end{aligned}$$

where the last line comes from applying the probability of an outcome in a fair experiment. Thus, $P(E_i) = 1/3$ for $i = 1, 2$. Most readers will have known this answer intuitively.

Example 4.5: Applying Axiom III, Example 2

Consider the same experiment described in the previous example. However, let's ask a slightly different question: what is the probability that either the value on the first die is less than 3 **or** the value on the second die is less than 3. (This could also include the case that both are less than 3.) Mathematically, we write this as $P(E_1 \cup E_2)$ using the events already defined.

Since E_1 and E_2 correspond to events on completely different dice, it may be tempting to apply Axiom III like:

$$\begin{aligned} P(E_1 \cup E_2) &= P(E_1) + P(E_2) \\ &= \frac{1}{3} + \frac{1}{3} \\ &= \frac{2}{3}. \end{aligned}$$

However, it is easy to see that this thinking is somehow not correct. For example, if we defined events G_i to be the event that the value on die i is less than 5, this approach would imply that

$$\begin{aligned} P(G_1 \cup G_2) &= P(G_1) + P(G_2) \\ &= \frac{2}{3} + \frac{2}{3} \\ &= \frac{4}{3}. \end{aligned}$$

Hopefully, you recognize that this is not an allowed value for a probability! Let's see what went wrong. We can begin by estimating $P(E_1 \cup E_2)$ using simulation:

```
import numpy as np
import numpy.random as npr

num_sims = 100_000

# Generate the dice values for all simulations:
die1 = npr.randint(1, 7, size=num_sims)
die2 = npr.randint(1, 7, size=num_sims)

# Each comparison will generate an array of True/False value
E1occurred = die1 < 3
E2occurred = die2 < 3

# Use NumPy's union operator (|) to return True where either array is True:
Eoccurred = E1occurred | E2occurred

# NumPy's count_nonzero function will count 1 for each True value
# and 0 for each False value
print("P(E1 or E2) =~", np.count_nonzero(Eoccurred) / num_sims)
```

```
P(E1 or E2) =~ 0.55938
```

The estimated probability is about 0.56, which is lower than predicted by trying to apply Axiom III. The problem is that Axiom III does not hold for events E_1 and E_2 because they are not disjoint: both can occur at the same time. Let's enumerate everything that could happen by writing the outcomes of die 1 and die 2 as a tuple, where (j, k) means that die 1's outcome was j and die 2's outcome was k.

The tables below show the outcomes with selected events highlighted in color and bold. The Python code to generate all of these tables is available online at fdsp.net/4-4. Let's start by printing all outcomes with the outcomes in event E_1 highlighted in blue:

```
Outcomes in E1 are in bold and blue:
(1, 1)   (1, 2)   (1, 3)   (1, 4)   (1, 5)   (1, 6)
(2, 1)   (2, 2)   (2, 3)   (2, 4)   (2, 5)   (2, 6)
(3, 1)   (3, 2)   (3, 3)   (3, 4)   (3, 5)   (3, 6)
(4, 1)   (4, 2)   (4, 3)   (4, 4)   (4, 5)   (4, 6)
(5, 1)   (5, 2)   (5, 3)   (5, 4)   (5, 5)   (5, 6)
(6, 1)   (6, 2)   (6, 3)   (6, 4)   (6, 5)   (6, 6)
```

We can easily modify this to highlight the events in E_2 in green:

```
Outcomes in E2 are in bold and green:
(1, 1)   (1, 2)   (1, 3)   (1, 4)   (1, 5)   (1, 6)
(2, 1)   (2, 2)   (2, 3)   (2, 4)   (2, 5)   (2, 6)
(3, 1)   (3, 2)   (3, 3)   (3, 4)   (3, 5)   (3, 6)
(4, 1)   (4, 2)   (4, 3)   (4, 4)   (4, 5)   (4, 6)
(5, 1)   (5, 2)   (5, 3)   (5, 4)   (5, 5)   (5, 6)
(6, 1)   (6, 2)   (6, 3)   (6, 4)   (6, 5)   (6, 6)
```

You should already see that the set of outcomes in E_1 overlaps with the set of outcomes in E_2. To make that explicit, let's highlight the outcomes that are in both E_1 and E_2 in red:

```
(1, 1)   (1, 2)   (1, 3)   (1, 4)   (1, 5)   (1, 6)
(2, 1)   (2, 2)   (2, 3)   (2, 4)   (2, 5)   (2, 6)
(3, 1)   (3, 2)   (3, 3)   (3, 4)   (3, 5)   (3, 6)
(4, 1)   (4, 2)   (4, 3)   (4, 4)   (4, 5)   (4, 6)
(5, 1)   (5, 2)   (5, 3)   (5, 4)   (5, 5)   (5, 6)
(6, 1)   (6, 2)   (6, 3)   (6, 4)   (6, 5)   (6, 6)
```

So, does this mean that we cannot use Axiom III to solve this problem? No. We just have to be more careful. Let's highlight all the outcomes that belong to $E_1 \cup E_2$ with a yellow background.

```
Outcomes in  E1 union E2 are on a yellow background:
(1, 1)   (1, 2)   (1, 3)   (1, 4)   (1, 5)   (1, 6)
(2, 1)   (2, 2)   (2, 3)   (2, 4)   (2, 5)   (2, 6)
(3, 1)   (3, 2)   (3, 3)   (3, 4)   (3, 5)   (3, 6)
(4, 1)   (4, 2)   (4, 3)   (4, 4)   (4, 5)   (4, 6)
(5, 1)   (5, 2)   (5, 3)   (5, 4)   (5, 5)   (5, 6)
(6, 1)   (6, 2)   (6, 3)   (6, 4)   (6, 5)   (6, 6)

Number of outcomes in E1 OR E2 is 20
```

If an event is written in terms of a set of K **outcomes** $o_0, o_1, \ldots, o_{K-1}$, and the experiment is fair and has N total outcomes, then Axiom III can be applied to calculate the probability as

$$\begin{aligned}
P(E) &= P\left(\{o_0, o_1, \ldots, o_{K-1}\}\right) \\
&= P\left(o_0\right) + P\left(o_1\right) + \ldots + P\left(o_{K-1}\right) \\
&= \frac{1}{N} + \frac{1}{N} + \ldots + \frac{1}{N} && \text{(total of } K \text{ terms)} \\
&= \frac{K}{N}.
\end{aligned}$$

We believe that this experiment is fair and that any of the 36 total outcomes is equally likely to occur. The form above is general to any event for a fair experiment, and it is convenient to rewrite it in terms of set cardinalities as

$$P(E) = \frac{|E|}{|S|}.$$

Applying this to our example, we can easily calculate the probability we are looking for as

$$\begin{aligned} P(E_1 \cup E_2) &= \frac{|E_1 \cup E_2|}{|S|} \\ &= \frac{20}{36} \\ &= \frac{5}{9}. \end{aligned}$$

The calculated value matches our estimate from the simulation:

```
5 / 9
```

```
0.5555555555555556
```

The key to making this work is that we had to realize several things:

- E_1 and E_2 are not outcomes. They are events, and they can occur at the same time.
- The outcomes of the experiment are the combination of the outcomes from the individual rolls of the two dice.
- The composite experiment is still a fair experiment. It is easy to calculate probabilities using Axiom III and the properties of fair experiments once we determine the number of outcomes in the event of interest.

However, we can see that the solution method is still lacking in some ways:

- It only works for fair experiments.
- It requires enumeration of the outcomes in the event – this may be challenging to do without a computer and may not scale well.

Some of the difficulties in solving this problem come from not having a larger toolbox; i.e., the axioms provide a very limited set of equations for working with probabilities. In the next section, we explore several corollaries to the axioms and show how these can be used to simplify some problems in probability.

Terminology review and self-assessment questions

Interactive flashcards to review the terminology introduced in this section and self-assessment questions are available at fdsp.net/4-4, which can also be accessed using this QR code:

4.5 Corollaries to the Axioms of Probability

Corollaries are results that can be proven from more fundamental theorems or properties. In this case, we are interested in what additional properties or relations we can develop for the probability measure based on the Axioms of Probability.

Let $A \in \mathcal{F}$ and $B \in \mathcal{F}$. Then the following properties of P can be derived from the axioms and the mathematical structure of $\mathcal{F}$:

Corollary 1. Let $\overline{A}$ denote the complement of A; i.e., $\overline{A}$ contains every outcome in S that is not in A. Then

$$P\left(\overline{A}\right) = 1 - P\left(A\right).$$

Proof:

The proof uses Axioms II and III, as well as properties of sets. First, note that $A \cup \overline{A} = S$ and that A and $\overline{A}$ are disjoint by definition. Then by Axioms II and III,

$$\begin{aligned} P(S) &= 1 \\ P(A \cup \overline{A}) &= 1 \\ P(A) + P(\overline{A}) &= 1 \\ P(A) &= 1 - P(\overline{A}). \end{aligned}$$

Example 4.6: A Pair of Die Values Less Than 3, Take 1

A fair six-sided die is rolled twice, and the top faces are recorded. What is the probability that neither roll is less than 3?

Let E_i be the event that the outcome of roll i is less than 3. Then we are asked to find $P\left(\overline{E_1} \cap \overline{E_2}\right)$. By DeMorgan's rules,

$$P\left(\overline{E_1} \cap \overline{E_2}\right) = P\left(\overline{E_1 \cup E_2}\right).$$

We can apply Corollary 1 to get

$$P\left(\overline{E_1 \cup E_2}\right) = 1 - P\left(E_1 \cup E_2\right).$$

But we already found the probability on the right-hand side to be 5/9 in Section 4.4. Thus, the probability we are looking for is $1 - 5/9 = 4/9$.

Corollary 2. $P(A) \leq 1$

As previously noted, this restriction is not included in the axioms.

Proof:

By Corollary 1, we have

$$P(A) = 1 - P(\overline{A}).$$

By Axiom I, $P(\overline{A}) \geq 0$, so it must be that $P(A) \leq 1$.

Corollary 3. $P(\emptyset) = 0$

Proof:

By Corollary 1, we have

$$P(\emptyset) = 1 - P(\overline{\emptyset}).$$

But $\overline{\emptyset} = S$. Thus,

$$\begin{aligned} P(\emptyset) &= 1 - P(S) \\ &= 1 - 1 \\ &= 0. \end{aligned}$$

Corollary 4. If $A_0, A_1, \ldots, A_{n-1}$ are pairwise mutually exclusive, then

$$P\left(\bigcup_{k=1}^{n-1} A_k\right) = \sum_{k=1}^{n-1} P(A_k).$$

Proof: The proof is by induction and is omitted.

Corollary 5. $P(A \cup B) = P(A) + P(B) - P(A \cap B)$

Proof:

The proof requires a bit of work with sets and applying Axiom III. It is based on the Venn diagram for the event $A \cup B$ that is shown in Fig. 4.2. Note that the regions $A \cap \overline{B}$, $A \cap B$, and $B \cap \overline{A}$ are disjoint.

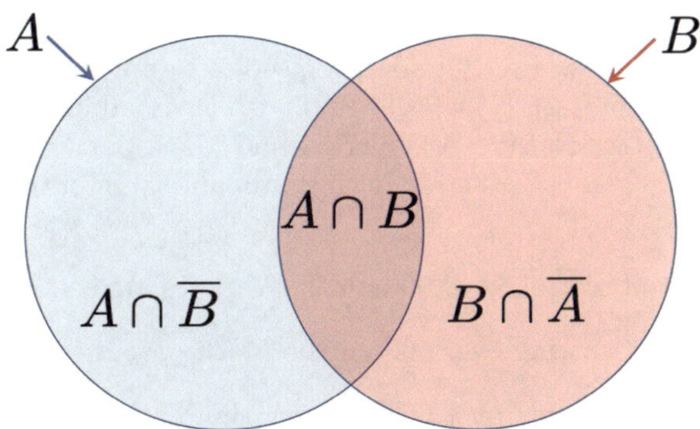

FIGURE 4.2
Illustration of decomposition of union of A and B into 3 disjoint regions.

In addition, we can write

$$\begin{aligned} A \cup B &= \left(A \cap \overline{B}\right) \cup \left(A \cap B\right) \cup \left(B \cap \overline{A}\right), \\ A &= \left(A \cap \overline{B}\right) \cup \left(A \cap B\right), \text{ and} \\ B &= \left(B \cap \overline{A}\right) \cup \left(A \cap B\right). \end{aligned}$$

Applying Axiom III to each of these identities yields

$$P(A \cup B) = P\left(A \cap \overline{B}\right) + P\left(A \cap B\right) + P\left(B \cap \overline{A}\right),$$
$$P(A) = P\left(A \cap \overline{B}\right) + P\left(A \cap B\right), \text{ and}$$
$$P(B) = P\left(B \cap \overline{A}\right) + P\left(A \cap B\right).$$

We can write the last two equations as

$$P\left(A \cap \overline{B}\right) = P(A) - P\left(A \cap B\right), \text{ and}$$
$$P\left(B \cap \overline{A}\right) = P(B) - P\left(A \cap B\right).$$

Substituting into the remaining equation yields

$$P(A \cup B) = \left[P(A) - P\left(A \cap B\right)\right] + P\left(A \cap B\right) + \left[P(B) - P\left(A \cap B\right)\right],$$

which simplifies to the desired result.

Example 4.7: Applying Unions and Intersections of Events to Questions About SAT Scores

In the United States, most high school students applying for college take a standardized achievement test called the SAT, which is administered by the College Board. The main test consists of a Verbal and Math part, each of which is scored on a scale from 200 to 800. The following probability information is inferred from data found online[1].

- The probability of scoring over 600 on the Verbal part is 0.24.
- The probability of scoring over 600 on the Math part is 0.25.
- The probability of scoring over 600 on both the Math and Verbal parts is 0.16.

(a) If a college requires that a student score over 600 on at least one of the Math and Verbal parts of the SAT to be eligible for admission, what is the probability that a randomly chosen student will meet the college's SAT criterion for admission?

Define the following notation:

- V = event that Verbal score > 600
- M = event that Math score > 600

We are looking for $P(V \cup M)$. We can apply Corollary 5,

$$\begin{aligned} P(V \cup M) &= P(V) + P(M) - P(V \cap M) \\ &= 0.24 + 0.25 - 0.16 \\ &= 0.33. \end{aligned}$$

About 1/3 of students who take the SAT will meet the college's criterion.

(b) What is the probability that a randomly chosen student does not score over 600 on either part of the SAT?

This probability can be written as $P(\overline{V} \cap \overline{M})$. It is the complement of the event in part (a):

$$\begin{aligned} P(\overline{V} \cap \overline{M}) &= P\left(\overline{V \cup M}\right) \\ &= 1 - P(V \cup M) \\ &= 0.67. \end{aligned}$$

(c) What is the probability that a randomly chosen student scores over 600 on the Math part but scores 600 or less on the Verbal part?

This probability can be written as $P(M \cap \overline{V})$. I will provide a purely mathematical answer, but it is helpful to draw a Venn diagram to visualize this scenario. Note that

$$\begin{aligned} P(M) &= P(M \cap V) + P\left(M \cap \overline{V}\right) \\ P\left(M \cap \overline{V}\right) &= P(M) - P(M \cap V) \\ &= 0.25 - 0.16 \\ &= 0.09. \end{aligned}$$

[1]Data on correlation from https://eportfolios.macaulay.cuny.edu/liufall2013/files/2013/10/New_Perspectives.pdf . Data on mean and variance is from https://blog.prepscholar.com/sat-standard-deviation .

Example 4.8: A Pair of Die Values Less Than 3, Take 2

A fair six-sided die is rolled twice. What is the probability that either of the rolls is a value less than 3?

As before, let E_i be the event that the top face on roll i is less than 3, for $i = 1, 2$. Referring back to Section 4.4, note that it is much easier to calculate the number of outcomes in $E_1 \cap E_2$ than to count the number of items in $E_1 \cup E_2$. (Intersections are always no bigger than the smallest constituent set, where as unions are no smaller than the largest of the constituent sets.)

The intersection is shown as the red and bolded outcomes in the following table:

(1, 1)	**(1, 2)**	(1, 3)	(1, 4)	(1, 5)	(1, 6)
(2, 1)	**(2, 2)**	(2, 3)	(2, 4)	(2, 5)	(2, 6)
(3, 1)	(3, 2)	(3, 3)	(3, 4)	(3, 5)	(3, 6)
(4, 1)	(4, 2)	(4, 3)	(4, 4)	(4, 5)	(4, 6)
(5, 1)	(5, 2)	(5, 3)	(5, 4)	(5, 5)	(5, 6)
(6, 1)	(6, 2)	(6, 3)	(6, 4)	(6, 5)	(6, 6)

We see that $|E_1 \cap E_2| = 4$, which means that $P(E_1 \cap E_2) = 4/36 = 1/9$. Then we can calculate the desired probability as

$$\begin{aligned} P(E_1 \cup E_2) &= P(E_1) + P(E_2) - P(E_1 \cap E_2) \\ &= \frac{2}{6} + \frac{2}{6} - \frac{1}{9} \\ &= \frac{5}{9}. \end{aligned}$$

Corollary 6. If $A \subset B$, then $P(A) \leq P(B)$.

Proof:

It may be helpful to refer to the Venn diagram in Fig. 4.3 for some intuition:

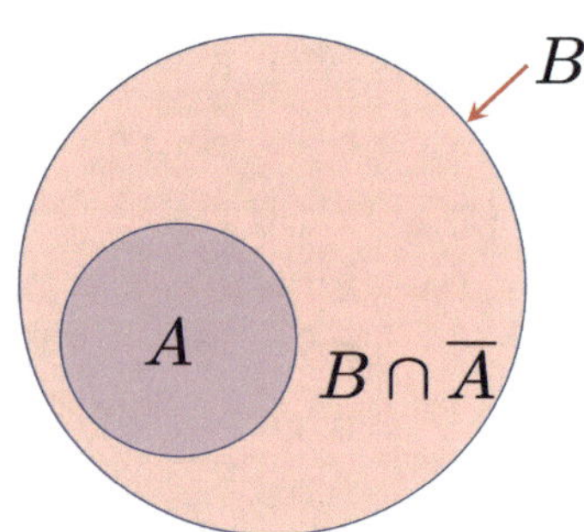

FIGURE 4.3
Venn diagram for A a subset of B, showing the decomposition of B and A into disjoint sets.

Note that $A \subset B$ implies that $A \cap B = A$. Then

$$\begin{aligned} P(B) &= P(B \cap A) + P\left(B \cap \overline{A}\right) \\ &= P(A) + P\left(B \cap \overline{A}\right) \\ P(A) &= P(B) - P\left(B \cap \overline{A}\right). \end{aligned}$$

Since $P\left(B \cap \overline{A}\right) \geq 0$, $P(A) \leq P(B)$.

Corollary 7.

$$\begin{aligned} P\left(\bigcup_{k=0}^{n-1} A_k\right) = &\sum_{k=0}^{n-1} P(A_j) - \sum_{j<k} P(A_j \cap A_k) + \cdots \\ &+ (-1)^{(n-1)} P(A_0 \cap A_1 \cap \cdots \cap A_{n-1}). \end{aligned}$$

Here the ellipses indicate that the pattern should be continued until exhaustion: add all single events, subtract off all intersections of pairs of events, add in all intersections of 3 events, ...

Proof: The proof is by induction and is omitted.

This list of Corollaries is not unique; nor is it meant to be comprehensive. Rather, these represent some common tools that we will use in our work on probability.

Terminology review and self-assessment questions

Interactive flashcards to review the terminology introduced in this section and self-assessment questions are available at fdsp.net/4-5, which can also be accessed using this QR code:

4.6 Combinatorics

"Counting is the religion of this generation... Anybody can count..."
– Gertrude Stein

For fair experiments with a finite sample space S, we used Axiom III of the Axioms of Probability to show that the probability of an event E is simply

$$P(E) = \frac{|E|}{|S|}.$$

Thus, the problem of calculating $P(E)$ is simplified to counting the cardinalities of S and E. As most people have learned to count as young children, this sounds like a simple exercise. However, in practice, this is often quite challenging. In fact, this general problem space is rich enough that this branch of mathematics has its own name:

> Definition
>
> **combinatorics**
> The mathematics of counting.

WARNING

One of the biggest mistakes made by people learning the material in this section is to try to mix probabilities and combinatorics when computing the probability of an event. When solving most problems with combinatorics, the solution should only consist of these three steps:

1. Find $|S|$ using combinatorics. This is usually not difficult.
2. Find $|E|$ using combinatorics. This may be very challenging, even if the textual description of an event is simple.
3. Calculate

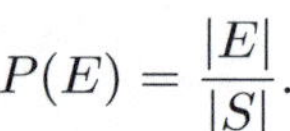

$$P(E) = \frac{|E|}{|S|}.$$

No probability is computed until the last step. This approach is illustrated when calculating probabilities in the examples below.

For the data science topics covered in this book, we do not encounter many very challenging counting problems. However, if you look back over the past chapters, you may discover that several problems have already been introduced that could be solved by counting:

- **(A)**: In Section 2.1, we asked "If you only observe 6 heads on the 20 flips, should you reject the idea that the coin is fair?". We solved this problem by simulating flipping a fair coin 20 times and determining the probability of observing 6 or fewer heads. However, if we record the ordered heads and tails outcomes, this is a fair experiment, and so we could solve for this probability by counting the number of events with 6 or fewer heads.
- **(B)**: In Section 3.6, we conducted a bootstrap test where we pooled the data for the US states and then performed bootstrap sampling to create random samples under the null hypothesis. When determining how many random draws we might use, it is useful to know how many ways there are to partition the data in this way.
- **(C)**: In Section 4.3, the relative frequencies were shown for the sum of two fair 6-sided dice, as used in Monopoly. If we record the ordered set of top faces of the dice, then this is a fair experiment, and we can calculate the exact probabilities for the sum of the faces using combinatorics.

The first and third examples are types of compound experiments, which consist of a sequence of subexperiments. Suppose there are K subexperiments, and the sample space for the ith subexperiment is S_i. Then the sample space for the compound experiment is

$$S = \left\{(s_0, s_1, \ldots, s_{K-1}) \mid s_0 \in S_0, s_1 \in S_1, \ldots s_{K-1} \in S_{K-1}\right\}.$$

This notation for S is tedious. We introduce the *Cartesian product* operator to simplify the notation for the sample space:

Definition

Cartesian product

The *Cartesian product* of two sets A and B is denoted $A \times B$ and is defined by

$$A \times B = \{(a,b) \mid a \in A \text{ and } b \in B\}.$$

That is, it is the set of all two-tuples with the first element from set A and the second element from set B.

We can form the sample space for our repeated experiment through repeated application of the Cartesian product to the individual sample spaces:

$$S = S_0 \times S_1 \times \ldots \times S_{K-1}.$$

We start by seeing how we can use Python to calculate probabilities by enumerating and then counting sample spaces and events:

4.6.1 Enumerating sample spaces and events using IterTools

Let's start by enumerating S and showing how it can be used to calculate probabilities. We will use the Python `itertools` library, which is distributed as part of standard Python distributions, to enumerate S. Begin by importing this library:

```
import itertools
```

Example 4.9: (A) Monopoly Dice

Consider applying combinatorics to the Monopoly dice problem. The sample spaces for the two dice are the same. In Python, we can define them using simple ranges. The `itertools` library has a `product()` function to carry out the Cartesian product over these two ranges:

```
S0 = range(1, 7)
S1 = range(1, 7)

S = itertools.product(S0, S1)
for s in S:
    print(s, ' ', end='')
```

```
(1, 1)  (1, 2)  (1, 3)  (1, 4)  (1, 5)  (1, 6)  (2, 1)  (2, 2)  (2, 3)  (2,␣
↪4)  (2, 5)  (2, 6)  (3, 1)  (3, 2)  (3, 3)  (3, 4)  (3, 5)  (3, 6)  (4, 1) ␣
↪(4, 2)  (4, 3)  (4, 4)  (4, 5)  (4, 6)  (5, 1)  (5, 2)  (5, 3)  (5, 4)  (5,␣
↪5)  (5, 6)  (6, 1)  (6, 2)  (6, 3)  (6, 4)  (6, 5)  (6, 6)
```

WARNING

Note that the `itertools` functions generally provide an *iterator* for the resulting set. Iterators produce values one at a time until exhausted. Unlike looping over a range or list, you cannot execute the loop again using the iterator once it has reached the end. Also, the number of items to be iterated over cannot be directly determined – you must iterate over all of the items to determine how many there are.

When the number of items being iterated over is small, the iterator can be used to directly create a list of these items. As expected from our previous examples, there are 36 items in the sample space:

```
Slist = list(itertools.product(S0, S1))
len(Slist)
```

```
36
```

We can find the probability for the sums of the dice faces by iterating over S and counting the number of times each sum occurs. It is easy to see that the sum of the faces will be between 2 and 12. We initialize a list with 13 zeros (from 0 to 12) to store the counts.

```
counts = [0] * 13
for s in Slist:
    counts[sum(s)] += 1

print('sum:', '# ways of occurring')
for c in range(2, 13):
    print(f'{c:>2} : {counts[c]}')
```

```
sum: # ways of occurring
 2 : 1
 3 : 2
 4 : 3
 5 : 4
 6 : 5
 7 : 6
 8 : 5
 9 : 4
10 : 3
11 : 2
12 : 1
```

The right-hand column is the cardinality of the event described by the left-hand column. As the events partition the sample space (i.e., they are disjoint and cover everything in the sample space), the sum of the right-hand column is equal to the cardinality of S:

```
sum(counts), len(Slist)
```

```
(36, 36)
```

If we let E_i denote the event that the sum of the dice faces is i, then $P(E_i) = |E_i|/|S|$, where the values of $|E_i|$ are given in the table above. Thus, the probabilities are:

```
probs = [0] * 13

print('sum: probability')
for c in range(2, 13):
    probs[c] = counts[c] / len(Slist)
    print(f'{c : >2} : {probs[c] : .4f}')
```

```
sum: probability
 2 :  0.0278
 3 :  0.0556
 4 :  0.0833
 5 :  0.1111
 6 :  0.1389
 7 :  0.1667
 8 :  0.1389
 9 :  0.1111
10 :  0.0833
11 :  0.0556
12 :  0.0278
```

Fig. 4.4 compares the relative frequencies from simulating the dice to the analytical results.The results match quite closely. (Code to generate this figure is available online at fdsp.net/4-6.)

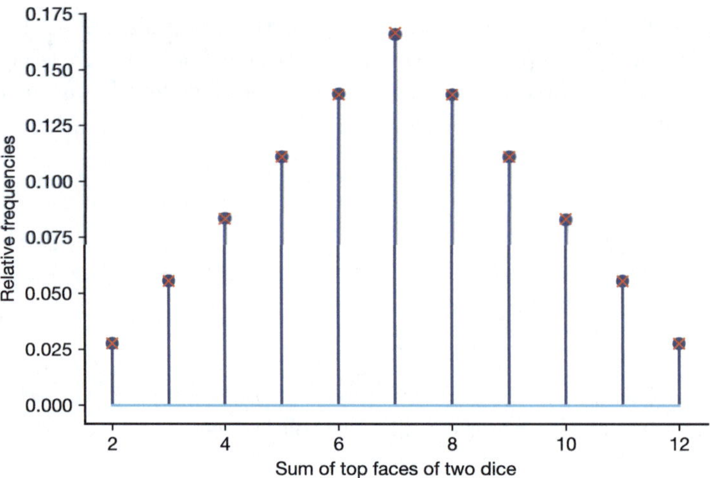

FIGURE 4.4
Comparison of analytical and theoretical results for probabilities of sum of two fair, six-sided dice.

Example 4.10: (B) Bootstrap Sampling

In the example in Chapter 3, we performed null hypothesis testing using bootstrap resampling. This is performed by pooling all the data and then repeatedly creating new groups by sampling with replacement from the pooled data. The sizes of the new groups are equal to the sizes of the groups in the original comparison.

A reasonable question to ask when performing bootstrap resampling is: **How many ways are there to resample the data using bootstrap sampling?**

It turns out that for the full set of 50 US states, the number of ways that two groups of size 25 can be created via bootstrap resampling is too large to even iterate over in Python.

Instead, we consider the smaller example of data from 6 states partitioned into two groups of size 3. It is convenient to represent the pooled data as $P_B = [0, 1, 2, 3, 4, 5]$. The actual data values are not important for counting:

```
PB = range(6)
```

Now we use `itertools` to iterate over all the groups of size 3 that can be created by sampling without replacement. This is a compound experiment with identical sample spaces for each component experiment. We can create an iterator for the sample space of one of these groups using `itertools.product()` by passing `PB` and the keyword argument `repeat` with the number of items in the groups:

```
S3 = itertools.product(PB, repeat=3)
countB1 = 0
for s in S3:
    countB1 += 1
```

(continues on next page)

(continued from previous page)

```
print('Number of ways to choose a group of size 3 under bootstrap sampling is',
      countB1)
```

```
Number of ways to choose a group of size 3 under bootstrap sampling is 216
```

Note that we are not finished. We are interested in the number of ways to choose both groups of size 3. This is a compound experiment in which each of the individual experiments has 216 outcomes. In other words, the second group has 216 outcomes for **each** outcome of the first group:

```
S3 = itertools.product(PB, repeat=3)
countB2 = 0
for s in S3:
    S3_2 = itertools.product(PB, repeat=3)
    for s in S3_2:
        countB2 += 1
print('No. of ways to choose TWO groups of size 3 under bootstrap sampling ',
      f'is {countB2}')
```

```
No. of ways to choose TWO groups of size 3 under bootstrap sampling is 46656
```

If we are running a simulation to randomly draw groups, then it makes little sense to use more than 46,656 draws because:

1. We could just iterate over all of the 46,656 groups. (This is called an *exact permutation test* and is considered in Section 5.4.)
2. As the number of random draws gets large (close to 46,656), the number of draws that are repeats of other random draws in the simulation will increase. Thus, we are really not gaining new information by further increasing the number of draws.

Example 4.11: (C) Flipping a Fair Coin 20 Times

Now consider flipping a fair coin 20 times and determining the probability of an outcome less than or equal to 6. Each of the 20 subexperiments has the same sample space. Using H to denote heads and T to denote tails, we can refer to these sample spaces as:

```
Si = ['H', 'T']
```

As before, we can create an iterator for the sample space of the compound experiment using `itertools.product` by passing `Si` and the keyword argument `repeat` with the number of repetitions as follows:

```
Sdice = itertools.product(Si, repeat=20)
```

We can count the cardinality of the sample space and the event that the number of heads is 6 or less simultaneously while looping over the outcomes in the sample space:

```
Sdice = itertools.product(Si, repeat=20)

Scount = 0
Ecount = 0

for s in Sdice:
    Scount += 1
    if s.count('H') <= 6:
        Ecount += 1

print('|E|=', Ecount, '    |S|=', Scount)
```

```
|E|= 60460     |S|= 1048576
```

Thus, the probability of seeing 6 or fewer heads is:

```
print(f'P(6 or fewer heads) = {Ecount / Scount: .4f}')
```

```
P(6 or fewer heads) =  0.0577
```

Compare this value with the estimated probability found via simulation in Section 2.4. The two values are very close, so the simulation did a good job of estimating this probability (at least with 1,000,000 iterations). Note that you probably don't want to go through all 1,048,576 outcomes by hand. Moreover, if the number of coin flips were increased significantly, it may be challenging to even iterate over them. This motivates us to develop mathematical methods for counting the cardinalities of sample spaces and events without enumerating them.

4.6.2 Determining cardinalities of sample spaces and events mathematically

We start with a basic result on counting in the context of sample spaces for compound experiments. If S is a set that can be written as a Cartesian product,

$$S = S_0 \times S_1 \times \ldots \times S_{K-1},$$

then the cardinality of S is the product of the cardinalities of the sets in the Cartesian product:

$$|S| = |S_0| \cdot |S_1| \cdot \ldots \cdot |S_{K-1}|.$$

For instance, for the Monopoly dice problem, $|S_0| = |S_1| = 6$, so $|S| = |S_0| \cdot |S_1| = 6 \cdot 6 = 36$.

This is an example of drawing items from a set and recording the ordered sequence of outcomes, where each item is placed back into the set before the next draw. This is called sampling **with replacement** and **with ordering**. For a set of k draws from N items, sampling with replacement means that each $|S_i|$ in the expression above is equal to N. Thus, we have the following result:

Sampling with Replacement and with Ordering

The number of ways to choose k items from a set of N items with replacement and with ordering is

$$\underbrace{N \cdot N \cdot \ldots \cdot N}_{k \text{ times}} = N^k.$$

For flipping a fair coin 20 times, $N = 2$ and $k = 20$, so $|S| = 2^{20} = 1,048,576$.

Example 4.12: A: Monopoly Dice

Let's consider how to find the probability for a particular value of the sum of the dice. Let E_{10} be the event that the sum of the numbers on the top faces of the two dice is 10. To find $P(E_{10})$, we have to determine $|E_{10}|$. Note that if we know the value of the first die, then the value of the second die is determined. Moreover, not all values of the first die can result in a sum of 10. So we just need to determine what values of the first die **can** result in sums of 10. The smallest value of the first die that can result in a sum of 10 will occur when the second die has a value of 6. So, the first die must be at least 4. Clearly, if it is larger than 4, there will be a value of the second die that results in a sum of 10. From this we conclude that $|E_{10}| = 3$. To be explicit,

$$E_3 = \{(4,6),\ (5,5),\ (6,4)\}.$$

Then $P(E_3) = 3/36 = 1/12$:

```
1 / 12
```

```
0.08333333333333333
```

(Reader, please compare this with the probability found through enumeration in the version of this example using `itertools`.)

Example 4.13: (B) Bootstrap Resampling

Consider again bootstrap resampling from a pool of 6 data points to two groups of cardinality 3. This a problem of sampling with replacement and with ordering, so the total number of groups is:

```
(6 ** 3) * (6 ** 3)
```

```
46656
```

This is the same result we found via `itertools`.

Now consider bootstrap resampling data from all 50 US states into groups of 25. We saw for the pooled data of size 6 that it did not make sense to draw from it more than about 50,000 times. Should we be concerned about having a similar problem when working with the full data set?

The total number of groups that can be created using the full data set is

```
(50 ** 25) * (50 ** 25.0)
```

```
8.881784197001254e+84
```

(I purposefully put a decimal on one of the 25s so that the result would be shown in scientific notation rather than as an extremely long integer.)

For reference, compare this number to the number of atoms in the observable universe: https://en.wikipedia.org/wiki/Observable_universe#Matter_content%E2%80%94number_of_atoms. According to the article, the number of atoms in the observable universe is less than 10^{80}, which is less than the number of groups we can create via bootstrap resampling. Thus, there is no concern that our resampling simulation will exhaust the number of groups when dealing with the full data set.

Example 4.14: C: Flipping a Fair Coin 20 Times

This experimental setup seems even easier than the Monopoly dice problem because of the small size of the subexperiment sample spaces, but enumerating the event that the number of heads is 6 or fewer turns out to be much more challenging and will require us to introduce some new mathematical tools. Before we get to that, let's answer a few questions that will help us build to our ultimate result:

Let H_i be the event that there are **exactly** i heads on the 20 flips.

First, what is $|H_0|$? There is exactly one way to get zero heads. All of the 20 flips were tails.

Next, what is $|H_1|$? There is exactly one head in the 20 flips. It can either be on the first flip, the second flip, ..., or finally the 20th flip. In other words, there are 20 different places for the heads to be, so $|H_1| = 20$.

Now, what is $|H_2|$? This is where things start to get challenging and interesting. We will solve this in two ways. The first way will get us the answer. The second way will help lead us to a general solution for H_i.

Counting H_2: Way 1

We can count $|H_2|$ in much the same way that we counted H_1. For convenience, let's consider the flips in order. For each place the first heads occurs, we will have multiple places that the second heads could occur. I.e., if the first heads is on flip 0, then the second heads can be on flips 1 through 19. But if the first heads is on flip 18, the second heads has to be on flip 19. The total number of outcomes in H_2 can therefore be written as

$$
\begin{aligned}
|H_2| &= \sum_{i=0}^{18} \sum_{j=i+1}^{19} 1 \\
&= \sum_{i=0}^{18} [19 - (i+1) + 1] \\
&= 19 \cdot 20 - \sum_{i=1}^{19} i \\
&= 380 - \frac{19 \cdot 20}{2} \\
&= 190.
\end{aligned}
$$

(The last summation is a standard form.)

We could have used Python to calculate this sum:

```
total = 0
for i in range(0, 19):
    for j in range(i + 1, 20):
        total += 1
print(total)
```

```
190
```

Exercise

Extend the approach described above to give a formula for the $|H_3|$. (For this purpose, I recommend you use three summation terms, although a more general and sophisticated solution for skilled programmers can be created using recursion.) Use Python to evaluate the sum. Check your answer using the self-assessment quiz at fdsp.net/4-6.

This approach will get increasingly tedious and challenging to write and compute as we consider H_i for larger i.

Counting H_2: Way 2

Consider a second approach in which we try to count how many ways we could create the result (i.e., an n-tuple) of the flips:

- The first heads can go in any of the 20 places.
- The second heads can go in any of the remaining 19 places.

Then the total number of results is $20 \cdot 19 = 380$.

But this result does not match the one we just computed in detail! Why?

The reason is that we have **overcounted**. For instance, let's mark the first value we choose by underlining it. Here are two outcomes we will find this way (values not shown are all T):

$$(\underline{H}, T, T, H, T, \ldots)$$
$$(H, T, T, \underline{H}, T, \ldots)$$

Our counting mechanism has created two different representations (*orderings*) of the same outcome (i.e., where there is a heads on rolls 0 and 3). The number we will count in this way is twice the total number because for any particular outcome, there are two different orders by which we could have created it (i.e., by putting the first H in the leftmost position or by putting the first H in the rightmost position). So, we have to divide by two to get: $|H_2| = 20 \cdot 19/2 = 190$, which agrees with our previous result.

Let's consider how we can rewrite this value to make it extensible to find H_i for $i > 2$. We first introduce the number of ways that a set of objects can be ordered:

Definition

permutation

A *permutation* is an ordering (or reordering) of a set of objects.

Given n distinct objects, it is not hard to calculate the number of permutations possible. Consider drawing the objects one at a time, until all objects have been drawn, to create the ordered set of objects:

- There are n ways to choose the first object.
- Then there are $n-1$ ways to choose the second object from the remaining set.
- Then there $n-2$ ways to choose the third object from the remaining set.
- ...
- On the final (nth) draw, there is only one object remaining in the set.

The number of permutations of n distinct objects is written as $n!$, which is read "n factorial" (en fact-or-ee-ul). The rules for Cartesian products can be applied to calculate

$$n! = n(n-1)(n-2)\cdots(2)(1).$$

In Python, I recommend you use the factorial function from SciPy.special. If the argument is 20 or less, I would recommend passing the keyword parameter `exact=True` to get back an integer solution.

```
from scipy.special import factorial
```

Then the number of ways that 20 unique objects can be arranged is

```
factorial(20, exact=True)
```

```
2432902008176640000
```

Now consider our equation for choosing 2 items out of 20 with ordering: $20 \cdot 19$. We can rewrite this equation using factorial notation as $20!/18!$. And it is easy to extend this equation to choosing k items with ordering. Because each time we choose an item (here, we are choosing the *positions* for the Hs), we remove it from the set, we call this sampling without replacement and with ordering:

Sampling without Replacement and with Ordering

The number of ways to choose k items from a set of N items without replacement and with ordering is

$$(n)(n-1)(n-2)\cdots(n-k+1) = \frac{n!}{(n-k)!}.$$

As we saw before, the ordered result overcounts the number of outcomes for this problem. What we really want is the unordered set of locations for the Hs. For a given unordered set of k locations, the number of orderings will be the number of permutations for k unique items, which is just $k!$. Since every unordered set of k items will show up $k!$ times in the ordered list, we can find the number of unordered sets of locations by dividing by $k!$.

This is an example of sampling without replacement and *without* ordering:

Sampling without Replacement and without Ordering

The number of ways to choose k items from a set of N items without replacement and without ordering is

$$\binom{N}{k} = \frac{N!}{(N-k)!\,k!},$$

which is read as "N choose k" and is known as the *binomial coefficient*.

To find the binomial coefficient in Python, I recommend that you use the `comb()` function from SciPy.special.

Example 4.15: C – continued

Then we can calculate H_2 as

$$\binom{20}{2} =$$

```
from scipy.special import comb
comb(20, 2)
```

```
190.0
```

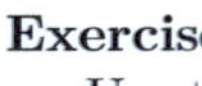

Exercise

Use the `comb()` *function to calculate the cardinalities and probabilities in Combinatorics Question Set 2 online at* fdsp.net/4-6.
(Recall that $P(H_i) = |H_i|/|S|$, and we found $|S|$ further above.)

These basics about combinatorics will be useful when we conduct permutation tests in Chapters 5 and 11 and in understanding some random variables in Chapter 8.

Terminology review and self-assessment questions

Interactive flashcards to review the terminology introduced in this section and self-assessment questions are available at fdsp.net/4-6, which can also be accessed using this QR code:

4.7 Chapter Summary

The primary purpose of this chapter is to formalize our language and techniques for dealing with random phenomena. Our framework for doing that is a probability space $(S, \mathcal{F}, P)$, which consists of:

- A sample space S, which contains all of the outcomes for a random experiment.
- An event class, $\mathcal{F}$, which contains sets of events to which we will assign probability.
- A probability measure, P, which is a set function that assigns probabilities to the events. The event class has to be a σ-algebra, which basically means that we can combine any events using set operations and the result will still be in the event class. The properties of the probability measure come from the three Axioms of Probability.

We showed how to calculate probabilities for fair experiments using counting. However, counting the number of outcomes in an event is sometimes nontrivial, and I introduced several counting techniques from combinatorics to help with this. We will see some of these same formulas again when we introduce random variables in Chapter 8.

Now that we have better terminology and tools, we take a deep dive into null hypothesis testing with resampling in Chapter 5.

Access a list of key take-aways for this chapter, along with interactive flashcards and quizzes at fdsp.net/4-7, which can also be accessed using this QR code:

5

Null Hypothesis Tests

Does additional education after an undergraduate degree generally increase family wealth? Does it make you more likely to be a millionaire? How can we get the data to answer this type of question, and how can we be confident that the answers we get are not caused by randomness in the underlying data? In this chapter, we conduct statistical tests to give insight into these issues.

In Chapter 2 and Chapter 3, we introduced our first statistical tests. In each of these chapters, the examples ended with a test to determine whether some observed effect was "significant" in the sense that it is unlikely to have occurred under some default condition. In Chapter 3, we began to formalize this by defining this default condition as a *null hypothesis.* We then introduced null hypothesis significance testing (NHST) as a way to conduct a statistical test to determine if an observed result can simply be explained by randomness in the data and limitations in the sample size.

This chapter provides a deeper and more detailed discussion of NHST and different ways to conduct these tests using resampling. To better understand such tests and how they are applied in this book, I start with a discussion of the different types of studies and experiments usually encountered in statistics.

5.1 Statistical Studies

In Chapter 3, we analyzed data about socioeconomic factors and COVID rates across the United States to answer research questions about whether these socioeconomic factors are associated with differences in COVID rates. This was our first example of a *statistical study*:

> Definition
>
> **statistical study**
>
> A means to answer a research question using data.

For conciseness, we will just use the term *study* to refer to a statistical study. This section introduces different types of studies and the trade-offs between them. A lot of new terminology is introduced to help characterize different types of studies. Studies can be classified as either *experimental* or *observational*:

DOI: 10.1201/9781003324997-5

Definition

experimental study,
experiment

A study in which the investigator controls one or more of the variables.

Definition

observational study

A study in which data is collected about variables of interest for some participants without any attempt to control the variables for the participants. In particular, if some of the variables correspond to treatments, there is no attempt to randomize which participants receive which treatments.

5.1.1 Experiments

The *gold standard* experiment is a *randomized controlled trial (RCT)*. RCTs are most often used in the medical field, so the variable being controlled is typically called the *treatment*:

Definition

randomized control trial (RCT)

An experiment involving participants who are **randomly assigned** to either a *control group* or one or more *treatment groups*. The control group receives no treatment or the standard treatment. The treatment group receives the novel treatment that is to be evaluated.

Typically, *blinding* is used so that the researchers do not even know which participant got which treatment(s) until the treatment and data collection portions of the RCT are complete.

Definition

blinding

A technique that prevents some of the researchers in an experimental study from knowing which patients received which treatment until the study is complete and the data is to be analyzed.

Blinding is used to prevent researchers in an experimental study from unintentionally providing differentiated treatment to patients in different treatment groups. Because researchers can directly compare the effect of a novel treatment to the effect of the standard treatment without potential biases in which participant received which treatment, statistical significance in an RCT is usually believed to indicate a *causal relationship*. That is, if the novel treatment shows a statistically significant beneficial outcome, then the presumption is that the novel treatment *causes* the beneficial outcome.

Although RCTs are the gold standard, they are also the most expensive and challenging to conduct. Moreover, it is not possible to carry out RCTs for many phenomena because of ethical reasons. For instance, if we believe that a chemical may cause cancer, we cannot ethically give that chemical to some randomly chosen participants in a study.

5.1.2 Observational studies

To motivate the use of observational studies, consider the question: does smoking tobacco cause cancer? If we believe that smoking causes cancer, then we cannot ethically assign people to smoke or not smoke at random. Similarly, because many people have a strong preference about smoking, a random assignment to smoke or not smoke is not likely to be maintained. Instead, we must rely on observational studies. Using observational data, we can only determine whether cancer and smoking are associated; we cannot directly infer that smoking tobacco causes cancer. Any observed association may be attributable to:

- People who have cancer are more likely to smoke, for instance, to relieve pain.
- Smoking causes cancer.
- There may be some other factor, such as genetics, that predisposes people to both cancer and the desire to smoke.

Remember from Section 3.6 that the null hypothesis was first proposed by Ronald A. Fisher? Fisher was an avid smoker and argued that the strongly observed association between smoking cigarettes and cancer could be explained by factors other than smoking causing cancer. (See, for instance, his 1957 letter to the *British Medical Journal*, "Alleged Dangers of Cigarette-Smoking," available at https://www.york.ac.uk/depts/maths/histstat/fisher269.pdf).

Occasionally, circumstances result in a situation in which a particular treatment is applied to a portion of a population in such a way that the result can be treated as if it were applied to a random subset of the participants. For instance, from the 1970s to the 1990s, twelve states had their entire prison population under court control because of litigation that alleged overcrowding. The effect of this court control was that the prison population grew more slowly than in other states. Most of these states had nearby states with similar demographics that were not under such a court order. Thus, the court orders controlling prison population growth created a *natural experiment*:

Definition

natural experiment

An experiment in which some participants are exposed to a novel treatment, while others are exposed to a control treatment, in a way that approximates the random assignment of a randomized control trial.

For the example of the court controlling prison population growth, the resulting natural experiment was used by Steven Levitt (co-author of the book *Freakonomics*) to estimate the effect of incarceration rates on crime rates; see "The effect of prison population on crime rates: Evidence from prison overcrowding litigation," *The Quarterly Journal of Economics*, vol. 111, no. 2, pp. 319–351.

5.1.3 Population studies

Much of the realm of statistics is about studying the effect of something on a particular group of people. We generally will identify some group of people who share some common characteristic as the *population* to be studied:

> Definition
>
> **population**
>
> *All members* of a group that share some common characteristic.

A study on a population is called a *population study*:

> Definition
>
> **population study**
>
> In a *population study*, data are gathered about a group (often people) whose members share some common characteristic(s).

For example, the Nurse's Health Study (NHS) is one of the most well-known population studies. Although the study has expanded over time, the original study focused on "married registered nurses, aged 30 to 55 in 1976, who lived in the 11 most populous states" (i.e., these are the "common characteristics" of this population). The NHS is an observational study. Note that we often want the population to have some characteristics that are **not the same** so that we can assess the effects of such characteristics. For instance, the NHS originally focused on the health effects of contraception and/or smoking.

Note that the NHS identifies a population ("married registered nurses..."), but that does not mean that it collects data for *every* member of the population. Instead, data is collected from a *sample* of the total population:

> Definition
>
> **sample**
>
> In a population study, a *sample* is a subset of the population for which data is collected.

Samples from a population can be created in different ways. In many cases, the ideal is *random sampling*:

> Definition
>
> **random sampling**
>
> In *random sampling*, the sample is chosen randomly from the population.

Random sampling avoids biases from choosing samples in other ways. A common alternative that tends to be very biased is called *convenience sampling*:

Definition

convenience sampling

In *convenience sampling*, the sample is chosen based on access to a subset of the population.

For instance, running a poll on social media uses convenience sampling, but that sample is biased toward the types of people who use social media.

Population studies can be further categorized based on when data about a population is collected. In particular, studies often either collect data at a single point in time or else collect data about participants across time:

Definition

cross-sectional study

A study that collects data about some population at a single instance in time.

Cross-sectional studies are usually observational studies. The term *cross-sectional* refers to a variety of components that make up a whole. The use of this term may apply to cross-sectional studies in two ways:

1. A cross-sectional study should have a sample that includes the different types of members that make up a population.
2. Cross-sectional studies are often used to analyze differences among the groups that make up a population.

Since a cross-sectional study focuses on a single time, it cannot be used to analyze trends in the population over time.

By contrast, a *longitudinal study* can be used to track trends:

Definition

longitudinal study

A study that involves collecting repeated observations over time for a population.

Like cross-sectional studies, longitudinal studies are usually observational studies.

When a longitudinal study collects observations on the same participants at each time, the group of participants is called a *cohort*, and the study is a *longitudinal cohort study*:

Definition

longitudinal cohort study

A study that involves collecting repeated observations over time for the same set of members (called a cohort) of a population.

The NHS is an example of a longitudinal cohort study. Additional examples of large longitudinal studies in the United States include the National Longitudinal Surveys:

https://www.bls.gov/nls/, which are conducted by the Bureau of Labor Statistics. As of 2022, this study consists of three different cohorts, each of which consists of approximately 10,000 participants.

Note:

The data from a longitudinal study at any particular observation time is cross-sectional data.

5.1.4 Prospective versus retrospective studies

Studies can also be classified based on the relation between 1) when and why a research question is asked, and 2) when, why, and how data was collected. Based on the relation between these factors, a study may be classified as either *prospective* or *retrospective*:

Definition

prospective study

A study in which the research question is formulated *first* and used to formulate the design of the data collection (including not only what data is to be collected, but from what subjects, and how the data is to be collected).

Definition

retrospective study

A study that tries to answer a research question by using data that has already been collected.

Results from prospective studies are generally more reliable because the study can be designed to help eliminate biases. By contrast, retrospective studies may be abused by first observing some effect in the collected data and then using the study to carry out a test to confirm the significance of the observed effect. This is called *post hoc analysis*:

Definition

post hoc analysis

Retrospective analysis involving computing new statistics or searching for new effects that were not identified or hypothesized in formulating the study.

The problem with *post hoc* analysis is that there are many different effects that **could** be observed in the data. Suppose that none of the possible effects is actually caused by underlying differences in the sample population, but each may give an effect that appears to be statistically significant with some small probability, q. Then if there are N possible independent effects that could be observed, the probability of observing at least one effect that appears to be significant is approximately $1-(1-q)^N$, which is approximately Nq for small q.. For instance, if there are 250 possible effects, each of which has a 0.1% probability of occurring at a level that appears significant, then the overall probability of observing some significant effect is over 22%.

Retrospective studies are also prone to *selection bias*:

> Definition
>
> **selection bias**
>
> *A type of bias that occurs when the sample is chosen in a way that causes it to* differ from a random sample from the population being studied.

In a retrospective study, the sample was chosen before the research question was formulated. Thus, it is not possible to ensure that the way the sample was chosen accurately reflects the population for which the research question is being asked.

Although retrospective studies may be prone to biases and the dangers of *post hoc* analysis, they are still very useful and practical. Governments often fund large studies of their populations to better understand their populations and trends in those populations over time. Studies such as the General Social Survey (GSS) collect a wide variety of data over many years. For instance, the GSS celebrated its 50th anniversary in 2022 and collects data on over 500 variables. Surveys were conducted approximately annually and had approximately 1500 participants per survey.

As previously mentioned, as of 2022 the NLS has three active cohorts, dating back to 1979, and each of these has many variables. For instance, NLSY79 has tens of thousands of survey variables across twelve different categories (see the NLS Investigator: https://www.nlsinfo.org/investigator. The five categories with the most variables are shown below, with the number of variables shown in parentheses:

- Employment (51982)
- Income, Assets & Program Participation (4324)
- Health (3330)
- Household, Geography & Demographics (2945)
- Sexual Activity, Pregnancy & Fertility (1215)

This large number of variables allows researchers to investigate many different research questions. In the next sections, we will use NLSY79 to explore one of the questions mentioned earlier in this section: **What is the effect on family wealth of schooling beyond undergraduate college?**

Terminology review and self-assessment questions

Interactive flashcards to review the terminology introduced in this section and self-assessment questions are available at fdsp.net/5-1, which can also be accessed using this QR code:

5.2 General Resampling Approaches for Null Hypothesis Significance Testing

In Section 3.6, I introduced Null Hypothesis Significance Testing (NHST), and we used resampling to conduct NHSTs on whether socioeconomic factors among the states are associated with a significant difference in mean COVID rates. Consider again *why* we perform resampling for NHST: we have observed some difference between the groups that is captured in a test statistic, and we want to determine the probability of seeing such an extreme value of the test statistic under the null hypothesis. As discussed in Section 3.6, the test statistic comes from a random distribution, which is known as the *sampling distribution*:

Definition

sampling distribution (test statistic)

For some statistic of sample data, the *sampling distribution* characterizes the possible values of that statistic and the mapping of probability to those values, under some underlying assumption(s) on the population. The sampling distribution varies with the size(s) of the group(s) in the sample.

If we know the sampling distribution of the test statistic, we can potentially determine the probability of seeing such an extreme value of the test statistic analytically. Since we do not know the sampling distribution, we generally have two choices:

1. We can use a *model-based* approach, in which we assume that under H_0, the test statistic can be modeled as coming from some known random distribution, and we infer the parameters of the distribution from the data. This approach is discussed more in Chapter 9.
2. We can use a *model-free* approach, in which we approximate the sampling distribution of the test statistic by resampling from the data.

This chapter focuses on the model-free approach using resampling.

Let's consider a generic NHST resampling problem. We have data about some variable, and we are interested in whether there is a difference in that variable across groups. Use the groups to partition the data into sets $\mathcal{A}$ and $\mathcal{B}$. For each group, compute the value of the same summary statistic, and denote the computed values by $T_{\mathcal{A}}$ and $T_{\mathcal{B}}$, respectively. For example, the summary statistic is often the average (or sample mean). Suppose that we observe some difference in the summary statistics of the form $T_{\mathcal{A}} > T_{\mathcal{B}}$. Then we can conduct a generic NHST via resampling as follows:

Generic NHST via Resampling

1. Choose a significance level, α, where $\alpha > 0$ and $\alpha << 1$ (i.e., α is much less than 1). The lower the value, the more strict our test is (in some sense that we will discuss further in the next section).

2. Determine the conditions to be tested. We have observed a difference in some summary statistic. Create the test statistic $\Delta = T_{\mathcal{A}} - T_{\mathcal{B}}$. Then our resampling test will be to determine the **probability of seeing a**

test statistic "at least as extreme" as Δ under H_0. (I am leaving this a bit open-ended right now because I want to explore this step further below.)

3. Under H_0, the two populations have the same random distribution for the variable of interest. Collect all the data for that variable from both populations into a single set of *pooled data*: $\mathcal{P} = \mathcal{A} \cup \mathcal{B}$.

4. Under H_0, the assumption is that $\mathcal{A}$ and $\mathcal{B}$ are just random samples from some larger population and the observed difference Δ is created by the fact that $\mathcal{A}$ and $\mathcal{B}$ are samples of limited size from that population. To determine the probability of such a large difference occurring, we conduct an N-iteration simulation to estimate the probability of seeing a difference as extreme as Δ. Then in each iteration i, we:

1. Create new samples $\mathcal{A}_i$ and $\mathcal{B}_i$, which we call *test samples*, based on H_0. This implies that we should draw the test samples $\mathcal{A}_i$ and $\mathcal{B}_i$ from the original population; however, we do not have access to the original population. So, we use **resampling and draw the new samples from** $\mathcal{P}$. (Again, I am purposefully leaving this open-ended so that we can explore this more below.)
2. Calculate the summary statistic for the test samples, $T_{\mathcal{A}_i}$ and $T_{\mathcal{B}_i}$.
3. Calculate and store the difference between the summary statistics for the test data, $\Delta_i = T_{\mathcal{A}_i} - T_{\mathcal{B}_i}$. We will call this the *test difference.*
4. Evaluate what the test samples tell us about the statistical significance of the observed difference in the data.

This procedure is quite general – you should be able to use it as a skeleton for any NHST using resampling. However, the procedure above is purposefully vague about several things:

1. What does a difference "at least as extreme" as Δ mean?
2. How should we draw test samples from the pooled data, P?
3. How should we evaluate statistical significance using the test differences, $\Delta_i,\ i = 0, 1, \ldots, N-1$?

Each of these is considered in the following sections. However, to make the discussion more concrete, let's use the National Longitudinal Survey of Youth 1979 (NLSY79) data set to try to answer the question "What is the effect on family wealth of schooling beyond undergraduate college?" This question is too vague to be used as a research question because it does not clearly define "the effect". We will consider two questions:

1. For people with a college education, is post-baccalaureate education associated with an increase in median net family wealth?
2. For people with a college education, do those with post-baccalaureate education have a higher probability of becoming millionaires?

First, we consider why and how the NLSY79 might be used to answer these questions. The NLSY79 is a longitudinal survey of 9,964 participants who were 14–22 when interviewed in 1979. The survey contains detailed information on participants' education and family wealth. In particular, the 2016 survey contains a variable T56845.00 TNFW_TRUC, which is the total net family wealth[1], with the top 2% of values *topcoded*:

Definition

topcoding

The process in which some distinctive high values in data are replaced by a representative value, usually for the purposes of protecting sensitive information. *For instance, some percentage of the highest family wealth values in a data set* may be replaced by the mean or median of that group because those wealth values are sensitive and may also be used to uniquely identify the participants.

In this case, *topcoded* means that the top 2% of family wealth values are each replaced with the average of all the family wealth values in the top 2%. This helps protect the personal information of the participants who belong to that group.

The participants in the NLSY79 cohort were 51–59 years old in 2016. Although most people in the cohort were likely still working in 2016, it is likely that most of the people in the cohort would have already been through the majority of their working life. So any overall effect of post-baccalaureate education is likely to be observable in the participants.

There are also reasons that NLSY79 may not be able to answer these questions. In particular, the US economy has shifted to be more service-based and focused on technology than in the past decades. Thus, new college graduates who are considering using statistical results from NLSY79 may find themselves in very different economic circumstances.

Although NLSY79 is a longitudinal survey, we will be using the data for 2016 as cross-sectional data. This is a huge survey with thousands of variables, but we will work with the following small subset that can be used to answer our research question:

Reference Number	Question Name/ Description	Variable Title	Survey Year
R00001.00	CASEID	IDENTIFICATION CODE	1979
T55976.00	Q11-GENHLTH_4B	RESPONDENT GENDER	2016
T56845.00	TNFW_TRUNC	FAMILY NET WEALTH (TRUNC)	2016
T99000.00	HGC_EVER	HIGHEST GRADE EVER COMPLETED	XRND

Two notes about the data in this table:

1. NLSY79 is a longitudinal cohort study, so each respondent has a unique IDENTIFICATION CODE that was established in the survey's first year. This is why IDENTIFICATION CODE has Survey Year equal to 1979.
2. Because respondents did not provide their highest grade completed in every survey year, I elected to use the HIGHEST GRADE EVER COMPLETED variable, which is collected across all survey years. According to the NLS Glossary (see https://www.nlsinfo.org, XRND stands for "cross-round", which implies that the data is collected from multiple survey years. (In fact, the NLS Glossary references the highest grade ever reported for the NLSY97 cohort as an example.)

[1] Note that the total family wealth variable is a *created* or *computed* variable, meaning that participants did not provide total family wealth directly; instead, the value is computed from answers to other survey questions about assets and liabilities.

The NLSY79 data for the variables mentioned above was retrieved from NLS Investigator and can be loaded as follows:

```
import pandas as pd
df = pd.read_csv('https://www.fdsp.net/data/nls.csv')
print(f'len(df) = {len(df)}')
df.head()
```

	R0000100	T5597600	T5684500	T9900000
0	1	-5	-5	12
1	2	2	-3	12
2	3	2	115000	12
3	4	2	112850	14
4	5	-5	-5	18

The column names for the data frame are the Reference Numbers with the decimal points removed. Since these are difficult to interpret, let's rename the columns to more human-friendly versions. This can be done using the `rename()` method of the dataframe. We pass this method a Python dictionary that provides the map from each original column name to its new value. Here is the renaming dictionary I will use:

```
remap = {'R0000100' : 'CASE_ID',     'T5597600' : 'GENDER',
         'T5684500' : 'NET_WEALTH', 'T9900000' : 'HIGHEST_GRADE_EVER' }
```

To rename columns in the dataframe, we can pass `remap` using the keyword argument `columns`. We also specify to perform this remapping in place; i.e., it will change the columns of `df` directly:

```
df.rename(columns=remap, inplace=True)
df.head(3)
```

	CASE_ID	GENDER	NET_WEALTH	HIGHEST_GRADE_EVER
0	1	-5	-5	12
1	2	2	-3	12
2	3	2	115000	12

Note that some of the entries in the dataframe are less than 0. For instance, for `CASE_ID==1`, both the `GENDER` and `NET_WEALTH` entries are negative. When I downloaded this data from NLS Investigator, I chose to include a *codebook* that provides information on how to interpret the results: Codebook for NLSY79 Data on Wealth vs Education: https://www.fdsp.net/data/nls.cdb.

Definition

codebook

A guide that provides information about the meanings of labels for different features, as well as how the values are encoded for those features.

A value of -5 indicates that the respondent was not interviewed in that year. A value of -3 is an "Invalid Skip", which indicates that the respondent should have answered the question but did not. We will filter the data to preserve only those entries that have valid values for `NET_WEALTH` and `HIGHEST_GRADE_EVER`. This is an example of *data cleaning*:

Definition

data cleaning

A process to deal with incorrect, duplicate, missing, corrupt, or incomplete data.

We will only use those rows of the dataframe where the value of `NET_WEALTH` is either non-negative or less than -5 and the value of `HIGHEST_GRADE_EVER` is non-negative. We can combine these conditions in `df.query()` using an ampersand (&), which represents the logical *and* operator and a "pipe" (|), which represents the logical *or* operator. The resulting query is:

```
df2=df.query('HIGHEST_GRADE_EVER >= 0 & (NET_WEALTH>=0 | NET_WEALTH<-5)')
print(f'len(df2) = {len(df2)}')
df2.head(3)
```

```
len(df2) = 6008
```

	CASE_ID	GENDER	NET_WEALTH	HIGHEST_GRADE_EVER
2	3	2	115000	12
3	4	2	112850	14
5	6	1	95825	16

Note that the number of rows has dropped from 12,686 to 6,008.

Before carrying out any statistical tests, we will do some initial data exploration. Fig. 5.1 shows a scatter plot of net wealth as a function of highest grade ever completed. Note the markers at the very top of the plot – those are the top-coded values.

To progress toward answering our research question, let's extract the data for two groups:

- We will consider students with 16 or 17 years of education as having undergraduate education and place their data into a dataframe called `undergrad`.
- We will consider students with 18 or more years of education as having post-baccalaureate or graduate education and place their data into a data frame called `grad`.

```
undergrad = df2.query('HIGHEST_GRADE_EVER >= 16 & HIGHEST_GRADE_EVER <=17')
grad = df2.query('HIGHEST_GRADE_EVER >= 18')
```

A scatter plot for these two groups is shown in Fig. 5.2. At first glance, the data for the two groups looks very similar. Part of this is because the high values of net wealth have been top-coded, as discussed previously. If we look at the maximum value for each group, we will see they have the same top-coded value (approximately $5.5 million):

```
undergrad['NET_WEALTH'].max(), grad['NET_WEALTH'].max()
```

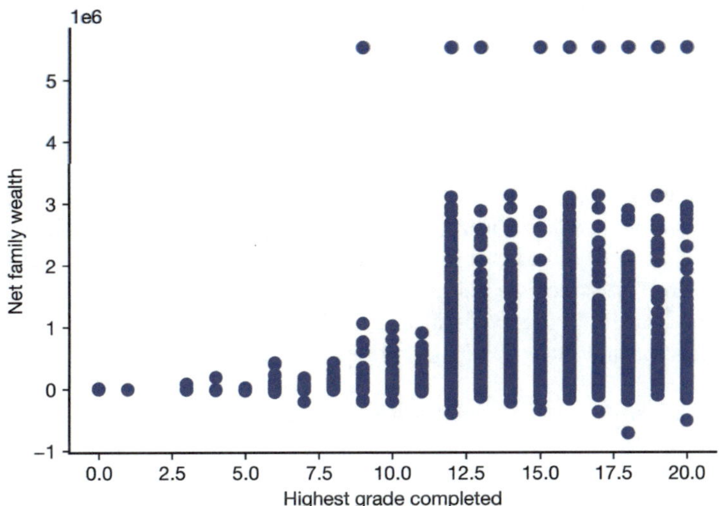

FIGURE 5.1
Net family wealth versus highest grade ever completed based on 2016 data from the NLS79 survey.

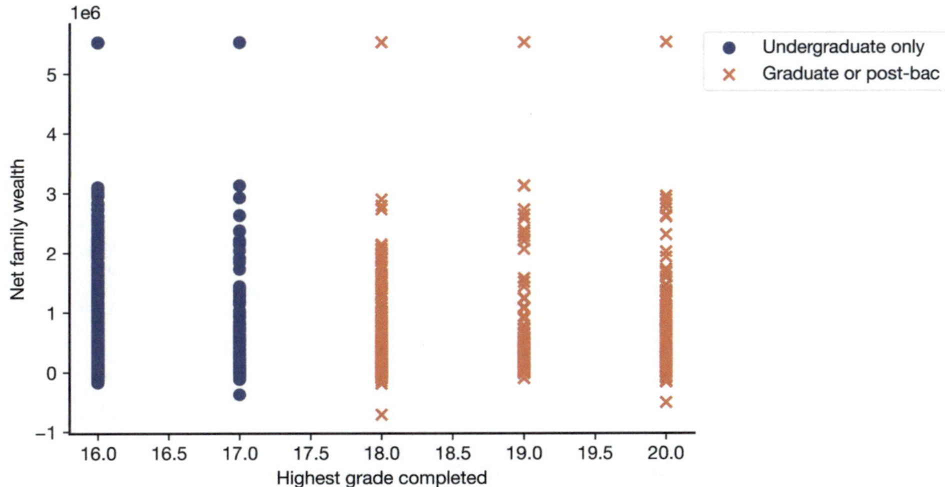

FIGURE 5.2
Net family wealth versus highest grade completed for groups with undergraduate or post-baccalaureate educations, 2016 data from NLS79 survey.

```
(5526252, 5526252)
```

Let's check the median wealth in each of these groups:

```
undergrad['NET_WEALTH'].median(), grad['NET_WEALTH'].median()
```

```
(356325.0, 416000.0)
```

You may wonder why I chose to use the *median* here instead of the average (or sample mean). The reason is that the top coding may cause the average value to be inaccurate. In particular, suppose that the undergrad group has billionaires, but the grad group does not. Both sets of values are top coded to the same value, but the effects of those values on the actual sample means for the two groups would likely be very different.

The median value for the `grad` group is higher than the `undergrad` group by:

```
delta_median = grad['NET_WEALTH'].median() - undergrad['NET_WEALTH'].median()
print(delta_median)
```

```
59675.0
```

Let's also consider the relative frequency of millionaires in each group:

```
R_undergrad = len(undergrad.query('NET_WEALTH >= 1_000_000')) / len(undergrad)
print(f'{R_undergrad * 100: .1f}% millionaires')
```

```
22.0% millionaires
```

```
R_grad = len(grad.query('NET_WEALTH >= 1_000_000')) / len(grad)
print(f'{R_grad * 100: .1f}% millionaires')
```

```
27.4% millionaires
```

The relative frequency of millionaires in the `grad` group is higher than in the `undergrad` group by:

```
delta_M_freq = R_grad - R_undergrad
print(f'{delta_M_freq: 0.3f}')
```

```
0.054
```

The difference is a positive value. We will always achieve this by putting the group with the larger summary statistic first in the difference equation.

We can use NHST to answer the following questions:

1. Is the observed increase in median wealth ($\Delta = \$59,675$) between the undergrad and grad groups statistically significant?
2. Is the observed increase in the relative frequency of millionaires between the undergrad and grad groups ($\Delta \approx 0.054$) statistically significant?

Note that for both cases, we need to sample from the same pooled data: the net family wealth from both the `undergrad` and `grad` groups. Since these correspond to all students with at least 16 years of education, we can create the pooled data as follows:

```
pooled = df2.query('HIGHEST_GRADE_EVER >= 16')
```

In the next sections, we answer both these questions by applying different approaches to resample from this pooled data, to determine whether a result is "at least as extreme" as the observed one, and to determine statistical significance.

Terminology review and self-assessment questions

Interactive flashcards to review the terminology introduced in this section and self-assessment questions are available at fdsp.net/5-2, which can also be accessed using this QR code:

5.3 Calculating p-Values

The simplest and most common test for statistical significance is to estimate the probability of a random sample from the population having a test difference "at least as extreme" as the observed difference in summary statistics under H_0. Let's express this mathematically. Let E be the event that an arbitrary sample test difference Δ_i is "at least as extreme" as Δ. Then we use the notation $P(E \mid H_0)$ to denote the probability that E occurs **under the null hypothesis**. (The vertical bar "|" denotes a conditional probability, and we delve more into the meaning of these types of probabilities in Chapter 6.) In practice, we denote $P(E \mid H_0)$ by p, which is called the "p-value".

To estimate p using resampling, we use a simulation of the form described in the last section and estimate p by the relative frequency

$$p \approx r_E(N) = \frac{n_E(N)}{N},$$

where $n_E(N)$ is the number of times E is observed in N trials. The only remaining issue is how to determine whether E occurred based on the simulated test differences, Δ_i.

5.3.1 One-Tail or Two-Tail Tests

NHSTs are classified as one-tailed or two-tailed (also called one-sided or two-sided) based on the way that "at least as extreme as Δ" is interpreted. Let's illustrate the different approaches using the statistical questions from the last section:

1. Is the observed difference in median wealth ($\Delta = \$59,675$) between the undergrad and grad groups statistically significant?
2. Is the observed difference in the relative frequency of millionaires between the undergrad and grad groups ($\Delta \approx 0.054$) statistically significant?

Let's use the first one of these as an example to understand one-tailed versus two-tailed tests, let $\tilde{U}_i$ and $\tilde{G}_i$ denote the **observed median values** for the ith undergrad and grad test samples. Recall that we previously defined $\Delta_i = \tilde{G}_i - \tilde{U}_i$.

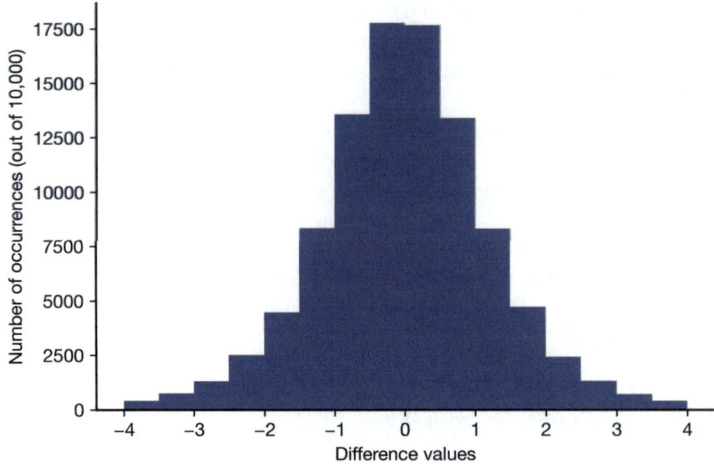

FIGURE 5.3
Example histogram for NHST test statistic.

In many statistical tests, the test statistic under H_0 is equally likely to be positive or negative, with most of the values centered around zero. (In Chapters 8 and 9, we will introduce formal terminology for this. We say the difference has zero mean, and its distribution is symmetric around the mean.) Fig. 5.3 shows an example of the type of histogram we expect to encounter.

In this histogram, the values with the higher counts correspond to those with a higher probability of occurring. The way the probabilities map to the values is called the *distribution* of the random values. (Again, this is discussed in more detail in Chapter 8.) We can define the *mode* for a distribution of random values:

> Definition
>
> **mode(s) (of a random distribution)**
>
> The value(s) with the maximum probability or probability density.

We can also classify a distribution based on how many modes it has. The most common case, and the main one we will utilize in this book, is when a distribution is *unimodal*:

> Definition
>
> **unimodal distribution**
>
> A distribution of random values that has a single mode.

Because the total probability in the distribution must be 1, for most unimodal distributions the probabilities will decrease as the values get farther from the mode of the distribution. We can then define a *tail probability*:

Definition

tail probability

For a unimodal distribution in which the probabilities decrease as some function of the distance from the mode, a *tail probability* is the probability of being at least some distance from the mode.

There are two types of tail probabilities: one-sided tails and two-sided tails. Let M_Δ be the mode of the distribution of the test difference. One-sided tails can either be:

Definition

right tail,
upper tail

For a random value Δ that has a distribution with mode M_Δ, the *right tail* or *upper tail* is of the form $P(\Delta \geq M_\Delta + d)$ for some value $d > 0$.

Definition

left tail,
lower tail

For a random value Δ that has a distribution with mode M_Δ, the *left tail* or *lower tail* is of the form $P(\Delta \leq M_\Delta - d)$ for some value $d > 0$.

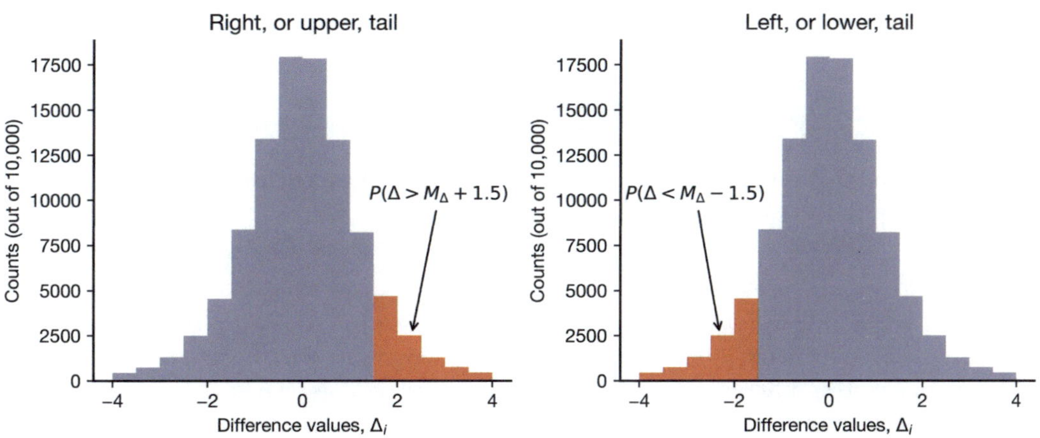

FIGURE 5.4
Example of regions corresponding to tail probabilities.

These two types of tails are shown in the diagrams in Fig. 5.4. A *two-sided tail* simply includes the probability of both upper and lower tails:

Definition

two-sided tail

For a random value Δ that has a distribution with mode M_Δ, a *two-sided tail* is of the form $P(|\Delta - M_\Delta| \geq d)$ for some value $d > 0$.

A two-sided tail is the union of a left tail and a right tail. Fig. 5.5 illustrates a two-sided tail.

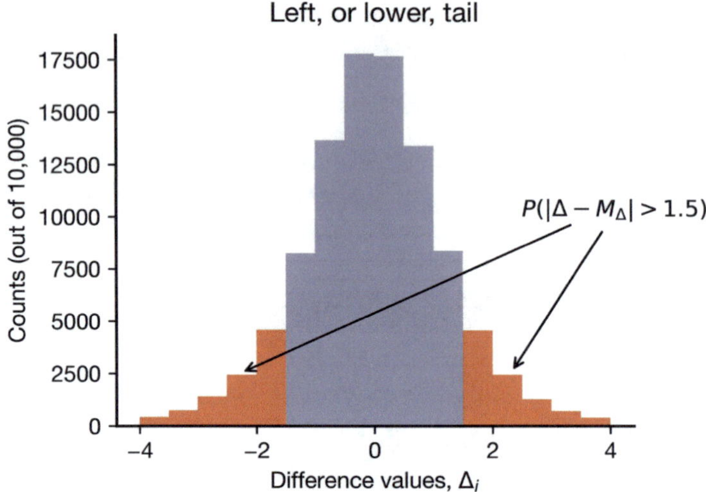

FIGURE 5.5
Example of regions corresponding to two-tail probability.

One-tailed Test

A one-tailed test is **concerned about the particular direction of the observed difference** in test statistics between the two groups.

A one-tailed test on median wealth estimates the probability that the grad group's median wealth is at least \$59,675 higher than the undergrad group's median wealth under H_0. For a given test difference Δ_i, we say that E occurs if $\Delta_i \geq 59{,}675$. Thus, the one-tailed test estimate is $P(\Delta_i \geq 59{,}675 \mid H_0)$.

If we collected all the values of Δ_i in our simulation into a NumPy array `diffs`, then we can calculate the relative frequency of the observed difference `Delta_i` exceeding a threshold `thresh`, by

- comparing the array to the threshold `diffs >= thresh`
- counting the number of `True` values; since `False` is counted as zero, we can do this and the above step in Python as `np.count_nonzero(diffs >= thresh)`
- dividing the number of values where `diffs >= thresh` by the number of simulated test differences

Because I am not ready to discuss how to draw samples from the pooled data yet, let's illustrate this idea by generating an array of 100 random values in the interval $[-0.5, 0.5)$ to represent an array of simulated differences. We estimate the probability that the simulated differences are bigger than 0.9. Run this code a few times and observe the outcomes. Increase the number of simulated differences to 100,000. What do you observe?

```
import numpy as np
import numpy.random as npr

num_sims=100
thresh = 0.4

diffs = npr.rand(num_sims) - 0.5 # Random values on [-0.5, 0.5)
np.count_nonzero( diffs >= thresh) / num_sims
```

```
0.13
```

Two-tailed Test

A two-tailed test is only concerned that the test statistics of the two groups differ by at least as much as the observed difference Δ; it is **not concerned with the direction of the difference.**

A two-tailed test on median wealth estimates the probability that the difference in wealth between the two groups is at least \$59,675. More specifically, E occurs if either $\tilde{G}_i - \tilde{U}_i \geq 59{,}675$ or $\tilde{U}_i - \tilde{G}_i \geq 59{,}675$. These cases are mutually exclusive, and we can write the combined event more concisely as $|\tilde{G}_i - \tilde{U}_i| \geq 59{,}675$, which is the same as $|\Delta_i| \geq 59{,}675$, where the vertical bars indicate absolute value. Thus, the two-tailed test estimates

$$P\left(|\Delta_i| \geq 59{,}675 \;\middle|\; H_0\right).$$

In this expression, the first pair of vertical bars indicates absolute value, whereas the last vertical bar indicates that we are calculating this probability under the null hypothesis, H_0.

Given an array called `diffs` of all the values of Δ_i in our simulation, we can calculate the relative frequency of the absolute value of the difference `Delta_i` exceeding a threshold `thresh` by:

- comparing the absolute value of the array to the threshold: `np.abs(diffs) >= thresh`
- counting the number of `True` values; since `False` is counted as zero, we can do `np.count_nonzero( np.abs(diffs) >= thresh)`
- dividing the number of values where `np.abs(diffs) >=thresh` by the number of simulated test differences

The Python code is just a minor modification of that for the one-tailed test: we just take the *absolute value* of the test differences before comparing them to the threshold:

```
num_sims=1000
thresh = 0.4

diffs = npr.rand(num_sims) - 0.5
np.count_nonzero( np.abs(diffs) >= thresh) / num_sims
```

```
0.23
```

Note that the p-value for the two-tailed test is always higher than for the one-tailed test because the two-tailed probability is the union of two one-sided probabilities.

5.3.2 Choosing Between a One-Tailed or Two-Tailed Test

Whether to use a one-tailed or two-tailed test depends primarily on the initial research question. If the initial research question specifies a particular direction, then a one-tailed test is justified. For instance, if the initial research question is "Does post-baccalaureate education increase median net family wealth?", then a one-tailed test is justified. However, if the initial research question asks only about a difference in effects, then a two-tailed test should be used. For instance, if the initial research question is "Is there a difference in net family wealth between those with an undergraduate education and those with post-baccalaureate education?", then a two-tailed test should be used. Because a two-tailed test produces higher p-values, it is less likely to be smaller than the threshold, α. Thus, a two-tailed test is a more conservative test—it is less likely to find statistical significance than a one-tailed test. When in doubt, it is best to choose a two-tailed test.

Terminology review and self-assessment questions

Interactive flashcards to review the terminology introduced in this section and self-assessment questions are available at fdsp.net/5-3, which can also be accessed using this QR code:

5.4 How to Sample from the Pooled Data

In Section 3.6, I introduced one way to sample from the pooled data, bootstrap sampling. In this section, we will delve deeper into ways to sample from the pooled data and the different implications of these different ways. In general, sampling can either be *with replacement* or *without replacement*. Performing resampling with replacement is called *bootstrapping*:

Definition

bootstrapping

A resampling technique that approximates sampling from the distribution of a population by drawing samples **with replacement** from the original data.

A non-random alternative is *permutation testing*:

Definition

permutation testing

A statistical technique that uses every mapping (i.e., permutation) from the original data to the test samples.

In many cases, the number of permutations is too large, and the test can be approximated by sampling from the possible permutations. In practice, this is implemented by sampling from the original data **without replacement**, an approach called *Monte Carlo permutation testing*:

Definition

Monte Carlo permutation testing

A resampling technique that approximates a permutation test by drawing samples **without replacement** from the original data.

It is easiest to see the differences among these approaches if we use a small data set with unique values. Consider whether a professor's course grades increased after the professor was promoted. The table below shows the average course grade for a graduate course taught at the University of Florida over six semesters with the same professor:

Year	Grade
2013	74.1
2014	74.5
2015	79.4
2016	79.0
2018	78.4
2019	79.3

Note that the grades are all distinct. Also, the professor did not teach that particular course in 2017. The professor was promoted between the 2014 and 2015 courses. Let's compare the average of the course grades pre- and post-promotion:

```
import numpy as np
grades1=np.array([74.1, 74.5])
grades2=np.array([79.4,79.0, 78.4,79.3])
```

```
np.mean(grades1), np.mean(grades2)
```

```
(74.3, 79.025)
```

We observe that the average of the course grades increased from 74.3 before the professor's promotion to 79.025 after the professor's promotion. We can define the test statistic as the difference between these averaged course grades:

```
np.round(np.mean(grades2) - np.mean(grades1),3)
```

```
4.725
```

We will pool the data and then consider different resampling approaches below. The pooled data can be created by concatenating the two vectors of data using `np.hstack()`:

```
pooled = np.hstack( (grades1, grades2) )
```

5.4.1 Bootstrap Sampling

As previously discussed, bootstrapping approximates drawing values from the sampling distribution of the test statistic by randomly sampling with replacement from the pooled data. This means that a single value in the pooled data may appear multiple times in a test sample. As in Section 3.6, we will use `npr.choice()` to sample from the pooled data; by default, `npr.choice()` samples *with replacement*.

The code cell below contains a simulation that draws 10 bootstrap samples of the data. In each iteration, we construct a sample version **for each group**, and we print out the grades in each group, as well as a list of the repeated values across the two sample groups:

```
import numpy.random as npr
num_sims = 10

# Don't worry about how to do all this fancy f-string formatting!
print(f'{"iter":^6} {"pre-promotion":^16} {"post-promotion":^24} '
      + f'{"repeated values":^20}')
print(f'{"----":^6} {"-"*14:^16} {"-"*22:^24}  {"-"*22:^20}')

for sim in range(num_sims):
  sample1 = npr.choice(pooled, len(grades1) )
  sample2 = npr.choice(pooled, len(grades2) )

  # Find the repeated values
  vals, counts = np.unique(np.hstack( (sample1, sample2) ),
                           return_counts=True)

  print(f'{sim:^6} {str(sample1):^16} {str(sample2):^24} '
        + f'{ str(vals[np.where(counts >1 )]):^20}')
```

```
 iter   pre-promotion       post-promotion          repeated values
 ----   --------------   ----------------------   ----------------------
  0      [79.3 79. ]     [74.1 74.5 74.1 74.1]          [74.1]
  1      [79.  74.5]     [79.3 79.  79.3 79. ]       [79.  79.3]
  2      [78.4 74.1]     [74.1 78.4 74.5 79.4]       [74.1 78.4]
  3      [79.3 79.4]     [74.5 74.5 78.4 74.1]          [74.5]
  4      [74.1 74.5]     [79.  74.5 79.4 79.3]          [74.5]
  5      [74.5 79.3]     [79.4 79.  79.  78.4]          [79.]
  6      [79.3 79. ]     [74.1 74.1 79.3 79. ]     [74.1 79.  79.3]
  7      [74.5 74.1]     [74.1 79.3 79.3 78.4]       [74.1 79.3]
  8      [78.4 74.5]     [79.3 74.1 78.4 74.5]       [74.5 78.4]
  9      [74.5 79. ]     [74.1 79.4 74.1 79.3]          [74.1]
```

It is likely that every group has at least one repeated value. In fact, the probability of having no repeats in an iteration is about 1.5×10^{-2}. On the other hand, the probability of having at least one row with no repeats in the table is approximately 0.144. So, if you run this repeatedly, you should see some rows with no repeats occasionally.

To conduct an NHST, we will use bootstrapping to approximate drawing from the sampling distribution of the test statistic. Our test statistic is the difference in averages of the two samples, so to create the bootstrap samples of the test statistic, we will calculate the averages for each test sample and then calculate the difference:

```
num_sims = 10

print(f'{"iter":^6} {"pre-promotion avg":^18} {"post-promotion avg":^24}  '
      + f'{"sample test stat":^20}')
print(f'{"----":^6} {"-"*18:18} {"-"*22:^24}  {"-"*22:^20}')
for sim in range(num_sims):
  sample1 = npr.choice(pooled, len(grades1) )
  sample2 = npr.choice(pooled, len(grades2) )

  print(f'{sim:^6} {sample1.mean():^18.3} {sample2.mean():^24.3}' \
        + f'{ sample2.mean() - sample1.mean():^20.3}')
```

```
 iter  pre-promotion avg    post-promotion avg      sample test stat
 ----  ------------------  ----------------------  ----------------------
  0          79.0                 77.5                    -1.53
  1          74.3                 78.9                     4.58
  2          79.3                 78.0                    -1.3
  3          74.3                 76.8                     2.53
  4          78.7                 77.8                    -0.875
  5          78.7                 78.0                    -0.65
  6          79.0                 78.0                    -0.975
  7          79.2                 79.2                     0.0
  8          76.8                 79.1                     2.38
  9          76.8                 78.6                     1.88
```

Note that we get positive, negative, and zero values for the sample test statistic. This should be expected under the null hypothesis, as the values from the two groups are coming from the same distribution. We can further explore this by looking at the distribution of the values of the sample test statistic using a histogram. The following code generates a histogram from 100,000 sample test statistics created using bootstrapping:

```
import matplotlib.pyplot as plt

num_sims = 100_000
diffs = np.zeros(num_sims)

for sim in range(num_sims):
  sample1 = npr.choice(pooled, len(grades1) )
  sample2 = npr.choice(pooled, len(grades2) )
  diffs[sim] = sample2.mean() - sample1.mean()
```

(continues on next page)

(continued from previous page)

```
plt.hist(diffs, bins = 100);
```

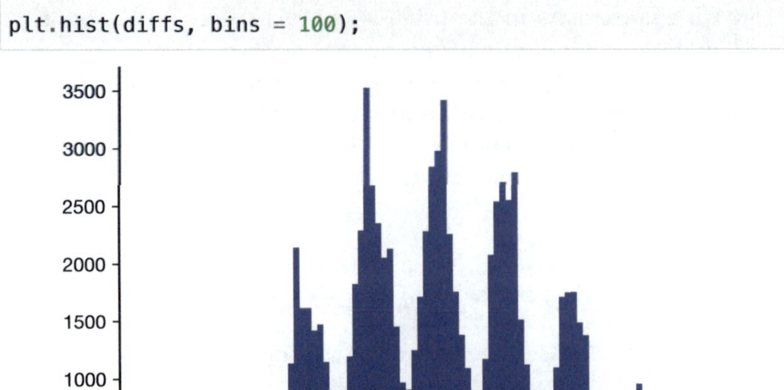

Although the histogram is not completely symmetric, it should be clear that the actual distribution of the bootstrap test statistic is symmetric around 0. An NHST can be conducted using this distribution by determining the probability of being in a tail determined by the observed value of the test statistic. Suppose we wish to conduct a one-tailed test. We can estimate the probability of seeing a value as high as the observed difference in averages by measuring the relative frequency of such events, as shown in the following code:

```
num_sims = 100_000
count = 0

observed_diff = np.round(np.mean(grades2) - np.mean(grades1),3)

for sim in range(num_sims):
  # Generate test samples and sample value of test statistic
  sample1 = npr.choice(pooled, len(grades1) )
  sample2 = npr.choice(pooled, len(grades2) )
  sample_diff = sample2.mean() - sample1.mean()

  # One-sided test: count if sample value of test statistic is
  # at least as large as observed test difference
  if sample_diff >= observed_diff:
    count += 1

print(f'Relative frequency of events with difference >= '
      +f'{observed_diff:.3f} is {count/num_sims}')
```

```
Relative frequency of events with difference >= 4.725 is 0.01268
```

If we are using $\alpha = 0.05$ as a threshold for statistical significance, then the result is statistically significant. Even with this small data set, the difference is big enough that we **reject the null hypothesis at the $p < 0.05$ level**.

5.4.2 Permutation Tests

When we carried out the bootstrap test, we observed that almost every row had repeated values of the data. This is attributable to resampling with replacement, and the probability of at least one repeated value increases as the size of the data increases. There is also nothing preventing the sample groups from repeating in different iterations, although the probability of repeated sample groups decreases with the size of the data.

If our goal is to conduct a statistical test that directly breaks any dependence on the assignment of the data to the two groups, then it makes more sense to consider simply shuffling, or *permuting*, the data among the groups, with no repetitions allowed. When the data is small, we can simply try every permutation of the data across the two groups. This is a permutation test.

A detailed discussion of exact and Monte Carlo permutation tests, along with code and results for this data set, is available on the book's website at fdsp.net/5-4. The p-value for the exact test is 0.067 and the p-value for the Monte Carlo permutation test is 0.065. In both cases, we fail to reject the null hypothesis at the $\alpha = 0.05$ level. Permutation tests are usually more conservative than bootstrap tests, in the sense that they are less likely to reject the null hypothesis.

5.4.3 Which Test to Use?

Some factors you should consider when determining whether to use a bootstrap or permutation test:

- The bootstrap test is generally the most popular test, although this varies with the particular research community.
- The goal of the bootstrap test is to approximate drawing values from the sampling distribution of the test statistic. If the reason for resampling is focused on the sampling distribution of the test statistic, then a bootstrap test should be used. We will give an example in the next section.
- If the goal of resampling is to break up any dependence on the data from the assignment to the underlying groups, then a permutation test makes the most sense. For instance, if each group represents a treatment, then the goal of resampling may be to break up the mapping of data to treatments; this would merit the use of a permutation test.
- When the data is small, an exact permutation test can consider every possible mapping of data to groups.
- Permutation tests are generally more conservative than bootstrap tests in the sense that they generally produce larger p-values and are thus less likely to reject the null hypothesis and indicate statistical significance.

In the remainder of this book, we will primarily use bootstrap tests because of their simplicity and popularity. An exception is that we will use an exact permutation test for categorical data in Chapter 11. In the following section, we will apply bootstrap tests on the questions we formulated about the impact of post-baccalaureate work on family wealth. In addition to exploring these questions, we will investigate two issues: 1) further exploring the meaning of p-values, and 2) creating confidence intervals for the test statistic using bootstrap sampling.

Terminology review and self-assessment questions

Interactive flashcards to review the terminology introduced in this section and self-assessment questions are available at fdsp.net/5-4, which can also be accessed using this QR code:

5.5 Example Null Hypothesis Significance Tests

Let's return to the example introduced in Section 5.2. We will use the NLSY79 data to try to answer research questions about whether post-baccalaureate education increases family wealth. In particular, we asked two questions:

1. Is the observed increase in median wealth ($\Delta = 59,675$) between the undergrad and grad groups statistically significant?
2. Is the observed increase in the relative frequency of millionaires between the undergrad and grad groups ($\Delta \approx 0.054$) statistically significant?

The previous sections should have helped you understand how to calculate p-values, one- and two-tailed tests, and different approaches to sample from the pooled data. With this knowledge, let's perform statistical tests to try to answer these questions.

Let's create the pooled data by using `np.hstack()` to concatenate the wealth data for the `undergrad` and `grad` groups into a single vector:

```
pooled = np.hstack( ( undergrad, grad) )
```

For research question 1, the test statistic will be the observed difference in medians for the two groups. Let's compute this difference and store it in a variable:

```
delta1 = grad.median() - undergrad.median()
print(f'Observed difference in median net family wealth was {delta1}')
```

```
Observed difference in median net family wealth was 59675.0
```

For research question 2, we are computing the proportion of families with net wealth over $1,000,000. Because we will be computing these proportions repeatedly during our statistical test, let's make a helper function to do the computation:

```
def find_proportion(data, thresh=1_000_000):
    return np.sum(data >= thresh) / len(data)
```

The test statistic is the difference in observed values of the proportion of millionaires between the grad and undergrad populations. Let's compute that difference and store it in a variable:

```
delta2 = find_proportion(grad) - find_proportion(undergrad)
print(f'Observed difference in proportion of millionaires is {delta2:.3f}')
```

```
Observed difference in proportion of millionaires is 0.054
```

Now we are ready to plan our resampling tests:

First, we need to decide whether to perform a one-tailed or two-tailed test. Because the two research questions focus on whether post-baccalaureate education is associated with an increase in net family wealth, a one-tailed test makes sense. If instead the questions asked whether post-baccalaureate education affects net family wealth, then a two-tailed test would be more appropriate.

Second, we need to decide how we will do resampling. Because we are going to perform an NHST on sample values of the test statistic, a bootstrap test is justified and is the test that is most commonly used in this type of scenario. Thus, we will use a bootstrap test. Note, however, that a Monte Carlo permutation test is not wrong because we are performing a test where we are trying to remove the effect of a difference in treatment (the highest grade completed).

5.5.1 One-tailed Bootstrap Test on Difference of Medians

Our simulation will follow the form in Section 5.2. In each iteration, we:

- Draw new groups from the pooled data to represent randomly chosen `undergrad` and `grad` groups. The samples are drawn *with replacement* using the default behavior of `npr.choice()`.
- Find the sample value of the test statistic, compare it to the observed value, and increment a counter if it is as extreme as the observed value. Because we are conducting a one-tailed test, we will increment the counter if the sample value of the test statistic is *at least as large* as the observed value.

At the end of the simulation, we print out the relative frequency of the test statistic being as large as the value observed in the data.

Choosing the number of simulation iterations for NHST

To accurately estimate the probability of some event via simulation, we usually need to conduct enough simulations to capture around 100 occurrences of that event. If the event occurs with probability p, we need approximately $100/p$ iterations in our simulation.

However, for an NHST, we do not necessarily need to accurately estimate the probability of the event – we only need to determine if the probability is close to our significance threshold. If the probability is very close to our threshold, then we want enough simulations to determine whether the probability is above or below the threshold. Since the significance threshold is usually 0.05 or 0.01, then **we need around** $100/(0.01) = 10,000$ **simulations.**

Here is a simulation based on the discussion above:

```
import numpy.random as npr

num_sims = 10_000
count=0

```

(continues on next page)

(continued from previous page)

```
# Define these to avoid calling a function to get them in every iteration
undergrad_len = len(undergrad)
grad_len = len(grad)

for sim in range(num_sims):
  # Create random groups under H0 using bootstrap sampling:
  undergrad_sample = npr.choice(pooled, undergrad_len)
  grad_sample = npr.choice(pooled, grad_len)

  # Calculate the sample value of the test statistic
  sample_diff = np.median(grad_sample) - np.median(undergrad_sample)

  # Increment the counter based on the one-tailed test
  if sample_diff >= delta1:
    count+=1

print(f'The p-value is {count/num_sims}')
```

```
The p-value is 0.0758
```

Since $p \approx 0.076$ is over our threshold of 0.05, we fail to reject the null hypothesis. This does not mean that the observed difference in median net wealth is not real; it only means that there is not enough data to be confident that the difference is not a result of random effects and limited sample sizes.

Important!

Always report the measured p-value along with the conclusion. It is not sufficient to just report $p > 0.05$ because the specific p value gives additional insight into the results of the statistical test.

5.5.2 One-tailed Bootstrap Test on Proportion of Millionaires

The simulation for research question 2 is almost identical to that for research question 1. The only changes are in calculating the test statistic and the threshold to which the test statistic is compared.

```
num_sims = 10_000
count=0

# Define these to avoid calling a function to get them in every iteration
grad_len = len(grad)
undergrad_len = len(undergrad)

for sim in range(num_sims):
  # Create random groups under H0 using bootstrap sampling:
  undergrad_sample = npr.choice(pooled, undergrad_len)
```

(continues on next page)

(continued from previous page)

```
    grad_sample = npr.choice(pooled, grad_len)

    # The test statistic for research question 2 is the difference in
    # proportions of millionaires
    sample_diff = find_proportion(grad_sample) - \
                  find_proportion(undergrad_sample)

    # The test statistic is compared to the difference in proportions of
    # millionaires that was observed in the original data
    if sample_diff >= delta2:
      count+=1
print(f'The p-value is {count/num_sims}')
```

```
The p-value is 0.0106
```

Since the p-value is less than our specified significance threshold of 0.05, we say that we **reject the null hypothesis at the $p < 0.05$ level, since $p \approx 0.01$.** Thus, although the observed difference in median net wealth among the groups was not found to be statistically significant, the observed difference in the proportions of millionaires was found to be statistically significant.

> Note:
>
> When applying multiple NHSTs using the same data set, a stricter criteria should be applied to ensure statistical significance at a given level. However, the details are beyond the scope of this book. Observing a value of $p \approx 0.01$ is still sufficient to ensure statistical significance at the $p < 0.05$ level.

One criticism (of many) of NHSTs is that they throw away a lot of information about the test by only calculating and reporting a single number, the p-value. In the next section, I show how we can use the entire set of sample values of the test statistic to create confidence intervals, which capture more information about the statistical test.

Terminology review and self-assessment questions

Interactive flashcards to review the terminology introduced in this section and self-assessment questions are available at fdsp.net/5-5, which can also be accessed using this QR code:

5.6 Bootstrap Distribution and Confidence Intervals

Alternatives to reporting p-values include providing a characterization of the full distribution of the test statistic under the null hypothesis and reporting a confidence interval, which provides a range that characterizes the most likely values of the test statistic under the null hypothesis. We start with the distribution of the test statistic under the null hypothesis.

5.6.1 Bootstrap Distribution

If we perform bootstrap sampling and create sample values of the test statistic, then the set of those values can be used to characterize the distribution of the test statistic. The resulting distribution is called the *bootstrap distribution*:

Definition

bootstrap distribution

When we apply bootstrap sampling to create sample values of some statistic, each sample of the statistic is a random value. The *bootstrap distribution* characterizes the possible values of that statistic and the mapping of probability to those values.

(The term *distribution* is defined more precisely when we define random variables in Chapter 8.) Whereas the sampling distribution is based on samples from the original random distribution, the bootstrap distribution is based on resampling from a fixed set of data. In this chapter, we will study the bootstrap distribution using histograms.

We will revisit our test on median wealth using the NLSY79 data set to illustrate these ideas. To estimate the bootstrap distribution, let's modify our previous bootstrapping simulation to save every sample value of the test statistic. An efficient way to do this is to create an array of zeros (to pre-allocate all the needed storage space) and then fill in the appropriate value of the array at each iteration of the simulation:

```
import numpy.random as npr

num_sims = 10_000

# Pre-allocate an array to store the test statistic values
test_statistics = np.zeros(num_sims)

# Define these to avoid calling a function to get them in every iteration
grad_len = len(grad)
undergrad_len = len(undergrad)

for sim in range(num_sims):
  undergrad_sample = npr.choice(pooled, undergrad_len)
  grad_sample = npr.choice(pooled, grad_len)

  # Calculate the sample value of the test statistic
  sample_diff = np.median(grad_sample) - np.median(undergrad_sample)

  # Now we store the value of the test statistic
  # instead of doing a one-tailed test
  test_statistics[sim] = sample_diff
```

Given the array of test statistics, let's start by generating a histogram with 40 bins. Because so many of the values are large, let's plot in thousands of dollars, which we will abbreviate $K. In addition, I have drawn a vertical line at the test statistic value observed in the original data using the Matplotlib function `plt.axvline()`:

```
plt.hist(test_statistics/1000, bins=40)
plt.xlabel('Test statistic value ($K)')
plt.ylabel('Count')

# Draw a vertical line at the observed value of the test statistic
plt.axvline(59.675, color='C1');
```

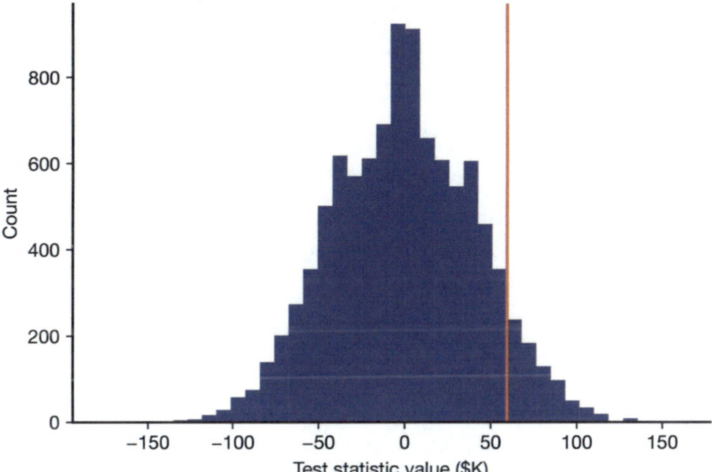

This is the shape of the bootstrap distribution. We can immediately make a few observations:

- The distribution is (approximately) symmetric around zero. This should be expected because under the null hypothesis, the data from each group are coming from the same distribution. Thus, any differences are equally likely to be positive or negative.
- Almost all of the values are in the interval $[-150, 150]$, with the majority of the values in $[-100, 100]$.

The p-value for the one-tailed test is the proportion of values in the sampling distribution that are to the right of the red line. It is hard to infer the value of p from a traditional histogram. However, we can pass a couple of keyword parameters to `plt.hist()` that will make it easier to find the p-value:

- `cumulative = True` will make the values accumulate from left to right, and
- `density = True` will normalize the histogram. With the `cumulative = True` option, the y-axis value corresponding to any point on the x-axis will be the cumulative relative frequency (the proportion of values that are less than or equal to that x-axis value).

```
plt.hist(test_statistics/1000, bins=40, cumulative = True, density = True)
```

The code to plot this type of histogram follows. The full plot with labels is shown in Fig. 5.6a. If we zoom in to the upper right quadrant, we get the plot shown in Fig. 5.6b. The proportion less than 59,675 is approximately 0.92, which means that the probability of the test statistic being greater than 59,675 under the null hypothesis is approximately 0.08.

We will revisit the bootstrap distribution again in Section 9.7 when we have the tools we need to better understand random distributions and their characterization.

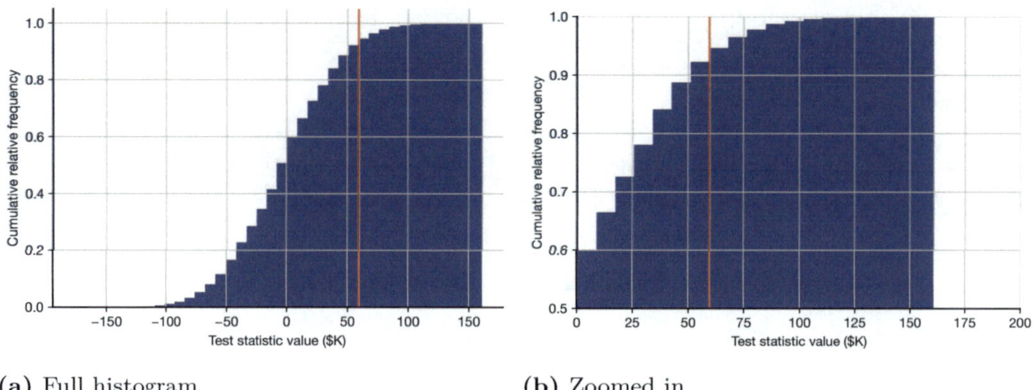

(a) Full histogram (b) Zoomed in

FIGURE 5.6
Cumulative, normalized histogram showing bootstrap distribution of test statistic for median family wealth between people with undergraduate and graduate educations. The vertical line shows the observed value in the original data.

5.6.2 Confidence Intervals

Confidence intervals provide a method of summarizing information about the distribution of the test statistic under the null hypothesis in a way that is different than p-values. Confidence intervals are often suggested as a better alternative to p-values. However, we will see that they are actually very similar to p-values and, like p-values, are often subject to misinterpretation. In this section, I define confidence intervals, show different ways to calculate confidence intervals, and discuss their interpretation. We begin with a definition:

Definition

confidence interval (CI)

Given samples from a random distribution and a confidence level c%, the c% CI for some parameter is an interval that will contain the true value of the parameter c% of the time if the sampling process were repeated many times.

A typical confidence level is 95%, and we will use that value in this section.

For the purposes of this chapter, the observed value of the test statistic is the parameter of interest. We construct a CI under the null hypothesis, and we can assign statistical significance to the result if it does **not** belong to the CI for the null hypothesis. If the data came from the null hypothesis, then the observed value will not lie within the CI 95% of the time that data like this is drawn from the underlying distribution.

WARNING

A common *misinterpretation* is that given a particular c% CI is that it has a c% chance of containing the true parameter value. This is incorrect because c% refers to the proportion of randomly drawn CIs that will contain the true parameter value. Thus, the c% CI is a measure of how reliable the CI estimation process is, not how reliable a particular CI is.

There are different ways to generate CIs depending on the application. In the case of null hypothesis testing, the distribution of the test statistic is centered around zero. This knowledge allows us to create a c% CI by finding the interval that contains c% of the bootstrapped test statistic values. (This method is sometimes called the *percentile method* and may not be safe to apply directly in other applications.)

It is easiest to understand how to create a CI using the percentile method by looking at the normalized cumulative histogram. For a 95% CI, we wish to find the interval that excludes 2.5% of the data in the left tail and 2.5% of the data in the right tail. The left threshold is where the normalized cumulative histogram takes on the value 0.025, and the right threshold is where that histogram takes on the value $1 - 0.025 = 0.975$. These thresholds are shown as horizontal lines in the figure below:

```
plt.hist(test_statistics/1000, bins=40, cumulative = True, density = True)
plt.xlabel('Test statistic value ($K)')
plt.ylabel('Cumulative relative frequency')
plt.axhline(y=0.025, color='C1', linestyle='-')
plt.axhline(y=0.975, color='C1', linestyle='-');
```

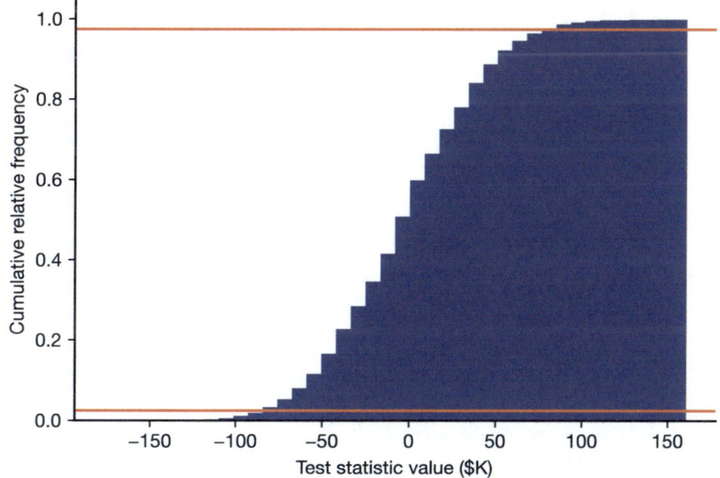

By looking at where the horizontal lines intersect the surface of the histogram, we can estimate that the 95% confidence interval is approximately $[-85, 80]$. Fig. 5.7 shows zoomed-in views that make it easier to better determine the confidence interval boundaries. From Fig. 5.7, we can see that the 95% confidence interval is approximately $[-84, 77]$. If we wanted to use this to determine statistical significance, we would note that the confidence interval contains the observed value of the test statistic ($59.7K); therefore, the result is not statistically significant.

Using the normalized cumulative histogram is intuitive but also error-prone, and its resolution is limited by the number of bins used in the histogram. Fortunately, we can find these thresholds in a very simple way. We will just find the values that are 2.5% of the way through the sorted data and 97.5% of the way through the sorted data. These are called the 2.5% and 97.5% *percentiles*, and NumPy has a `np.percentile()` function that will return the percentile values without us having to sort the data and choose the appropriate indices. We call `np.percentile()` with two arguments: 1) the array of values and 2) a list of the percentiles to compute. For example, we can find the 2.5% and 97.5% percentiles in the array `test_statistics` as follows:

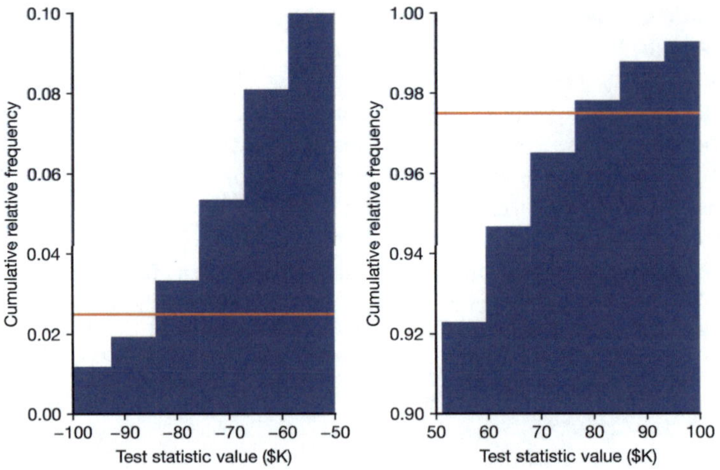

FIGURE 5.7
Zoomed in cumulative, normalized histogram for test statistic computed from median wealth of NLSY79 participants with undergrad versus post-baccalaureate education.

```
CI = np.percentile(test_statistics, [2.5, 97.5])
print(f'The 95% CI is {np.round(CI/1000,1)} in $K')
```

```
The 95% CI is [-80.  82.5 ] in $K
```

Again, the 95% CI under the null hypothesis contains the observed value of the decision statistic, so the result is not statistically significant.

> Note:
>
> Percentiles are closely related to quantiles. For data, the quantiles give the value that is a certain proportion of the way through the data. Thus, the 2.5 percentile value is the same as the 0.025 quantile.

SMALL DATA WARNING

Confidence intervals created using the bootstrap with small data may be too narrow. For instance, Hesterberg[2] indicates that for highly skewed data, bootstrapping may be worse than more conventional approaches when the number of samples is less than or equal to 34.

[2]Hesterberg, Tim C., "What Teachers Should Know About the Bootstrap: Resampling in the Undergraduate Statistics Curriculum". *The American Statistician*, vol. 69:4 (2015): 371-386.

Terminology review and self-assessment questions

Interactive flashcards to review the terminology introduced in this section and self-assessment questions are available at fdsp.net/5-6, which can also be accessed using this QR code:

5.7 Types of Errors and Statistical Power

This section introduces a few important terms and related notation for binary hypothesis testing. I will present these in terms of NHSTs, but the ideas can also be applied to other binary hypothesis tests that we consider later. The focus in this section is on the *terminology* and *meaning* of these terms, but I leave the math until Chapter 9 and Chapter 10.

5.7.1 Types of Errors

Let's review and introduce notation for an NHST. The null hypothesis is H_0, and the alternative hypothesis is H_a. Let $\widehat{H_0}$ denote the event that we do *not* reject the null hypothesis, and let $\widehat{H_a}$ be the event that we reject the null hypothesis. Then four different scenarios can occur, as shown in the matrix below:

Ground truth		Decision: $\widehat{H_0}$ fail to reject null hypothesis	Decision: $\widehat{H_a}$ reject null hypothesis
H_0	null hypothesis is true	True Negative	False Positive *Type I Error*
H_a	alternative hypothesis is true	False Negative *Type II Error*	True Positive *Power*

FIGURE 5.8
Error matrix showing combinations of ground truth and decisions for null hypothesis significance testing (NHST).

The rows of Fig. 5.8 correspond to different possible realities. If the ground truth is H_0, the null hypothesis is actually true; for NHST, this typically means that the two groups of data represent samples from a single distribution. If the ground truth is H_a, then the alternative hypothesis is true; for NHST, this means that the two groups of data come from distributions that differ in some way.

The columns of Fig. 5.8 correspond to different possible decisions from the NHST. If $\widehat{H_0}$, then the null hypothesis is not rejected. In the cells of the matrix, we will label results corresponding to $\widehat{H_0}$ as *negative* to indicate that there is no difference between the groups (in the test statistic). If $\widehat{H_a}$, then the null hypothesis is rejected. In the cells of the matrix, we will label results corresponding to $\widehat{H_a}$ as *positive* to indicate that there is a real difference between the groups.

When the decision matches the ground truth, that result is said to be *True*; if the decision does not match the ground truth, the result is said to be *False*.

Then the entries in the cells of the matrix show the combination of these effects:

- The top left cell corresponds to the null hypothesis being true (H_0) and not rejected ($\widehat{H_0}$), so this is a *True Negative*.
- The top right cell corresponds to the null hypothesis being true (H_0) but rejected ($\widehat{H_a}$), so this is a *False Positive*.
- The bottom left cell corresponds to the alternative hypothesis being true (H_a) but the null hypothesis being accepted ($\widehat{H_0}$), so this is a *False Negative*.
- The bottom right cell corresponds to the alternative hypothesis being true (H_a) and the null hypothesis being rejected ($\widehat{H_a}$), so this is a *True Positive*.

Type I and Type II Errors

Note that two of the cells in Fig. 5.8 correspond to errors, which we have called *false positive* and *false negative*. These are also commonly referred to as Type I and Type II errors:

Definition

Type I error

A Type I error is a **false positive** and is sometimes denoted by the Greek letter α (alpha). For NHST, a Type I error occurs if the null hypothesis is actually true, but it is rejected.

For NHST, the significance threshold α is the acceptable probability of Type I error. It is the acceptable probability of falsely indicating significance by rejecting H_0 when H_0 is actually true.

Definition

Type II error

A Type II error is a **false negative** and is sometimes denoted by the Greek letter β (beta). For NHST, a Type II error occurs if the alternative hypothesis is actually true, but the null hypothesis is not rejected.

One of the key principles of NHST is that it requires no knowledge of the alternative hypothesis. Thus, under NHST it is not possible to quantify the probability of failing to reject H_0 when H_a is actually true. However, designing experiments (such as choosing the sample size) often requires us to make some assumptions about H_a and the *power* of the test.

How to remember the meaning of Type I and Type II errors?

1. Remember that both of these are types of errors that appear in the NHST error matrix. The entries in that matrix are either True results or False results, and errors correspond to False results.
2. Remember that entries in the NHST error matrix are either Positive (indicating an effect) or Negative.

3. The previous two points will help you remember that errors are either False Positive or False Negative.
4. Finally, here are three ways to remember the relation between Type I/Type II and False Positive/False Negative[3]:
    1. Map Positive to True and Negative to False. Then False Positive has 1 False and thus is a Type I error. False Negative has 2 Falses and so is a Type II error.
    2. Recall the story of the boy who cried wolf, and treat the normal case of no wolf as the null hypothesis. The first time he cried wolf, the townspeople made a Type I error: there was no wolf, and they believed him that there was a wolf – this was a False Positive. The second time he cried wolf was a Type II error: there was a wolf, but the townspeople believed that there was no wolf – this was a False Negative.
    3. Recall that we can only evaluate the probability of False Positive under NHST. Thus, it makes sense that these are the Type I errors. Determining the probability of Type II errors requires information about both H_0 and H_a.

5.7.2 Statistical Power

We start by defining the *power* of a statistical test:

> Definition
>
> **statistical power**
>
> The probability of rejecting the null hypothesis when the alternative hypothesis is true.

If the probability of Type II error is β, then the power of the test is $1 - \beta$. Power is often used in experimental design and, in particular, used to choose sample sizes. However, just like the probability of Type II error, determining power requires knowing some characteristics about how the random distribution of the underlying data is different under H_a in comparison to H_0. For instance, in Section 3.4, we introduced the average, or mean, of a sample. The underlying distributions also have associated means. If we know something about the difference in means, we may be able to estimate how large the sample size must be to ensure that the null hypothesis will be rejected with a high probability. How to choose the necessary sample sizes based on α, β, and another parameter called effect size, is discussed in Section 9.8.

Terminology review and self-assessment questions

Interactive flashcards to review the terminology introduced in this section and self-assessment questions are available at fdsp.net/5-7, which can also be accessed using this QR code:

[3]The first two of these are from a question about power on StackExchange Cross Validated: https://stats.stackexchange.com/questions/1610/is-there-a-way-to-remember-the-definitions-of-type-i-and-type-ii-errors. The Wolf analogy may be originally attributable to Patrick Collison, who published it in this tweet about the boy who cried wolf and Type I and II errors: https://twitter.com/patrickc/status/976833754864943105.

5.8 Chapter Summary

This chapter provided a detailed discussion of null hypothesis significance testing (NHST) and how to use resampling to implement it. Different resampling approaches were presented, including bootstrap tests, permutation tests, and Monte Carlo permutation tests. One-tail and two-tail tests were also presented. Some important terminology used in statistical testing was introduced, including Type I and Type II errors and power. We built simulations to perform NHSTs to evaluate whether the observed effects of post-graduate education on family wealth were statistically significant.

It is important to note that there has been a lot of discussion in the research community about the use (and abuse) of NHSTs. There are two main ways that the community has chosen to deal with this:

1. *New Statistics* replace p-values with other metrics, such as confidence intervals, and try to provide more nuanced discussions of significance than using a single threshold.
2. *Bayesian Methods* try to estimate the probabilities associated with the phenomena of interest given that we observed the data.

Understanding Bayesian methods requires knowledge of one of the most important tools of probability – *conditional probability.* Conditional probabilities allow us to decompose problems in different ways and precisely formulate questions about the dependence between events. Chapter 6 covers the basics of independence, dependence, and conditional probability. Chapter 7 provides an introduction to Bayesian methods for statistical analysis.

Access a list of key take-aways for this chapter, along with interactive flashcards and quizzes at fdsp.net/5-8, which can also be accessed using this QR code:

6

Conditional Probability, Dependence, and Independence

Multiple random phenomena can depend on each other in ways that can be obvious or subtle. *Conditional probability* is the study of how to model such dependence and use these models to solve problems and make optimal decisions. This is one of the most important concepts in probability. In fact, it is often said that "all probabilities are conditional," which means that we should always understand probability through the concepts taught in this chapter. Some examples of how we can use conditional probabilities are:

- In null hypothesis significance testing, we are asking whether an observed difference in a summary statistic depends on the underlying grouping of the data. This dependence can be best expressed in terms of conditional probabilities and conditional expectations.
- In wireless communications, we receive a noisy signal and need to determine which information symbol was transmitted. The noisy signal depends on the transmitted information, and thus we can use tools from conditional probability to make optimal decisions about the transmitted symbol.
- In compound experiments, the later experiments often depend on the results of the earlier experiments. For instance, consider drawing cards from a deck. If the first card is an ace, then the probability that the second card will be an ace is different than if the first card had not been an ace.

In this chapter, I introduce conditional probability and the tools for working with conditional probabilities. These tools will often help us answer questions that are challenging to understand without conditional probability. Below I give a few quick examples of simple problems that people often find challenging to answer correctly:

Balls and boxes/urns

A classic set of probability problems involves pulling colored balls from an urn. I have found that many students are not familiar with the word *urn* (which is like a large vase), so I will use a box instead. To make the problem interesting, the box must contain more than two balls, and the balls must be of at least two different colors.

Answer the interactive questions about this in the **Interactive Quiz About Balls in a Box** at fdsp.net/6, which can also be accessed using this QR code:

If you are surprised or confused by the answers, don't worry. That is normal, and even after we solve these in detail, it may still take you some time to build intuition and deeper understanding of these problems.

DOI: 10.1201/9781003324997-6

The Monty Hall problem

Some readers may be familiar with this problem. However, even if you have seen it before, you may still not understand the mathematics behind solving it. This problem became famous when it appeared in the "Ask Marilyn" column of *Parade* magazine, which was distributed with many Sunday newspapers. In this column, Marilyn vos Savant answered tricky questions, and she provided an answer to this problem that provoked a lot of surprise and correspondence.

The problem is based on an old TV game show called *Let's Make a Deal*, which was hosted by Monty Hall. (The problem setup varies somewhat from how the actual TV game worked.) Here is a slightly paraphrased version of the problem from Parade magazine:

You are on a game show, and you are given the choice of three doors:

- Behind one door is a car
- Behind the other doors are goats

You pick a door, and the host, who knows what's behind the doors, opens another door, which he knows has a goat. The host then offers you the option to switch doors. Does it matter if you switch? If switching changes your probability of getting the prize, what is the new probability?

Answer the interactive questions in the **Interactive Quiz for the Monty Hall Problem** at fdsp.net/6, which can also be accessed using this QR code:

If you do not understand how to get the answer, don't worry – we will solve this problem later in the chapter.

6.1 Simulating and Counting Conditional Probabilities

We will begin to explore conditional probabilities through two methods that we have been using in previous chapters: simulation and counting. As with the questions in the introduction, the following example has a simple setup but quickly creates many interesting questions to explore:

6.1.1 The Magician's Coin, Part 1

Suppose you attend a magic show. The magician shows you two coins: a two-headed coin and a fair coin. She asks you to pick one of the coins at random and flip it, observing only the top face. If the outcome of that first flip is heads, does that affect the probability that heads would come up if you flip that coin again?

We will first answer a simpler question: **What is the probability of getting heads when one of the coins is chosen at random and flipped?**

I will show you how to easily analyze this probability in the next section, but let's first build a simple simulation to estimate this probability.

To facilitate learning, I build the simulation in three steps below. All of them leverage the `random` library, so let's start by importing that:

```
import random
```

Step 1. Let's use `random.choice()` to choose a random coin from the list `{['fair', '2head'{]}}` and print it out for a very small simulation. (There are **much** more computationally efficient ways to implement this simulation, but the goal here is to make the simulation as **clear as possible** to the reader.)

```
def choose_coin(num_sims=10):
    coins=['fair','2head']
    for sim in range(num_sims):
        coin=random.choice(coins)
        print(coin, end=' ')
```

```
choose_coin()
```

```
2head 2head fair 2head 2head fair 2head 2head 2head 2head
```

Step 2. Now, we can check the outcome of the coin choice and randomly choose one of the faces of the chosen coin. If you are following along in a separate Jupyter notebook, I suggest you copy the function from above, rename it, and then add the rest of the code (the part after the comment in the function below):

```
def choose_and_flip(num_sims=10):
    coins=['fair','2head']
    for sim in range(num_sims):
        coin=random.choice(coins)
        # Delete the print statement and add the following:
        if coin=='fair':
            faces=['H','T']
        else:
            faces=['H','H']
        value=random.choice(faces)
        print(f'({coin:5}, {value})', end= '   ')
        if sim == 4:
          print()
```

```
choose_and_flip()
```

```
(fair , H)   (2head, H)   (fair , T)   (2head, H)   (2head, H)
(2head, H)   (fair , T)   (fair , T)   (fair , H)   (fair , T)
```

By inspection, you should see that the relative frequency of heads is more than 1/2. Hopefully, you already expected this by intuition. You may even already have an idea of what the probability of heads is on one flip of a randomly chosen coin.

Step 3. Finally, let's estimate the probability of heads by determining the relative frequency of `H`. Outside the simulation loop, we initialize a counter called `num_heads` to zero. In a simulation iteration, we increment the counter every time the outcome is `H`.

Important: We need to drastically increase the number of simulated coin flips, so we will stop printing the outcomes inside the `for` loop.

```
def one_flip(num_sims=100_000):
    coins=['fair','2head']

    num_heads=0
    for sim in range(num_sims):
        coin=random.choice(coins)
        if coin=='fair':
            faces=['H','T']
        else:
            faces=['H','H']
        value=random.choice(faces)

        # Count the number of heads, regardless of coin type
        if value=='H':
            num_heads+=1

    print("Prob. of H is approximately", num_heads/num_sims)
```

```
one_flip()
```

```
Prob. of H is approximately 0.74796
```

How can we calculate this probability using equally likely outcomes? Let's create a table that enumerates all the outcomes:

Coin	Face 1	Face 2
Fair	**Heads**	Tails
Two-Headed	**Heads 1**	**Heads 2**

Here, I separately list the two different heads outcomes for the two-headed coin because I want the outcomes to be equally likely. Since the coins are equally likely to be chosen, and the two faces of each coin are equally likely, the resulting outcomes of the combined experiment are also equally likely.

Let H_1 denote the event that the top face of the coin on the first flip is heads. From the table, $|H_1| = 3$ and $|S| = 4$. Since the outcomes are assumed to be equally likely, $P(H_1) = |H_1|/|S| = 3/4$, which matches our simulation.

6.1.2 The Magician's Coin, Part 2

The reader may wonder what happens if we ask about the second flip instead. Let H_i be the event that the ith flip of the coin is heads. If you do not already know what $P(H_2)$ will be, take a moment to think about what the simulation will look like – with careful thought, you should be able to know $P(H_2)$.

Let's build the simulation:

```
def two_flips(num_sims=100_000):
    coins=['fair','2head']
```

(continues on next page)

(continued from previous page)

```
    num_heads=0
    for sim in range(num_sims):
        # Note that the coin is chosen once -- the same coin
        # is used for both flips!!!
        coin=random.choice(coins)
        if coin=='fair':
            faces=['H','T']
        else:
            faces=['H','H']
        value1=random.choice(faces)
        value2=random.choice(faces)
        if value2=='H':
            num_heads+=1

    print(f'Prob. of H2 is approximately {num_heads/num_sims: .3f}')
```

```
two_flips()
```

```
Prob. of H2 is approximately  0.749
```

Did you guess this answer? It should not be a surprise – we now flip twice as many coins and then throw half of them away (i.e., we never do anything with `value1`)! So, the simulation is just a less efficient version of `one_flip()`. If we don't have any other information about the coin (say, from observing the results of some of the flips), then the probability of heads on any flip, $P(H_i)$, is equal to $P(H_1)$: there are three possible heads outcomes out of four total outcomes.

Comments on things not observed

Some students of probability have problems grasping the meaning of some of the probabilities discussed up to this point. For instance, students often ask questions like:

- How can we evaluate the probability of getting heads if we don't know which coin was chosen?
- How can we ask about the probability of getting heads on the ith flip, if we don't even know what the outcomes of flips 1 through $i - 1$ were?

It is normal to ask these questions, and we will explore this idea more as we learn about conditional probability. The short answer is that we can ask any question we like, but we may not always know how to answer it. A longer answer is that if we want to ask about the probability of some event A that depends on some possible events B_i, we can find the probabilities of A for each of the possible B_i that could have occurred and then combine them appropriately to find the probability of A. We will define all this precisely and introduce the appropriate mathematics later in this section.

6.1.3 The Magician's Coin, Part 3

Now, let's return to our motivating question: if the coin is flipped once and comes up heads, what is the probability that it will be heads on a second flip (of the same coin)?

If you want to take a guess, you can do so using the interactive quiz labeled **Magician's Coin Quiz 1** at fdsp.net/6-1, which can also be accessed using this QR code:

Again, let's simulate the answer before trying to find it analytically. This simulation will be a bit different than the ones we have created before. In the previous simulations, every simulation iteration affects the probability being estimated. In this simulation, that will not be the case. We will simulate two coin flips, but **we are only interested in those simulation iterations in which the coin came up heads on the first flip**. We will need two counters to estimate the probability of getting heads on the second flip given that the first flip was heads:

- First, we need to count how many times the first flip was heads.
- Second, for those iterations in which the first flip was heads, we need to count how many times the second flip was also heads. Then the probability we are looking for is the ratio of the second counter to the first counter.

To implement these in Python, we note that both these counters are only updated under certain conditions. Moreover, the second counter is only updated if the condition for the first counter holds **and** another condition holds (the second flip was also heads). These conditions will be checked with `if` statements, and the condition-within-a-condition will be in the form of a *nested if statement.* Much like the nested for loops introduced in *Chapter 1*, a *nested* `if` statement occurs when an `if` statement is entirely enclosed within the suite of another `if` statement.

The function below implements this counting scheme to estimate the probability of getting heads on the second flip of a coin when the first flip was heads:

```
def two_flips(num_sims=10000):
    coins=['fair','2head']
    heads_count1=0
    heads_count2=0

    for sim in range(num_sims):

        #Choose a coin at random and set up the faces accordingly
        coin=random.choice(coins)
        if coin=='fair':
            faces=['H','T']
        else:
            faces=['H','H']

        # Flip the coin (choose a face at random) twice:
        coin1_face=random.choice(faces)
        coin2_face = random.choice(faces)

        # Check if the first flip was H
        if coin1_face =='H':
            heads_count1+=1
            # If the first flip was H, check if the second flip was H
            if coin2_face == 'H':
```

(continues on next page)

(continued from previous page)

```
            heads_count2+=1

    print('Prob. of heads on second flip given heads on first flip is',
          f'{heads_count2/heads_count1: .3f}')
```

```
two_flips()
```

```
Prob. of heads on second flip given heads on first flip is  0.835
```

Exercise

Modify the simulation for the Magician's Coin to answer the interactive quiz labeled **Magician's Coin Quiz 2** at fdsp.net/6-1, which can also be accessed using this QR code:

6.1.4 Discussion

Many people are surprised that the outcome of the first coin flip affects the probabilities for the second coin flip. The reason is that their intuition tells them that the coin flips should be independent. If you are one of those people, here is a way to help fix your intuition in cases like these:

Replace the problem with a more extreme version of the same problem! Suppose we considered the probability of the coin coming up heads on the next flip given that *the coin was heads on the first 1,000,000 flips.* If you heard that the coin came up heads on 1,000,000 flips, what would you have to assume? You would have to assume that the selected coin was the two-headed coin. So, the probability of heads on the next flip will be 1 (or extremely close to that). If you agree that the probability increases toward 1 as we observe 1,000,000 consecutive heads, then what about 100,000 heads? 1000 heads? 100 heads? It must be that if you see a lot of heads, the probability of the magician having selected the two-headed coin increases. So, shouldn't it increase if even one head is observed? By how much? We will answer that later in this chapter.

6.2 Conditional Probability: Notation and Intuition

In the last section, we introduced the *concept* of conditional probability: we want to know the probability of an event **given that another event has occurred**. To make our work on conditional probability easier and to communicate effectively with others, we will introduce standard mathematical notation for this type of conditional probability.

A conditional probability is indicated if there is a vertical bar "|" inside the arguments of the probability measure. (Users of Unix/BSD-type operating systems and programmers often refer to the | symbol as "pipe" – in the context of conditional probability, we will read it as "given".) If there is a | symbol, then the parts of the argument on each side of the | should both be events. The part that follows the | symbol is the conditioning event (i.e., the event that is given to have happened).

Let's make this more concrete by formalizing the notation for the Magician's Coin problem. As before, we let H_i be the event that the coin came up heads on flip i. Then if we want to know the conditional probability that the coin came up heads on the second flip given that it came up heads on the first flip, we can write that probability in mathematical notation as

$$P(H_2 \mid H_1).$$

When we see such a statement, we will read the mathematical formulation (i.e., without referring to the definitions of H_1 and H_2) as "the conditional probability of H_2 given H_1". If we want to additionally refer to the definition of H_i, then we would read that probability as some variation of "the conditional probability of getting heads on the second flip of the coin given that the first flip was heads".

Note:

- Whatever follows the | symbol is the conditioning information.
- There cannot be more than one | symbol. I.e., this type of notation is nonsense:
  ✗ $P(H_3 \mid H_2 \mid H_1)$
- If we want to know the probability of H_3 given that both H_1 and H_2 occurred, then we can use the intersection operator $\cap$ to express the event that both H_1 and H_2 occurred:
  ✓ $P(H_3 \mid H_1 \cap H_2)$

Now that we have appropriate mathematical notation, we can work through a simple example. Here, we use basic counting arguments to calculate the indicated probabilities without yet providing a mathematical definition for conditional probability.

Example 6.1: Computer Lab

A computer lab contains

- two computers from manufacturer A, one of which is defective
- three computers from manufacturer B, two of which are defective

A user sits down at a computer at random.

Let the properties of the selected computer be denoted by a two-letter code, where the first letter is the manufacturer, and the second letter is D for a defective computer and N for a non-defective computer. Since there would be two BD outcomes with this labeling and the elements in a set must be unique, we denote the two defective computers from manufacturer B as BD_1 and BD_2. Then the sample space is

$$S = \{AD, AN, BD_1, BD_2, BN\}$$

Now we define the following events:

- E_A is the event that the user's computer is from manufacturer A.

- E_B is the event that the user's computer is from manufacturer B.
- E_D is the event that the user's computer is defective.

Use basic counting to answer the interactive questions in **Quiz 1: Non-conditional Probabilities** at fdsp.net/6-2, which can also be accessed using this QR code:

Now consider how these probabilities change if you are given some information about the selected computer. For instance, suppose you observe that the computer is from manufacturer A. Given that information, does that change your answer for the probability that the computer is defective?

We can denote this probability by $P(E_D \mid E_A)$. If you don't immediately know the value of this probability, we can use the following approach. Given E_A, the only possible outcomes are AD and AN. Thus given E_A, we can consider that the outcomes come from a new, smaller (conditional) sample space:

$$S_{|A} = \{AD, AN\}.$$

All the outcomes were equally likely before conditioning. Given information that simply restricts the possible outcomes to two of the outcomes will not change the outcomes from being equally likely in the new sample space.

The event E_D in the conditional sample space $S_{|A}$ is $E_D = \{AD\}$, and

$$P(E_D) = P(\{AD\}) = \frac{|\{AD\}|}{|S_{|A}|} = \frac{1}{2}.$$

Use the counting approach for conditional probabilities to answer the interactive questions in **Quiz 2: With Conditional Information** at fdsp.net/6-2, which can also be accessed using this QR code:

This approach of finding the set of outcomes and calculating the probabilities by taking advantage of equally likely outcomes is useful for simple problems, but it has several requirements for it to be valid:

1. The set of outcomes has to be finite.
2. The outcomes have to be equally likely.
3. The conditioning event has to only restrict the set of outcomes to a subset of the sample space; it cannot make some of the remaining outcomes more likely than others.

These requirements are often not met, and so we need a more systematic way to define and calculate conditional probabilities. That is the subject of the next section.

6.3 Formally Defining Conditional Probability

In Section 6.2, we introduced the idea of a *conditional sample space*. Suppose we are interested in the conditional probability of A given that event B occurred. A Venn diagram for a general set of events A and B is shown in Fig. 6.1.

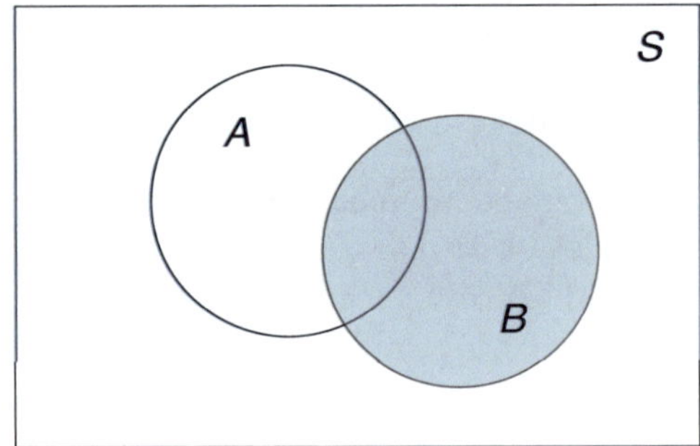

FIGURE 6.1
Venn diagram of two generic events A and B, where the event B is shaded to indicate that it is known to have occurred.

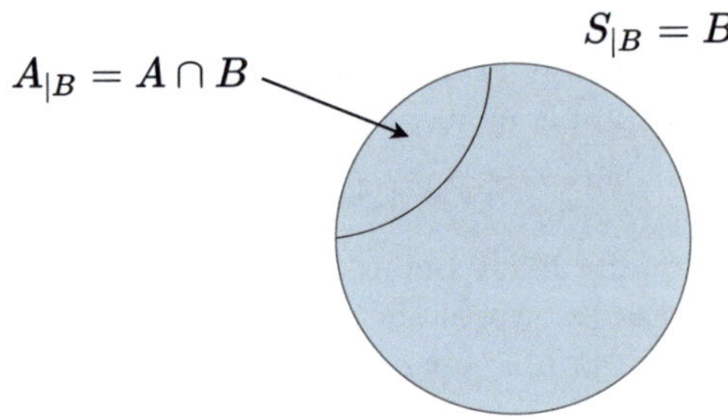

FIGURE 6.2
Venn diagram of two generic events A and B, where the event B is shaded to indicate that it is known to have occurred.

If the original sample space is S, then the set of outcomes that can have occurred given that B occurred can be thought of as a "conditional sample space",

$$S_{|B} = S \cap B = B.$$

Note:

The reason that I have put "conditional sample space" in quotation marks is that although this concept is useful to understand where the formula for calculating conditional probabilities comes from, it is also misleading in that we are not restricted to only calculating conditional probabilities for events that lie within $S_{|B}$. In fact, our assumption in drawing the Venn diagrams is that A is an event that is not wholly contained in B. Further below, we show that conditioning on an event B induces a new *conditional probability measure* on the **original sample space and event class**.

Now, given that B occurred, the only possible outcomes in A that could have occurred are those in $A_{|B} = A \cap B$. Then the "conditional sample space" $S_{|B}$ and a corresponding conditional event $A_{|B}$ are shown in Fig. 6.2.

Based on Fig. 6.2, we make the following observations under the condition that B is known to have occurred:

- If A and B are disjoint, then there are no outcomes of A contained in the event B. Thus if B occurs, A cannot have occurred, and thus $P(A \mid B) = 0$ in this case.

- If $B \subset A$, then the intersection region $A \cap B = B$. In other words, every outcome in B is an outcome in A. If B occurred, then A must have occurred: $P(A \mid B) = 1$.
- Under the condition that B occurred, only the outcomes in A that are also in B are possible. Thus, $P(A \mid B)$ should be proportional to $P(A \cap B)$ (i.e., the smaller region in Fig. 6.2).

These observations lead to the following definition of *conditional probability*:

Definition

conditional probability

The conditional probability of an event A given that an event B occurred, where $P(B) \neq 0$, is

$$P(A \mid B) = \frac{P(A \cap B)}{P(B)}.$$

Now suppose we have a probability space $(S, \mathcal{F}, P)$ and an event B with $P(B) \neq 0$. Then we define a new probability space $(S, \mathcal{F}, P(\ \mid B))$, where $P(\ \mid B)$ is the conditional probability measure given that B occurred. To be more precise, we define $P(\ \mid B)$ on the event class $\mathcal{F}$ using the original probability measure $P()$ as follows:

For each $A \in \mathcal{F}$,

$$P(A \mid B) = \frac{P(A \cap B)}{P(B)}.$$

To claim that the triple $(S, \mathcal{F}, P(\ \mid B))$ defined above is a probability space, we need to verify that the conditional probability measure $P(\ \mid B)$ **satisfies the axioms** in this probability space:

1. Axiom 1 is that the probabilities are non-negative. Let's check:

$$P(A|B) = \frac{P(A \cap B)}{P(B)}.$$

We are already given that $P(B) > 0$, and $P() \geq 0$ for all events in $\mathcal{F}$. Since $\mathcal{F}$ is a σ-algebra, $A \cap B \in \mathcal{F}$ and so $P(A \cap B) \geq 0$. Thus, $P(A \mid B)$ is a non-negative quantity divided by a positive quantity, and so $P(A \mid B) \geq 0$.

2. Axiom 2 is that the probability of S (the sample space) is 1. Let's check:

$$P(S \mid B) = \frac{P(S \cap B)}{P(B)} = \frac{P(B)}{P(B)} = 1.$$

3. Axiom 3 says that if A and C are disjoint events in $\mathcal{F}$, then the probability of $A \cup C$ is the sum of the probability of A and the probability of C. Let's check if this still holds for our conditional probability measure:

$$\begin{aligned} P(A \cup C \mid B) &= \frac{P[(A \cup C) \cap B]}{P[B]} \\ &= \frac{P[(A \cap B) \cup (C \cap B)]}{P[B]}. \end{aligned}$$

Note that $A \cap C = \emptyset \Rightarrow (A \cap B) \cap (C \cap B) = (A \cap C) \cap B = \emptyset$, so

$$\begin{aligned} P(A \cup C \mid B) &= \frac{P[A \cap B]}{P[B]} + \frac{P[C \cap B]}{P[B]} \\ &= P(A \mid B) + P(C \mid B). \end{aligned}$$

The important thing to notice here is that the new conditional probability measure $P(\ |B)$ satisfies the axioms with the original sample space and event class – we are not restricted to applying $P(\ |B)$ to those events that lie within the smaller "conditional sample space", $S_{|B}$.

Exercise

Consider again the problem with five computers in a lab, with sample space denoted by

$$S = \{AD, AN, BD_1, BD_2, BN\},$$

and the following events:

- E_A is the event that the user's computer is from manufacturer A.
- E_B is the event that the user's computer is from manufacturer B.
- E_D is the event that the user's computer is defective.

Use the formula for conditional probability,

$$P(A \mid B) = \frac{P(A \cap B)}{B}$$

to calculate the probabilities specified in the Self Assessment Questions (see link below). (*It is easier to solve these using intuition/counting, but I encourage you to practice using the formula in the definition, which we will need to use in more complicated scenarios soon.*) Submit your answers as a fraction or a decimal with at least two digits of precision.

Terminology review and self-assessment questions

Interactive flashcards to review the terminology introduced in this section and self-assessment questions are available at fdsp.net/6-3, which can also be accessed using this QR code:

6.4 Relating Conditional and Unconditional Probabilities

As you may have seen from the Magician's Coin example, conditional probability is an area of probability where our intuition about probabilities may start to fail. One of my goals in writing this book is to leverage my experience teaching related courses to help you identify where you may already have some incorrect intuition about probability. To this end, consider this simple question:

How does having conditional information change a probability? Answer the full question at fdsp.net/6-4, which can also be accessed using this QR code:

To see how conditional information can change probabilities, let's consider a few examples:

Example 6.2: Relating Conditional and Unconditional Probabilities: Two Flips of a Fair Coin

Suppose a **fair** coin is flipped twice, and let H_i denote the event that the coin came up heads on flip i. Then clearly, $P(H_i) = 1/2$ and $P(H_1 \cap H_2) = 1/4$. From the definition of conditional probability,

$$(H_2 \mid H_1) = \frac{P(H_1 \cap H_2)}{P(H_1)} = \frac{1/4}{1/2} = 1/2.$$

So in this case, $P(H_2 \mid H_1) = P(H_2)$, and **conditional information did not change the probability**. This is expected because the outcome of the second flip should be independent of what happened on the first flip.

Example 6.3: Relating Conditional and Unconditional Probabilities: Repeated Draws from a Box of Balls

Suppose I have a box containing two balls, one of which is white and one of which is black. Balls are drawn consecutively from the box **without replacement**. Let W_i be the event that the ith draw is a white ball. If we are told that the first draw was a white ball, then clearly the remaining ball is black, so $P(W_2 \mid W_1) = 0$. Let's check using the definition of conditional probability. From our previous examples, we know that $P(W_i) = 1/2$, and $P(W_1 \cap W_2) = 0$ because there is no way to draw white balls out on both draws. Thus,

$$P(W_2 \mid W_1) = \frac{P(W_1 \cap W_2)}{P(W_1)} = \frac{0}{1/2} = 0.$$

So, in this case, $P(W_2 \mid W_1) < P(W_2)$, and **conditional information reduced the probability**.

In fact, this will happen whenever the two events being evaluated are disjoint. Let A and B be disjoint events with $P(A) > 0$ and $P(B) > 0$. Then

$$P(A \mid B) = \frac{P(A \cap B)}{P(B)} = \frac{P(\emptyset)}{P(B)} = 0.$$

If we know that two events are disjoint and one of the events occurred, then the other event cannot have occurred.

Example 6.4: Relating Conditional and Unconditional Probabilities: Other Events for Two Flips of a Fair Coin

Consider again the scenario where a fair coin is flipped twice and the ordered results are noted. Let E_i be the event that the number of heads observed is i, and let G be the event that at least one heads occurred.

From counting, we see $P(E_1) = 1/2$ and $P(G) = 3/4$. Let's find $P(G \mid E_1)$ using the definition of conditional probability. We need to calculate $P(G \cap E_1)$. We can write out the relevant events in terms of their outcomes as $E_1 = \{(H,T),\ (T,H)\}$ and $G = \{(H,T),\ (T,H),\ (H,H)\}$. Then $G \cap E_1 = E_1$. So,

$$P(G \mid E_1) = \frac{P(G \cap E_1)}{P(E_1)} = \frac{P(E_1)}{P(E_1)} = 1.$$

So, in this case, $P(G \mid E_1) > P(G)$, and **conditional information increased the probability**.

In general, if we have events A and B where B is a subset of A, then $A \cap B = B$. Then if $P(A) < 1$ and $P(B) > 0$,

$$P(A \mid B) = \frac{P(A \cap B)}{P(B)} = \frac{P(B)}{P(B)} = 1,$$

and $P(A \mid B) > P(A)$.

In conclusion:

Important!

There is no general answer to how conditional information will change the probability of an event. The conditional probability may be greater than, less than, or equal to the unconditional probability, depending on the relation between the two events.

6.5 More on Simulating Conditional Probabilities

Simulation is a key technique used in this book to explore random phenomena. Thus, it is important to understand the simulation techniques we use for different scenarios. Consider again the Magician's Coin problem, and let H_i denote the event that the coin comes up heads on the ith flip. In Section 6.1, we investigated $P(H_2 \mid H_1)$, the probability of getting heads on a second flip of the chosen coin given that heads was observed on the first flip. If we apply the definition of conditional probability from Section 6.3, then we can calculate this probability as

$$P(H_2 \mid H_1) = \frac{P(H_1 \cap H_2)}{P(H_1)}. \tag{6.1}$$

However, in Section 6.1, we did not estimate either $P(H_1 \cap H_2)$ or $P(H_1)$. Instead, we counted the proportion of times that we observed the outcome H_2 when H_1 was the outcome of the first flip. Let's see why these produce the same result.

Consider estimating the probabilities on the right-hand side of (6.1) using relative frequencies. We will use the following notation for the counters we have in the simulation:

N is the total number of simulation trials, N_1 is the number of simulation trials in which the outcome of flip 1 was heads (i.e., the number of times H_1 occurred), and N_{12} is the number of simulation trials in which the outcomes of both flip 1 and flip 2 were heads (i.e., the number of times $H_1 \cap H_2$ was observed).

In our original simulation for this scenario, the counter that incremented inside the nested `if` statement is equivalent to N_{12} because it is only updated when the outer `if` statement detects that the first flip was an `H` and the inner `if` statement detects that the second flip was an `H`.

Now, we can estimate the probability of H_1 using its relative frequency,

$$P(H_1) = \frac{N_1}{N}.$$

Similarly, we can estimate the probability of $H_1 \cap H_2$ using its relative frequency,

$$P(H_1 \cap H_2) = \frac{N_{12}}{N}.$$

If we substitute these into (6.1), we get

$$P(H_2 \mid H_1) = \frac{N_{12}/N}{N_1/N} = \frac{N_{12}}{N_1}.$$

I.e., the result is the same as the approach taken in Section 6.1. In general, we can directly estimate the conditional probability of an event B given an event A using nested `if` statements to:

- Count the number of trials in which A occurred. Call this number N_A.
- Within those trials in which A occurred, count the number of trials in which B also occurred. Call this number N_{AB}.
- Estimate the conditional probability of B occurring given that A occurred as

$$P(B \mid A) \approx \frac{N_{AB}}{N_A}.$$

WARNING

When designing simulations to estimate conditional probabilities, you must be especially careful in selecting the number of trials to be simulated.

In simulations of single events, if we know the approximate probability of the event, we can use that probability to determine the number of trials needed to estimate it. For example, if we are estimating a probability $P(A)$ that we believe is close to 0.1, then we can estimate that the event A occurs approximately once in every $1/P_A = 10$ trials. If we capture 100 instances of A occurring, we will get a reasonable estimate of $P(A)$, so we might use this information to simulate $100/P_A = 1000$ trials.

This approach will not work to estimate a conditional probability!

It is not sufficient to use only a conditional probability to estimate the number of trials needed.

Let's use an example to show why this is the case. Suppose A and B are events for which we wish to estimate the conditional probability $P(B \mid A)$. Consider the scenario where $P(A) = 10^{-3}$ and $P(A \cap B) = 9 \times 10^{-4}$. Then

$$P(B \mid A) = \frac{P(A \cap B)}{P(A)} = \frac{9 \times 10^{-4}}{10^{-3}} = 0.9.$$

If we used only the knowledge of $P(B \mid A)$, then we might use $100/P(B \mid A) \approx 111$ trials. But remember that in the simulation, we estimate the conditional probability as

$$P(B \mid A) \approx \frac{N_{AB}}{N_A}.$$

We can estimate[1] the values of these counters as follows. N_A will be approximately equal to the number of trials times $P(A)$, so it will be approximately equal to $111P(A) = 0.11$. In other words, most of the time when we run a simulation with 111 trials, we will not see any event A occur. The estimated value of counter N_{AB} will be the estimated number of trials in which we see both A and B occur, which will be approximately equal to $111P(A \cap B) \approx 0.1$. Again, in most simulation runs, we will not see $A \cap B$ occur at all. We may either not be able to estimate $P(B \mid A)$ at all (because the denominator is zero), or our estimate may be very inaccurate.

Important!

!

If we only use the conditional probability information, we may severely underestimate the number of trials required to observe a sufficient number of events.

⇒ When determining a sufficient number of trials for a simulation of a conditional probability $P(B \mid A)$, we must determine the required number of trials from the (estimated) value of $P(A \cap B)$, if known.

If no information about $P(A \cap B)$ is available, the value of N_{AB} should be checked at the end of the simulation to ensure that a sufficient number of events (at least 10; 100 is better) were observed.

Because of this effect, some people will code their simulation to run until N_{AB} reaches a specific threshold. This will bias the estimate of the $P(B \mid A)$ for reasons that are beyond the scope of this book. However, the bias will be small if the threshold on the number of events is large (>100).

6.6 Statistical Independence

The word "independence" generally means free from external control or influence. We will apply the concept of independence to many random phenomena, and the implication of

[1] We will formalize the meaning of these estimates later in the book when we talk about expected values/ensemble means.

independence is generally the same as the definition above: phenomena that are independent cannot influence each other.

We have already been applying the concept of independence throughout this book when we assume that the outcome of a coin flip, die roll, or simulation iteration does not depend on the values seen in other trials of the same type of experiment. However, now we have the mathematical tools to define the concept of independence precisely.

6.6.1 Conditional probabilities and independence

Based on the discussion above, try to answer the following question about what independence should mean for conditional probabilities. (Don't worry if you don't intuitively know the answer – you can keep trying if you don't get it right at first!)

Answer the interactive questions in **Quiz 1: Conditional probabilities and independence** at fdsp.net/6-6, which can also be accessed using this QR code:

If B is independent of A, then knowledge of A occurring should not change the probability of B occurring. I.e., if we are *given* that A occurred, then the conditional probability of B occurring should equal the unconditional probability:

$$P(B \mid A) = P(B).$$

Let's see the implications of this by substituting the formula for $P(B \mid A)$ from the definition:

$$\begin{aligned} \frac{P(A \cap B)}{P(A)} &= P(B) \\ \Rightarrow P(A \cap B) &= P(A)P(B). \end{aligned} \tag{6.2}$$

Now we might ask: if B is independent of A, does that imply that A is independent of B? Let's assume that (6.2) holds and apply the result to the definition for $P(A \mid B)$, assuming that $P(B) > 0$:

$$\begin{aligned} P(A \mid B) &= \frac{P(A \cap B)}{P(B)} \\ &= \frac{P(A)P(B)}{P(B)} \\ &= P(A). \end{aligned}$$

So if $P(B \mid A) = P(B)$, then $P(A \mid B) = P(A)$.

6.6.2 Formal definition of statistically independent events

A definition for statistically independent events that satisfies all the forms of independence discussed above and that can deal with events with probability zero is as follows:

Definition

statistically independent (two events)

Given a probability space $(S, \mathcal{F}, P)$ and two events $A \in \mathcal{F}$ and $B \in \mathcal{F}$, A and B are *statistically independent* if and only if (iff) $P(A \cap B) = P(A)P(B)$.

If the context is clear, we often write "independent" instead of "statistically independent" or write *s.i.*, a commonly used abbreviation. If events are not s.i., then we say they are *statistically dependent*.

Note:

Please take time to study the definition of *statistically independent* carefully. In particular, note the following:

- Up to this point, we have only defined statistical independence and statistical dependence for events.
- Probabilities **are not** something that are statistically independent or dependent.
- The "if and only if" statement means that the definition applies in both directions:
  - If events A and B are statistically independent, then the probability of the intersection of the events factors as the product of the probabilities of the individual events, $P(A \cap B) = P(A)P(B)$.
  - If we have events A and B for which $P(A \cap B) = P(A)P(B)$, then A and B are statistically independent.

6.6.3 When can we assume independence?

Statistical independence is often assumed for many types of events. However, it is important to be careful when applying such a strong assumption because events can be coupled in subtle ways. For example, consider the Magician's Coin example. Many people assume that the event of getting heads on the second flip of the chosen coin will be independent of the outcome of the first flip of the coin. However, we have seen that this assumption is wrong! So, when can we assume that events will be independent?

Important!

Events can be assumed to be statistically independent if they arise from completely separate random phenomena.

In the case of the Magician's Coin, this assumption is violated in a subtle way. If we knew that the two-headed coin was in use, then we would know the results completely. What is subtle is the fact that observing the outcome of the first flip may give some information about which coin is in use (although we won't be able to show this until Section 7.2).

Examples that are assumed to result from separate random phenomena are extensive:

- **Devices to generate randomness in games:** Independence can usually be assumed for different flips of a fair coin or rolls of a fair die.
- **Failures of different devices in systems:** Mechanical and electrical devices fail at random, and the failures at different devices are often assumed to be independent; examples include light bulbs in a building or computers in a lab.
- **Characteristics of people unrelated to any grouping of those people:** For example, for a group of people at a meeting, having a March birthday would generally be independent events across any two people.

Let's apply statistical independence to find a simpler way to solve a problem that was introduced in Section 4.4.

Example 6.5: A Pair of Dice Values Less Than 3 – Take 3

A fair six-sided die is rolled twice. What is the probability that the value of either roll is less than 3?

As before, let E_i be the event that the top face on roll i is less than 3, for $i = 1, 2$. We assume that different rolls of the die are independent, so E_1 and E_2 are independent.

As in Section 4.5, we can use Corollary 5 of the Axioms of Probability to write

$$P(E_1 \cup E_2) = P(E_1) + P(E_2) - P(E_1 \cap E_2).$$

Before, we had to enumerate $E_1 \cap E_2$ over the sample space for the combined rolls of the die to determine $P(E_1 \cap E_2)$. Now, we can apply statistical independence to write $P(E_1 \cap E_2) = P(E_1)P(E_2)$, yielding

$$\begin{aligned} P(E_1 \cup E_2) &= P(E_1) + P(E_2) - P(E_1)P(E_2) \\ &= \frac{1}{3} + \frac{1}{3} - \left(\frac{1}{3}\right)\left(\frac{1}{3}\right) \\ &= \frac{5}{9}. \end{aligned}$$

Exercises

Answer the interactive questions in **Quiz 2: Applying Statistical Independence to Unions** at fdsp.net/6-6, which can also be accessed using this QR code:

Note:

If A and B are s.i. events, then the following pairs of events are also s.i.:

A and $\overline{B}$; $\overline{A}$ and B; $\overline{A}$ and $\overline{B}$.

I.e., if the probability of an event A occurring does not depend on whether some event B occurs, then it cannot depend on whether the event B does not occur. This probably

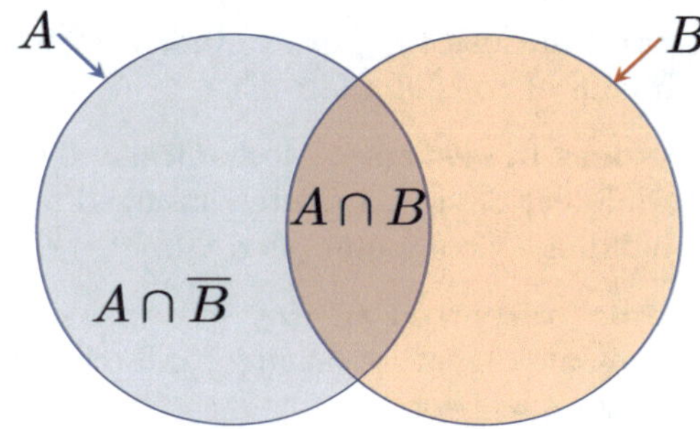

FIGURE 6.3
Venn diagram showing relation of A, $A \cap \overline{B}$, and $A \cap B$.

matches your intuition. However, we should verify it. Let's check the first example. We need to evaluate $P(A \cap \overline{B})$ to see if it factors as $P(A)P(\overline{B})$. Referring to the Venn diagram in Fig. 6.3, we can see that A consists of the union of the disjoint parts, $A \cap B$ and $A \cap \overline{B}$. So we can write $P\left(A \cap \overline{B}\right) = P(A) - P(A \cap B)$.

Then by utilizing the fact that A and B are s.i., we have

$$\begin{aligned} P\left(A \cap \overline{B}\right) &= P(A) - P(A \cap B) \\ &= P(A) - P(A)P(B) \\ &= P(A)\left[1 - P\left(B\right)\right] \\ &= P(A)P\left(\overline{B}\right). \end{aligned}$$

So, if A and B are s.i., so are A and $\overline{B}$. The other expressions can be proven through similar manipulation. This is important because we often use this fact to simplify solving problems. We start with a simple example to demonstrate the basic technique.

Example 6.6: A Pair of Dice Values Less Than 3 – Take 4

A fair six-sided die is rolled twice. What is the probability that the value of either roll is less than 3?

As before, let E_i be the event that the top face on roll i is less than 3, for $i = 1, 2$, and E_1 and E_2 are s.i. then

$$\begin{aligned} P(E_1 \cup E_2) &= 1 - P\left(\overline{E_1 \cup E_2}\right) \\ &= 1 - P\left(\overline{E_1} \cap \overline{E_2}\right) \\ &= 1 - P\left(\overline{E_1}\right) P\left(\overline{E_2}\right) \\ &= 1 - \left[1 - P\left(E_1\right)\right]\left[1 - P\left(E_2\right)\right] \\ &= 1 - \left[1 - \left(\frac{2}{6}\right)\right]\left[1 - \left(\frac{2}{6}\right)\right] \\ &= \frac{5}{9}. \end{aligned}$$

Of course, for this simple example, it is easiest to directly compute $P\left(\overline{E_1}\right)$, but the full approach shown here is a template that is encountered often when dealing with unions of s.i. events.

To see the power of this method, we first need to define s.i. for more than two events:

Definition

statistically independent (any number of events)

Given a probability space S, $\mathcal{F}$, P, a collection of events $E_0, E_1, \ldots, E_{n-1}$ in $\mathcal{F}$ are *statistically independent* if and only if

$$\begin{aligned} P(E_i \cap E_j) &= P(E_i)P(E_j), & \forall i \neq j \\ P(E_i \cap E_j \cap E_k) &= P(E_i)P(E_j)P(E_k), & \forall i \neq j \neq k \\ &\vdots \\ P(E_0 \cap E_1 \cap \ldots \cap E_{n-1}) &= P(E_0)P(E_1)\cdots P(E_{n-1}). \end{aligned}$$

It is not sufficient to just check that the probability of every pair of events factors as the product of the probabilities of the individual events. That defines a weaker form of independence, called *pairwise statistical independence*:

Definition

pairwise statistically independent

Given a probability space S, $\mathcal{F}$, P, a collection of events $E_0, E_1, \ldots, E_{n-1}$ in $\mathcal{F}$ are *pairwise statistically independent* if and only if

$$P(E_i \cap E_j) = P(E_i)P(E_j), \quad \forall i \neq j.$$

If we want to find the probability of the union of s.i. events, we can use complements to convert the unions to intersections, and the resulting general form looks like

$$P\left(\bigcup_i E_i\right) = 1 - \prod_i \left[1 - P\left(E_i\right)\right].$$

It may be helpful to interpret this as follows. The complement of the event that any of a collection of events occurring is that none of those events occurs; thus, the probability that any of a collection of events occurs is one minus the probability that none of those events occurs.

Compare the simplicity of this approach to the form for directly solving for the probability of unions of events (Corollary 7 from Section 4.5:

$$P\left(\bigcup_{k=0}^{n-1} A_k\right) = \sum_{k=0}^{n-1} P\left(A_j\right) - \sum_{j<k} P\left(A_j \cap A_k\right) + \cdots$$
$$+ (-1)^{(n-1)} P\left(A_0 \cap A_1 \cap \cdots \cap A_{n-1}\right).$$

Now apply this approach to solve the following practice problems. Answer the interactive questions in **Quiz 3: More on Independence and Probabilities of Unions** at fdsp.net/6-6, which can also be accessed using this QR code:

6.6.4 Relating statistically independent and disjoint events

Before reading below, try to answer the question in **Quiz 4: Relating Statistical Independent and Disjoint Events** at fdsp.net/6-6, which can also be accessed using this QR code:

Suppose A and B are events that are both disjoint and statistically independent.

- Since A and B are disjoint, $A \cap B = \emptyset$, which further implies $P(A \cap B) = P(\emptyset) = 0$.
- Since A and B are s.i., $P(A \cap B) = P(A)P(B)$.

Combining these, we have that $P(A \cap B) = P(A)P(B) = 0$, which can only occur if either or both of $P(A) = 0$ or $P(B) = 0$.

Thus, events **cannot be both statistically independent and disjoint unless at least one of the events has probability zero**.

To gain some further insight, consider the condition for events to be disjoint, $A \cap B = \emptyset$. This condition implies that if A occurs, then B cannot have occurred, and vice versa. Thus, knowing that either A or B occurred provides a lot of information about the other event. Thus, A and B cannot be independent if they are disjoint, except in the special case already identified.

Terminology review and self-assessment questions

Interactive flashcards to review the terminology introduced in this section and self-assessment questions are available at fdsp.net/6-6, which can also be accessed using this QR code:

6.7 Conditional Probabilities and Independence in Fair Experiments

It may be helpful to consider conditional probabilities and independence in the context of a fair experiment. Recall that for a fair experiment, the (unconditioned) probability of an event E can be computed as

$$P(E) = \frac{|E|}{|S|}.$$

Let's see what this means in terms of conditional probabilities and independence by using an example. Consider again the example where a die is rolled two times and the outcome is the tuple of ordered values on the top faces. As before, we let E_i be the event that the value on face i is less than 3. We previously showed that $P(E_1) = P(E_2) = 1/3$. Let T_i be the event that the total (sum) value shown on the two faces is i. For instance, we can write T_7 in terms of the outcomes of the composite experiment as

$$T_7 = \{(1,6),\ (2,5),\ (3,4),\ (4,3),\ (5,2),\ (6,1)\}.$$

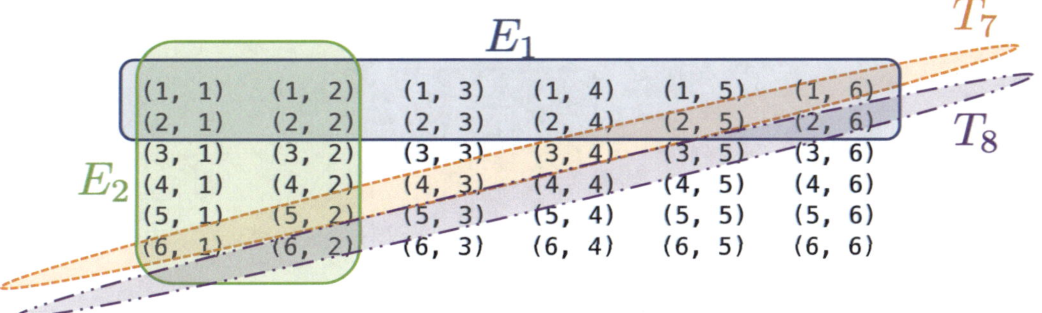

FIGURE 6.4
Diagram showing outcomes included in events E_1, E_2, and F.

Then

$$P(T_7) = \frac{|T_7|}{|S|} = \frac{6}{36} = \frac{1}{6}.$$

Now we will consider some conditional probabilities and see how the computation simplifies for fair experiments. In addition, we will see the implications of statistical independence for fair experiments. In Fig. 6.4, we illustrate four events that we will discuss: E_1, E_2, T_7, and T_8.

First, consider some generic events (i.e. representing any possible event) A and B for this experiment. Then the conditional probability of A given B can be written as

$$\begin{aligned} P(A \mid B) &= \frac{P(A \cap B)}{P(B)} \\ &= \frac{\frac{|A \cap B|}{|S|}}{\frac{|B|}{|S|}} \\ &= \frac{|A \cap B|}{|B|}. \end{aligned}$$

It is useful to interpret this as follows: given that B occurs, $|A \cap B|$ is the number of outcomes in A that are within B. Then the conditional probability of A given B is just the proportion of outcomes of A in B. This verifies the approach we took when we introduced conditional probability: the conditional probability of A given B can be computed as the probability of A within the reduced sample space, $S_{|B}$.

To make this more concrete, let's apply it to several of the events we have introduced.

Example 6.7: $P(E_1 \mid E_2)$

From Fig. 6.5, we can see that there are 4 outcomes of E_1 within E_2, and there are 12 total outcomes in E_2, so the proportion of outcomes of E_1 in E_2 is $P(E_1 \mid E_2) = 4/12 = 1/3$.

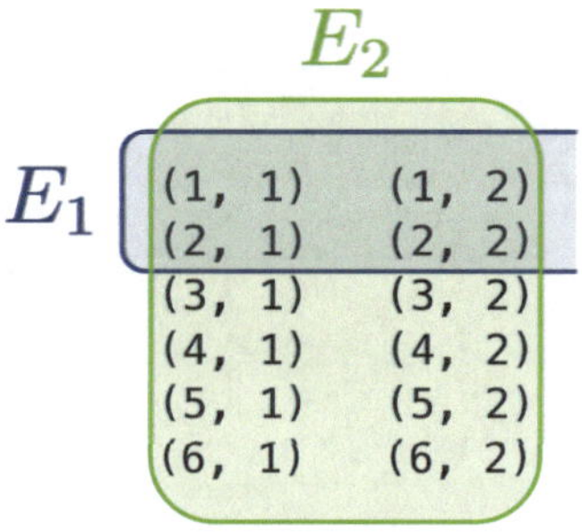

FIGURE 6.5
Diagram showing the event E_2 and the outcomes of E_1 that are within E_2.

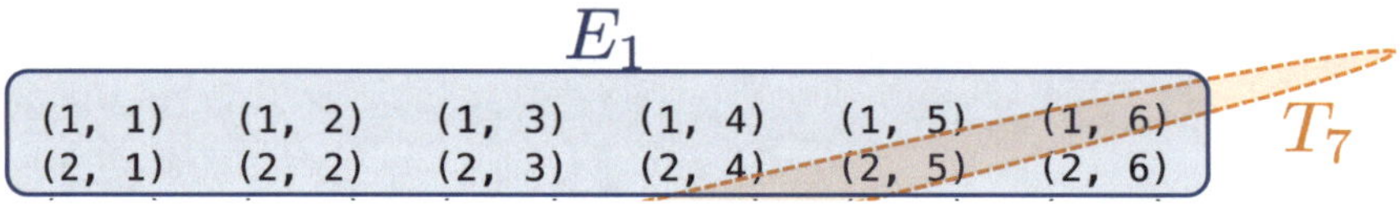

FIGURE 6.6
Diagram showing the event E_1 and the outcomes of T_7 that are within E_1.

Example 6.8: $P(T_7 \mid E_1)$ and $P(E_1 \mid T_7)$

These probabilities can be calculated by referring to Fig. 6.6.

There are 2 outcomes of T_7 out of the 12 total outcomes that make up E_1, so the conditional probability of T_7 given E_1 is $P(T_7 \mid E_1) = 2/12 = 1/6$.

There are 2 outcomes of E_1 within the 6 total outcomes of T_7, so $P(E_1 \mid T_7) = 2/6 = 1/3$.

Example 6.9: $P(T_8 \mid E_2)$ and $P(E_2 \mid T_8)$

We will use Fig. 6.7 for reference.

By inspection, we see that the proportion of outcomes of T_8 in E_2 is 1/12, so $P(T_8 \mid E_2) = 1/12$. Similarly, the proportion of outcomes of E_2 in T_8 is 1/5, so $P(E_2 \mid T_8) = 1/5$.

Now, let's consider which pairs of events that we have discussed are independent and what that means in terms of the outcomes.

- We calculated that $P(E_1 \mid E_2) = 1/3 = P(E_1)$, so E_1 is independent of E_2.
- We calculated that $P(T_7 \mid E_1) = 1/6 = P(T_7)$, so T_7 is independent of E_1.
- We calculated that $P(E_1 \mid T_7) = 1/3 = P(E_1)$, so E_1 is independent of T_7 (which we also could infer from the previous step).
- We calculated that $P(T_8 \mid E_2) = 1/12$, but $P(T_8) = 5/36$, so T_8 is *not* independent of E_2.

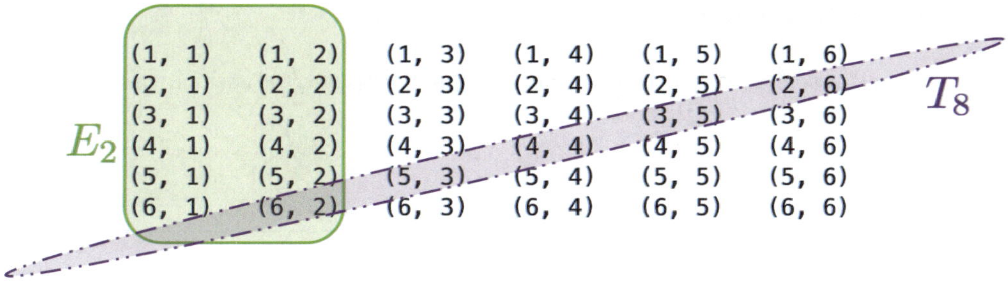

(1, 1)	(1, 2)	(1, 3)	(1, 4)	(1, 5)	(1, 6)
(2, 1)	(2, 2)	(2, 3)	(2, 4)	(2, 5)	(2, 6)
(3, 1)	(3, 2)	(3, 3)	(3, 4)	(3, 5)	(3, 6)
(4, 1)	(4, 2)	(4, 3)	(4, 4)	(4, 5)	(4, 6)
(5, 1)	(5, 2)	(5, 3)	(5, 4)	(5, 5)	(5, 6)
(6, 1)	(6, 2)	(6, 3)	(6, 4)	(6, 5)	(6, 6)

E_2 T_8

FIGURE 6.7
Diagram showing the events E_2 and T_8.

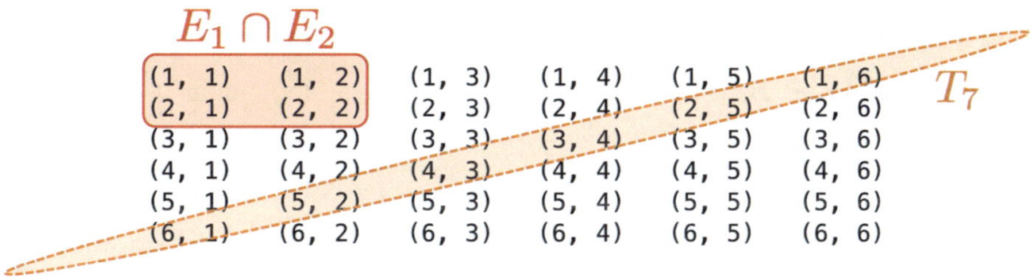

$E_1 \cap E_2$

(1, 1)	(1, 2)	(1, 3)	(1, 4)	(1, 5)	(1, 6)
(2, 1)	(2, 2)	(2, 3)	(2, 4)	(2, 5)	(2, 6)
(3, 1)	(3, 2)	(3, 3)	(3, 4)	(3, 5)	(3, 6)
(4, 1)	(4, 2)	(4, 3)	(4, 4)	(4, 5)	(4, 6)
(5, 1)	(5, 2)	(5, 3)	(5, 4)	(5, 5)	(5, 6)
(6, 1)	(6, 2)	(6, 3)	(6, 4)	(6, 5)	(6, 6)

T_7

FIGURE 6.8
Diagram showing the events $E_1 \cap E_2$ and T_7.

- We calculated that $P(E_2 \mid T_8) = 1/5$, but $P(E_2) = 1/3$, so E_2 is *not* independent of T_8 (again, we could have inferred this from the previous step).

So, what is required for independence for fair experiments?

In a fair experiment, for an event A to be independent of an event B, the proportion of outcomes of A in B must equal the proportion of outcomes of A in S (the original sample space). (Or we could also test if the proportion of outcomes of B in A is equal to the proportion of outcomes of B in S.)

Please note that all these computations involving proportions do not necessarily hold (and usually do not hold) if the experiment is not fair!

Now, note that E_1 and E_2 are independent. And we showed that T_7 and E_1 are independent. It is not hard to see that T_7 and E_2 are also independent. Combined, we can see that E_1, E_2, and T_7 are *pairwise statistically independent*. So, are these events independent? We only need to check if one more condition holds: does $P(E_1 \cap E_2 \cap T_7)$ equal $P(E_1)P(E_2)P(T_7)$? We have

$$\begin{aligned} P(E_1)P(E_2)P(T_7) &= \left(\frac{1}{3}\right)\left(\frac{1}{3}\right)\left(\frac{1}{6}\right) \\ &= \frac{1}{54}. \end{aligned}$$

Now consider $E_1 \cap E_2 \cap T_7$. We can find the outcomes in this event by performing the leftmost intersection first, as $(E_1 \cap E_2) \cap T_7$, as shown in Fig. 6.8.

The events $E_1 \cap E_2$ and T_7 are disjoint, so there are no outcomes in $E_1 \cap E_2 \cap T_7$, and $P(E_1 \cap E_2 \cap T_7) = 0 \neq P(E_1)P(E_2)P(T_7)$.

Thus, E_1, E_2, and T_7 are an example of events that are pairwise s.i. but not s.i.

Self-assessment questions

Interactive self-assessment questions are available at fdsp.net/6-7, which can also be accessed using this QR code:

6.8 Conditioning and (In)dependence

Many books do not discuss how conditioning can affect the dependence or independence of events. However, understanding this is key to techniques we will apply to analyze several types of problems.

Let A, B, and C be events. Consider the following "simple" questions:

- Is it possible for A and B to be independent but become dependent if we are told that C occurred?
- Is it possible for A and B to be dependent but become independent if we are told that C occurred?

It is surprising to many people studying probability that the answer to both these questions is *yes*. To discuss this carefully, we need to extend our concept of independence:

> Definition
>
> **conditional independence (events)**
>
> Events A and B are *conditionally independent* given an event C if and only if A and B are independent under the conditional probability measure $P(\cdot \mid C)$. I.e., A and B are conditionally independent if and only if
>
> $$P(A \cap B \mid C) = P(A \mid C)\, P(B \mid C).$$

Now, let's consider some examples.

Example 6.10: Conditional Independence for Dice Example

Consider again the example from Section 6.7, in which a fair die is rolled two times and the ordered values on the top faces are noted. Four events were defined, and these are illustrated in Fig. 6.4.

We previously showed that E_1, E_2, and T_7 are pairwise statistically independent. However, note that

$$E_1 \cap E_2 \cap T_7 = \emptyset,$$

so

$$P(E_1 \cap E_2 \cap T_7) = 0.$$

Thus,

$$P\left(E_1 \cap E_2 \mid T_7\right) = \frac{P(E_1 \cap E_2 \cap T_7)}{P(T_7)} = \frac{0}{1/6} = 0.$$

But

$$P(E_1 \mid T_7) = P(E_2 \mid T_7) = \frac{P(E_i \cap T_7)}{P(T_7)} = \frac{2/36}{1/6} = \frac{1}{3}.$$

Since

$$P(E_1 \cap E_2 \mid T_7) \neq P(E_1 \mid T_7)\, P(E_2 \mid T_7),$$

E_1 and E_2 are **not conditionally independent** given T_7, even though they are independent if no information about T_7 is known.

It is relatively easy to see why this is the case for this example. From the picture, we can see that if T_7 occurs, then it is not possible for both E_1 and E_2 to occur. Thus, if we know that T_7 occurred, then knowing E_1 occurred tells us that E_2 did not occur. Without knowing whether T_7 occurred, knowledge of whether E_1 occurred does not affect the probability that E_2 occurred.

Example 6.11: Conditional Independence in the Magician's Coin Problem

Consider the Magician's Coin problem, which was introduced in Section 6.1. Considering the two sides of the two-headed coin as "H1" and "H2", we can enumerate all the possible outcomes as shown below:

Coin				
Fair	**(H, H)**	**(H, T)**	**(T, H)**	**(T, T)**
Two-Headed	**(H1, H1)**	**(H1, H2)**	**(H2, H1)**	**(H2, H2)**

As before, let H_i denote the event that the coin came up heads on flip i. Let F denote the event that the coin is fair. It is easy to see that

$$P(H_1 \cap H_2) = \frac{5}{8}$$

because there are five total outcomes that have heads on both flips. As before, we can see that $P(H_1) = P(H_2) = 3/4$.

Since

$$P(H_1 \cap H_2) = \frac{5}{8} \neq P(H_1)P(H_2) = \frac{3}{4},$$

events H_1 and H_2 are **dependent**. However, if we are given F, we have

$$P(H_1 \cap H_2 \mid F) = \frac{1}{4},$$

and

$$P(H_1 \mid F) = P(H_2 \mid F) = \frac{1}{2}.$$

Since

$$P(H_1 \cap H_2 \mid F) = \frac{1}{4} = P(H_1 \mid F)\, P(H_2 \mid F) = \left(\frac{1}{2}\right)\left(\frac{1}{2}\right),$$

we can see that H_1 and H_2 are independent given F.

Question: Are H_1 and H_2 independent given $\overline{F}$?

In fact, it is possible for events A and B to be independent given an event C and dependent given an event $\overline{C}$. However, such an example is outside the scope of this book.

Terminology review and self-assessment questions

Interactive flashcards to review the terminology introduced in this section and self-assessment questions are available at fdsp.net/6-8, which can also be accessed using this QR code:

6.9 Chain Rules and Total Probability

Chain rules and total probability use conditional probability to decompose the probability of an event. The goal is to express unknown probabilities of events in terms of probabilities that we already know.

6.9.1 Using conditional probability to decompose events: Part 1 – chain rules

In working with probability, we often need to find probabilities of intersections of events, while only knowing the probabilities of individual events and conditional probabilities of combinations of the events. *Chain rules* are often helpful in these cases:

Definition

chain rules

Express the probability of an intersection of events in terms of a sequence of conditional probabilities involving those events.

This is easiest to understand for the case of two events, A and B. From the definition of conditional probability, we can write $P(A \mid B)$ as

$$\begin{aligned} P(A \mid B) &= \frac{P(A \cap B)}{P(B)} \\ \Rightarrow P(A \cap B) &= P(A \mid B)\, P(B), \end{aligned}$$

and we can write $P(B \mid A)$ as

$$\begin{aligned} P(B \mid A) &= \frac{P(A \cap B)}{P(A)} \\ \Rightarrow P(A \cap B) &= P(B \mid A)\, P(A). \end{aligned}$$

After manipulating the expressions as shown, we get two *different* formulas for expressing $P(A \cap B)$. These are **chain rules** for the probability of the intersection of two events. Such rules are often used when:

- Two events are dependent on each other, but the relation is simple if the outcome of one of the experiments is known.

- The events are at two different points in a system, such as the input and output of a system.

Example 6.12: Draws from a Card Deck

A simple example of the former is in card games. Two cards are drawn (without replacement) from a deck of 52 cards (without jokers). What is the probability that they are both Aces? Let A_i be the event that the card on draw i is an Ace. Then the most natural way to apply the chain rule is to write

$$P(A_1 \cap A_2) = P(A_2 \mid A_1) P(A_1).$$

The probability of getting an Ace on draw 1 is 4/52=1/13 because there are 4 Aces in the deck of 52 cards. The probability of getting an Ace on the second draw *given that the first draw was an Ace* is $3/51 = 1/17$. This is because, after the first draw, there are 3 Aces left in the remaining deck of 51 cards. Thus,

$$P(A_1 \cap A_2) = P(A_2 \mid A_1) P(A_1) = \left(\frac{1}{17}\right)\left(\frac{1}{13}\right) = \frac{1}{221}.$$

As a check, we can compare with a solution using combinatorics. There are

$$\binom{4}{2} = 6$$

ways to choose the Aces from the four total Aces. There are

$$\binom{52}{2} = \frac{52!}{50!2!} = 1326$$

ways to choose two cards from 52. So,

$$P(A_1 \cap A_2) = \frac{6}{1326} = \frac{1}{221},$$

which matches our answer using conditional probability.

The solution using conditional probability is usually much more intuitive for learners who are new to probability, but being able to use both techniques is a powerful method for checking your work.

Review Questions

Answer the interactive questions in **Quiz 1: Using Chain Rules** at fdsp.net/6-9, which can also be accessed using this QR code:

Example 6.13: Chain Rule Example for the Magician's Coin

For the Magician's Coin problem, what is the probability that the fair coin is selected and it comes up heads on the first flip? This is an example of a system where one unobserved event (the choice of coin) affects the entire sequence of outputs. This is an example of a system with *hidden state*, and we will delve into this general concept more in Section 7.2. When we have such problems, we usually need to decompose them in terms of the probabilities of the hidden state and the conditional probabilities of the output given the hidden state. Let H_i denote the event that the coin comes up heads on flip i. Let F be the event that the fair coin was chosen. We are looking for $P(F \cap H_1)$, which we can write as

$$P(F \cap H_1) = P(H_1 \mid F) P(F).$$

If there is one fair coin and one two-headed coin, $P(F) = 1/2$. Given that the coin is fair, $P(H_1 \mid F) = 1/2$. So,

$$P(F \cap H_1) = \left(\frac{1}{2}\right)\left(\frac{1}{2}\right) = \frac{1}{4}.$$

Note that it is generally **not helpful to write the probability using the other form of the chain rule:**

$$P(F \cap H_1) = P(F \mid H_1) P(H_1).$$

Neither $P(H_1)$ nor $P(F \mid H_1)$ are probabilities that can be known directly from the problem statement. Thus, although the expression is valid mathematically, it is not helpful in solving this problem.

Review Questions

Answer the interactive questions in **Quiz 2: Chain Rules and the Magician's Coin** at fdsp.net/6-9, which can also be accessed using this QR code:

The chain rule can be easily generalized to more than two events. The easiest way is to write probabilities in terms of conditional probabilities that are expressed as fractions (as in the definition of probability), such that the denominator of one fraction cancels with the numerator of the next fraction to make sure the expression does not change when it is rewritten. This will make more sense with an example for rewriting the probability of the intersection of 3 events (A, B, and C):

$$\begin{aligned} P(A \cap B \cap C) &= \frac{P(A \cap B \cap C)}{P(B \cap C)} \cdot P(B \cap C) \\ &= \frac{P(A \cap B \cap C)}{P(B \cap C)} \cdot \frac{P(B \cap C)}{P(C)} \cdot P(C) \\ &= P(A \mid B \cap C) P(B \mid C) P(C). \end{aligned}$$

This decomposition assumes that we know the probability of A given that B and C have occurred and that we know the probability of B given C. Such dependence occurs naturally in many systems, but the particular decomposition will depend on what we know about these probabilities. We could just have easily written

$$P(A \cap B \cap C) = P(C \mid A \cap B) P(B \mid A) P(A).$$

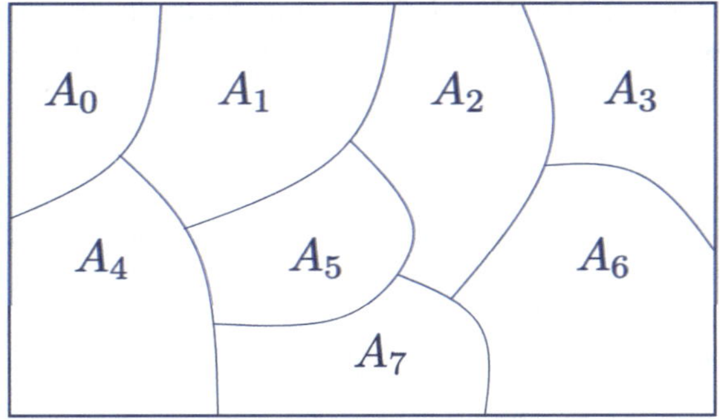

FIGURE 6.9
Example partition of the sample space.

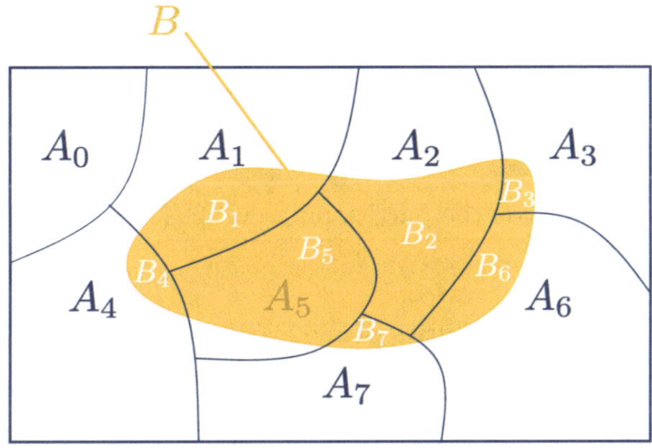

FIGURE 6.10
Example event B superimposed on partition A_0, $A_1, \ldots, A_7$ of the sample space.

6.9.2 Using conditional probability to decompose events: Part 2 – partitions and total probability

We previously introduced the concept of partitions in Section 3.3 as a way to take a collection of data and break it into separate, disjoint groups. Now, we are ready to apply this concept to events, which are sets of outcomes. In particular, we will most often partition the sample space, S:

> Definition
>
> **partition (of the sample space)**
>
> A collection of events $A_1, A_2, \ldots$ *partitions* the sample space S if and only if
>
> $$S = \bigcup_i A_i, \text{ and } A_i \cap A_j = \emptyset, \quad i \neq j.$$

For example, the collection of disjoint events A_0, $A_1, \ldots, A_7$ shown in Fig. 6.9 is a partition for S. Now consider how we can use a partition A_0, $A_1, \ldots, A_{n-1}$ of S to decompose any event $B \subseteq S$. In Fig. 6.10, an example event B is shown on top of our example partition. Note that we do not require that B have a nonempty intersection with every partition event (e.g., in this example $B \cap A_0 = \emptyset$). Then we can use our partition to decompose B into smaller subsets as

$$B_i = B \cap A_i, \quad i = 0, 1, \ldots, n-1,$$

as shown in Fig. 6.10.

Note that $A_i \cap A_j = \emptyset$ also implies that the set of events $\{B_j\}$ are disjoint. This should be intuitive from Fig. 6.10, but a mathematical proof is included at fdsp.net/6-9. Since $\{A_i\}$ covers the entire sample space, the union of the $\{B_j\}$ is equal to the original event B. A formal proof is provided online at fdsp.net/6-9. These two properties imply that $B_0, B_1, \ldots, B_{n-1}$ are a partition for B. If we want to express the probability of B, we can write

$$\begin{aligned} P(B) &= P\left(\bigcup_i B_i\right) \\ &= \sum_i P(B_i) \\ &= \sum_i P(B \cap A_i)\,. \end{aligned}$$

Now suppose that we choose the partitioning events $\{A_i\}$ such that:

- We know the probabilities $P(A_i)$.
- We know the conditional probabilities of the event B given that A_i occurred, $P(B \mid A_i)$.

Applying the chain rule, we can write $P(B \cap A_i) = P(B \mid A_i)\,P(A_i)$ for each i. Putting this all together, we get the *Law of Total Probability*:

Definition

Law of Total Probability

Given an event $B \subseteq S$ and a partition of S denoted by $A_1, A_2, \ldots$,

$$P(B) = \sum_i P(B \mid A_i)\,P(A_i).$$

The law of total probability is often used in systems where there is either:

- *hidden state*, or
- random inputs and outputs, where the output is dependent on the input.

Note that these are not necessarily disjoint; systems can have both hidden state and random inputs and outputs. For the Magician's Coin problem, we would usually treat the choice of coin as a hidden state because it does not change over time but is an unobservable property of the system that influences the system's outputs. When applying chain rules or the Law of Total Probability in such problems, the conditioning will usually be on the different possibilities of the hidden state or unknown input.

Example 6.14: The Magician's Coin

As before, a magician has a fair coin and a two-headed coin in her pocket. Let H_i denote the event that the outcome of flip i is heads. We can use total probability to answer more complicated questions regarding the probabilities of the outputs:

(a) $P(H_1)$

As mentioned above, we can condition on the hidden state, which is whether the coin is fair (F) or not ($\overline{F}$).

Note:

One thing that is often confusing to learners of probability is determining what is actually a partition of S. If you have a set of events that are both *disjoint* and *one of those events must occur* (i.e., the events cover the sample space), then they are a partition.

In this case, F and $\overline{F}$ are complements, so they are disjoint. Moreover, either the coin is fair (F) or it is not ($\overline{F}$), so one of these events must occur. Therefore, the events $F, \overline{F}$ partition S.

Applying the Law of Total Probability,

$$\begin{aligned} P(H_1) &= P(H_1 \mid F)\, P(F) + P\left(H_1 \mid \overline{F}\right) P(\overline{F}) \\ &= \left(\frac{1}{2}\right)\left(\frac{1}{2}\right) + \left(1\right)\left(\frac{1}{2}\right) \\ &= \frac{3}{4}. \end{aligned}$$

(b) $P(H_1 \cap H_2)$

We can use the same partition as above to write:

$$P(H_1 \cap H_2) = P(H_1 \cap H_2 \mid F)\, P(F) + P\left(H_1 \cap H_2 \mid \overline{F}\right) P(\overline{F}).$$

However, now we need to know the probabilities $P(H_1 \cap H_2 \mid F)$ and $P\left(H_1 \cap H_2 \mid \overline{F}\right)$. When the coin is fair, the events H_1 and H_2 are conditionally independent, so $P(H_1 \cap H_2 \mid F) = P(H_1 \mid F)\, P(H_2 \mid F)$. When the coin is two-headed, it always comes up heads, so $P\left(H_1 \cap H_2 \mid \overline{F}\right) = 1$. Then

$$\begin{aligned} P(H_1 \cap H_2) &= P(H_1 \mid F)\, P(H_2 \mid F)\, P(F) + (1)P(\overline{F}) \\ &= \left(\frac{1}{2}\right)\left(\frac{1}{2}\right)\left(\frac{1}{2}\right) + \left(1\right)\left(\frac{1}{2}\right) = \frac{5}{8}. \end{aligned}$$

(c) $P(H_2 \mid H_1)$

By calculating this probability, we can assess whether getting heads on flip 2 is independent of getting heads on flip 1, and if so, how information about the value of the first flip changes the probabilities for the second flip. We can directly apply the definition of conditional probability to calculate this from the answers to parts (a) and (b):

$$P(H_2 \mid H_1) = \frac{P(H_1 \cap H_2)}{P(H_1)} = \frac{5/8}{3/4} = \frac{5}{6}.$$

If we did not know H_1, then $P(H_2) = P(H_1) = 3/4$ (you should verify this using the Law of Total Probability with H_1 as the hidden state). So knowing that heads occurred on the first flip increases the probability that heads will occur on the second flip.

If this surprises you (again), then recall that we can anticipate this if we take it to extremes. What if I told you that heads occurred on the first 1000 flips? Then you would surely think that the magician had chosen the two-headed coin and expect the probability of getting heads on the 1001st flip to be close to 1. If heads is observed on one flip and the same coin is flipped again, we should expect the probability of getting heads on the second flip to be larger than if we did not know the outcome of the first flip.

In fact, after observing a single outcome of heads, the probability of having chosen the two-headed coin should increase to more than 1/2. It does – but we will need some new tools that we will develop in Chapter 7 before we can calculate the new probability.

Review Questions

Answer the interactive questions in **Quiz 3: Total Probability and the Magician's Coin** at fdsp.net/6-9, which can also be accessed using this QR code:

Example 6.15: Two Random Selections

At the beginning of this chapter, the following question was asked: A box contains two white balls and one black ball. I reach into the box and remove one ball. Without inspecting it, I place it in my pocket. I withdraw a second ball. What is the probability that the second ball is white?

This question stumps many people learning probability, but the answer turns out to be simple. Let W_i be the event that a white ball is chosen on draw i. We are asking about W_2. Then $P(W_2) = P(W_1) = 2/3$. However, this is not intuitive because our brain tells us that the probabilities for the second draw must depend on what happened on the first draw – which is unknown in this case.

We can formally show this using the Law of Total Probability. The conditioning event will be the unobserved result of the first draw. The partition is $W_1, \overline{W_1}$. Then

$$P(W_2) = P\left(W_2 \mid W_1\right) P(W_1) + P\left(W_2 \mid \overline{W_1}\right) P\left(\overline{W_1}\right).$$

If a white ball is drawn first (W_1), then there is one white ball and one black ball left, so $P\left(W_2 \mid W_1\right) = 1/2$. If a black ball is drawn first ($\overline{W_1}$), then there are two white balls left, so $P\left(W_1 \mid \overline{W_1}\right) = 1$. Then

$$P(W_2) = \left(\frac{1}{2}\right)\left(\frac{2}{3}\right) + \left(1\right)\left(\frac{1}{3}\right) = \frac{2}{3},$$

which is equal to $P(W_1)$. We can use similar math to show that $P(W_3) = 2/3$, also. In the absence of information about what happened on previous draws, the probability of getting a white ball on any draw is equal to the original proportion of white balls in the box.

Example 6.16: The Monty Hall Problem

You are on a game show, and you are given the choice of three doors:

- Behind one door is a car.
- Behind the other doors are goats.

You pick a door, and the host (who knows what is behind the doors) opens another door, which he knows has a goat. The host then offers you the option to switch doors. Does it matter if you switch? If switching changes your probability of getting the prize, what is the new probability?

Here is a simple solution. Let W_i be the event that you are winning on choice i. We will analyze the probability of winning for the case where you never switch and the case where you switch.

Case 1: Never Switch: Consider first if you never switch. Because there are two goats and the host always shows you a goat, the fact that he reveals a goat to you does not change the probability that you have selected the car. Thus, for this case, $P(W_2) = P(W_1) = 1/3$.

Case 2: Always Switch: It may be tempting to think that either:

1. Switching doesn't matter because the host was always going to show you a goat, so your new choice is just as likely to be a car as it was before, or
2. After the host reveals a goat, there is one goat and one car, so the probability of choosing the car when you switch is $1/2$.

It turns out that neither of these is correct! Let's condition on what happens on choice 1:

$$P(W_2) = P(W_2 \mid W_1) P(W_1) + P\left(W_2 \mid \overline{W_1}\right) P(\overline{W_1}).$$

The probability of winning on choice 1 does not depend on whether you switch after that choice, so $P(W_1) = 1/3$, and $P(\overline{W_1}) = 2/3$.

Now, consider what happens on the second choice. There are two possibilities:

- If you are winning on the first choice (W_1), then the two doors you have not chosen both contain goats. The host reveals one of the goats, and you will switch to the other goat. Thus, $P(W_2 \mid W_1) = 0$.
- If you are not winning on the first choice ($\overline{W_1}$), then one of the doors you have not chosen has a goat and the other has the car. The host shows the goat that is not behind your door. Then if you switch, you will switch to the door with the car. Thus, $P\left(W_2 \mid \overline{W_1}\right) = 1$.

Putting this all together, if you always switch then

$$
\begin{aligned}
P(W_2) &= P\left(W_2 \mid W_1\right) P(W_1) + P\left(W_2 \mid \overline{W_1}\right) P(\overline{W_1}) \\
&= \left(0\right)\left(\frac{1}{3}\right) + \left(1\right)\left(\frac{2}{3}\right) = \frac{2}{3}.
\end{aligned}
$$

Why is the probability not 1/2 using the reasoning above about one car and one goat left? Because you do not randomly choose between the two doors. The host is revealing information when he reveals the goat because he cannot choose your door and he cannot choose the door with the car. If you were to randomly choose between the two doors after the goat is revealed, the probability would be 1/2.

Terminology review and self-assessment questions

Interactive flashcards to review the terminology introduced in this section and self-assessment questions are available at fdsp.net/6-9, which can also be accessed using this QR code:

6.10 Chapter Summary

In this chapter, I introduced fundamental notation and definitions for conditional probabilities. I demonstrated how to estimate conditional probabilities through simulation, and I pointed out a common misunderstanding in determining how many trials of an experiment must be simulated to estimate a conditional probability.

I introduced the concepts of statistical independence and conditional independence and gave examples to illustrate these concepts. Then chain rules and the Law of Total Probability were introduced, and several tricky problems were analyzed using these tools.

Conditional probability is fundamental to many things data scientists and engineers do. In the next chapter, we build on these basic concepts to develop techniques for making optimal decisions in systems where the outputs are nondeterministic.

Access a list of key take-aways for this chapter, along with interactive flashcards and quizzes at fdsp.net/6-10, which can also be accessed using this QR code:

7

Introduction to Bayesian Methods

In many systems, we cannot directly observe some phenomena that we are interested in, but we have some observations that are related to that phenomena. We want to use the observations to find the probabilities for the phenomena of interest or to make an optimal decision about that phenomena. For example:

- Suppose a magician has a fair coin and a two-headed coin in their pocket. They choose one at random and flip it three times. If it comes up heads every time, what is the probability that it is the two-headed coin?
- A 40-year-old woman's mammogram indicates a possible breast cancer tumor. Given data on the accuracy of the mammogram and prevalence of breast cancer in 40-year-old women, what is the probability that this woman has breast cancer?
- In a binary communication system, 0s and 1s are transmitted over a noisy channel. Suppose we know the probability that a given bit is a 0, and we know the probabilities associated with the noisy channel. Then given an observation of the channel's output, what is the optimal decision?

To answer these questions, we need to introduce several fundamental techniques for working with conditional probabilities. In particular, we need to use Bayesian techniques, which provide methods for taking probability information about the phenomena of interest (such as a system input or hidden state) and conditional probability information about how the observations depend on that phenomena, and use those to calculate conditional probabilities of the phenomena given the observations.

7.1 Bayes' Rule

Consider again a motivating example from the Introduction:

> **A 40-year-old woman's mammogram indicates a possible breast cancer tumor. Given data on the accuracy of the mammogram and prevalence of breast cancer in 40-year-old women, what is the probability that this woman has breast cancer?**

This problem can be modeled as a *stochastic system*:

> Definition
>
> **stochastic system**
>
> A system with one or more inputs and one or more outputs, for which the outputs are not a deterministic function of the inputs.

DOI: 10.1201/9781003324997-7

This question would be hard to answer for most people who are new to data science. Some of the reasons for this are common to many word problems involving random phenomena:

Note:

One of the biggest challenges for students learning probability is solving a word problem. This usually requires multiple steps, including the following:

1. Identifying and finding the information provided in the problem
2. Defining mathematical notation for mathematical objects, such as events and probabilities
3. Formulating the problem using mathematical notation
4. Choosing the appropriate techniques, such as corollaries and the tools of conditional probability
5. Identify any missing information that will be needed to solve the problem and determine how to find this information
6. Performing mathematical manipulation to solve the problem

Note that mathematical manipulation is the last step. For the first two steps, some keys to success are:

1. Identify all of the phenomena (inputs, outputs, state) that may be random. Note that we should identify phenomena that may be random **even if we are given a particular outcome, event, or realization for that phenomenon**.

2. Introduce mathematical notation and define it precisely.

In this problem, we use randomness to model our lack of knowledge. For example, for the woman under consideration, we do not know whether she has breast cancer or not. Thus, whether the woman actually has cancer is one random phenomenon. Although we are told that the mammogram indicates a possible cancer tumor, in general, it can be modeled as a random phenomenon that depends on whether the woman being tested has breast cancer. These two are the only random phenomena in this problem, so we can now introduce notation.

- Let C be the event that a woman develops breast cancer in a given year.
- Let D be the event that a mammogram detects possible breast cancer in a woman in a given year.

Note:

One key to success in the next step is as follows:

3. If asked to find a probability, determine if it will be a conditional probability. If the statement of the desired probability uses the words "given that", then it is usually expressing a conditional probability, but other words, such as "if", can also imply conditional information. In addition, be sure to interpret the desired probability in the context of all of the information given in the problem.

For example, this problem asks, "Given data on the accuracy of the mammogram and prevalence of breast cancer in 40-year-old women, what is the probability that the woman has breast cancer?". This statement may introduce confusion for two reasons. First, the part that reads "Given data on the accuracy of the mammogram and prevalence of breast cancer in 40-year-old women" sounds like conditional information because it begins with "Given". However, the following "data" is not intended to be something random in this experiment. Instead, this statement is intended to say that we will need some probability information about these events.

Second, the remainder of the statement does not read like a conditional probability. Thus, some readers may translate the desired probability to $P(C)$. However, that is not the probability that this problem is asking about! The reason is that we need to interpret that sentence in terms of the other information given in the problem. In this case, the first sentence states, "A 40-year-old woman's mammogram indicates a possible breast cancer tumor". In other words, we are **given** that the mammogram came back positive for possible breast cancer, which in our mathematical notation is D. Thus, the probability that this problem is seeking is $P(C|D)$, which is the *conditional* probability that the woman has cancer *given* that her mammogram detected a possible breast cancer tumor.

To begin to interpret this probability, the reader needs to understand that medical tests are never perfect. When a mammogram detects a possible breast cancer tumor, that does *not* necessarily mean that a woman has breast cancer. If a mammogram detects breast cancer and a woman does not actually have breast cancer, that is a Type I error (or false alarm). The probability of a false alarm is $P(\overline{C}|D)$. You can find information on false alarm rates for mammograms on the Susan G. Komen Foundation's web pages at

https://www.komen.org/breast-cancer/screening/mammography/accuracy/

Based on the information on that page, let $P(D|\overline{C}) = 0.1$.

If a woman does have cancer, a mammogram should detect it most of the time. The associated probability is called the *sensitivity* of the test:

> Definition
>
> **sensitivity**
>
> In a detection test, the probability of detecting some presence or effect, given that presence or effect is actually there.

(Sensitivity is similar to *power* in an NHST, except that we are truly detecting an effect and not just rejecting the null hypothesis.) From the same Susan G. Komen Foundation web page, the sensitivity of mammograms is $P(D|C) = 0.87$.

> Note:
>
> As a reminder, we should not expect $P(D|\overline{C})$ and $P(D|C)$ to add to 1 because they are computed using different conditional probability measures.

If a woman does have cancer, but the mammogram fails to detect it, it is called a *miss*. The probability of miss is $P(\overline{D}|C) = 1 - P(D|C) = 0.13$

The probabilities $P(D|C)$, $P(\overline{D}|C)$, $P(D|\overline{C})$, and $P(\overline{D}|\overline{C})$ represent the probabilities of observing an effect given some true state (or input). These are called *likelihoods*:

Definition

likelihood (discrete-input stochastic system)

Consider a stochastic system with a discrete set of possible input events $\{A_0, A_i, \ldots\}$ and a discrete set of possible output events $\{B_0, B_1, \ldots\}$. Then the ***likelihoods*** are the conditional probabilities of the output events given the input events:

$$P\left(B_j \mid A_i\right).$$

Now we have a lot of different probabilities, but none of these are of the form $P(C|D)$. Instead, we only have probabilities like $P(D|C)$, $P(D|\overline{C})$, etc. (the likelihoods). The probability $P(C|D)$ is the probability of the true state (or input) given the observed effect. It is called an *a posteriori* or *posterior* probability, meaning that it is **after the observation or measurement**:

Definition

***a posteriori* probability (discrete stochastic system)**

Consider a stochastic system with a discrete set of possible input events $\{A_0, A_i, \ldots\}$ and a discrete set of possible output events $\{B_0, B_1, \ldots\}$. Then the *a posteriori probabilities* are the conditional probabilities of the input events given the observed outputs:

$$P\left(A_i \mid B_j\right).$$

A posteriori probabilities are sometimes abbreviated APPs.

We need to develop a new formula to calculate *a posteriori* probabilities from the likelihoods (and some additional information).

7.1.1 Deriving Bayes' Rule

Bayes' Rule is a technique for expressing probabilities of the form $P(X|Y)$ in terms of probabilities of the form $P(Y|X)$ and some additional information. Thus, Bayes' rule can be used to find the *a posteriori* probabilities for our motivating problem.

We begin by expressing the desired probability using the definition of conditional probability:

$$P(C|D) = \frac{P(C \cap D)}{P(D)}.$$

We can apply a chain rule to express $P(C \cap D)$ in terms of a likelihood:

$$P(C|D) = \frac{P(D|C)P(C)}{P(D)}.$$

Note that in addition to the likelihood, we also need $P(C)$. $P(C)$ is the probability of cancer without taking into account the mammogram. $P(C)$ is called an *a priori* probability:

Definition

a priori probability (discrete stochastic system)

Consider a stochastic system with a discrete set of possible input events $\{A_0, A_i, \ldots\}$. Then the *a priori probabilities* $P(A_i)$ are the probabilities of the input events before any output is observed or measured.

A priori probabilities are sometimes called *priors* because they are the probabilities of the inputs *prior* to the output being observed or measured.

We are still left with $P(D)$, which we don't know. However, we can apply the Law of Total Probability to express $P(D)$ in terms of the likelihoods and *a priori* probabilities. Note that C and $\overline{C}$ form a partition of S, so

$$P(D) = P(D|C)P(C) + P(D|\overline{C})P(\overline{C}).$$

Finally, we have our formula for the desired *a posteriori* probability,

$$P(C|D) = \frac{P(D|C)P(C)}{P(D|C)P(C) + P\left(D|\overline{C}\right)P\left(\overline{C}\right)}.$$

This last formula is a form of **Bayes' Rule**:

Definition

Bayes' Rule (discrete stochastic system)

Consider a stochastic system with a discrete set of possible input events $\{A_0, A_i, \ldots\}$ and a discrete set of possible output events $\{B_0, B_1, \ldots\}$. Then the *a posteriori* probabilities $P(A_i|B_j)$ can be written in terms of the likelihoods $P(B_j|A_i)$ and the *a priori* probabilities $P(A_i)$ as

$$P(A_i|B_j) = \frac{P\left(B_j\,|A_i\right)P\left(A_i\right)}{\sum_i P\left(B_j\,|A_i\right)P\left(A_i\right)}.$$

Note that the form of the numerator and the summands in the denominator are the same.

7.1.2 Motivating Example – Part 2

Now that we have applied Bayes' Rule to our problem, we have a formula for $P(C|D)$ in terms of the likelihoods (which we know) and the *a priori* probabilities (which we do not yet know). The *a priori* probabilities for this problem are $P(C)$ and $P\left(\overline{C}\right)$. Consider $P(C)$ – this is the probability that our 40-year-old woman has breast cancer without knowing the outcome of the mammogram.

As with the other probabilities, we can estimate $P(C)$ from data on the Susan G. Komen foundation website at https://www.komen.org/breast-cancer/risk-factor/age/#fig2-1 The graph in Fig. 7.1 shows data on the annual cancer rates for women by age, extracted from the graph on that page. Based on this figure, $P(C) \approx 140/100{,}000 = 1.4 \times 10^{-3}$, and $P\left(\overline{C}\right) = 1 - P(C)$.

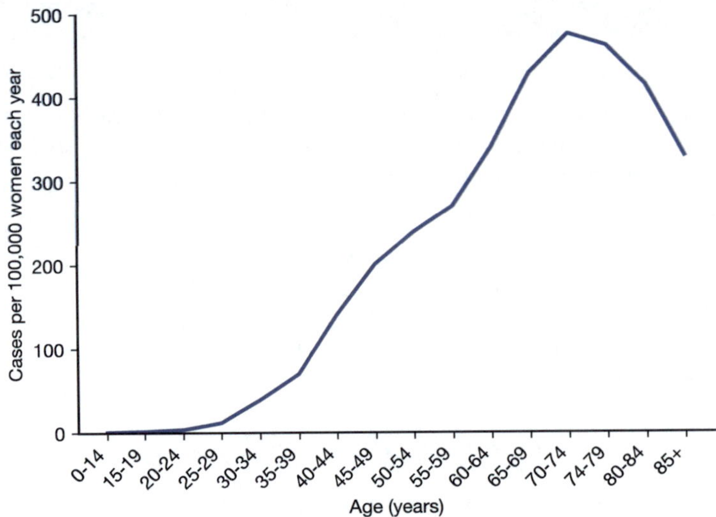

FIGURE 7.1
Annual cancer rates by age for women in the United States.

Now we have enough data to calculate the conditional probability that a 40-year-old woman has cancer given that she has a mammogram that detects breast cancer:

$$\begin{aligned} P(C|D) &= \frac{P(D|C)P(C)}{P(D|C)P(C) + P\left(D|\overline{C}\right)P\left(\overline{C}\right)} \\ &= \frac{(0.87)(1.4 \times 10^{-3})}{(0.87)(1.4 \times 10^{-3}) + [(0.1)\left(1 - 1.4 \times 10^{-3}\right)]}. \end{aligned}$$

We use Python to actually perform the calculation:

```
PC       = 1.4e-3
PNotC    = 1-PC
PD_C     = 0.87
PD_NotC = 0.1

PC_D= ( PD_C*PC )/ (PD_C*PC + PD_NotC * PNotC)

print('The probability a 40-year-old woman actually has cancer when she has a')
print(f'positive mammogram test is approximately {PC_D:.3f}')
```

```
The probability a 40-year-old woman actually has cancer when she has a
positive mammogram test is approximately 0.012
```

Is this result surprising to you? The test only has a false positive rate of 10%; i.e., it only indicates cancer when cancer is not present 10% of the time. This is an example of the *base rate fallacy*:

Definition

base rate fallacy

Many people have incorrect intuition about dependent phenomena when given both likelihoods and *a priori* probabilities. The tendency is to focus on the likelihoods, which describe the relation among the phenomena, and neglect the *a prioris*, which give some **base rate** at which one of the phenomena occurs.

The base rate fallacy is especially problematic when the base rate is particularly close to 0 or 1. The base rate fallacy can often be avoided by applying Bayes' rule to determine the *a posteriori* probabilities. For more examples illustrating the base rate fallacy, see the Wikipedia page https://en.wikipedia.org/wiki/Base_rate_fallacy.

For this example, the probability that a woman in her 40s has breast cancer is very low, so most women getting a mammogram will not have cancer. Thus, even a small false alarm rate results in a large number of women for whom cancer is incorrectly detected.

It is easiest to interpret this result by treating the probabilities as rates and determining the number of women for which cancer is correctly detected and the number of women for which cancer is falsely detected. We will focus our interpretation on the USA, so let's start by getting some data on how many 40-year-old women are in the US. Type "US population by age and gender" into Google. You can then select a source for your data – I would suggest the US Census Bureau, which took me to this page:

Age & Sex Tables:

https://www.census.gov/topics/population/age-and-sex/data/tables.html

By clicking on "All", I got to a page that provides the information we want for 2019:

Age and Sex Composition in the United States: 2019:

https://www.census.gov/data/tables/2019/demo/age-and-sex/2019-age-sex-composition.html

If you download the Excel spreadsheet linked to "Table 1. Population by Age and Sex: 2019" and open it in Microsoft Excel, you should see the spreadsheet shown in Fig. 7.2. According to this data, there were approximately 9,956,000 women from age 40 to 44 (5 years) in 2019. For convenience, let's estimate that there are now approximately 2 million women of age 40 – the exact number is not important in this analysis. Then the number of 40-year-old women who develop breast cancer in a year is approximately

$$2 \times 10^6 \cdot \frac{140}{100,000} = 2800.$$

The number without cancer is $2 \times 10^6 - 2800 = 1,997,200$.

According to the Kaiser Family Foundation, approximately 72% of women over age 40 have had a mammogram in the past two years:

https://www.kff.org/womens-health-policy/state-indicator/mammogram-rate-for-women-40-years/

The percentage is probably lower for 40 year olds, so let's estimate the probability that a 40-year-old woman has a mammogram in a year as 1/4. If this rate applies equally to women

	A	B	C	D	E	F	G
1							
2	Table 1. Population by Age and Sex: 2019						
3	(Numbers in thousands. Civilian noninstitutionalized population.[1])						
4							
5	Age	Both sexes		Male		Female	
6		Number	Percent	Number	Percent	Number	Percent
7	All ages	324,356	100.0	159,028	100.0	165,328	100.0
8	Under 5 years	19,736	6.1	10,094	6.3	9,642	5.8
9	5 to 9 years	20,212	6.2	10,328	6.5	9,884	6.0
10	10 to 14 years	20,827	6.4	10,650	6.7	10,177	6.2
11	15 to 19 years	20,849	6.4	10,545	6.6	10,304	6.2
12	20 to 24 years	21,254	6.6	10,716	6.7	10,538	6.4
13	25 to 29 years	23,277	7.2	11,792	7.4	11,485	6.9
14	30 to 34 years	21,932	6.8	10,935	6.9	10,997	6.7
15	35 to 39 years	21,443	6.6	10,629	6.7	10,814	6.5
16	40 to 44 years	19,584	6.0	9,628	6.1	9,956	6.0
17	45 to 49 years	20,345	6.3	9,993	6.3	10,351	6.3
18	50 to 54 years	20,355	6.3	9,930	6.2	10,425	6.3

FIGURE 7.2
US Census Bureau data on age and sex composition in the United States, 2019.

who are ultimately determined to have cancer and those who do not, then the number of 40-year-old women who have a mammogram in a year is 500,000. Applying the probability of cancer for this age group implies that there will be approximately 700 women who have breast cancer, and approximately 499,300 women who do not have breast cancer.

The number of women who have cancer and who will have cancer detected by a mammogram is approximately

$$0.87(700) = 609.$$

The number of women who do not have cancer and who will have cancer detected by a mammogram is approximately

$$0.1(499,300) = 49,930.$$

Are you surprised? For every 40-year-old woman who has breast cancer correctly detected by a mammogram, there are almost 82 who have breast cancer **incorrectly detected** by a mammogram.

The total number of women for which breast cancer is detected is

$$609 + 49,930 = 50,539.$$

Thus, the proportion of women for whom a mammogram detects breast cancer who actually have breast cancer is

$$\frac{609}{50,539} \approx 0.012,$$

which is the same result we got before. Using explicit numbers makes this easier to understand, but the important part of the math is exactly the same.

Terminology review and self-assessment questions

Interactive flashcards to review the terminology introduced in this section and self-assessment questions are available at fdsp.net/7-1, which can also be accessed using this QR code:

7.2 Bayes' Rule in Systems with Hidden State

When a system has some form of memory, we refer to the current value of that memory as the *state* of the system. If that memory cannot be observed, then we call that *hidden state*:

Definition

hidden state

An internal state (for example, memory contents) of a system that can affect outputs but that is not directly observable.

Systems with hidden states are different from systems with unknown inputs in that the hidden state can affect more than one output: depending on how the state evolves, it may affect one output, a few outputs, or all outputs. In fact, we have already seen a system with hidden state: the Magician's Coin problem. In this problem, the hidden state is which coin was chosen, and the chosen coin affects all future outputs of the system. Although the Magician's Coin does not have both hidden state and unknown input(s), many have both.

In this section, I revisit the Magician's Coin problem to better expose why the observations affect the probabilities of future events in the way they do. This gives us a chance to use Bayes' Rule to explore the probabilities of the hidden state, and it will also give us an opportunity to apply our knowledge about conditional independence.

Recall the simple question we started with: **If the coin comes up heads on the first flip, what is the probability that it comes up heads on the second flip?** As before, let H_i be the event that the chosen coin comes up heads on flip i. Also, we assume that the magician is equally likely to choose between a fair coin and a two-headed coin. Let F be the event that the chosen coin is fair.

We are asked to find $P(H_2|H_1)$. In the previous section, we solved this using the definition of conditional probability and the Law of Total Probability as follows:

$$P(H_2|H_1) = \frac{P(H_1 \cap H_2)}{P(H_1)},$$

where

$$P(H_1) = P(H_1|F)P(F) + P(H_1|\overline{F})P(\overline{F}) = 3/4,$$

and

$$P(H_1 \cap H_2) = P(H_1 \cap H_2|F)P(F) + P(H_1 \cap H_2|\overline{F})P(\overline{F}) = 5/8.$$

So, $P(H_2|H_1) = 5/6$. This is the easiest way to solve for $P(H_2|H_1)$. However, it is non-intuitive because it tells us nothing about why knowing H_1 affects the probability of H_2. We previously argued that every time we see an outcome of heads, we should be more likely to believe that we have the two-headed coin (i.e., the probability that we have the two-headed coin should increase). Only now do we have the tools to test this hypothesis:

Consider $P(\overline{F}|H_1)$, the probability of having selected the two-headed coin after the first outcome is observed to be heads. We do not have outcomes in this form, but we do have outcomes of the forms $P(H_1|\overline{F})$ and $P(\overline{F})$. In fact, these probabilities are a likelihood and an *a priori* probability for this system. So, we can solve for the desired probability using Bayes' Rule:

$$P\left(\overline{F}\,|H_1\right) = \frac{P\left(H_1\,|\overline{F}\right)P(\overline{F})}{P(H_1)}.$$

I did not expand the denominator because we have already found $P(H_1)$ using the Law of Total Probability. We already know the probabilities in the numerator, so we can calculate

$$P\left(\overline{F}\,|H_1\right) = \frac{(1)(1/2)}{P(3/4)} = \frac{2}{3}.$$

Thus, after observing a single heads, the probability that the magician has selected the two-headed coin goes from 1/2 to 2/3.

Now, I show how we can apply our knowledge of the updated state probabilities to find the probability of H_2. Because of the multiple conditioning that arises, this may be confusing at first. To help with this, I'm going to first hide the conditioning on H_1 by defining

$$\tilde{P}(H_2) = P(H_2|H_1).$$

Recalling that conditioning on H_1 creates a new (conditional) probability measure, $\tilde{P}$ is a different name for that probability measure.

Now, we can apply the Law of Total Probability using $\{F, \overline{F}\}$ as a partition:

$$\tilde{P}\left(H_2\right) = \tilde{P}\left(H_2|F\right)\tilde{P}\left(F\right) + \tilde{P}\left(H_2|\overline{F}\right)\tilde{P}\left(\overline{F}\right).$$

Next, we substitute back in that $\tilde{P}(A) = P(A|H_1)$ for any event A. Recall that conditioning on two events results in a single conditioning on the intersection of the two events. Thus, the resulting expression is

$$P(H_2|H_1) = P(H_2|H_1 \cap F)P(F|H_1) + P\left(H_2\,|H_1 \cap \overline{F}\right)P\left(\overline{F}\,|H_1\right).$$

We know $P(\overline{F}|H_1) = 2/3$, and $P(F|H_1) = 1 - P(\overline{F}|H_1) = 1/3$. But this is where things get tricky – what are the other two probabilities? We have to apply conditional independence to find these. If we know the true state, F or $\overline{F}$, then information about what happened on the first flip cannot affect the outcome of the second flip. In other words, H_2 is *conditionally independent* of H_1 given either F or $\overline{F}$. Thus,

$$P(H_2|H_1) = P(H_2|F)P(F|H_1) + P\left(H_2\,|\overline{F}\right)P\left(\overline{F}\,|H_1\right).$$

Finally, we are ready to solve:

$$P(H_2|H_1) = \left(\frac{1}{2}\right)\left(\frac{1}{3}\right) + (1)\left(\frac{2}{3}\right) = \frac{5}{6},$$

which is the same answer we got above. However, we have a new perspective now. Given H_1:

- The magician will have the fair coin with probability 1/3, and another flip of that coin will yield heads with probability 1/2.
- The magician will have the two-headed coin with probability 2/3, and another flip of that coin will yield heads with probability 1.

Terminology review and self-assessment questions

Interactive flashcards to review the terminology introduced in this section and self-assessment questions are available at fdsp.net/7-2, which can also be accessed using this QR code:

7.3 Optimal Decisions for Discrete Stochastic Systems

We are ready to tackle the third example from this chapter's introduction:

Example: Binary Communication System

> In a binary communication system, 0s and 1s are transmitted over a noisy channel. Suppose we know the probability that a given bit is a 0, and we know the probabilities associated with the noisy channel. Then given an observation of the channel's output, what is the optimal decision?

Let's start by considering one particular instantiation of this problem:

7.3.1 Binary Communication System

Suppose we are given a system that has binary inputs and ternary outputs. Let's use A_0 and A_1 to denote the input events and B_0, B_1, and B_2 to denote the output events. The channel is completely specified by giving the likelihoods, $P(B_j|A_i)$ for all $i \in \{0,1\}$ and $j \in \{0,1,2\}$.

Let $p_{ij} = P(B_j|A_i)$. Note the order of the i and j. This is the probability of transitioning from input i to output j, and these are also called *channel transition probabilities.* It is helpful to visualize the channel transition probabilities/likelihoods on a diagram, where we label the arrow connecting input event A_i with output event B_j by p_{ij}. An example of such a diagram is shown in Fig. 7.3. For example, from this diagram, we can read that $p_{00} = 5/8$, $p_{01} = 1/4$, and $p_{12} = 3/4$.

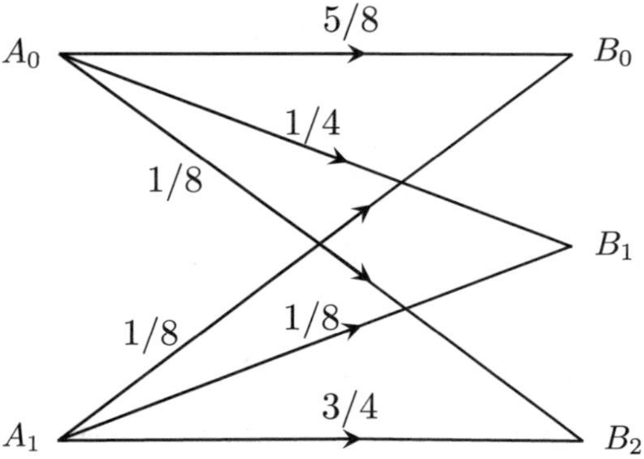

FIGURE 7.3
Example channel transition diagram for channel with two inputs and three outputs.

It is convenient to collect the channel transition probabilities into an array, such that the (i, j)th entry in the array is P_{ij}. Let's import NumPy and create an array with these channel transition probabilities:

```
import numpy as np

P=np.array([
    [5/8, 1/4, 1/8],
    [1/8, 1/8, 3/4]
])

print(P)
```

```
[[0.625 0.25  0.125]
 [0.125 0.125 0.75 ]]
```

Note that the probabilities with the same conditioning should add to 1. In other words,

$$\sum_{j \in \{0,1,2\}} P(B_j|A_i) = 1, \quad i = 0, 1.$$

In the diagram, this corresponds to the transition probabilities that emerge from the same input. In the matrix **P**, this corresponds to all of the entries in a row. We can tell NumPy to sum the array across the column by using the `np.sum()` function and passing the `axis=1` argument to tell it to sum across the second axis (the columns):

```
np.sum(P, axis=1)
```

```
array([1., 1.])
```

Note that the probabilities for a particular B_j (i.e., merging into a particular output) do not necessarily sum to 1.

$$\sum_{i \in \{0,1\}} P(B_j|A_i) =?, \quad j = 0, 1, 2.$$

This corresponds to the column sums of the matrix **P**, and we can get these column sums by using the `np.sum()` function and passing the `axis=0` argument to tell NumPy to sum across axis 0 (the rows):

```
np.sum(P, axis=0)
```

```
array([0.75 , 0.375, 0.875])
```

7.3.2 Decision Problem and Decision Rules

The *decision problem* for the binary communication systems is concisely described as follows: given the observed output, determine which input was sent. A *decision rule* tells how to choose an input given an observed output. In general, a decision rule may result in a randomized choice for the input, but we only consider deterministic decision rules.

We can turn this into an *optimal decision problem* if we specify criteria to be optimized. Here are two common criteria:

1. Maximize the likelihood of the input.

2. Choose the input that minimizes the probability of error.

The optimum decision under criterion 1 is called the *maximum likelihood (ML)* decision. In this problem, the likelihoods are of the form $P(B_j|A_i)$. Let $\widehat{A}_i$ denote the event that the receiver decides that input was A_i. Then the ML decision rule given B_j is observed is

$$\widehat{A}_i, \text{ where } i = \arg \max_{i \in \{0,1\}} P(B_j|A_i).$$

The values of $P(B_j|A_i)$ are given in Fig. 7.3. For each output, the ML rule selects the input that has the largest likelihood, which corresponds to the arrow with the largest probability merging into that output in Fig. 7.3. Similarly, the ML decision corresponds to the row number with the largest probability for each column of the transition probability matrix, **P**. To get the index of the largest value in each column, we can use the `np.argmax()` function and pass the `axis=0` keyword parameter to tell NumPy to maximize over the rows. (Note that `np.max()` returns the maximum value, whereas `np.argmax()` returns the index of the maximum value.)

Thus, the ML decisions are as follows:

```
np.argmax(P, axis=0)
```

```
array([0, 0, 1])
```

We can read these decisions as follows. The value in position j is the decision given B_j is received. Then the decision rules are as follows:

- Given B_0 is received, decide A_0.
- Given B_1 is received, decide A_0.
- Given B_2 is received, decide A_1.

Unfortunately, the ML solution does not necessarily minimize the probability of error. For instance, suppose that we know that 0 is sent with probability 1; i.e., $P(A_0) = 1$. Then the ML rule will make an error whenever B_2 is received. Let's use total probability to calculate the probability of each B_j:

$$P(B_j) = \sum_{i \in \{0,1\}} P(B_j|A_i)P(A_i).$$

We start by setting up a NumPy vector to hold the *a priori* probabilities:

$$[P(A_0), P(A_1)].$$

Then we use a nested for loop to calculate all the values of $P(B_j)$. The outer loop iterates over the values of j. The inner loop (for a given value of j) carries out the sum over the values for i:

```
aprioris = np.array([1,0])
for j in range(3):
    pBj=0
    for i in range(2):
        pBj+=P[i,j]*aprioris[i]

    print(f'P(B{j})= {pBj}', end= '      ')
```

```
P(B0)= 0.625      P(B1)= 0.25      P(B2)= 0.125
```

Let E be the event that an error occurs (i.e., the decision differs from the transmitted symbol). Then for this simple example, $P(E) = P(B_2) = 0.125$. We know that it is suboptimal because we could just use the decision rule "Always decide A_0" and get error probability 0.

We can guess that there must be some value q_0 such that:

- if $P(A_0) < q_0$, the ML rule performs better, and
- if $P(A_0) > q_0$, always deciding A_0 performs better.

Let's build a simulation to test this. First, we will see how to efficiently generate the events A_0 and A_1 given any probabilities $P(A_0)$ and $P(A_1)$ such that $P(A_0) + P(A_1) = 1$. We again use NumPy's `npr.choice()` function, but we now pass it the probability information as a keyword parameter. Take a look at the help for `npr.choice()`. It accepts a `p` parameter that is the associated probabilities for the items that are being selected from. Thus, if we want to output a 0 with probability $P(A_0) = 0.75$ and a 1 with probability $P(A_1) = 0.25$, we can simulate 1000 such events as follows:

```
sim_values = npr.choice([0,1], 1000, p = [0.75,0.25])
```

We can check that we are achieving the desired probabilities by comparing the relative frequencies to the probabilities we passed as arguments:

```
np.sum(sim_values==0)/1000
```

```
0.771
```

```
np.sum(sim_values==1)/1000
```

```
0.229
```

The relative frequencies of 0.771 and 0.229 are close to the desired probabilities of 0.75 and 0.25. Some variation is expected, since we only simulated 1000 events.

Now we are ready to build a function to carry out the simulation. I introduce a new Python concept here: we can pass a function as an argument to another function. This allows us to create one simulation and test it for multiple different decision rules. The parameter called `decision_rule` will take a function to make decisions based on the channel observations and the likelihoods and *a priori* probabilities. The following code will simulate the error probability:

```
def sim2to3 (decision_rule, P, PA0, num_sims = 100_000, verbose = False):
    # Create all the input events at the same time:
    inputs = npr.choice([0, 1], num_sims, p = [PA0, 1-PA0])

    # Create an array to determine the channel outputs
    outputs = np.zeros(num_sims)

    # Create an array to store the decisions
```

(continues on next page)

(continued from previous page)

```
    decisions = np.zeros(num_sims)

    # There are more efficient ways of doing this using NumPy, but
    # individually determining each output for each input should make
    # this easier to understand for most learners
    for sim in range(num_sims):
        input_bit = inputs[sim]
        # Choose observation according to transition probabilities
        # for given input bit:
        observation = npr.choice([0, 1, 2], p = P[input_bit])

        # Now pass this observation to the decision_rule function:
        decisions[sim] = decision_rule(observation, P, PA0)

    # Finally, calculate the error probability. An error occurs
    # whenever the decision is not equal to the true input
    errors = np.sum(inputs != decisions)

    error_prob = errors/num_sims
    if verbose:
        print( f'The error probability is approximately {error_prob:.2f}')
    return error_prob
```

Now let's create and test our decision rule functions. An easy one is to always decide 0:

```
def always_decide0 (observation, P, PA0):
    return 0
```

This decision rule should result in zero errors when $P(A_0) = 1$:

```
sim2to3(always_decide0, P, 1, verbose=True);
```

```
The error probability is approximately 0.00
```

The error probability should increase as $P(A_0)$ decreases:

```
sim2to3(always_decide0, P, 0.8, verbose=True);
```

```
The error probability is approximately 0.20
```

Now let's implement our ML decision rule as a function:

```
def ML (observation, P, PA0):
    #Here I selected the  column and then did argmax, but the other way also works
    return np.argmax(P[:,observation])
```

Let's try this for a few values of $P(A_0)$:

```
sim2to3(ML, P, 1, verbose = True);
```

```
The error probability is approximately 0.12
```

```
sim2to3(ML, P, 0.8, verbose = True);
```

```
The error probability is approximately 0.15
```

The ML rule does worse than the "always decide A_0" rule for $P(A_0) = 1$, as expected. The ML rule performs better when $P(A_0) = 0.8$.

Let's create one more decision rule function: always decide A_1:

```
def always_decide1 (observation, P, PA0):
    return 1
```

The following code calculates the error probabilities for these three decision rules as a function of $P(A_0)$. The resulting error probabilities are plotted in Fig. 7.4. (Code to generate this figure is available online at fdsp.net/7-3.)

```
input_probs = np.linspace(0,1,21)

pe_always0 = []
pe_always1 = []
pe_ML = []

for PA0 in tqdm(input_probs):
    pe_always0 += [sim2to3(always_decide0, P, PA0)]
    pe_always1 += [sim2to3(always_decide1, P, PA0)]
    pe_ML += [sim2to3(ML, P, PA0)]
```

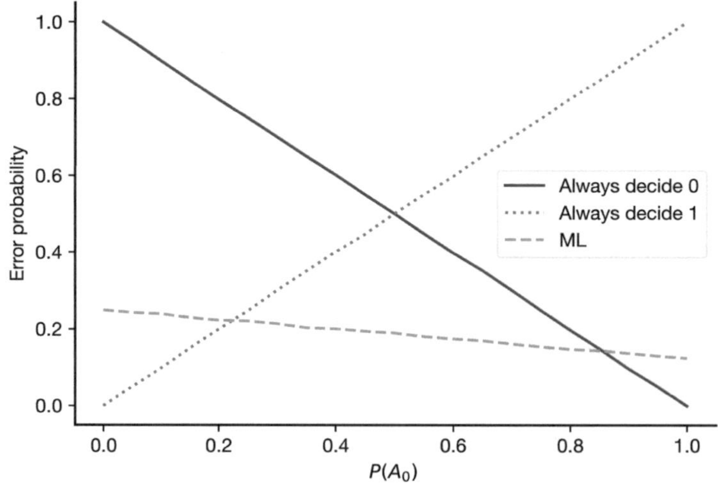

FIGURE 7.4
Error probabilities for a binary communication system for three different decision rules, as a function of the *a priori* probability $P(A_0)$.

From Fig. 7.4, we can see that each of these three different decision rules minimizes the probability of error over some specific range of $P(A_0)$. For low $P(A_0)$, it is best to always decide A_1. For medium $P(A_0)$ (in the range $0.2 < P(A_0) < 0.85$), it is best to apply the ML rule, and for larger $P(A_0)$, it is best to always decide A_0. And other decision rules are possible. We need a principled way to determine the optimum decision rule to minimize the error probability.

7.3.3 Optimal Decisions to Minimize the Probability of Error: The MAP Rule

As before, let $\widehat{A}_i$ denote the event that the decision is A_i. Then an error occurs if the decision differs from the actual input event. Let E denote the error event, so

$$E = \left(\widehat{A}_0 \cap A_1\right) \cup \left(\widehat{A}_1 \cap A_0\right).$$

Note that $\left(\hat{A}_0 \cap A_1\right)$ and $\left(\hat{A}_1 \cap A_0\right)$ are mutually exclusive (since A_0 and A_1 are complementary events). Thus

$$P(E) = P\left(\hat{A}_0 \cap A_1\right) + P\left(\hat{A}_1 \cap A_0\right).$$

We can minimize the error probability if we minimize the conditional error probability given the observation B_j for every B_j:

$$P(E|B_j) = P\left(\hat{A}_0 \cap A_1 \left|B_j\right.\right) + P\left(\hat{A}_1 \cap A_0 \left|B_j\right.\right).$$

In general, we could make a decision rule that is probabilistic, meaning that given B_j, we choose $\hat{A}_0$ with some probability and $\hat{A}_1$ with some probability. However, that is not necessary: the optimum decision rule is deterministic. Thus, only one of the two terms in the summation will be retained; i.e., given B_j only one of $\hat{A}_0$ and $\hat{A}_1$ will occur. Suppose the decision rule given B_j is to always decide the most likely input was 0; then $\hat{A}_0$ occurs, and $P(E) = P(A_1 \left|B_j\right.)$. Conversely, suppose the decision rule given B_j is to always decide the most likely input was 1; then $\hat{A}_1$ occurs, and $P(E) = P(A_0 \left|B_j\right.)$.

Since we wish to minimize the $P(E)$, the decision rule given B_j is received should be:

- $\hat{A}_0$ if $P(A_1 \left|B_j\right.) < P(A_0 \left|B_j\right.)$, and
- $\hat{A}_1$ if $P(A_1 \left|B_j\right.) > P(A_0 \left|B_j\right.)$, and
- either $\hat{A}_0$ or $\hat{A}_1$ if $P(A_1 \left|B_j\right.) = P(A_0 \left|B_j\right.)$.

Note:

Although we didn't prove that the optimal decision rule is deterministic, it is not hard to see that any probabilistic decision rule would have an error probability that is a linear combination of the probabilities $P(A_1 \left|B_j\right.)$ and $P(A_0 \left|B_j\right.)$. The minimum value of a line on a closed interval is at one of the endpoints (i.e., one of the decisions has probability 1, and the other has probability 0), so the decision rule that minimizes the error probability is deterministic.

The decision rule that minimizes the error probability can be summarized as "choose the input that maximizes the *a posteriori* probability given the observation B_j". Mathematically, the minimum error probability rule is

$$\hat{A}_i, \text{ where } i = \arg\max_{i \in \{0,1\}} P(A_i|B_j).$$

Since we choose the input that maximizes the *a posteriori* probability (APP), we call this a *maximum a posteriori (MAP)* decision rule. We also use the following notation when the decision rule is between two possible inputs:

$$P(A_0 | B_j) \underset{1}{\overset{0}{\gtrless}} P(A_1 | B_j). \tag{7.1}$$

This is interpreted as follows:

- When the top inequality holds, the decision is A_0.
- When the bottom inequality holds, the decision is A_1.

As in Section 7.1, the APPs are not given in the problem formulation, but we can use Bayes' rule to find the APPs from the likelihoods and the *a priori* probabilities:

$$P(A_i|B_j) = \frac{P(B_j|A_i)P(A_i)}{\sum_i P(B_j|A_i)P(A_i)}.$$

If only the MAP decision is needed, then (7.1) can be simplified:

$$\begin{aligned} P(A_0 | B_j) &\underset{1}{\overset{0}{\gtrless}} P(A_1 | B_j) \\ \frac{P(B_j|A_0)P(A_0)}{\sum_i P(B_j|A_i)P(A_i)} &\underset{1}{\overset{0}{\gtrless}} \frac{P(B_j|A_1)P(A_1)}{\sum_i P(B_j|A_i)P(A_i)} \\ P(B_j|A_0)P(A_0) &\underset{1}{\overset{0}{\gtrless}} P(B_j|A_1)P(A_1). \end{aligned}$$

The APPs can be computed by implementing Bayes' rule in Python as follows for $P(A_0) = 1/5$ and $P(A_1) = 4/5$:

```
aprioris = np.array([1/5,4/5])
for j in range(3):
    pBj = 0
    for i in range(2):
        pBj += P[i,j]*aprioris[i]
        print(f'P(B{j}) = {pBj:.2f}: ', end = '')
    for i in range(2):
        print(f'P(A{i}|B{j}) = {P[i,j]*aprioris[i]/pBj: .2f}', end = '     ')
    print()
    print()
```

```
P(B0) =  0.23: P(A0|B0) =  0.56     P(A1|B0) =  0.44

P(B1) =  0.15: P(A0|B1) =  0.33     P(A1|B1) =  0.67

P(B2) =  0.63: P(A0|B2) =  0.04     P(A1|B2) =  0.96
```

From these example *a prioris*, we see that the MAP rule is **not any of the three rules previously introduced**! Let's create a MAP decision rule function:

```
def MAP(observation, P, PA0):
    # Take the jth column and multiply it elementwise by the
    # a priori probability vector
    scaled_apps = P[:,observation] * np.array([PA0, 1-PA0])
    return np.argmax(scaled_apps)
```

The following code simulates the performance of all of the decision rules discussed for different values of the *a priori* probability $P(A_0)$. The resulting error probabilities are plotted in Fig. 7.5.

```
input_probs = np.linspace(0,1,21)

pe_always0 = []
pe_always1 = []
pe_ML = []
pe_MAP = []

for PA0 in tqdm(input_probs):
    pe_always0 += [sim2to3(always_decide0, P, PA0)]
    pe_always1 += [sim2to3(always_decide1, P, PA0)]
    pe_ML += [sim2to3(ML, P, PA0)]
    pe_MAP += [sim2to3(MAP, P, PA0)]
```

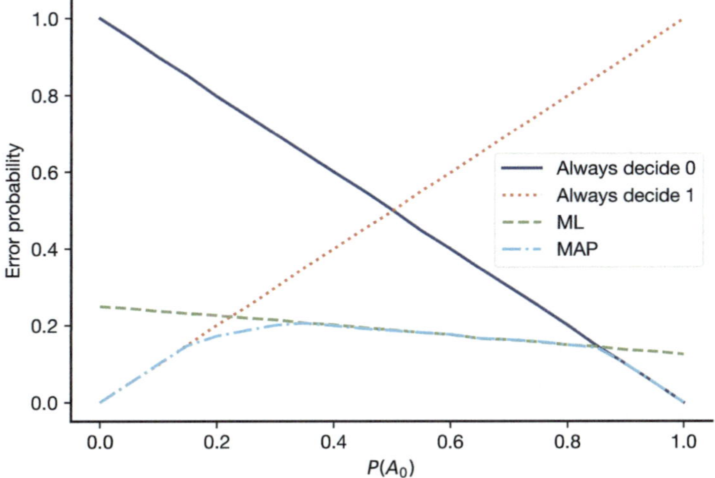

FIGURE 7.5
Error probabilities for a binary communication system for four different decision rules, as a function of the *a priori* probability $P(A_0)$.

As seen in Fig. 7.5, the MAP rule achieves the lowest error probability for all values of $P(A_0)$. However, this requires the receiver to know the *a priori* values of the inputs. If the *a priori* probabilities are not known, then the ML decision rule is usually used.

Terminology review and self-assessment questions

Interactive flashcards to review the terminology introduced in this section and self-assessment questions are available at fdsp.net/7-3, which can also be accessed using this QR code:

7.4 Bayesian Hypothesis Testing

In Chapters 2–5, we introduced null hypothesis significance testing (NHST) via resampling. In NHST, we assign significance to some observed statistic if the value of the statistic has a very low probability of occurring under some testable null hypothesis. This approach is very common, but it also has some problems:

- NHST cannot utilize information from other sources about the probability of a hypothesis being true.
- NHST can only evaluate the probability of seeing such an extreme value of a statistic under the null hypothesis; it cannot assign probabilities to the null and alternative hypotheses.
- Moreover, NHST is not easily generalized to handle multiple possible hypotheses.
- The result of an NHST (the p-value) is often misinterpreted, and p-values are often subject to manipulation through "p hacking".

NHST can be considered a *frequentist* technique because it only involves determining the proportion of trials for which we see a certain type of value under the null hypothesis.

An alternative is to use a *Bayesian hypothesis test* in which the goal is to estimate the *a posteriori* probability of the possible hypotheses. To simplify the introduction to this approach, let's use a variation on our first statistical test, which we introduced in Section 2.4. Since biased coins only exist in probability problems, let's consider whether a die might be loaded instead:

You find a 6-sided die. Suppose you roll the die four times, and it comes up 1 on every roll. You would like to conduct a statistical test to determine if it could be a fair die. If it were a fair die, then there is one outcome of "all 1s" among the 6^4 possible outcomes, meaning that the probability of observing all 1s is

```
print(f'{1 / 6 ** 4:.6f}')
```

```
0.000772
```

Under the NHST approach, we determine the probability of seeing an outcome that is at least this extreme under the null hypothesis. The typical null hypothesis would be that the die is fair. However, under a Bayesian approach, we can find the *a posteriori* probability that this is a fair die given the observed outcomes of the die rolls.

Unlike NHST, we have to establish a model for all the possible behaviors of the die and assign probabilities to this model. Let's start with the simplest possible model that might fit our observations: the die is either fair or it is loaded to always return 1.

Now, let's establish some notation. Let:

- F = event the die is fair
  (Then $\overline{F}$ = the event that the die is loaded to always land showing 1.)

- E = event that the die comes up 1 on four consecutive rolls

A Bayesian statistical test estimates the APP $P(F|E)$. As usual, we do not have direct knowledge of this type of probability. In addition, we know from our previous work in this chapter that we need to know the *a priori* probabilities to calculate *a posteriori* probabilities. For the model we have chosen, we need to know $P(F)$ and $P(\overline{F}) = 1 - P(F)$.

Given the *a prioris*, we can find $P(F|E)$ in two ways:

1. We can run a simulation of the system.
2. We can calculate it using Bayes' rule.

Although approach 2 is tractable for this simple problem, let's apply approach 1, since this approach is applicable to a larger class of problems.

We need the *a priori* probabilities, but there is no way for us to know the true *a prioris*. So what to do? We have to choose the *a prioris*. How? There are two typical approaches:

1. **Uninformative prior**: We choose a prior that assumes as little information about the inputs as possible. For discrete inputs, this will often be equally likely probabilities.
2. **Informative prior**: We choose a prior based on other knowledge of the problem.

Below, we consider both of these approaches and compare the effect of these assumptions on the *a posteriori* probabilities and on our conclusions:

7.4.1 Uninformative Prior

Suppose that we have two possible cases: the die is a normal fair die, or the die is loaded to always come up 1. For the uninformative prior, we assign probability 1/2 to each of these possibilities. Then we can construct a simulation to determine the probability that the die is fair given that it came up 1 on four consecutive rolls, when the die is chosen according to this uninformative prior:

```
import random

# How many sets of die rolls to simulate
num_sims = 1_000_000

# This is the number of rolls of the die and the target for the event E
rolls = 4

# Set up some counters. As we saw before, when we estimate
# a conditional probability, we generally need two counters:
event_count = 0
fair_count = 0

# We have two types of die:
dietypes = ["fair", "loaded"]

for sim in range(num_sims):

    # Choose a die at random and set up the sample space for the die roll:
```

(continues on next page)

(continued from previous page)

```
    dietype = random.choice(dietypes)
    if dietype == "fair":
        faces = [1,2,3,4,5,6]
    else:
        faces = [1]

    # Now roll the die the chosen number of times:
    face_values = random.choices(faces, k=rolls)
    # Count how many 1s were observed
    num_1s = face_values.count(1)

    # Check if E occurred and update counter if it did
    if num_1s == rolls:
        event_count += 1

        # Now check that the event F occurred when E occurred
        # and update the counter:
        if dietype == "fair":
            fair_count += 1


print(f'Prob. of fair die given that {rolls} ones were observed is',
      f'{fair_count / event_count :.2g}')
```

```
Prob. of fair die given that 4 ones were observed is 0.00078
```

Given the uninformative prior, the probability that the die is fair is very small.

From this result, you can make the argument that the chance that the die was fair was very small indeed. But would you be willing to show this result to your boss if your job were on the line? What could go wrong?

The most likely thing to go wrong is that your boss would say: "**How many loaded dice have you ever seen?**" I have never (to my knowledge) seen a loaded die, but I know they exist. We can use this knowledge to choose a prior other than the uninformative prior. We consider how to handle that next.

7.4.2 Informative prior

Given our prior experience, a randomly chosen die has a very small probability of being loaded. So, the *a priori* probabilities should not be equal to 0.5. The *a posteriori* probabilities will depend on the particular values chosen for the *a prioris*. However, if you bias these probabilities too much, it will give someone viewing your results yet another point of criticism. So, for instance, if you say that the probability of finding a loaded die is 1 in 1,000,000, then you may be forced to defend why you chose that particular value. But, if you assume that the probability of finding a loaded die is 1/100 and still find a reasonable probability that the die is fair, then that is much easier to defend. To make assessing this tradeoff easier, let's create a function that can simulate the *a posteriori* probabilities for any choice of the probability of a loaded die.

The `random.choices()` function has a `weights` keyword parameter that can be used to change the probability of the choices according to input probabilities or other weightings.

```
def bayes_die(prob_loaded = 0.5, rolls = 4, num_sims = 1_000_000):
    event_count = 0
    fair_count = 0

    # We have two types of die:
    dietypes = ["fair", "loaded"]

    for sim in range(num_sims):

        # Choose a die at random and set up the sample space for the die roll:
        dietype = random.choices(dietypes,
                                 weights=[1 - prob_loaded, prob_loaded])[0]
        if dietype == "fair":
            faces = [1, 2, 3, 4, 5, 6]
        else:
            faces = [1]

        # Now roll the die the required number of times:
        dies = random.choices(faces, k=rolls)
        # Count how many ones were observed
        num1s = dies.count(1)

        # Check if E occurred and update counter if it did
        if num1s == rolls:
            event_count += 1

            # If the event F occurred when E occurred, update the counter:
            if dietype == "fair":
                fair_count += 1

    return fair_count / event_count
```

Let's start by checking the output of our function for the uninformative prior, which uses the default value of `prob_2tails = 0.5`:

```
bayes_die()
```

```
0.0007174188906985342
```

Now, we can check how the answer varies if we set the probability of finding a loaded die to 1/100:

```
bayes_die(prob_loaded = 1/100)
```

```
0.07271533120660537
```

The *a posteriori* value of the die being fair will increase when the probability of finding a loaded die decreases:

```
bayes_die(prob_loaded = 1/1000)
```

```
0.4407158836689038
```

Even though the chance of getting all ones on four rolls of a fair die is very small, the *a posteriori* probability of the die being fair can be large if the *a priori* probability of the die being loaded is small.

If we return to what we could be confident in telling our boss, we can say that if the probability of finding a loaded die is less than 1/100, then the probability that the die is fair is greater than 0.07. Although this probability is still less than 1/2, it is above the 5% acceptable probability of false alarm that we used in NHST.

Example 7.1: Prior to Make Die Equally Likely to Be Fair or Loaded

What prior makes it equally likely for the die to be fair or loaded?

Let's try to analyze this using conditional probability. Let G_i be the event that the die's top face was 1 on roll i. If the die is fair, then the outcomes of the individual rolls are independent, so

$$P(E|F) = P\left(\bigcap_{i=0}^{3} G_i\right) = \prod_{i=0}^{3} P(G_i) = \left(\frac{1}{6}\right)^4 = \frac{1}{1296}$$

We know for the loaded die that $P\left(E|\overline{F}\right) = 1$. We are interested in finding the prior that makes $P(F|E) = P\left(\overline{F}|E\right)$, which implies that $P(F|E) = 1/2$.

To simplify the notation, let $P(F) = q$ and $P\left(\overline{F}\right) = 1 - q$. Then

$$\begin{aligned}
P(F|E) = \frac{1}{2} &= \frac{P(E|F)P(F)}{P(E|F)P(F) + P(E|\overline{F})P(\overline{F})} \\
&= \frac{(1/1296)q}{(1/1296)q + (1)(1-q)} \\
1 &= \frac{2q}{q + 1296 - 1296q} \\
1296 - 1295q &= 2q \\
q &= \frac{1296}{1297}.
\end{aligned}$$

To achieve equal APPs for the die being fair or loaded, the probability of the die being fair must be approximately:

```
1296/1297
```

```
0.9992289899768697
```

We can use our simulation to check our math:

```
bayes_die(prob_loaded = 1-0.999229)
```

```
0.5056962025316456
```

The simulation confirms that this choice of *a priori* probability achieves equal (up to the approximation level of the simulation) *a posteriori* probabilities for the die being fair or loaded.

7.4.3 Extending the set of *a prioris*

In practice, loaded dice do not have to return only a single value. Instead, they are usually biased to return some value more than others. Given the observed result, we may still assume that the die is biased to come up 1, but we can use this to introduce a new *model* for the die, and we have to choose a new prior.

However, this is where things may become confusing. The reason is that we can quantify the amount of loading of the die by the probability that the die comes up 1, $P(G)$. But now we want to assign probabilities to different biases for the loaded die. I.e., we have something like $P\left[P\left(G\right) = 0.8\right]$. And probabilities of this form are *a priori* probabilities for the loading. What we really want to estimate are *a posteriori* probabilities like $P\left[P\left(G\right) = 0.8 \,|E\right]$. To help simplify this confusing situation, let's define the loading as $L = P(G)$. Then we can write the *a priori* probabilities as $P(L = \ell)$ and the *a posteriori* probability $P\left(L = \ell|E\right)$ for different values of ℓ in the interval $[0, 1]$. Below, I only consider the case of an informative prior, but an example for an uninformative prior is given on the book's website at fdsp.net/7-4.

Let's again assume that at least 99% of the dice in circulation are fair dice (i.e., with loading $L = 1/6$). Then we still have to choose a model for how the remaining 0.01 probability will be distributed among the other loadings. Let's distribute it uniformly. Because we do not yet know how to estimate the *a posteriori* probability when the loading L is chosen randomly from the continuous range $[0, 1]$, we assign probabilities to discrete values separated by 0.01. We will let $P(L = 0.17) = 0.99$. There are 100 other biases, so we will assign each one a probability of $0.01/100 = 10^{-4}$. The following code shows how to generate one million points from this distribution. Fig. 7.6 shows a plot of the relative frequencies of the biases. The relative frequencies closely match the desired probabilities, so we are ready to carry out a statistical test using this informative prior.

```
bias_probs = np.full(101, 1e-4)
bias_probs[np.where(all_loadings == 0.17)] = 0.99
num_pts = 1_000_000
test = np.array(random.choices(all_loadings, weights=bias_probs, k=num_pts))
```

Below is a function `biased_die_i()` to simulate this scenario with the informative prior. It is based on the function `bayes_die()` above. The function replaces the previous parameter `prob_loaded` with a new keyword parameter `a_prioris` that will input the *a priori* probabilities on $[0, 1]$. The code is updated to draw a loading `L` based on the specified *a prioris* and then choose a random face according to the chosen value of `L`:

```
def biased_die_i(a_prioris, rolls=4, target=4, num_sims=100_000):
    """ Simulate randomly choosing the loading for a die (from 0 to 1 by 0.01) and
```

(continues on next page)

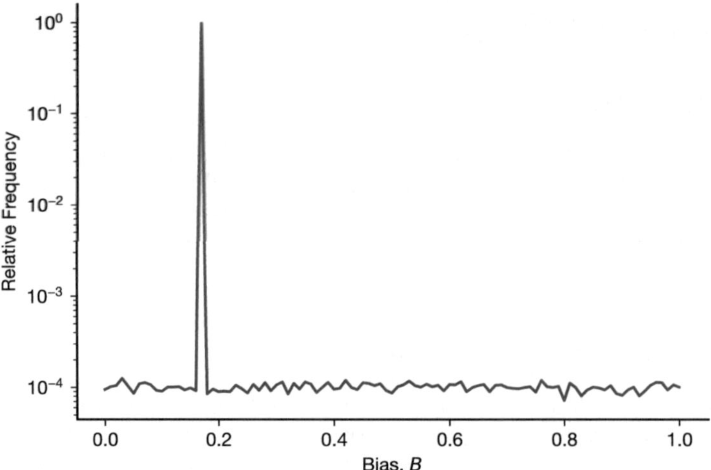

FIGURE 7.6
Relative frequencies of coin values observed under the approximation of a continuous, informative prior.

(continued from previous page)

```
    then rolling the die "rolls" times. Whenever it comes up tails "target" times,
    record that die's loading.  As before, the relative frequencies of the
    loadings in the resulting set approximate the /a posteriori/ probabilites
    of those biases."""

    # Generate the set of possible loadings
    all_loadings = np.arange(0, 1.01, 0.01)

    events = []
    for sim in range(num_sims):

        # Choose a loading for this die
        L = random.choices(all_loadings, weights=a_prioris)[0]

        # Now generate a value from a die with that loading and count
        # the number of 1s
        faces = random.choices([1,2,3,4,5,6], weights=[L] + [(1-L)/5]*5, k=rolls)
        num1s = faces.count(1)

        # Record the events with the target number of tails
        if num1s == target:
            events += [L]

    # Use the np.unique function (introduced in Ch. 2) to return
    # the values and their counts
    vals, counts = np.unique(events, return_counts=True)

    return events
```

Then we can simulate 100,000 events as shown:

```
events = biased_die_i(bias_probs)
```

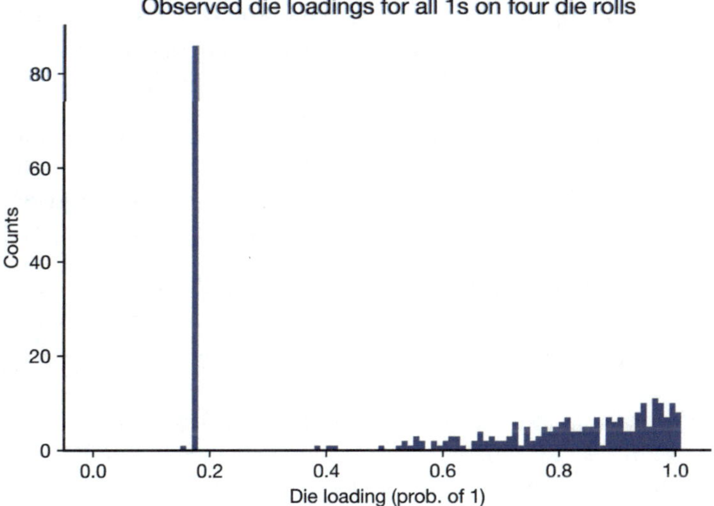

FIGURE 7.7
Histogram of loadings observed when four 1s were observed on four rolls of a die, informative prior.

Fig. 7.7 shows a histogram of the loadings observed when four 1s occurred on 4 rolls of a die. These frequencies are approximately proportional to the *a posteriori* probabilities (APPs). The figure indicates that there is a very high APP of the die having been the fair die; otherwise, the APPs increase as the loading gets closer to 1. By changing the weights, we can create a much more general *a priori* model than we had before. However, we have introduced a new problem:

How can we use the APPs in a statistical test – i.e., how can we make a decision about whether the die could be fair based on the *a posteriori* information?

A typical approach is to find whether the fair die ($L = 1/6$) is within a certain range of values where *almost all* of the *a posteriori* probability lies. We call such a region a *credible interval*:

Definition

credible interval

A $c\%$ *credible interval* is an interval of values that contains $c\%$ of the *a posteriori* probability.

Note that it is possible to choose different types of credible intervals. In this book, we will always generate *equal-tailed intervals*. For an equal-tail credible interval, the set of values below the credible interval has the same probability as the set of values above the credible interval. For more discussion on types of credible intervals, see

https://en.wikipedia.org/wiki/Credible_interval.

WARNING

Credible intervals are generated in Bayesian statistics using *a posteriori* probabilities. Be careful not to confuse them with *confidence intervals*, which can be generated in frequentist approaches, like NHSTs.

- Credible intervals have a straightforward interpretation: given the observation and a priori distribution, the probability that a value of interest lies within a $c\%$ credible interval is $c\%$.
- Confidence intervals do not have such a straightforward interpretation: a $c\%$ confidence interval implies that if an experiment were run again, generating independent results, there would be a $c\%$ chance the resulting confidence interval would contain the true value.
- Both confidence intervals and credible intervals can be generated using a model-based or model-free approach.

To find the $c\%$ credible interval, we can find the tail regions that contain $(1-c)/2\%$ of the probability. As in Section 5.6, we can use the normalized cumulative histogram to find the cumulative relative frequencies, except that now the histogram is of the *a posteriori* probabilities. The histogram is generated by the following code and shown in Fig. 7.8. I have included lines to determine a 95% credible interval. These are at $(100-95)/2 = 2.5\%$ and $100 - 2.5 = 97.5\%$, respectively.

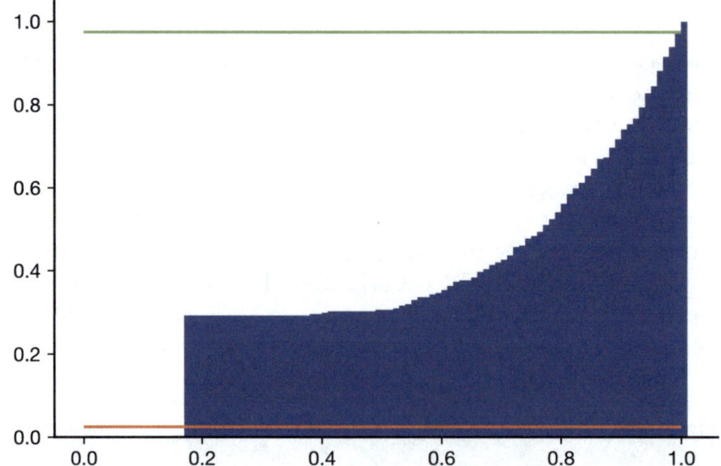

FIGURE 7.8
Cumulative normalized histogram of *a posteriori* relative frequencies for die loadings given that four heads were seen on four rolls, informative prior.

```
# Generate the set of possible biases
mybins = np.arange(0, 1.02, 0.01)

plt.hist(events2, bins=mybins, cumulative=True, density=True)
plt.hlines(0.025, 0, 1, color = 'C1')
plt.hlines(0.975, 0, 1, color = 'C2');
```

The sudden jump in the histogram at 0.17 indicates that the *a posteriori* probability at that point is approximately 0.32. Thus, with this informative prior, the fair die is within the 95% confidence interval. Similar to our approach for finding confidence intervals in Section 5.6, we can find the credible interval by finding the 2.5% and 97.5% percentiles of the *a posteriori* probabilities:

```
np.percentile(events2, [2.5, 97.5])
```

```
array([0.17 , 0.996])
```

Now we return to our original question: how can we use this to conduct a statistical test for whether the die could be fair? The approach we will use is to evaluate if the fair die is in a c% credible interval. Since 0.17 is included in the 95% confidence interval, we cannot reject the possibility that the die is fair. In fact, the results indicate that the *a posteriori* probability that the die is fair is approximately 0.32 under this model.

Terminology review and self-assessment questions

Interactive flashcards to review the terminology introduced in this section and self-assessment questions are available at fdsp.net/7-4, which can also be accessed using this QR code:

7.5 Chapter Summary

"Beliefs don't change facts. Facts, if you're rational, should change your beliefs".
– from Ricky Gervais - The Unbelievers Interview: https://youtu.be/iUUpvrP-gzQ.

This is the essence of Bayesian methods. We have some existing beliefs (*a priori* probabilities), but when presented with new facts, evidence, or data, Bayesian methods provide a systematic approach to updating our beliefs (*a posteriori* probabilities).

We began with deriving the core tool for performing these updates – Bayes' rule:

Bayes' Rule

For a stochastic system with a discrete set of possible input events $\{A_0, A_i, \ldots\}$ and a discrete set of possible output events $\{B_0, B_1, \ldots\}$, the *a posteriori* probabilities $P(A_i|B_j)$ can be written in terms of the likelihoods $(P(B_j|A_i))$ and the *a priori* probabilities $(P(A_i))$ as

$$P(A_i|B_j) = \frac{P(B_j|A_i)\,P(A_i)}{\sum_i P(B_j|A_i)\,P(A_i)}.$$

We applied Bayes' rule to show that for a 40-year-old woman who has a mammogram test come back positive, the probability that she actually has cancer is very small (around 0.01). This is an example of the *Base Rate Fallacy*, in which people tend to focus on the conditional probabilities of events (i.e., the false alarm rate for mammograms is low), while neglecting the base rates (i.e., *a priori* probabilities). In this case, it is very rare for a woman in her 40s to have breast cancer, and this strong *a priori* information results in the *a posteriori* probability of cancer being low, even when cancer has been detected on a mammogram.

I introduced the concept of *stochastic systems with hidden state*, in which there is some unknown memory or stored condition internal to the system that cannot be directly observed. Bayesian approaches allow us to update our beliefs about this hidden state after observing outputs from the system.

I introduced optimal decision problems, in which a system has an input and an output that depends on that input in some random way (for instance, because of noise). We studied a binary-input communication system as an example of this, and I showed two different approaches to making optimal decisions about the input given the observed outputs:

- **Maximum likelihood decision rules** have the advantage of only requiring information about the *likelihoods*, which are the conditional probabilities of the outputs given the inputs.
- **Maximum *a posteriori* decision rules** are Bayesian approaches that find the most probable input given the observed output; however, they require knowledge about both the likelihoods and the *a priori* probabilities of the inputs.

Finally, I briefly introduced Bayesian hypothesis testing and gave a simple example with different assumptions on the *a priori* probabilities. In particular, I introduced the concept of an *uninformative prior*, in which every input or system state is assumed equally likely to occur. We also considered an *informative prior*, which uses knowledge outside of the data to choose the *a priori* probabilities. I showed how to conduct both binary tests and nonbinary hypothesis tests. For the latter, I introduced the concepts of *credible intervals*, which are intervals that contain some specified percentage of the *a posteriori* probabilities. A hypothesis can be rejected in a Bayesian framework if it is not within the credible interval.

Bayesian approaches are especially powerful when our data does not come from discrete sets. However, we will need new approaches to model and analyze such random phenomena, and in the next chapter, we introduce random variables to help fulfill this need.

Access a list of key take-aways for this chapter, along with interactive flash-cards and quizzes at fdsp.net/7-5, which can also be accessed using this QR code:

8

Random Variables

In the previous chapters, when we have had to generate random phenomena, we have drawn values at random from a finite set. That is a very powerful technique that can be used to model many scenarios. However, in the real world, many phenomena are not only not finite, they are not countably infinite. For example, what is the lifetime of a computer CPU? Not only do we not know how to choose an upper bound, but the actual time is not limited in resolution to just days or even seconds – we usually just report such levels of detail for convenience.

More generally, much of the data that we work with will be numerical, especially in engineering and the hard sciences. All of these factors motivate the development of special approaches for working with numerically valued random phenomena. We call these *random variables*:

> Definition
>
> **random variable (informal definition)**
> A random phenomenon that takes on numeric values.

Because we are working with numerical phenomena, we can develop *functions* that allow us to better visualize how the probability is distributed and to calculate probabilities and other features that depend on the probabilities. We will use random variables as models for random phenomena and leverage these models to build analytical statistical tests.

8.1 Definition of a Real Random Variable

In the introduction, we provided an informal definition for a random variable. The informal definition helps build intuition about what we mean by a random variable. However, we want to develop random variables in such a way that we can take advantage of all of the properties of the *probability spaces* that we developed in Chapter 4. In this book, we only consider the case of **real** random variables, but random variables can also be complex-valued.

Let's start by introducing some notation that we will use for random variables and in our discussion below:

- We will generally use uppercase (capital) letters to represent a random variable, such as X, Y, or Z. We may choose the letter to represent a particular quantity, such as representing a random time by T or a random rate by R. For a generic random variable, we will usually use X.

DOI: 10.1201/9781003324997-8

- We want to ask about the probability that a random variable's value lies in some set. Until now, we have used P to denote either a probability or a probability measure. We will draw a distinction between these below, where we will write Pr() to denote a probability and $P()$ to denote a particular probability measure.
- We will have the need to provide notation for an arbitrary value that a random variable may take on. Such a value will be represented by the lowercase version of the letter used to denote the random variable. In other words, we can ask about the probability that the random variable X takes on a value x, which we can write as $\Pr(X = x)$. In particular, we can define functions this way; for instance, we can create a function of x that returns $\Pr(X = x)$ for each real x, like $p_X(x) = \Pr(X = x)$.

WARNING

Many people learning probability get confused by the notation of using an uppercase letter for a random variable and using a lowercase letter for the value of a random variable.
This confusion is amplified with handwritten math because many people use very similar letter styles for both the uppercase and lowercase versions of their letters.
Here are my recommendations to reduce this confusion:

- Write your random variables in a *san serif* style. Serifs are decorative strokes at the ends of letters, so write your uppercase letters in a block style without serifs.
- Write the lowercase versions with curvy *serifs*.

Some examples are below:

Random Variable	Value
X	x
Y	y
Z	z

I recommend crossing the stem of the uppercase Z to help distinguish it from the number two.

As mentioned above, we wish to be able to evaluate the probabilities that a random variable takes on certain sets of values. There are several immediate problems:

- How can we define a random variable so that we can evaluate the probability that it will take on a set of values in a way that is consistent with our work on probability?
- As in the case of event classes and probability measures, we cannot create a general function that will assign a meaningful probability to any arbitrary set of values that a

random variable might take on. Given this restriction, how can we choose reasonable sets of values to assign probability to, and how can we evaluate those probabilities?

Our approach may initially seem a bit strange, but it defines random variables in a way that directly leverages our work on probability spaces. Suppose we have a probability space $(S, \mathcal{F}, P)$, and we want to use the probability measure P from this probability space to determine the probabilities for a random variable X. Then if B is a set of real values, we want to translate $\Pr(X \in B)$ to some probability $P(E_B)$, where $E_B \in \mathcal{F}$ is an event in our probability space. To achieve this, we require that the outcomes (which do not have to be numbers) in E_B correspond in some way to the real values in B. Since we need to be able to determine what outcomes in E_B correspond to the set of reals B, there must be a **function** that maps from the outcomes in E_B to the real values in B. We define the random variable X as that function: it is a function that maps the outcomes in S to the real line.

Formally, we do not write a random variable as X but instead write it as $X(s)$ to indicate that it is a function of an outcome in the sample space. Note also that the function itself **is not random**. In other words, if we know that some particular value $s_1 \in S$ occurred, then $X(s_1)$ is a deterministic real number.

Note:

In the discussion below, we get into some fine details about random variables, and this discussion is more mathematically challenging than is usually included in a book at this level. I have chosen to include this level of detail for two reasons:

1. This careful approach establishes a solid foundation for further study of probability.
2. This approach will motivate our use of a particular function in working with random variables.

Do not be concerned if you do not understand every detail of these arguments: in practice, we will use a very pragmatic approach for working with random variables that does not require us to worry about the following details.

Thinking back to our work on probability spaces, you should recall that we cannot define $P(A)$ for all $A \subset S$ if S is uncountably infinite. We can only define $P(A)$ if $A \in \mathcal{F}$. Thus, we will need to introduce some additional restrictions on X to ensure that the sets of values B for which we define the probability $\Pr(X \in B)$ correspond to events $E_B \in \mathcal{F}$. There is no way to create a random variable that ensures that this will be true for every set of real values B, so we must restrict the sets $B_i \subset \mathbb{R}$ for which we will define $\Pr(X \in B_i)$.

Let $\mathcal{B}$ be a set of all of the subsets of $\mathbb{R}$ for which we will define $\Pr(X \in B)$ if $B \in \mathcal{B}$. We want to at least be able to ask questions like what is $\Pr(X = x)$ and what is $\Pr\{X \in [a, b)\}$, and we can easily see that we would like to create unions of such sets: for instance, if X represents the value on the top face of a six-sided die, we may want to determine the probability that the die is even. If A represents the age of people who have been hospitalized for COVID, we may wish to assess the impact on people outside of the usual working age by determining the probability that A is less than 18 or greater than or equal to 65. Thus, we can use the following generalization of the intervals:

Definition

Borel sets of $\mathbb{R}$

The collection of all countable unions, intersections, and complements of intervals.

The collection of Borel sets is called the *Borel σ-algebra* or the *Borel field* (technically, including the operators $\cup$ and $\cap$). For conciseness, we will use the terminology *Borel field* to refer to the collection of Borel sets of $\mathbb{R}$.

Note:

The Borel field for $\mathbb{R}$ can be formed from the countable unions, intersections, and complements of the half-open intervals $(-\infty, x)$ for all $x \in \mathbb{R}$.

To be able to define the probability $\Pr(X \in B)$ for every Borel set B, we must require that if we collect the set of outcomes that result in $X \in B$, then that set must be an event. In mathematical notation, we require that $\{s \,|X(s) \in B\} \in \mathcal{F}$.

Putting all of this together, we have our formal definition for a *random variable*:

Definition

random variable (formal definition)

Let $(S, \mathcal{F}, P)$ be a probability space. Then a random variable is a *function* that maps from the sample space S to the real line $\mathbb{R}$, such that if B is in the Borel field for $\mathbb{R}$, then $\{s \,|X(s) \in B\} \in \mathcal{F}$.

Since a random variable is a function, we have the following properties:

- A particular outcome $s \in S$ maps to exactly one value of $X(s)$.
- For a particular value $X = x$, there may be multiple values of s for which $X(s) = x$.
- For any two different values $x_1 \neq x_2$, the sets $\{s \,|X(s) = x_1\}$ and $\{s \,|X(s) = x_2\}$ are disjoint; otherwise, there must be some outcome $\tilde{s}$ that belongs to both sets and for which both $X(\tilde{s}) = x_1$ and $X(\tilde{s}) = x_2$. This is not possible if $X(s)$ is a function.

One important factor in classifying a random variable X is based on the number of values X can take on. Since X is a function of s, the set of values that X takes on is called the *range* of X:

Definition

range of a random variable

Let $(S, \mathcal{F}, P)$ be a probability space and let $X(s)$ be a random variable on that space. Then the *range* (or *image*) of $X(s)$, denoted by Range(X) is the set of values that $X(s)$ can take on, which is given by

$$\{X(s) \mid s \in S\}.$$

WARNING

At this point in our treatment of probability and random variables, it may be tempting to describe the range of $X(s)$ as the "set of values of $X(s)$ that have nonzero probability". However, we will soon discover that **this is not true in general** – in fact, we will find that for many of the random variables we commonly use, every value has probability zero!

Note that $\{s \,|\, X(s) \in B\}$ defines the set of outcomes that produce a set of values in the range of $X(s)$, so we abbreviate this as $X^{-1}(B)$.

Thus, we can write

$$\Pr(X \in B) = P\left[X^{-1}(B)\right].$$

Since the probability is evaluated using the probability measure P, for convenience of notation, I will just write $P(X \in B)$ in the remainder of this book.

To help make these ideas more clear, we consider some simple examples. In each example, we show a graphical depiction of the sample space for an experiment with directed arrows that indicate the mapping from the sample space to values of the random variable, which are shown on a number line.

Example 8.1: Random Variable from Flipping a Fair Coin Once

Let's create a binary random variable that has an equal probability of being 0 or 1. Such a random variable can be created using many different probability spaces, but let's pick a simple one that we are already familiar with. We will flip a fair coin. Then $S = \{H, T\}$ and

$$\mathcal{F} = 2^S = \{\emptyset, \{H\}, \{T\}, \{H, T\}\}.$$

The probability measure is given by

$$P(E) = \begin{cases} 0, & E = \emptyset \\ 1/2, & E = \{H\} \text{ or } E = \{T\} \\ 1, & E = \{H, T\} \end{cases}.$$

We can create our random variable as the function $X(s)$ given by

$$X(s) = \begin{cases} 0, & s = T \\ 1, & s = H \end{cases}.$$

This mapping is shown in Fig. 8.1.

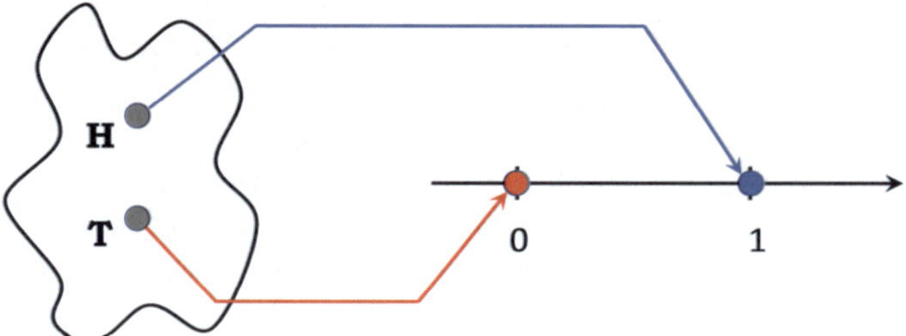

FIGURE 8.1
Illustration of mapping from sample space $S = \{H, T\}$ to random variable values $X(H) = 1$ and $X(T) = 0$ for Example 8.1.

Note:

This concrete example helps make an important point:

A random variable is defined on the sample space – it is a function whose argument is an **outcome**. In this case, the argument of the function is either H or T. Contrast this with the probability measure, which is defined on the event class – its argument is an **event**. The argument of the probability measure can be any of $\emptyset, \{H\}, \{T\}$, or $\{H, T\}$.

Let's implement our random variable as a function in Python. The function takes an argument which is one of the outcomes H or T and performs the mapping shown above:

```
def X(s):
    if s == 'T':
        return 0
    elif s == 'H':
        return 1
    else:
        raise Exception("The Random Variable X only takes inputs H or T")
```

Next, we will generate an outcome to simulate flipping a fair coin using `random.choice()` as we have before. Start by importing the random library and setting up the sample space:

```
import random
sample_space = [ 'H', 'T']
```

Let's draw 10 different random outcomes s from the sample space and output the corresponding value of the random variable:

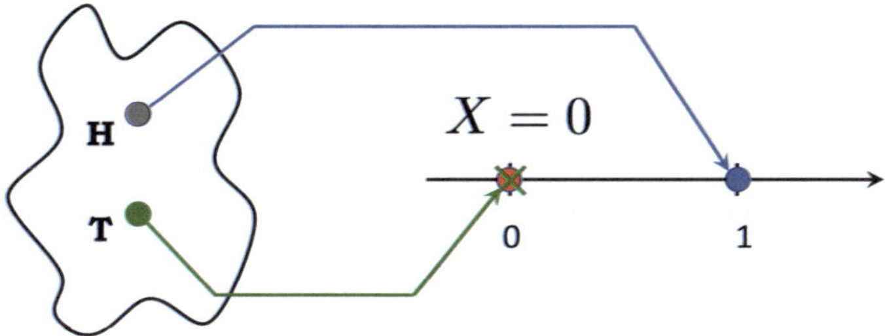

FIGURE 8.2
Mapping from value 0 back to outcome T for the random variable of Example 8.1.

```
print(f'{"i":^3}|{"s":^5}|{"X(s)":^7}')
print('-'*15)
for i in range(10):
    s = random.choice(sample_space)
    print(f'{i:^3}|{s:^5}|{X(s):^6}' )
```

```
 i |  s  | X(s)
---------------
 0 |  H  |  1
 1 |  T  |  0
 2 |  H  |  1
 3 |  H  |  1
 4 |  T  |  0
 5 |  H  |  1
 6 |  T  |  0
 7 |  H  |  1
 8 |  H  |  1
 9 |  H  |  1
```

Now we demonstrate how we can use the definition of the random variable and the underlying probability measure to evaluate probabilities for the random variable.

We start with the simplest case. Consider $X(s) = 0$, which is illustrated on the number line portion of the diagram in Fig. 8.2 by a green X.

Graphically, we can see that there is exactly one outcome $s \in S$ for which $X(s) = 0$; that is, $X(T) = 0$. Mathematically, we can find $P\left[X\left(s\right) = 0\right]$ as follows:

$$\begin{aligned} P\left[X(s) = 0\right] &= P\left[X^{-1}(0)\right] \\ &= P\big[\{T\}\big] \\ &= \frac{1}{2}. \end{aligned}$$

We can use our Python random variable to estimate this probability by generating a large number of values of the random variable and finding the relative frequency of the value 0:

```
num_sims = 10_000

counter=0
for sim in range(num_sims):
    s = random.choice(sample_space)

    if X(s) == 0:
        counter+=1

print(f'Pr( X = 0 ) is approximately {counter/num_sims}')
```

```
Pr( X = 0 ) is approximately 0.5002
```

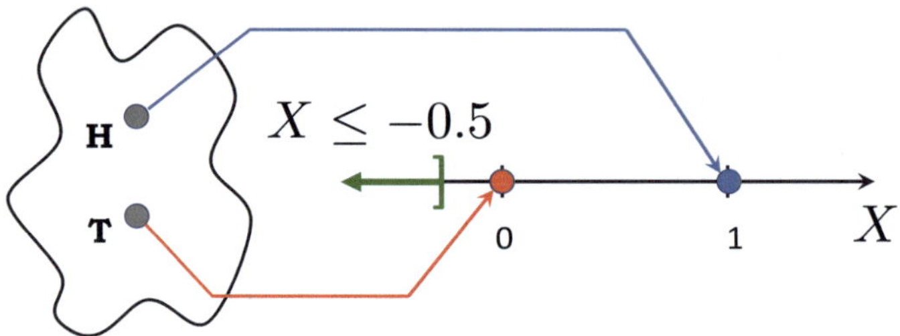

FIGURE 8.3
Mapping from $X \leq -0.5$ to null region in sample space for random variable of Example 8.1.

Now let's consider $X \leq -0.5$, which is shown by the green line in Fig. 8.3. The square brace at the point -0.5 indicates that the point -0.5 is included in the region, and the arrow indicates that the region continues to $-\infty$. We could write an equivalent form for this probability as

$$P\big(X \in (-\infty, -0.5]\big).$$

This region introduces an important concept: **we are not constrained to ask only about the probability of points in the range of $X(s)$. We can ask about the probability of any Borel set of $\mathbb{R}$.**

However, for this region, we can see that there are no points in S that yield any values of $X(s)$ in $(-\infty, -0.5]$. Thus, $P[X(s) \leq -0.5] = P[\emptyset] = 0$

Next consider the region $0 < X \leq 1$, which is shown in Fig. 8.4. The parenthesis at the point 0 indicates that it is not a part of this region. We can write the corresponding probability as $P(0 < X \leq 1)$ or as $P(X \in (0, 1])$.

As illustrated in the figure by the green line and the green shading on the point H in S, there is again only one $s \in S$ that gives a value of $0 < X \leq 1$. So, in this case, $P[0 < X \leq 1] = P[\{H\}] = 1/2$.

To simulate this probability, we will need to rewrite the condition $0 < X \leq 1$ into a form that more directly translates to Python. Note that this form is really a shorthand for two conditions:

- the left-hand side of the inequality is $0 < X$, but we more commonly flip that relation to write $X > 0$ when writing a single inequality, and
- the right-hand side of the inequality is $X \leq 1$. We are asking about the probability for values of X that satisfy both of these inequalities, so we may also write $X > 0 \cap X \leq 1$

The following simulation estimates the probability of X satisfying these conditions:

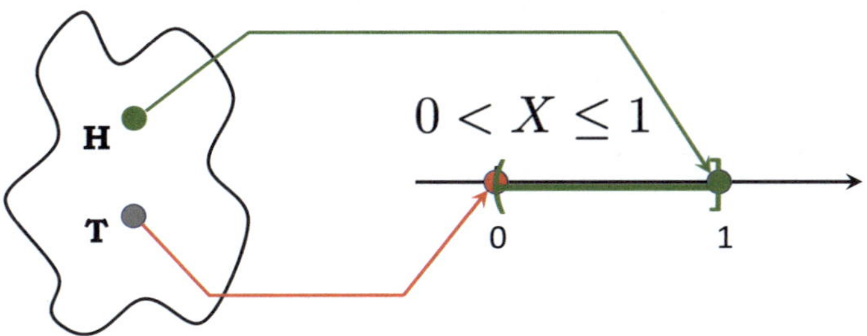

FIGURE 8.4
Mapping from region $0 < X \leq 1$ to the values in the sample space ($\{H\}$) for Example 8.1.

```
num_sims = 10_000

counter=0
for sim in range(num_sims):
    s = random.choice(sample_space)
    if X(s) >0 and X(s) <= 1:
        counter+=1

print(f'Pr( 0 < X <= 1 ) is approximately {counter/num_sims}')
```

```
Pr( 0 < X <= 1 ) is approximately 0.4893
```

We consider one last region of this random variable, $X < 1.5$, which is illustrated in Fig. 8.5.

In this case, there are two different values of $s \in S$ such that $X(s) < 1.5$. In particular, $X(T) = 0 < 1.5$ and $X(H) = 1 < 1.5$. If we let $B = (-\infty, 1.5)$, then $X^{-1}(B) = \{H, T\}$, and

$$\begin{aligned} P(X < 1.5) &= P\left[X \in B\right] \\ &= \left[X^{-1}(B)\right] \\ &= P\left[\{H, T\}\right] \\ &= 1. \end{aligned}$$

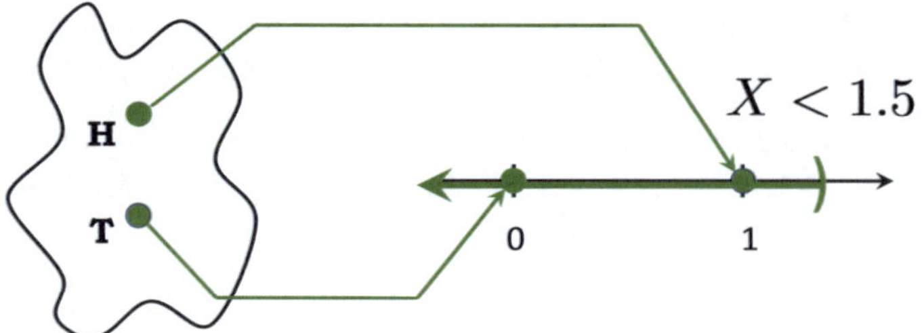

FIGURE 8.5
Mapping from region $X < 1.5$ to corresponding outcomes in sample space $\{H, T\}$ for Example 8.1.

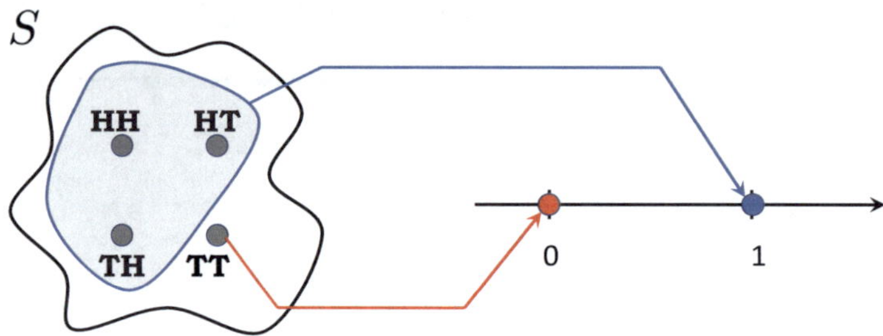

FIGURE 8.6
Illustration of mapping from sample space to real line for Example 8.2.

Example 8.2: Random Variable from Flipping a Fair Coin Twice

Let's create a binary RV Y from tossing a fair coin twice. We will use the following rule: the random variable is 1 if at least one heads occurs; otherwise, it is 0. Mathematically, we write the definition of $X(s)$ as

$$Y(s) = \begin{cases} 0, & s = TT \\ 1, & s \in \{HH, HT, TH\} \end{cases}$$

This relation is illustrated in Fig. 8.6, where all of the points within the blue region of S are mapped to 1.

From the figure and mathematical definition of $Y(s)$, we can see that $Y^{-1}(0) = \{TT\}$ and $Y^{-1}(1) = \{HH, HT, TH\}$. This is our first example where multiple outcomes map to the same value of a random variable. Thus even if we ask for a probability like $P[Y(s) = y]$, the underlying event may contain multiple outcomes, as in this case. Given that the coin is fair, it is easy to see that the probability of each outcome is 1/4, so $P[Y(s) = 0] = 1/4$ and $P[Y(s) = 1] = 3/4$. This is also our first example of a random variable with unequal probabilities, and it was built from a fair experiment, in which each outcome has an equal probability of occurring.

Now, let's implement Y as a function in Python. We will draw two values of the coin using `random.choices()`, which will return a list, so we will build the function `Y()` accordingly:

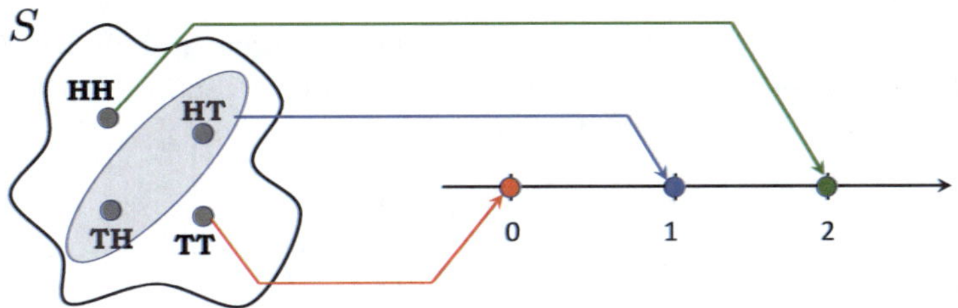

FIGURE 8.7
Illustration of mapping from sample space to real line for random variable $Z(s)$ in Example 8.3.

```
def Y(s):
    if s == ['T', 'T']:
        return 0
    elif s in [ ['T', 'H'], ['H', 'T'], ['H', 'H'] ]:
        return 1
    else:
        raise Exception('Must input a list of two elements, each of which may be
↪either "H" or "T"')
```

Now let's estimate the probability that $Y = 1$ via simulation:

```
num_sims = 10_000

sample_space = ['H', 'T']
counter=0
for sim in range(num_sims):
    s = random.choices(sample_space, k=2)
    if Y(s) == 1:
        counter+=1

print(f'P(Y=1) is approximately {counter/num_sims}')
```

```
P(Y=1) is approximately 0.7567
```

Example 8.3: Second Random Variable from Flipping a Fair Coin Twice

Let's create another RV Z using the same probability space as in Example 8.2; i.e., define $Z(s)$ using the same outcome. Let Z be the number of heads that occurred. Then the mathematical definition of $Z(s)$ is given by

$$Z(s) = \begin{cases} 0, & s = TT \\ 1, & s \in \{HT, TH\} \\ 2, & s = HH. \end{cases}$$

A graphical depiction of this mapping is shown in Fig. 8.7.

Note that even though Z is created from the same probability space, it takes on a different set of values and is not even a binary random variable. The point probabilities $P(Z = z)$ are easy to find:

$$P(Z = 0) = P(\{TT\}) = 1/4$$
$$P(Z = 1) = P(\{TH, HT\}) = 1/2$$
$$P(Z = 2) = P(\{HH\}) = 1/4$$
$$P(Z = z) = 0 \text{ if } z \notin \{0, 1, 2\}.$$

Example 8.4: Dependence Among Random Variables

Consider an example of what can happen when $Y(s)$ and $Z(s)$ are created from the same outcome. If you are given no information about Y, then we can see from above that $P(Z = 0) = 1/4$. Now suppose I told you that $Y = 0$. From above, we see that $Y^{-1}(0) = TT$ and $Z(TT) = 0$. So we know that $Z = 0$ with probability 1. Thus Y and Z are dependent random variables – we haven't defined exactly what this means yet mathematically, but we can see that they are dependent by this example.

To illustrate this concept, we implement the random variable Z as a function and conduct a simulation to estimate $P(Z = 0)$ and $P(Z = 0 \mid Y = 0)$, which corresponds to the description above that "I told you $Y = 0$".

```
def Z(s):
    if s == ['T', 'T']:
        return 0
    elif s in [ ['T', 'H'], ['H', 'T']]:
        return 1
    elif s == ['H', 'H']:
        return 2
    else:
        raise Exception('Must input a list of two elements, each of which may be␣
↪either "H" or "T"')
```

```
num_sims=100_000

count_z0 = 0
count_y0 = 0
count_z0_y0 = 0

for sim in range(num_sims):
    s = random.choices(sample_space, k=2)

    if Z(s) == 0:
        count_z0 += 1
```

(continues on next page)

(continued from previous page)

```
    if Y(s) == 0:
        count_y0 += 1
        if Z(s) == 0:
            count_z0_y0 += 1

print(f'P(Z=0) is approximately {count_z0/num_sims}')
print(f'P(Z=0|Y=0) is approximately {count_z0_y0/count_y0}')
```

```
P(Z=0) is approximately 0.25107
P(Z=0|Y=0) is approximately 1.0
```

As expected, these values closely match the results of our analysis.

All of these examples use discrete sample spaces, and the ranges of the random variables are all discrete, finite sets. These are called *discrete random variables*. The true power of working with random variables will be for the cases where these sets are uncountably infinite. However, before we get to those cases, the next section introduces some tools for working with discrete random variables and some common types of discrete random variables.

Terminology review and self-assessment questions

Interactive flashcards to review the terminology introduced in this section and self-assessment questions are available at fdsp.net/8-1, which can also be accessed using this QR code:

8.2 Discrete Random Variables

As mentioned in Section 8.1, we classify random variables based on their range. We do this because the mathematical tools we use to work such random variables vary depending on their range. We will start with the easiest type to understand, which is the *discrete random variable*:

> **Definition**
>
> **discrete random variable**
>
> A random variable is said to be a *discrete random variable* if its range is finite or countably infinite.

If X is a discrete random variable, then we can find the probability of any subset of its values by summing over them:

$$P(X \in A) = \sum_{x \in A} P(X = x) \text{ if } A \subset \text{Range}(X),$$

where we have implicitly taken advantage of the fact that the sets $\{s\,|X(s) = x\}$ are disjoint sets, as discussed in Section 8.1.

For any Borel set, B, we can express $\mathrm{P}(B)$ as a finite sum over the values of X that are in B and in the range of X, since all other values of X have probability zero:

$$\mathrm{P}(X \in B) = \sum_{x \in B \cap \mathrm{Range}(X)} \mathrm{P}(X = x) \text{ if } B \in \mathcal{B}.$$

Since both x and $\mathrm{P}(X = x)$ are real-valued, we define a function called the *probability mass function* to facilitate calculating these types of probabilities.

8.2.1 Probability Mass Functions

Definition

probability mass function (PMF)

For a discrete random variable, X, the *probability mass function* of X is the function $p_X(x)$ such that $p_X(x) = \mathrm{P}(X = x)$.

Note that

$$\begin{aligned}
\sum_{x \in \mathrm{Range}(X)} p_X(x) &= \mathrm{P}\left[X(s) \in \mathrm{Range}\,(X)\right] \\
&= \mathrm{P}\left[\{s\,|X(s) \in \mathrm{Range}\,(X)\,\}\right] \\
&= \mathrm{P}\left[s \in S\right] \\
&= 1.
\end{aligned}$$

As x ranges over all possible values of X, the events $\{s \mid X(s) = x\}$ are mutually exclusive and form a partition S.

When there is no confusion, we will drop the explicit mention of the range of X and just write

$$\sum_x p_X(x) = 1.$$

Similarly, for any set $A \in \mathrm{Range}(X)$, we will write

$$\mathrm{P}(X \in A) = \sum_{x \in A} p_X(x).$$

Given a functional definition of a random variable, we can find the probability mass function as follows:

- For each value $x \in \mathrm{Range}(X)$, find the set of outcomes for which $X(s) = x$, which we can write as $E_x = \{s\,|X(s) = x\,\}$. Then let $p_X(x) = \mathrm{P}(E_x)$.
- For each value of $x \notin \mathrm{Range}(X)$, let $p_X(x) = 0$.

We illustrate this with some examples.

Example 8.5: PMF For Binary Random Variable From Tossing a Fair Coin

Find the PMF for the binary random variable $X(s)$ defined in Example 8.1, which is illustrated in Fig. 8.1. Please refer to that example for the detailed mathematical formulation of the probability space and functional definition of the random variable.

For this random variable, $\text{Range}(X) = \{0, 1\}$. Hence, we calculate:

- $p_X(0) = \Pr[X(s) = 0] = \text{P}(\{T\}) = 1/2$, and
- $p_X(1) = \Pr[X(s) = 1] = \text{P}(\{H\}) = 1/2$.

Then the PMF for X is given by

$$p_X(x) = \begin{cases} \frac{1}{2}, & x \in \{0, 1\} \\ 0, & \text{o.w.} \end{cases}$$

Note that we often abbreviate "otherwise" in such formulas as "o.w.".

When we have a very simple PMF like this, we can implement it in Python for a single value of x using `if` statements:

```
def p_X1(x):
    if x == 0 or x == 1:
        return 1/2
    else:
        return 0
```

However, it is more useful to be able to return probabilities for a whole list or array of input values. We can use NumPy's `piecewise()` function to provide the appropriate values by conducting these tests on a vector of values (although a single value will also still work). The call signature for `np.piecewise()` is

```
Signature: np.piecewise(x, condlist, funclist, *args, **kw)
```

The `condlist` is a list of conditions, and the `funclist` is a **corresponding** list of functions. If condition i from `condlist` evaluates as true for an input value, then function i from `funclist` will be used to compute the output value. Here is how we can use it in action:

```
import numpy as np
def p_X(x):
    x = np.array(x).astype(float)
    return np.piecewise(x, [x==0, x==1], [0.5, 0.5, 0])
```

Here the line `x = x.astype(float)` is required to ensure that we can return an array of float values, even if the input contains all integers.

Because PMFs generally only have values at a small subset of the real numbers, we use a special type of plot in which the nonzero values are drawn as sticks with balls at the values. Because of the shape of these plots, these are called *lollipop plots* (especially when the independent axis contains categorical data). In Matplotlib, these are called *stem plots* and generated with the `plt.stem()` function. (Note that stem-and-leaf plots are something different).

To plot the PMF for X, we put the values in the range of X in a list or vector, and the corresponding PMF values in another list or vector and call `plt.stem()`. We can make the PMF easier to interpret by adjusting the limits of the axes. First, we increase the x-axis to help illustrate that the PMF is zero outside the range $[0, 1]$. Because we are plotting over a larger range, we also evaluate the PMF over a wider range, as shown for `x2` below. Second, we set the y-axis lower limit to zero and increase the upper limit so that it is easier to read off the maximum value of 0.5 for the PMF.

```
import matplotlib.pyplot  as plt

x2=[-1, 0, 1, 2]

plt.stem(x2, p_X(x2))

plt.xlim(-0.5, 1.5)
plt.ylim(0, 0.75)

plt.xlabel('$x$')
plt.ylabel('Probability mass function $p_X(x)$');
```

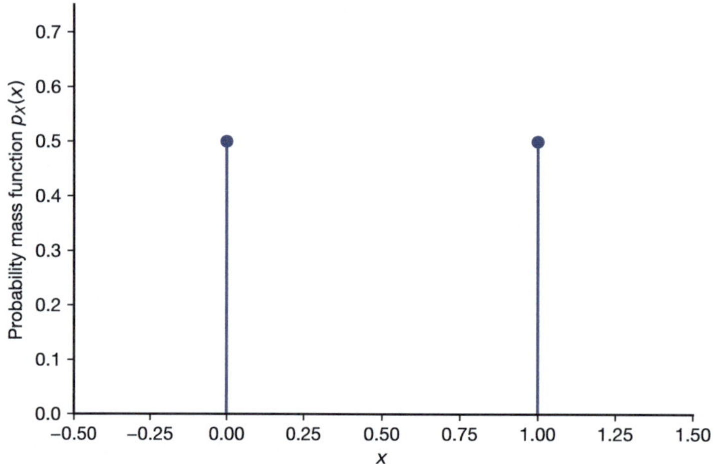

Example 8.6: PMF for Random Variable from Flipping a Fair Coin Twice

Next, we determine the PMF of the binary RV Y from Example 8.2, which is illustrated in Fig. 8.6.

The range of Y is the same as the range of X in Examples 8.1 and 8.5, $\text{Range}(Y) = \{0, 1\}$. However, the sample space and mapping from outcomes to values of the random variable are different than in those previous examples. From our previous work, we know that

- $p_Y(0) = \Pr[Y(s) = 0] = P(\{TT\}) = 1/4$, and
- $p_Y(1) = \Pr[Y(s) = 1] = P(\{HH, TH, HT\}) = 3/4$.

The PMF for Y is given by

$$p_Y(y) = \begin{cases} \frac{1}{4}, & y = 0 \\ \frac{3}{4}, & y = 1 \\ 0, & \text{o.w.} \end{cases}$$

Here is an implementation in Python:

```
def p_Y(y):
    y = np.array(y).astype(float)
    return np.piecewise(y, [y == 0, y == 1], [1/4, 3/4, 0])
```

Fig. 8.8 shows a plot of the PMF Y that was created using the same approach used for the PMF of X in Example 8.5.

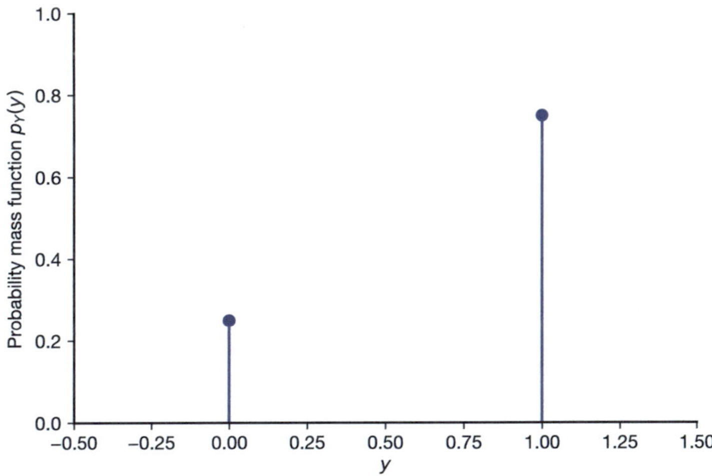

FIGURE 8.8
PMF for random variable Y from Example 8.2.

Example 8.7: PMF for Second Random Variable from Flipping a Fair Coin Twice

Now, consider the random variable Z from Example 8.3, which is illustrated in Fig. 8.7. This random variable is created using the same probability space as the random variable Y in Example 8.2 but has a different range, $\text{Range}(Z) = \{0, 1, 2\}$. From our previous work with this random variable,

- $p_Z(0) = P[Z(s) = 0] = P(\{TT\}) = 1/4$, and
- $p_Z(1) = P[Z(s) = 1] = P(\{HT, TH\}) = 1/2$, and
- $p_Z(2) = P[Z(s) = 2] = P(\{HH\}) = 1/4$.

Thus, the PMF of Z is

$$p_Z(z) = \begin{cases} \frac{1}{4}, & z \in \{0, 2\} \\ \frac{1}{2}, & z = 1 \\ 0, & \text{o.w.} \end{cases}.$$

The Python implementation follows. A graph of the PMF $p_Z(z)$ is shown in Fig. 8.9.

```
def p_Z(z):
    z = np.array(z).astype(float)
    return np.piecewise(z, [z == 0, z == 1, z == 2], [1/4, 1/2, 1/4, 0])
```

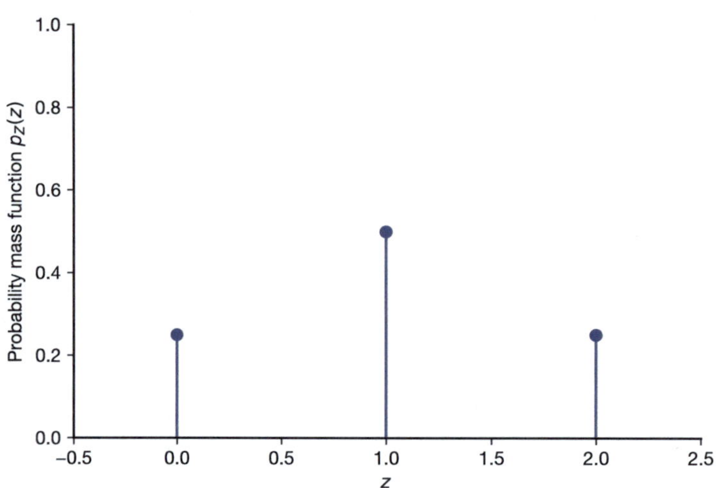

FIGURE 8.9
PMF of random variable Z from Example 8.3.

Example 8.8: PMF for Rolling a Fair 6-sided Die

In this example, we show how a random variable can be formed directly from the outcomes of a random experiment. In other words, for an experiment with a numerical outcome, we can let $X(s) = s$.

Create a probability space by rolling a fair 6-sided die and observing the top face. Let the event class be the power set of the sample space. Let the RV W be defined by $W(s) = s$ for all $s \in \{1, 2, 3, 4, 5, 6\}$.

Then $p_W(w) = P[W(s) = w] = P(\{w\})$ for $w \in \{1, 2, 3, 4, 5, 6\}$. Since this is a fair experiment, $P(\{w\}) = 1/6$.

The PMF for W is

$$p_W(w) = \begin{cases} \frac{1}{6}, & w = 1, 2, \dots, 6 \\ 0, & \text{o.w.} \end{cases}$$

A Python implementation is shown below:

```
def p_W(w):
    w = np.array(w).astype(float)
    return np.piecewise(w, [np.isin(w, [1,2,3,4,5,6])], [1/6, 0] )
```

Here, I used the `np.isin()` function to check each value of the w array to see if it is in the range of W. The PMF of W is shown in Fig. 8.10.

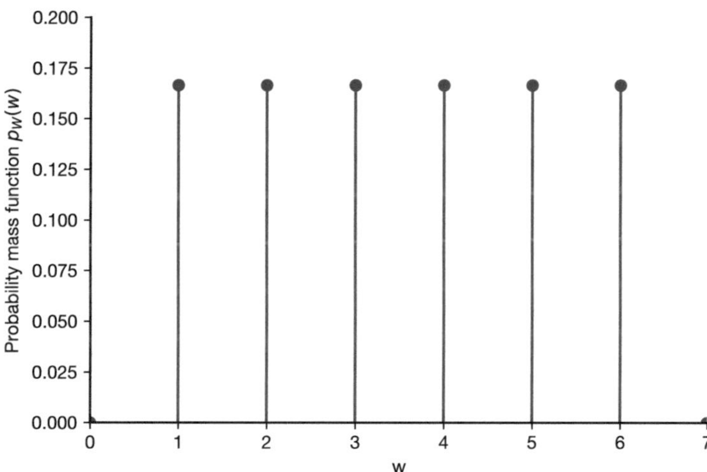

FIGURE 8.10
The PMF of the random variable W, which is equal to the top face on a fair six-sided die.

Example 8.9: PMF for Number of Coin Flips Until Heads Occurs

Recall that discrete random variables can have either a finite range or a countably infinite range. Let's create an example of the latter.

Flip a coin until the first heads occurs. Record the set of outcomes. Then the sample space is

$$S = \{H, TH, TTH, TTH, TTTH, \ldots\}.$$

It should be clear that there is no maximum number of flips until the first H occurs, and it should also be intuitive that the shorter sequences are more probable than the latter ones. We know that $P(H) = P(T) = 1/2$, and the results of different flips can be assumed to be independent. So

$$\mathrm{P}(H) = \frac{1}{2}$$
$$\mathrm{P}(TH) = \left(\frac{1}{2}\right)\left(\frac{1}{2}\right) = \frac{1}{4}$$
$$\mathrm{P}(TTH) = \left(\frac{1}{2}\right)\left(\frac{1}{2}\right)\left(\frac{1}{2}\right) = \frac{1}{8}$$

Let's introduce some notation. If a given sequence has $n-1$ T followed by one H, we will write it as $T^{n-1}H$. Then

$$\mathrm{P}(T^{n-1}H) = \frac{1}{2^n}.$$

Let N be a random variable that is equal to the number of flips until the first H occurs. Then if $n \in \{1, 2, 3, \ldots\}$,

$$\begin{aligned} \mathrm{P}\left[N(s) = n\right] &= \mathrm{P}\left[T^{n-1}H\right] \\ &= \left(\frac{1}{2}\right)^n. \end{aligned}$$

The PMF of N is

$$p_N(n) = \begin{cases} \left(\frac{1}{2}\right)^n, & n = 1, 2, 3, \ldots \\ 0, & \text{o.w.} \end{cases}$$

The implementation of this PMF as a Python function is tricky and beyond the scope of this book. A plot of the PMF is in Fig. 8.11.

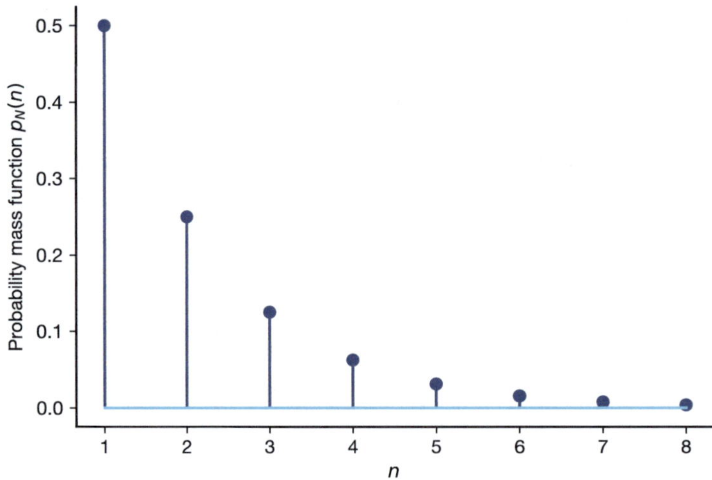

FIGURE 8.11
PMF of random variable N, which is the number of flips of a fair coin until heads occurs.

The probability mass function is easy to work with because:

- Its outputs are probabilities of the values that the random variable takes on.
- It is easy to calculate the probability of a discrete random variable taking on any of a small set of values by summing over the PMF outputs for those values.
- We know the range of the PMF is between 0 and 1 because its outputs are probabilities.

However, the probability mass function is not the best or most appropriate tool in all cases:

- It is not as convenient if we are interested in large intervals, such as if we want to evaluate $P(3 < N \leq 100)$. A computer can evaluate this range easily, but it is not easy to calculate by hand (even with a calculator).
- We will soon see that the PMF does not generalize to random variables for which the range is an uncountable set of values.

Thus, we are motivated to explore other functions that can address these last concerns. In the next section, we introduce the *cumulative distribution function (CDF)*, which instead of computing $\mathrm{P}(X = x)$, computes the probabilities of the form $\mathrm{P}(X \leq x)$. We will show that the CDF resolves the problems we identified with the PMF.

Terminology review

Interactive flashcards to review the terminology introduced in this section are available at fdsp.net/8-2, which can also be accessed using this QR code:

8.3 Cumulative Distribution Functions

We now introduce the *cumulative distribution function.* Recall that we wish to be able to assign probability to any of the Borel sets (of $\mathbb{R}$). In addition, the Borel sets can be formed from the countable unions, intersections, and complements of intervals of the form $(-\infty, x]$ for all real x. This motivates us to develop a function that assigns probabilities to all such sets. Noting that

$$\Pr\big(X \in (-\infty, x]\big) = \Pr(X \leq x),$$

we define the *cumulative distribution function* as follows:

> Definition
>
> **cumulative distribution function (CDF)**
>
> Let $(S, \mathcal{F}, P)$ be a probability space and X be a real RV on S. Then the **cumulative distribution function (CDF)** is the real function $F_X(x) = P(X \leq x)$.

As a reminder, $P(X \leq x)$ is an informal way of writing $P(\{s \,|X(s) \leq x\})$.

The cumulative distribution function may also just be called the *distribution function.* Since distribution functions evaluate to probability measures of some related events, they inherit many of the properties of probability measures.

WARNING

Be cautious about notation:

- The distribution function is written as an uppercase F; we will soon introduce a related function written as a lowercase f.
- We generally write the distribution function with a subscript that indicates which random variable it is associated with; i.e. $F_X()$ is the distribution function of the random variable X. Recall that random variables are written in uppercase.
- The argument of the distribution function refers to a value that the random variable can take on. When we are not referring to a specific value, we use the lowercase version of the random variable. Thus, we write $F_X(x)$ or $F_Z(z)$.
- I will write CDF in uppercase letters when abbreviating "cumulative distribution function" to remind the reader that this refers to the function written with an uppercase F, like $F_X(x)$.
- Some books call the CDF the probability distribution function, which they abbreviate as PDF. However, this is a source of much confusion for students when we introduce the probability density function, which is abbreviated pdf.

Note that if X is a discrete random variable, then

$$F_X(x) = P(X \leq x) = \sum_{k \leq x} p_X(k). \tag{8.1}$$

We now show how to find and plot the CDFs for some of our example random variables.

Example 8.10: CDF for Random Variable from Flipping a Fair Coin Once

Consider again the binary random variable X formed from flipping a fair coin, which we previously investigated in Examples 8.1 and 8.5.

To find the CDF, we can start by considering regions of the form $(-\infty, x]$ and plotting such a region onto a number line, as shown in Fig. 8.12. The shading on the number line indicates the region $(-\infty, x]$. Since that is an infinite region, the arrow shows that the region continues to $-\infty$. Also, note that the point x is included in the region, and a square bracket at that point indicates its inclusion.

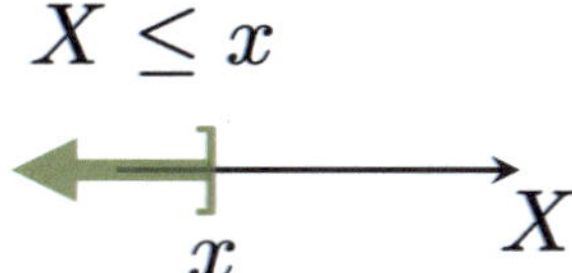

FIGURE 8.12
Region $X \leq x$.

WARNING

Many people learning probability initially have trouble understanding and drawing the different regions that we need to find the CDF. The primary confusion arises because we may often have **two** simultaneous inequalities involving x. These two inequalities have the following forms:

1. The CDF is $F_X(x) = P(X \leq x)$. Thus, **the first inequality that expresses the values of X relative to some threshold value x.** This inequality will always be represented by a shaded arrow that points left, toward $-\infty$. Thus, the first inequality will always be as shown in Fig. 8.12.

2. **The second inequality is on the value of x itself.** This inequality indicates what values of x we are considering when determining the value of the CDF. In terms of illustrating the shaded region that represents $X \leq x$, **the second inequality determines where we put the value x on the real line.**

The second inequality arises when we are trying to find the CDF because we often need to use different **cases** (in terms of values of x) because the region $X \leq x$ interacts with the probability of the random variable in different ways depending on the value of x.
So the problem is two-fold for students:

1. Two different inequalities are being considered simultaneously, and the meaning of each is sometimes confused.

2. It is confusing that we are illustrating regions like $X \leq x$ when the value of x is not a specific value.

It is hard to understand this from a general description – the best approach is to see how this works for specific examples.

For a general random variable U whose values fall within some finite range, $\text{Range}(U) \subseteq [a, b]$, the first case that we will usually consider will hold for any

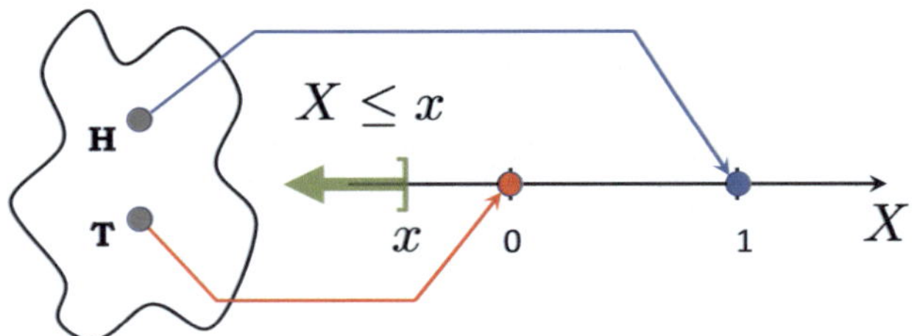

FIGURE 8.13
Region $X \leq x$ when $x < 0$ and relation to outcomes in S.

value of U less than a. Applying that to the random variable X that we are considering here, our first case is for any value of x less than 0. Because this case turns out to be very simple, we will refer to it as case 0:

Case 0:

For $x < 0$, we can illustrate the region $(-\infty, x]$ as shown on the number line in Fig. 8.13.

To find the probability that $X \in (-\infty, x]$ when $x < 0$, note that

$$P\big(X \in (-\infty, x]\big) = P(X \leq x) = F_X(x).$$

Then $F_X(x) = P(\{s\,|X(s) \leq x\})$. In other words, we wish to identify the outcomes s that give us values of X that are less than or equal to some value x, where we know that the value x is less than zero. Since the smallest value of $X(s)$ is $X(s) = 0$, there are no values of s that yield $X(s) \leq x$ when $x < 0$, so

$$F_X(x) = P(X \leq x) = 0, \quad x < 0.$$

Case 1

Next, we want to identify the next region of values of x for which either

- $F_X(x)$ will be constant (because all points in the region map back to the same outcomes in the sample space – this will be what happens for discrete random variables), or
- $F_X(x)$ can be represented by a single function of x for all x in that region – this is what happens for continuous random variables.

To understand what we mean by all points in the region mapping back to the same outcomes in the sample space, consider the following:

- Let A be the set of $s \in S$ such that $X(s) \leq 0$.
- Let B be the set of $s \in S$ such that $X(s) \leq 1/2$.

These two regions are illustrated in Fig. 8.14.

Both these regions contain a single point in Range(X); they contain $X = 0$. Thus $A = B = X^{-1}(0) = \{T\}$. Moreover, this will hold for **any** value of x in the range $[0, 1)$. So this yields our second value of $F(x)$:

$$F_X(x) = P\big(\{T\}\big) = \frac{1}{2}, \quad 0 \leq x < 1.$$

Case 2

Now we consider the next value where $F_X(x)$ changes. If $x = 1$, then $F_X(1) = P(X \leq 1) = 1$, since all of the values of X are less than or equal to 1. If we take any value of $x \geq 1$, then this will still hold because the region still includes the entire range of X, as shown in Fig. 8.15.

Since all values of $x \geq 1$ contain Range(X), then $X^{-1}\big((-\infty, x]\big) = S$. Thus,

$$F_X(x) = P(S) = 1, \quad x \geq 1.$$

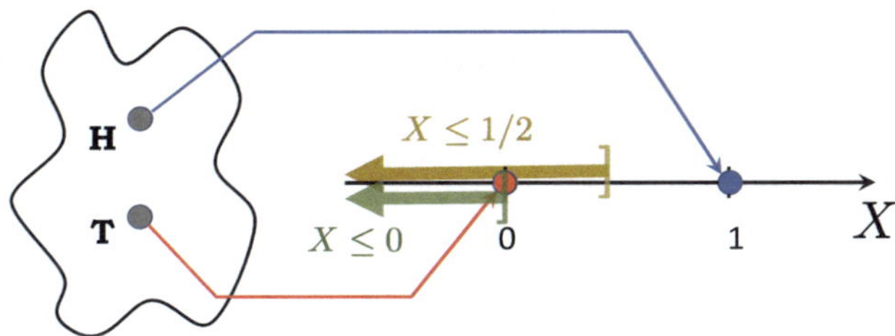

FIGURE 8.14
Regions $X \leq 0$ and $X \leq 1/2$ and relation to outcomes in S.

Putting It Together:

When asked to find the CDF for a random variable, you should provide a function that returns the probability $P(X \leq x)$ for every $x \in \mathbb{R}$. In this case, the function behaves differently for different values of x. This is very common for CDFs and other functions we will introduce later for working with random variables. Such functions are called *piecewise functions*:

Definition

piecewise function

A function that has a different functional relationship on different regions of its variable(s).

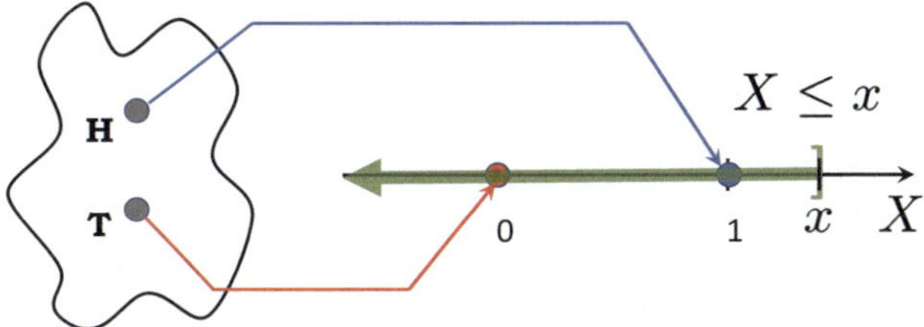

FIGURE 8.15
Region $X \leq x$ when $x \geq 1$ and relation to outcomes in S.

Example 8.10 (continued)

The function $F_X(x)$ can be specified piecewise as:

$$F_X(x) = \begin{cases} 0, & x < 0 \\ \frac{1}{2}, & 0 \leq x < 1 \\ 1, & x \geq 1. \end{cases}$$

Next, I present some useful techniques for evaluating and plotting distribution functions in Python. NumPy has a function `np.piecewise()` that can be used to define a piecewise function. Its call signature is

```
np.piecewise(x, condlist, funclist, *args, **kw)
```

We will only use the first three arguments:

- `x` is the value or array of values at which to evaluate the function.
- `condlist` is a list of conditions on `x` that specify the regions or conditions of the piecewise function.
- `funclist` is a list of functions of `x`, where each function matches with a corresponding condition in `condlist`.

The CDF above can be specified as:

```
import numpy as np

def F_X(x):
    return np.piecewise(x, [x<0, np.logical_and(x>=0, x<1), x>=1], [0, 1/2, 1] )
```

One part of this function requires some additional discussion. I needed to implement the condition $0 \leq x < 1$ for a NumPy array. However, because that is actually a combination of two conditions (i.e., $x \geq 0$ and $x < 1$), it cannot be implemented in a single logical command. Also, a single command like `x >=0` returns an array of Boolean values – see below for an example:

```
x=np.linspace(-2,2,5)
print (x, x>=0)
```

```
[-2. -1.  0.  1.  2.] [False False  True  True  True]
```

Thus, we want to combine the Boolean values from the two conditions using an elementwise and operator – that is the purpose of the `np.logical_and()`.

Since $F_X(x)$ is defined for all $x \in \mathbb{R}$, you might think that you can just call `plt.plot()` to plot it. However, the results may not be satisfying, as it will use sloped lines when it interpolates between points. We could reduce them by increasing the number of points, but a better way to resolve this problem is to use `plt.step()`, which changes the interpolation used by `plt.plot()`:

```
x = np.linspace(-1, 2, 61)
plt.step(x, F_X(x), where ='post')

plt.xlim(-1,2)
plt.xlabel('x')
plt.ylabel('$F_X(x)$');
```

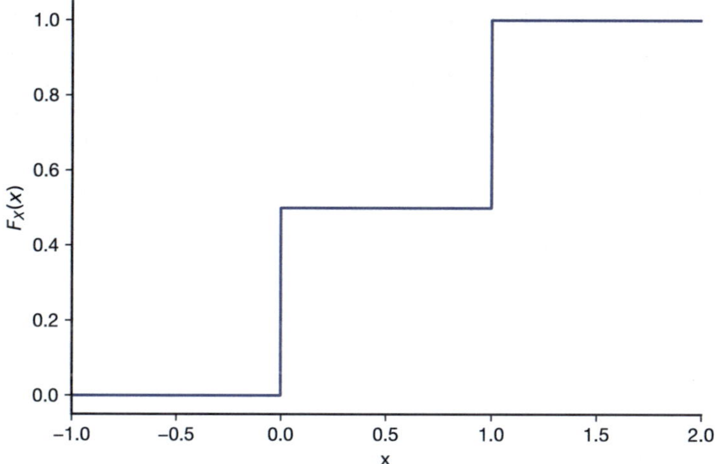

Note that we also pass `where = 'post'` to `plt.step()` because the probability is not achieved until we are at or past the values 1 and 2. This type of function is called a *staircase function*:

> Definition
>
> **staircase function**
>
> A staircase function is a piecewise-constant, nondecreasing function of its argument that has at most a countable number of pieces.

At each point that has probability greater than 0, the CDF will have a jump, and **the height of the jump is the probability at that point**. The CDF for any discrete random variable is a staircase function.

We can also find the CDF from the PMF using (8.1). Here is the PMF for X:

```
def p_X(x):
    x = np.array(x).astype(float)
    return np.piecewise(x, [x==0, x==1], [0.5, 0.5, 0])
```

We can find the CDF by first getting the values of $p_X(x)$ on a grid of points on the real line that spans the range **and** that includes all of the points in the range of X. We then evaluate the PMF at those values and then use `np.cumsum()` to return the cumulative sum up to each point, as shown in the following code:

```
x = np.linspace(-0.5, 1.5, int((1.5--0.5)*20)+1)
cdf_x = np.cumsum(p_X(x))
```

Now, we will jump to the last two examples from the previous sections. As we progress through this material, I will use more of a mathematical approach than a graphical and intuitive approach. If you are confused about a result, you are encouraged to draw diagrams such as those used in Example 8.1.

Example 8.11: CDF for Random Variable Representing Top Face on Roll of a Fair Die

Create a probability space by rolling a fair 6-sided die and observing the top face. Let the RV W be defined by $W(s) = s$ for all $s \in \{1, 2, 3, 4, 5, 6\}$.

Then $F_W(w) = P(W \leq w)$. Since Range$(W) = \{1, 2, 3, 4, 5, 6\}$, we know that these points will be important to find the values of the CDF. In fact, since all of the values of W that have nonzero probability are in Range(W), we can write

$$\begin{aligned} F_W(w) = P(W \leq w) &= P\big(W \in (-\infty, w) \cap \text{Range}(W)\big) \\ &= \sum_{k \in \{1,2,3,4,5,6\} \cap k < w} P(W = k) \\ &= \begin{cases} 0, & w < 1 \\ \frac{\lfloor w \rfloor}{6}, & 1 \leq w < 6 \\ 1, & w \geq 6 \end{cases}, \end{aligned}$$

where $\lfloor w \rfloor$ denotes the *floor* of w, which is the largest integer that is less than or equal to w. Intuitively, the CDF at any points $w \in \{1, 2, 3, 4, 5, 6\}$ is the cumulative probability at the points up to w. The use of the floor function allows us to create a function that returns the proper values at those points that are in-between those points where there are probability. The value of the CDF between the points that have probability will be the accumulated probability up to the lower endpoint.

Implementing this CDF in Python is a little tricky and therefore omitted. The CDF is shown in Fig. 8.16.

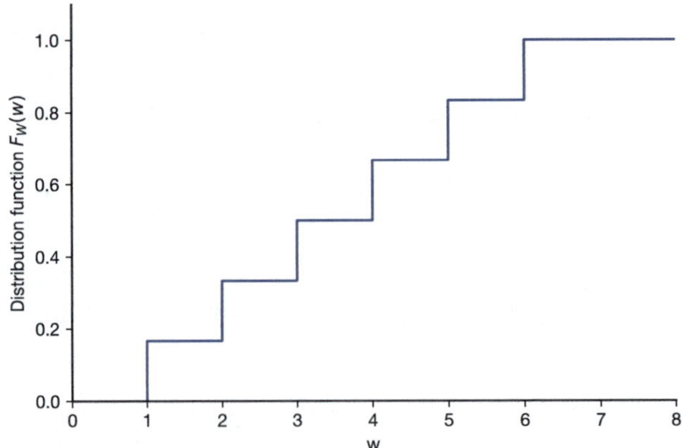

FIGURE 8.16
CDF for random variable representing value on top face of a fair six-sided die.

Example 8.12: CDF for Random Variable Representing Number of Coin Flips to First Heads

Flip a coin until heads occurs. Let N be a random variable that is equal to the number of flips until the first H occurs. A plot of the CDF $F_N(n)$ is shown in Fig. 8.17.

The CDF shows that the probability that the first heads is seen in 8 or fewer flips is over 0.99. We can easily check the last value of the vector `cdf_n`:

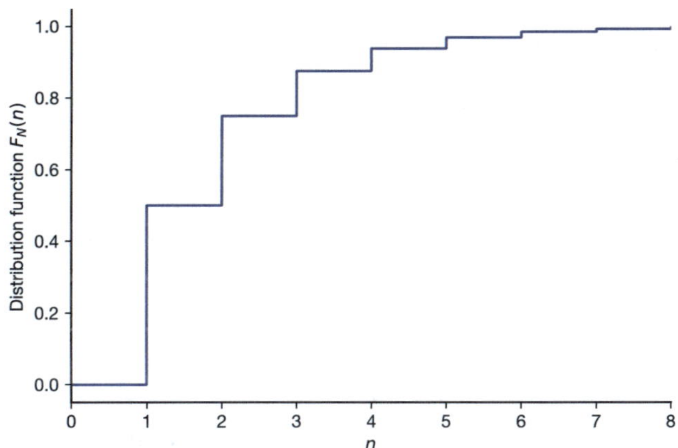

FIGURE 8.17
CDF for random variable representing number of flips of a fair coin until first heads.

```
n[-1], cdf_n[-1]
```

```
(8.0, 0.99609375)
```

8.3.1 Properties of CDFs

Some important properties of CDFs include:

1. $0 \leq F_X(x) \leq 1$

Proof: The distribution function returns a value from the probability measure of the underlying probability space.

2. $F_X(-\infty) = 0$ and $F_X(\infty) = 1$

Proof: The proof is beyond the scope of this book.

3. $F_X(x)$ is monotonically nondecreasing; i.e., $F_X(a) \leq F_X(b)$ iff $a \leq b$

Proof:

$$\begin{aligned} \mathrm{P}\{X \in (-\infty, b]\} &= \mathrm{P}\{X \in (-\infty, a] \cup X \in (a, b]\} \\ &= \mathrm{P}(X \in (-\infty, a]) + \mathrm{P}(X \in (a, b]), \end{aligned}$$

where the last step follows from the fact that $X(s)$ is a function, and hence disjoint values in the range of X must come from disjoint values in S. Thus,

$$F_X(b) = F_X(a) + \mathrm{P}(a < X \le b). \tag{8.2}$$

Since $\mathrm{P}(a < X \le b) \ge 0$, it must be that $F_X(b) \ge F_X(a)$.

4. $P(a < X \le b) = F_X(b) - F_X(a)$

Proof: Rewrite (8.2).

5. $P(X > x) = 1 - F_X(x)$

Proof:
If $A = \{s \in S \,|X(s) \le x\}$, then for all $X(s) > x$, it must be that $s \in \overline{A}$. Similarly, if $s \in \overline{A}$, then it must be that $X(s) > x$; otherwise, that s would have to belong to A. Thus, $\{s \in S \,|X(s) > x\} = \overline{A}$ and $A \cup \overline{A} = S$, so

$$\begin{aligned} P(X > x) &= P\left(\overline{A}\right) = 1 - P(A) \\ &= 1 - P(X \le x) = 1 - F_X(x). \end{aligned}$$

Probabilities of the form $P(X > x)$ arise often enough that this function is given a name:

Definition

survival function (SF)

Let $(S, \mathcal{F}, P)$ be a probability space and X be a real RV on S. Then the *survival function* is the real function $S_X(x) = P(X > x) = 1 - F_X(x)$.

6. When $F_X(x)$ has a jump discontinuity at a point $x = b$, then the height of the jump is $\Pr(X = b)$ and $F_X(b) = F_X(b^+)$, the limiting value from the right.

When the CDF of a random variable X has a jump discontinuity, there is a value of x where the CDF suddenly jumps up in value. For instance, the CDF $F_X(x)$ shown in Fig. 8.18 has a jump discontinuity at $x = 0$.

The height of a jump discontinuity at a point b is the difference between the limit from the left, written as $F_X(b^-)$, and the limit from the right, $F_X(b^+)$. Then

$$P(X = b) = F_X(b^+) - F_X(b^-).$$

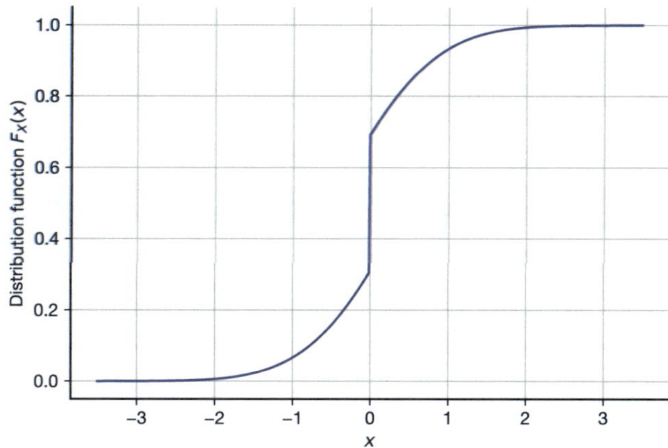

FIGURE 8.18
CDF $F_X(x)$ with jump discontinuity at $x = 0$.

For the example shown in the figure, the CDF jumps from $F_X(0^-) \approx 0.3$ to $F_X(0+) = 0.7$, so $P(X = 0) \approx 0.7 - 0.3 = 0.4$.

Since $F_X(b) = P(X \leq b)$, equality is included at b, and the value at the point $F_X(b) = F_X(b^+)$. For this example, $F_X(0) = 0.7$.

Proof: The proofs are outside the scope of this book.

In the following section, I introduce some common discrete random variables, show how to create and work with such random variables using the `scipy.stats` module, and discuss their application.

Terminology review

Interactive flashcards to review the terminology introduced in this section are available at fdsp.net/8-3, which can also be accessed using this QR code:

8.4 Important Discrete RVs

Discrete random variables are used in many statistics and engineering applications. We review the most common types below. We also show how to use the `scipy.stats` module to generate these random variables and work with the various functions that characterize the probability of these random variables. SciPy provides implementations of more than 120 types of random variables, including 19 types of discrete random variables (as of SciPy 1.10.1). To prepare for our use of `scipy.stats` we will import it as `stats`:

```
import scipy.stats as stats
```

8.4.1 Discrete Uniform Random Variable

In much of our previous work on probability, we considered fair experiments, where the experiment has a finite set of equally-likely outcomes. Here we extend this idea to a random variable:

> Definition
>
> **Discrete Uniform random variable**
>
> A random variable that has a finite number of values in its range.

For example, if we roll a fair die and let X be the number on the top face, then X is a Discrete Uniform random variable with

$$P_X(x) = \begin{cases} 1/6, & x \in \{1,2,3,4,5,6\} \\ 0, & \text{o.w.} \end{cases}$$

We can create a discrete uniform random variable in `scipy.stats` using `stats.randint()`. You should read the help for the `stats.randint` class:

```
?stats.randint
```

That help is long and complete, and it will be up-to-date with the latest version of `scipy.stats`. However, it is much more practical to refer to the web page scipy.stats.randint: https://docs.scipy.org/doc/scipy/reference/generated/scipy.stats.randint.html. I have included an image of the most important part below:

scipy.stats.randint

scipy.stats.randint = *<scipy.stats._discrete_distns.randint_gen object>* [source]

A uniform discrete random variable.

As an instance of the **rv_discrete** class, **randint** object inherits from it a collection of generic methods (see below for the full list), and completes them with details specific for this particular distribution.

Notes

The probability mass function for **randint** is:

$$f(k) = \frac{1}{\texttt{high} - \texttt{low}}$$

for $k \in \{\texttt{low}, \ldots, \texttt{high} - 1\}$.

randint takes `low` and `high` as shape parameters.

. . .

The range of random variables created with `stats.randint` consists of consecutive integers, and we specify the `low` and `high` values when creating a `stats.randint` object. As in most Python functions, the actual values that the random variable will take on will not include `high`; the actual highest value is `high - 1`.

Now let's see how to create and work with the Discrete Uniform distribution in Python. We will use an object-oriented (OO) approach in working with distributions in `scipy.stats`;

however, you will not need any prior OO knowledge to understand this. We will create an object for the desired distribution by calling the specified `scipy.stats` class as if it were a function. This is called a *constructor*, and we pass the desired distribution parameters as the arguments.

Let's create a random variable that represents the value on the top face when rolling a fair 6-sided die. It is conventional to use U for such a random variable if that does not conflict with other random variables' names. In Python, we can create an object to model this random variable as follows:

```
U = stats.randint(1, 7)
```

Alternatively, we can pass `low` and `high` as keyword parameters if we want to be more explicit:

```
U = stats.randint(low=1, high=7)
```

`U` is an *object*, and it has *methods* to work with the Discrete Uniform random variable with the given parameters. **Methods are just like functions, except that they belong to an object, and their behavior is affected by the internal attributes (i.e., properties) of the object.**

For instance, when we created `U`, we set its attributes to generate uniform values from 1 to 6 (inclusive). You can use Python's help function to see the methods of `U` (I have compressed the output for space reasons below):

```
help(U)
```

```
Help on rv_discrete_frozen in module scipy.stats._distn_infrastructure object:

class rv_discrete_frozen(rv_frozen)
 |  rv_discrete_frozen(dist, *args, **kwds)
 |
 |  Methods defined here:
 |
 |  logpmf(self, k)
 |  pmf(self, k)
 |  ----------------------------------------------------------------------
 |  Methods inherited from rv_frozen:
 |
 |  cdf(self, x)
 |  entropy(self)
 |  expect(self, func=None, lb=None, ub=None, conditional=False, **kwds)
 |  interval(self, confidence=None, **kwds)
 |  isf(self, q)
 |  logcdf(self, x)
 |  logsf(self, x)
 |  mean(self)
 |  median(self)
 |  moment(self, order=None, **kwds)
 |  ppf(self, q)
 |  rvs(self, size=None, random_state=None)
```

(continues on next page)

(continued from previous page)

```
 |  sf(self, x)
 |  stats(self, moments='mv')
 |  std(self)
 |  support(self)
 |  var(self)
```

A few of these should look familiar: `pmf()`, `cdf()`, and `sf()` refer to the same functions that we abbreviated in Section 8.2 and Section 8.3: the probability mass function, cumulative distribution function, and survival function, respectively. Each of the methods can be called by appending it to the object name after a period, followed by parentheses. Any parameters or values for the method should be given in the parentheses.

We can get the interval containing the range of a random variable in `scipy.stats` using the `support()` method:

```
U.support()
```

```
(1, 6)
```

WARNING

Note that `support()` returns the lowest value in the range and the highest value in the range, so be careful in using this method. If these values were used as arguments to create a new `stats.randint` object, that object would have a different range!

We can evaluate the PMF at any value:

```
U.pmf(3)
```

```
0.16666666666666666
```

These methods can also take lists or vectors as their arguments:

```
import numpy as np

uvals = np.arange(1, 7)
U.pmf(uvals)
```

```
array([0.16666667, 0.16666667, 0.16666667, 0.16666667, 0.16666667,
       0.16666667])
```

We can use the `rvs()` method to draw random values from this random variable. The argument is the number of random values to generate:

```
num_sims = 10_000
u = U.rvs(num_sims)
print(u[:20])
```

```
[4 4 4 2 1 5 6 1 2 1 2 1 3 4 3 2 3 4 4 1]
```

When working with discrete random variables, we can get the relative frequencies from `plt.hist()` by passing the `density=True` keyword argument, **provided the random variable is defined on the integers and bins of width 1 are used**. We will use bins of width 1 that are centered on each value of the random variable. The following code plots the normalized histogram with the PMF plotted on top using a stem plot:

```
newbins = np.arange(0.5, 7.5)
plt.hist(u, bins=newbins, density=True, color='C0',  edgecolor='black')
plt.stem(uvals, U.pmf(uvals), linefmt='C1')
plt.xlabel("$u$")
plt.ylabel("$P_U(u)$");
```

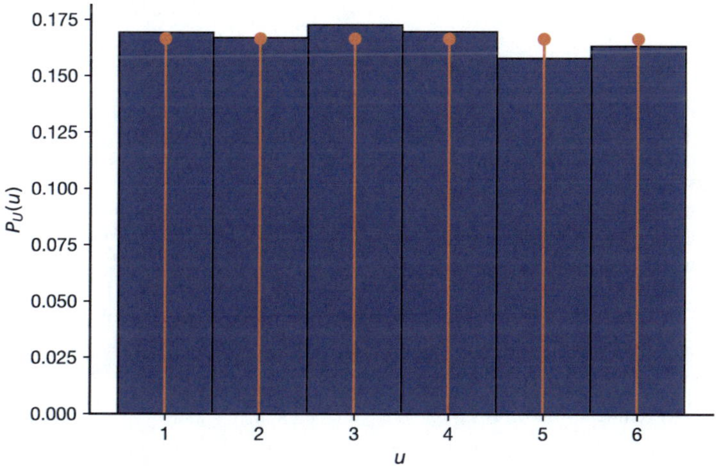

As expected, the relative frequencies match the PMF values closely. The following code plots the cumulative histogram (with both `cumulative = True` and `density = True`) along with the CDF:

```
uvals2 = np.arange(0, 7.1, 0.1)
plt.hist(u, cumulative=True, density=True, bins=newbins, alpha=0.5)
plt.step(uvals2, U.cdf(uvals2), where="post")

plt.xlabel("$u$")
plt.ylabel("$F_U(u)$");
```

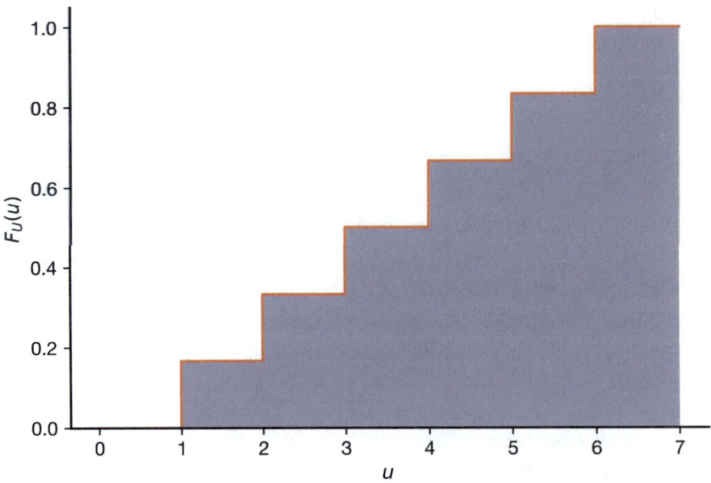

The cumulative histogram and CDF match almost exactly.

Exercise:

What happens if we plot the cumulative histogram without specifying the bins? Why do you think that happens?

8.4.2 Bernoulli Random Variable

The Bernoulli random variable is another of the simplest of the discrete random variables, and it is the simplest random variable that can take on unequal probabilities. We give an informal definition:

Definition

Bernoulli random variable

A Bernoulli random variable B takes on values of 0 or 1. It is specified by a parameter p such that $P(B = 1) = p$ and $P(B = 0) = 1 - p$.

Formally, we can define the Bernoulli random variable as follows:

Let $(S, \mathcal{F}, P)$ be a probability space. Let $A \in \mathcal{F}$ be an event and define $p = P(A)$; for instance, A may be an event corresponding to a "success". Then define the Bernoulli random variable B by

$$B = \begin{cases} 1, & s \in A \\ 0, & s \notin A \end{cases}$$

The PMF for a Bernoulli RV B is

$$p_B(b) = \begin{cases} p, & b = 1 \\ 1 - p, & b = 0 \\ 0, & \text{o.w.} \end{cases}$$

The random variables in Examples 8.1 and 8.2 are Bernoulli random variables with different values of p. Many other phenomena may be modeled as Bernoulli random variables.

Some engineering examples include: the value of an information bit, a value indicating whether a bit error has occurred, a failure indicator for some component of a system, or a detection indicator for a sensor.

We introduce the concept of the *distribution* of a random variable and some related notation below:

Definition

distribution (of a random variable)

The *distribution* of a random variable is a characterization of how the random variable maps sets of values to probabilities. For instance, the distribution may refer to a particular type of random variable along with whatever parameter(s) are required to completely specify the probabilities for that type of random variable.

Terminology and Notation for Distribution of a Random Variable

When the distribution refers to a particular type of random variable, we write either that a random variable has that distribution or that the random variable is *distributed* according to that type. For example, if X is a Bernoulli random variable with $p = P(X = 1)$, then we say that X has a Bernoulli(p) distribution or that X is distributed Bernoulli(p). Both of these statements have the same meaning, and we will denote this in shorthand notation as

$$X \sim \text{Bernoulli }(p).$$

We now show how to work with the Bernoulli random variable using `scipy.stats`. First, review the help page for `stats.bernoulli`:

```
?stats.bernoulli
```

For instance, the following Python code will create a Bernoulli random variable B_1 with probability $P(B_1 = 1) = 0.2$:

```
B1 = stats.bernoulli(0.2)
```

`B1` is now an object that represents a Bernoulli random variable with parameter $p = 0.2$. If you check its help page, you will see it has the same methods as for `randint` objects.

The following code shows the results of the `pmf()` method:

```
b = np.arange(-1, 3)
B1.pmf(b)
```

```
array([0. , 0.8, 0.2, 0. ])
```

Let's simulate 100,000 values of the Bernoulli(0.2) random variable and plot a normalized histogram (with `density=True`) of the values. We will also overlay the histogram with a stem plot of the PMF.

```
num_sims = 100_000

# Generate RVs
b = B1.rvs(num_sims)

# Plot histogram
mybins = np.arange(-0.5, 2.5, 1)
plt.hist(b, bins=mybins, density=True, alpha=0.5)

# Plot density
bvals = [0, 1, 2]
plt.stem(bvals, B1.pmf(bvals), linefmt='C1' );
```

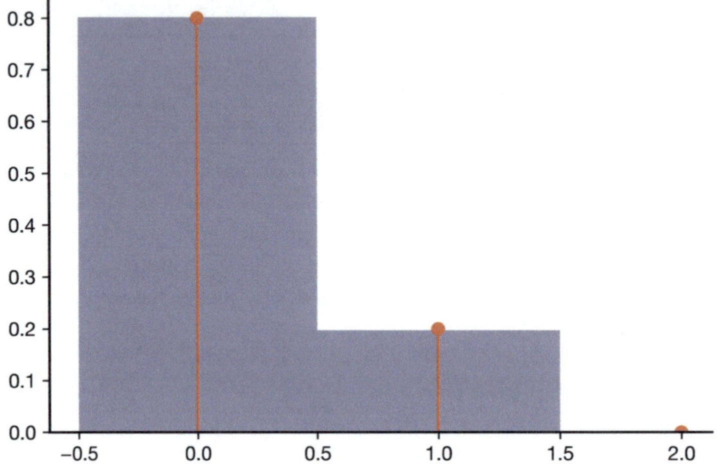

8.4.3 Binomial Random Variable

One of the next most common types of discrete random variables arises when we have N repeated, independent Bernoulli trials with identical probability of success p. Let B_2 be the number of successes. Then B_2 is a Binomial(N, p) random variable. For example, in our very first experiments in Section 2.1, we considered the probability of seeing six or fewer heads when a fair coin is tossed 20 times. We can model the number of heads as a Binomial (20, 0.5) random variable.

Note that we can also think of a Binomial (N, p) random variable as the **sum** of n independent Bernoulli(p) random variables.

In shorthand notation, we will write

$$B_2 \sim \text{Binomial } (N, p).$$

The PMF for a Binomial random variable is easily derived:

- For N trials with probability of success p, the probability of a particular ordering of k successes is $p^k(1-p)^{N-k}$.
- The number of different orderings of k successes and $N-k$ failures in N total trials is

$$\binom{N}{k}.$$

Thus, the probability of getting k successes on N independent Bernoulli (p) trials is

$$\binom{N}{k} p^k (1-p)^{N-k}.$$

Here, the variable k was used instead of b because k is used widely in practice to represent the number of successes for a Binomial random variable.

We summarize our definition of the Binomial random variable below:

Definition

Binomial random variable

A Binomial random variable B represents the number of successes on N independent Bernoulli trials, each of which has probability of success p. The probability mass function for the Binomial random variable is

$$p_B(k) = \begin{cases} \binom{N}{k} p^k (1-p)^{N-k}, & k = 0, 1, \ldots, N \\ 0, & \text{o.w.} \end{cases}$$

Some engineering examples include the number of bits in error in a data packet sent over a noisy communication channel or the number of defective items in a manufacturing run.

A binomial random variable can be created in `scipy.stats` using `stats.binom()`. The number of trials and probability of success must be passed as the arguments:

```
B2 = stats.binom(10, 0.2)
```

```
k = range(0, 11)
plt.stem(k, B2.pmf(k));
```

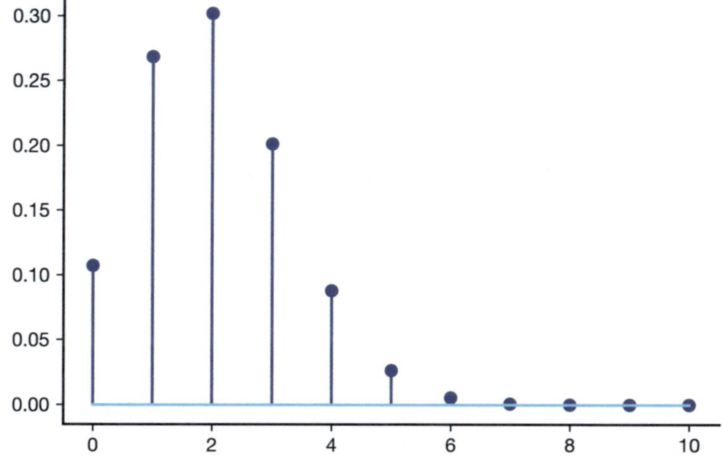

Intuitively, if p is small (such as $p = 0.2$ in the example above), then most of the probability mass will be around the small values of the random variable. Moreover, most people have the sense that if we conduct N trials with probability p, then the most likely outcomes will be around Np. We see that this is true for our example of $N = 10$ and

$p = 0.2$: the value with the highest probability is $Np = 2$. If we increase p to 0.6, we get the following PMF:

```
B3 = stats.binom(10, 0.6)
plt.stem(k, B3.pmf(k));
```

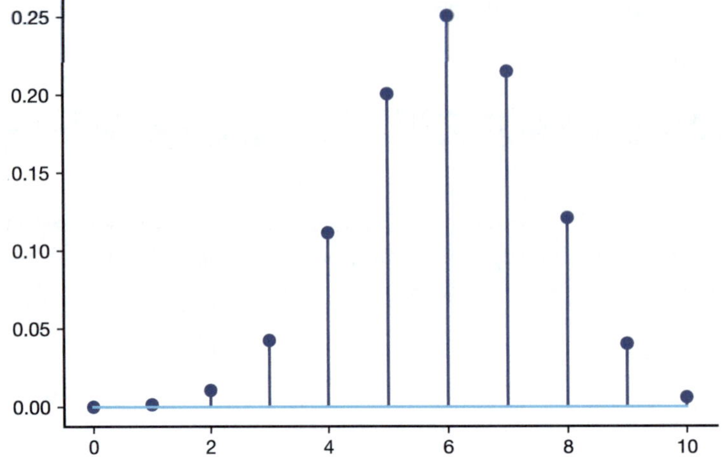

Again, $Np = 6$ is the most common value. Let's try $p = 0.75$, for which $NP = 7.5$ is not a possible value of the random variable:

```
B4 = stats.binom(10, 0.75)
plt.stem(k, B4.pmf(k));
```

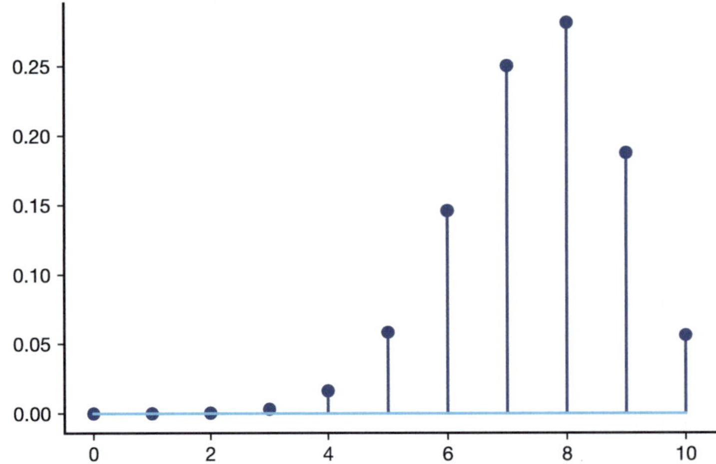

For the Binomial(10, 0.75) random variable, the value with the highest probability is 8, which is close to $Np = 7.5$.

Finally, let's generate some values from the Binomial(10, 0.75) distribution and compare a cumulative histogram (with `density=True`) to the CDF:

```
b4 = B4.rvs(100_000)
mybins = range(0, 12)
plt.hist(b4, bins=mybins, cumulative=True, density=True, alpha=0.3)
```

(continues on next page)

(continued from previous page)

```
plt.step(k, B4.cdf(k), where='post')
plt.xlim(0, 11);
```

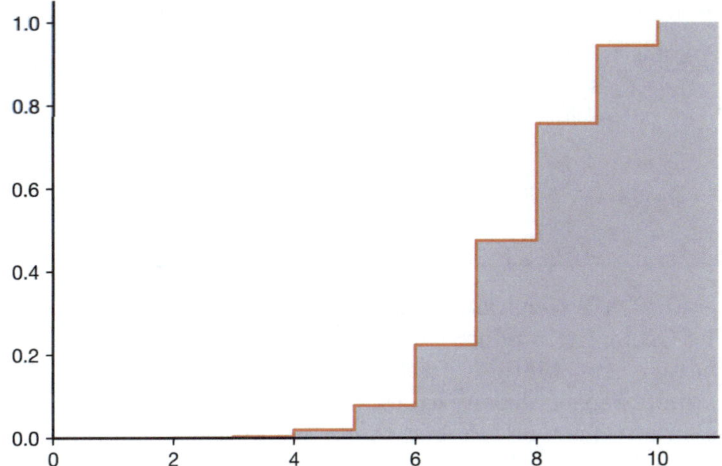

Note that the region where the CDF is increasing quickly is the region where most of the probability is concentrated (since the size of the jumps is equal to the probabilities at the location of the jumps).

Before we leave the binomial random variable, let's observe one more behavior. Consider how the PMF looks for a large number of trials. I have zoomed in on the section that contains most of the probability:

```
B5 = stats.binom(1000, 0.2)
b5vals = np.arange(0, 1001)
plt.stem(b5vals, B5.pmf(b5vals))
plt.xlim(150,250);
```

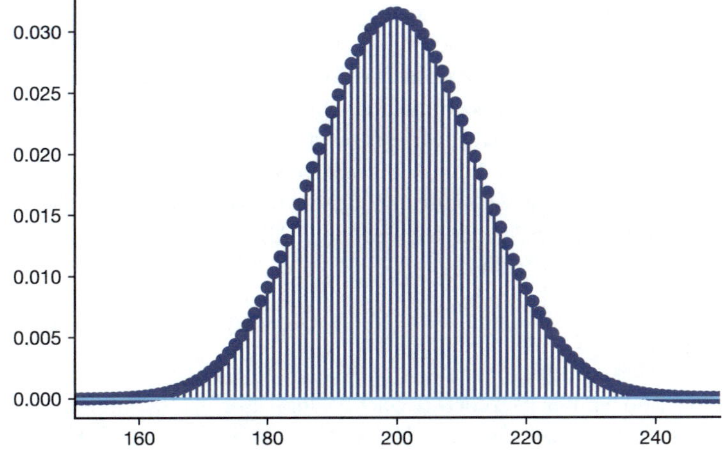

This shape is called a *bell shape* or *bell curve*. It plays an important role in probability and statistics, but we do not have the tools to explore it in detail yet.

8.4.4 Geometric Random Variable

Definition

Geometric random variable

If independent Bernoulli (p) trials are conducted until the first success, the number of trials required is a *Geometric random variable.* We write $X \sim$ Geometric(p), and the probability mass function for G is

$$p_G(k) = \begin{cases} p(1-p)^{k-1}, & k = 1, 2, \ldots \\ 0, & \text{o.w.} \end{cases}.$$

We already encountered such a scenario in Example 8.5, where we flipped a fair coin until the first heads. The Geometric random variable generalizes that example to handle Bernoulli trials with arbitrary probabilities. Unlike the Binomial, the Geometric random variable does not have a finite range; that is, we cannot specify any particular maximum number of Bernoulli trials that might be required to get the first success. However, since the range is the counting numbers $1, 2, 3, \ldots$, it is countably infinite.

It may seem that if the range of a random variable is an infinite set, then it will not be possible to assign a nonzero probability to every outcome. However, the Geometric random variable shows that this is not true. It is possible to assign nonzero probabilities to a discrete random variable with a countable number of outcomes, provided that the probabilities go to zero fast enough. Note that

$$\begin{aligned} \sum_{k=1}^{\infty} p(1-p)^{k-1} &= p \sum_{m=0}^{\infty} (1-p)^m \\ &= p\left(\frac{1}{p}\right) = 1, \end{aligned}$$

so the total probability assigned to all the outcomes sums to 1.

An example of this is the number of transmissions required for a packet to be successfully received when transmitted over a noisy channel and retransmitted whenever the received version is corrupted by noise. Another example is the number of people at a data science conference that a publisher must talk to before finding one who does not use Python.

A geometric random variable can be created in `scipy.stats` as `stats.geom()`, where the argument is the probability of success for the Bernoulli trials. See the help for a list of methods, which are similar to those available for the other discrete random variables that we have introduced. The code below creates a Geometric (0.2) random variable and draws 100,000 random values of that random variable.

```
G = stats.geom(0.2)
g = G.rvs(100_000)
```

A normalized histogram of the values is plotted on the left in Fig. 8.19, along with the PMF. The right plot in Fig. 8.19 shows the cumulative histogram and the CDF.

Rather than mathematically derive the CDF for a geometric random variable, we will determine it from a simple argument that is easy to remember. Consider instead the probability $P(G > k)$, which is the value of the survival function for G with argument k. If $G > g$, more than g trials are required **because there have been no successes in the**

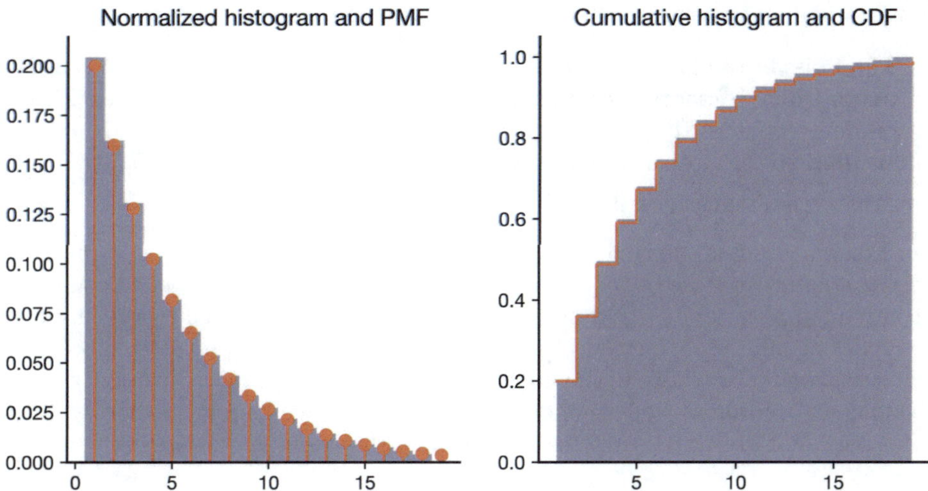

FIGURE 8.19
Histograms for samples from geometric distribution compared with theoretical PMF and CDF.

first g **trials**. Thus, we can calculate $P(G > g)$ as

$$\begin{aligned} P(G > g) &= \Pr(\text{no successes on } g \text{ consecutive independent Bernoulli}(p) \text{ trials}) \\ &= (1-p)^g. \end{aligned}$$

Then the CDF for G is

$$P(G \leq k) = 1 - P(G > k) = 1 - (1-p)^g.$$

8.4.5 Poisson Random Variable

Poisson is French and roughly pronounced "pwah - sahn". It is named after the French mathematician Siméon Denis Poisson (Wikipedia article on Siméon Denis Poisson: https://en.wikipedia.org/wiki/Siméon_Denis_Poisson).

The following story told by the famous Swiss psychiatrist Carl Jung in "Synchronicity: An Acausal Connecting Principle" about observations he made starting on April 1, 1949, has some interesting connections to the Poisson random variable:

> We have fish for lunch. Somebody happens to mention the custom of making an "April fish" of someone. That same morning I made a note of an inscription which read: "Est homo totus medius piscis ab imo". [Rough translation: It is man from the middle, fish from the bottom.] In the afternoon a former patient of mine, whom I had not seen for months, showed me some extremely impressive pictures of fish which she had painted in the meantime. In the evening I was shown a piece of embroidery with fish-like sea-monsters in it. On the morning of April 2 another patient, whom I had not seen for many years, told me a dream in which she stood on the shore of a lake and saw a large fish that swam straight towards her and landed at her feet. I was at this time engaged on a study of the fish symbol in history.

Seeing six fish in a 24-hour period seemed unusual. However, he recognizes that this must be assumed as a "meaningful coincidence" unless there is proof "that their incidence exceeds the limits of probability".

Now, you should be asking: what does this have to do with the Poisson random variable?

1. The Poisson random variable can help us answer questions like "**What is the probability of seeing six fish in 24 hours?**" and can be combined with other random variables to answer a question like "**What is the probably of seeing six fish in 24 hours at least sometime in a 20 year period?**"
2. *Poisson* is French for... **fish!**

The Poisson random variable is used to model phenomena that occur randomly over some fixed amount of time or space. For convenience of discussion and because it is the most common application, we will only consider periods of time, but everything we discuss below applies equally well to events that occur at some random rate over space.

A Poisson random variable is the number of occurrences given the average rate at which the phenomena occur and the length of time or area of space being considered. We will use the following notation for parameters associated with a random variable:

- λ is the average rate of occurrences over time or space,
- T is a length of time being considered, and
- $\alpha = \lambda T$ is the *average* number of occurrences over T.

If α is known, then it is not required to know λ and T separately.

Given the parameter(s), we can define a Poisson random variable in terms of its PMF:

Definition

Poisson random variable

A random variable X that models events that occur randomly over some fixed interval of time (or space). If α is the average number of events that occur over the interval, then the PMF of X is

$$p_X(x) = \begin{cases} \frac{\alpha^k}{k!} e^{-\alpha}, & k = 0, 1, \ldots \\ 0, & \text{o.w.} \end{cases}$$

Note that the range of a Poisson random variable is from 0 to ∞. The shape of the Poisson varies depending on its parameter, α. Consider first the PMF for a Poisson random variable with $\alpha = 0.5$:

```
P1 = stats.poisson(0.5)
p1vals = range(5)
plt.stem(p1vals, P1.pmf(p1vals));
```

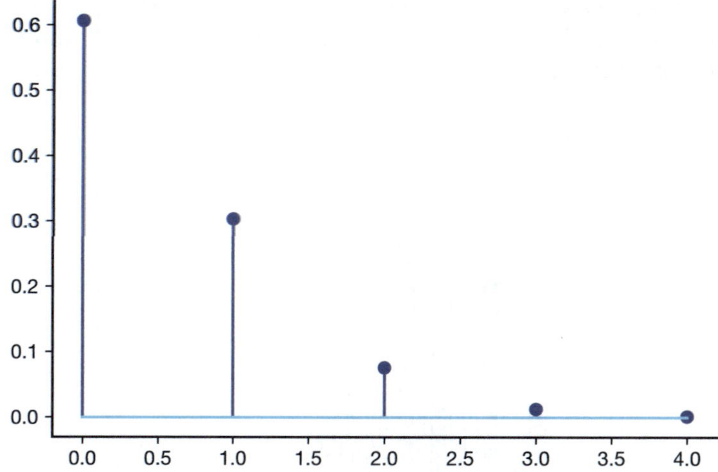

To experiment with the different shapes that the Poisson distribution can take on, download and run this notebook for an interactive plot that has a slider to see the PMF for different values of the parameter α:

https://www.fdsp.net/notebooks/important-discrete-rvs-widget.ipynb

There are basically three different cases:

- For $\alpha < 1$, the PMF is a strictly decreasing function of its argument,
- For $\alpha = 1$, then the PMF is a nonincreasing function of its argument: it has the same value at 0 and 1, and then decreases for all larger values,
- For $\alpha > 1$, the PMF first increases and then decreases. If α is an integer, then the PMF has the same value for arguments α and $\alpha - 1$.

Illustrations of the second two cases are shown in Fig. 8.20.

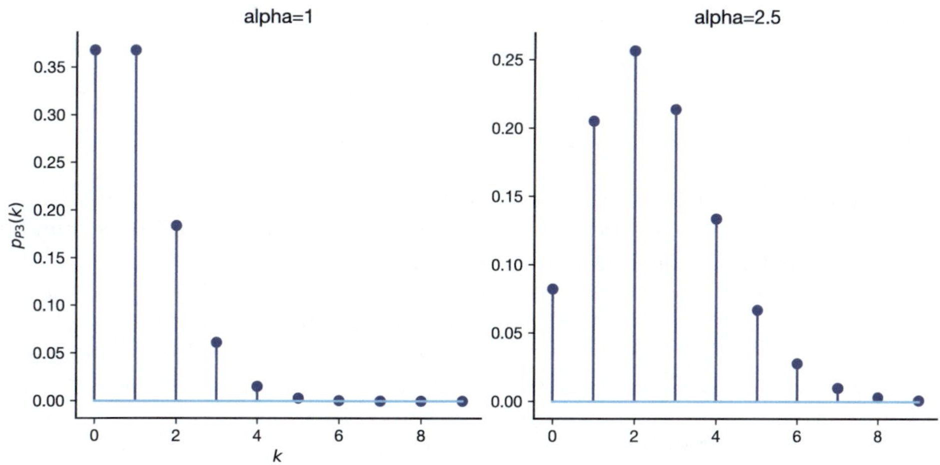

FIGURE 8.20
PMFs for Poisson random variables with $\alpha = 1$ and $\alpha = 2.5$.

For large α, the PMF takes on a familiar shape; for instance, the PMF for a Poisson random variable with $\alpha = 100$ is shown in Fig. 8.21. Interestingly, we get another bell shape, just as we saw with the Binomial random variable.

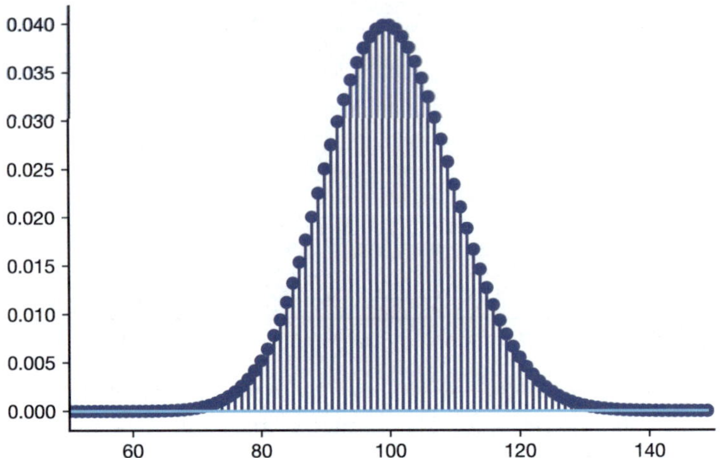

FIGURE 8.21
PMFs for Poisson random variable with $\alpha = 100$.

Now, let's see how to apply the Poisson random variable to a practical problem.

Example 8.13: An Unusual Number of Shark Attacks?

In this example, we will show how even minimal information provided in a news article can be used to conduct a statistical test when we can apply a model for the data. In the article "Man Bitten in Florida's 'Shark Capital of the World": https://www.theinertia.com/environment/man-bitten-in-floridas-shark-capital-of-the-world/, the town of New Smyrna, Florida is referred to as the "shark capital of the world". While mostly focusing on details of a particular shark attack in the beginning of 2022, the article provides the following information about shark bites in the same county in 2021:

> According to the International Shark Attack File (ISAF), Volusia County averages nine attacks per year but reported 17 attacks in 2021.

The much higher number of shark attacks than the average could either indicate a concerning trend or could be attributable to the randomness in the number of shark attacks that occur in a given year. How can we tell?

To try to get an answer to this, we have to first acknowledge that the article provides very little data: other than the number of shark attacks in 2021, we are only given the average number of shark attacks per year. With such limited data, we have two choices:

1. We can introduce a model, by which we mean that we assume that the data comes from a particular distribution that can be completely specified using the available information, or
2. We can find more detailed data, and use that to conduct a statistical test.

Let's use the model-based approach. Our previous discussion of fish may make us think of the Poisson distribution, but in general, we can ask several questions to determine whether the Poisson is a reasonable model:

- **Is there a specific maximum value that the data can take on? The answer should be no for the Poisson.** For instance, there is no way to upper bound the number of shark bites that might happen in a given year.
- **Do the individual occurrences occur randomly over time or space, and can we specify some rate of occurrence for the given data? The answer should be yes. If the rate is not given, we can usually estimate it from the data.** For the shark bite data, shark bites occur at random times.

Based on the information in the news article, let's assume that the number of shark attacks in Volusia County in a year is Poisson with $\alpha = 9$ representing the average number of shark attacks per year. Let's use Scipy.stats to create a Poisson object with this parameter:

```
S = stats.poisson(9)
```

Now we will see how to conduct a statistical test using this model. We can set up the following NHST:

- H_0: The observation comes from the Poisson (9) distribution.
- H_1: The observation comes from some other distribution.

We will use our usual p-value threshold of 0.05.

Under the NHST, we want to find the probability of seeing a result that is at least as extreme as the observed number under H_0. In this case, it makes sense to conduct a one-sided test, where we determine the probability of seeing 17 **or more** attacks in a year.

Note that $P(S \geq 17) = P(S > 16)$. This allows us to calculate this probability using the survival function:

```
S.sf(16)
```

```
0.011105909377583819
```

This probability is smaller than 0.05, so we reject H_0 under this test. The observed increase in shark bits in 2021 can be considered statistically significant at the $p < 0.05$ level.

However, the article might not have included this information if the result had not seemed significant in the first place. We might instead ask the probability of seeing such an extreme result at least once in a decade. We can model this as a binomial random variable with 10 trials, each of which has probability of "success" $P(S > 16)$. Let's call this binomial random variable S_2, and we will create a `scipy.stats` object to represent its distribution:

```
S2 = stats.binom(10, S.sf(16))
```

Then the probability of having at least one year with 17 or more shark attacks in a decade is $P(S_2 > 0)$, which is

```
S2.sf(0)
```

```
0.105669964152784
```

Thus there is more than a 10% chance that there will be at least one year with 17 or more shark attacks in Volusia County over a decade. Given this, we would not be able to reject the possibility that the 2021 result comes from the Poisson(9) distribution. It may just be attributable to the random nature of the number of shark attacks in a year.

Let's use this example to work with one more of our distributions. What is the probability that it is more than 10 years before there is another year with 15 or more shark attacks?

We can model the number of years until there is a year with 15 or more shark attacks as a Geometric random variable with probability of "success" $P(S > 14)$. Let S_3 denote this random variable, and we can create a Scipy.stats object to represent its distribution:

```
S3 = stats.geom(S.sf(14))
```

Then the probability that it is more than 10 years before there is another year with 15 or more shark attacks is $P(S_3 > 10)$, which is

```
S3.sf(10)
```

```
0.6547473480899506
```

There is over a 65% chance that we will not have a year with 15 or more shark attacks in the next decade.

8.4.6 Arbitrary Discrete RVs

We can create a discrete random variable with a finite set of integer values using the `stats.rv_discrete()` method:

```
?stats.rv_discrete
```

Note that this method implicitly assumes that the discrete random variable is defined on a subset of the integers.

Let's use this method to create a random variable based on Example 8.3, which uses two flips of a fair coin to create a random variable with values 0, 1, or 2. We need to pass a tuple to the constructor that includes the following:

- a list or vector containing the random variable's range, and
- a list or vector containing the probabilities of each point in the random variable's range.

The following code creates an object `A` with this distribution and uses that object to draw 100,000 random values from this distribution. The normalized histogram of the values is compared with the PMF of this random variable (using `A.pmf()`) in the left plot of Fig. 8.22. The normalized cumulative histogram of the values is compared with the CDF in the right plot of Fig. 8.22. As expected, the theoretical results almost exactly match the empirical histograms (the shaded regions).

```
range1 = [0, 1, 2]
probs1 = [1 / 4, 1 / 2, 1 / 4]

A = stats.rv_discrete(values=(range1, probs1))
a = A.rvs(size=100_000)
```

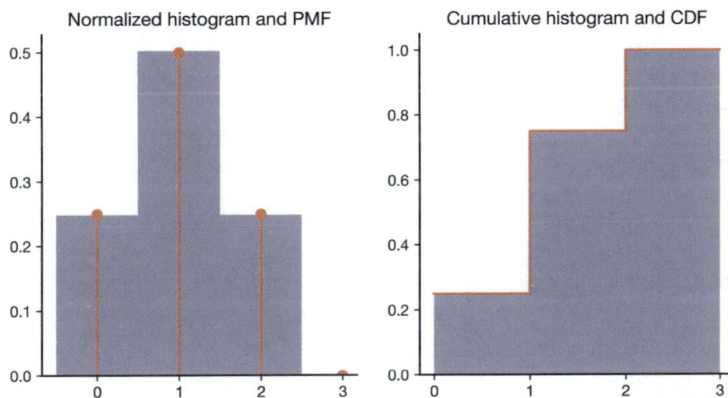

FIGURE 8.22
Histograms for samples from custom distribution compared with theoretical PMF and CDF.

Terminology review and self-assessment questions

Interactive flashcards to review the terminology introduced in this section and self-assessment questions are available at fdsp.net/8-4, which can also be accessed using this QR code:

8.5 Continuous Random Variables

The discrete random variables introduced in the last two sections can model many random phenomena of interest; however, there are also many phenomena that they cannot model. Let's begin with an example based on our study of the Poisson random variable in Section 8.4.

8.5.1 Motivating Example – Poisson Events

The Poisson random variable models the number of events that occur in some time period for a random phenomenon with events that occur randomly over time at some specified rate λ. However, the Poisson random variable does not tell us **when** those events occur.

Let's first try to get some insight into the distribution of the event times by simulating Poisson events. Let P be the number of Poisson events in an interval. Then given $P = p$ (i.e., given the number of Poisson events is some specific value p to be specified later), the p events should be equally likely to be anywhere in the observed interval. Let's simulate this scenario for the shark attack example from Section 8.4. The average number of shark attacks per year in Volusia County was given in the referenced article as 9. Let's treat the interval as 365 days instead of 1 year, so that the resulting times are not all small decimals.

The first thing to note is that the event times are **not discrete**. There is no need to limit our model of when events can occur to just the day, hour, or minute – the events can occur at any nonnegative real value in the observation interval. Let T_i, $0 \leq i \leq p-1$, be the random variable representing the time that event i occurs. If T_i is equally likely to be any time in that interval, then T is a *Uniform Continuous random variable*, which is usually just called a *Uniform random variable.*

We will begin using the Uniform random variable without formally defining it. Later, we will determine the important functions that characterize the Uniform random variable and will provide a careful definition at that point.

We can create a SciPy.stats object representing a continuous random variable on the interval $[0, 365]$ as follows:

```
import scipy.stats as stats

T = stats.uniform(0, 365)
```

Let's draw 10 random times from this interval:

```
T.rvs(10)
```

```
array([ 98.86230412, 279.39487612, 324.81884169,  96.58236547,
       110.24624772, 300.72179977, 129.08778775, 100.93429569,
       195.12634819, 236.44633485])
```

We can note a few things:

- Each random variable is drawn separately and independently, and the resulting values are not sorted.
- The values are not integers.
- Although the values are meant to represent values from $[0, 365]$, they are limited to values that can be represented by Python's `float` type, which has limited resolution.

Let's draw 100,000 random values from this distribution and plot the histogram of values drawn from this distribution:

```
import matplotlib.pyplot as plt

ts = T.rvs(100_000)
plt.hist(ts, bins = 100);
```

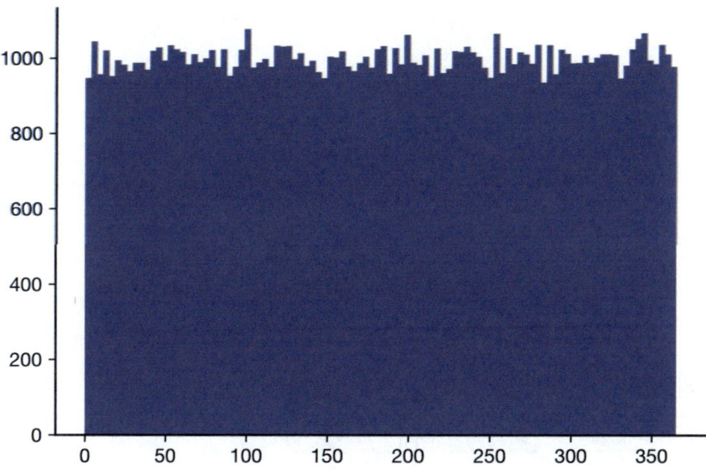

The relative frequencies are approximately equal for all of the bins, which we expect if the random values are equally likely to be anywhere in the interval.

Now let's plot the empirical CDF, which we can get by passing `density = True` and `cumulative = True` as parameters to `plt.hist()`.

Let's see if we can determine the CDF for T. If T is equally likely to be any value in $[0, 365]$, then if we choose any interval $[a, b] \subseteq [0, 365]$, then the probability that $T \in [a, b]$ should be proportional to the length $b - a$. Since the probability that T is any value in $[0, 365]$ is 1, it must be that

$$P\left(T \in [a, b]\right) = \frac{b - a}{365 - 0}, \text{ if}[a, b] \subseteq [0, 365].$$

In particular, note that $P(T \leq t) = P(T \in [0, t])$ if $t \in [0, 365]$. Thus

$$F_t(t) = \frac{t - 0}{365} = \frac{t}{365}, \quad 0 \leq t \leq 365,$$

which is shown overlaid on the empirical CDF below:

```
import numpy as np

# Empirical CDF
plt.hist(ts, bins = 100, density = True, cumulative = True, alpha=0.5);

# Analytical CDF
tvals = np.arange(0, 366, 1)
plt.plot(tvals, tvals/365)
plt.xlabel('$t$')
plt.ylabel('$F_T(t)$');
```

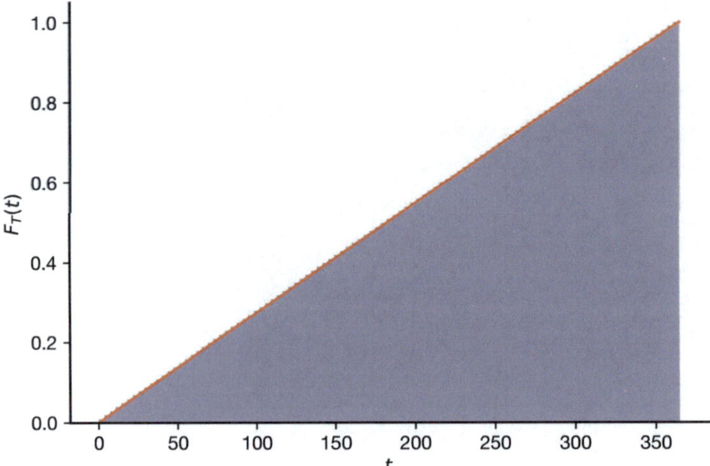

The empirical CDF and the analytical CDF match up almost exactly. So now we have a way to generate Poisson events: Generate a random value P according to the Poisson distribution. Then generate P random values from the Uniform continuous distribution on $[0, 365]$. Let's start by creating our SciPy.stats Python object with parameter 9:

```
P = stats.poisson(9)
```

Now let's create six different realizations of the Poisson events for a year. For each realization, we draw a random number of events. We use our usual notation that (upper case) `P` denotes the SciPy random variable object and (lower case) p denotes a particular value drawn from the random variable P. The figure below shows the Poisson events for these six different realizations:

```
rows=3
cols=2
fig, axs = plt.subplots(rows, cols)
for i in range(rows):
  for j in range(cols):
    p=P.rvs()
    t=T.rvs(p)
    axs[i, j].scatter(t, 0.1*np.ones(p), marker='v' )

    # Make it pretty
    axs[i, j].set_ylim(0,1)
    axs[i,j].spines[['top', 'left', 'right'] ].set_visible(False)
    axs[i,j].tick_params(axis='y', which='both', left=False,
                         right=False, labelleft=False)
    plt.tight_layout()
    plt.subplots_adjust(wspace=0.2)
```

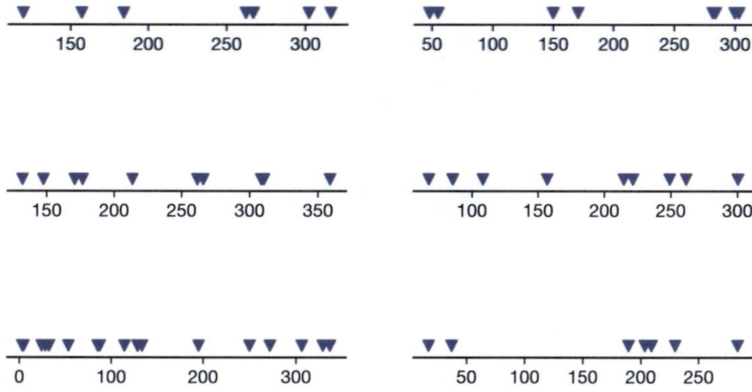

This approach will work to generate Poisson events. However, it cannot answer some simple questions like *what is the distribution for the first event time?* Let's start by finding out what the distribution for the first event time looks like by using simulation. Let's simulate 100,000 sets of Poisson events and store the *first* event time (i.e., the minimum of the randomly generated times) for each simulation iteration:

```
num_sims = 100_000
first_event_times=[]
for sim in range(num_sims):
  p = P.rvs() # a Poisson number of events
  if p > 0:
    ts = T.rvs(p)
    first_event_times += [ ts.min() ]
```

Now let's plot the normalized histogram and the normalized cumulative histogram for the first event times:

```
fig, axs = plt.subplots(1, 2, figsize=(8,4))
ax = axs[0]
ax.hist(first_event_times, bins=100, density=True)
ax.set_title('Normalized histogram')

ax = axs[1]
ax.hist(first_event_times, bins=100, density=True, cumulative=True);
ax.set_title('Normalized cumulative histogram');
```

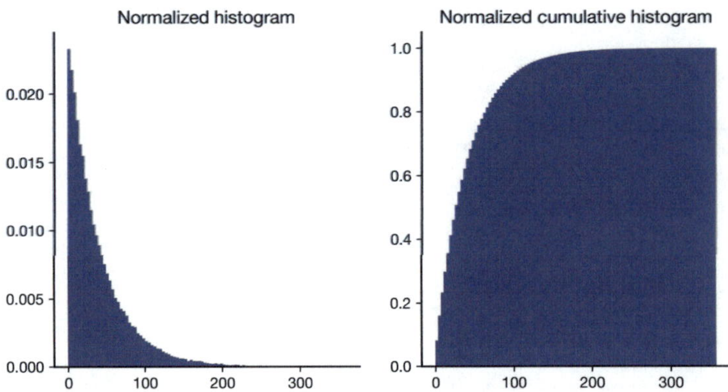

The histogram for the first event times looks very different than the histogram for a particular event's time. It clearly is not uniformly distributed on $[0, 365]$. Again, the empirical CDF appears to be a continuous function.

In fact, the Poisson random variable gives us enough information to find the CDF for the first event times. Let T_0 to be the time at which the first event occurs. Then suppose we want to find the probability that T_0 is greater than some specified value t. This will be true if there are no arrivals in the period $[0, t]$. For events that occur at rate λ, the number of events that occur in the period $[0, t]$ is a Poisson random variable with parameter $\alpha_t = \lambda t$; let's refer to this Poisson random variable as N_t. We can use the PMF for the Poisson random variable to give

$$\begin{aligned} P(T_0 > t) &= P(N_t = 0) \\ &= \frac{\alpha_t^0}{0!} e^{-\alpha_t} \\ &= e^{-\lambda t}. \end{aligned}$$

For our shark example, $\lambda = 9$ attacks per year or $\lambda = 9/365$ attacks per day. The figure below shows $P(T_0 > t)$ as a function of t:

```
t = np.linspace(0,365,366)
plt.plot(t, np.exp(- 9/365 * t))
plt.xlabel('t')
plt.ylabel('Pr(First event occurs after time $t$, $P(T_0 >t)$');
```

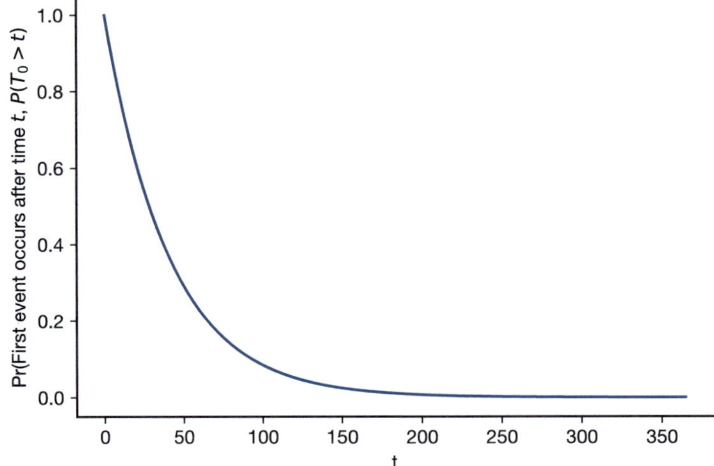

We can apply some intuition to see if this makes sense. If an average of 9 events occur, then it seems likely that the first event occurs in the first $365/9 \approx 40.6$ days. The figure confirms this intuition. Note also that the shape of this curve resembles the shape of the histogram for the data.

We can extend this analysis to get the CDF for T_0:

$$\begin{aligned} F_{T_0}(t) &= P(T_0 \leq t) \\ &= 1 - P(T > t) \\ &= 1 - e^{\lambda t}. \end{aligned}$$

Since $P(T_0 < 0) = 0$, we can summarize its CDF as

$$F_{T_0}(t) = \begin{cases} 1 - P(t) = 1 - e^{\lambda t}, & t \geq 0 \\ 0, & t < 0 \end{cases}.$$

The CDF is shown below for this example, plotted on the same graph as the empirical CDF:

```
# Empirical CDF
plt.hist(first_event_times, bins=100, density=True, cumulative=True, alpha=0.5);

# Analytical CDF
t = np.linspace(0, 365, 365)
plt.plot(t, 1 - np.exp(- 9/365 * t))

plt.xlabel('t')
plt.ylabel('$F_{T_0}(t)$');
```

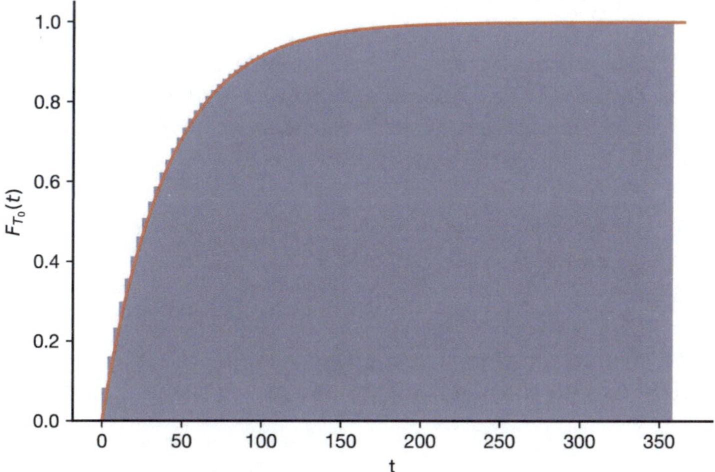

Again, the CDF is a continuous function, not a staircase function. This is a very different CDF than for the Uniform random variable. In fact, this is the CDF of an *Exponential random variable.*

8.5.2 Definition and Implications

For most purposes, the following definition of a continuous random variable will suffice:

Definition

continuous random variable

A random variable X that has an uncountably infinite number of values in its range and for which the cumulative distribution function (CDF) $F_X(x)$ is a **continuous function of** x.

You should already know what a continuous function is. Informally, it is easiest to think of how the function looks when plotted. A continuous function has no "breaks" in its graph, and we can imagine drawing it without having to pick up our pencil.

Mathematically, $F_X(x)$ is a continuous function at a point x_0 if and only

$$\lim_{\epsilon \to 0} F_X(x_0 - \epsilon) = F_X(x_0), \text{ and}$$
$$\lim_{\epsilon \to 0} F_X(x_0 + \epsilon) = F_X(x_0).$$

We say that the "limit from the left" and the "limit from the right" are equal to the value of $F_X()$ at x_0. Then $F_X(x)$ is a continuous function if and only if it is a continuous function at every point x.

If X is a continuous random variable, then the fact that $F_X(x)$ is a continuous random variable has an interesting implication. Recall from Section 8.3 that $P(X = x)$ is equal to the "height of the jump" in $F_X(x)$ at the point x. For continuous functions, there are no jumps! Mathematically, we saw that

$$P(X = b) = F_X(b^+) - F_X(b^-).$$

Since $F_X(x)$ is continuous, $F_X(x^+) = F_X(x) = F_X(x^-)$ for all x, and thus

$$P(X = x) = F_X(x^+) - F_X(x^-) = F_X(x) - F_X(x) = 0.$$

Since this holds for all x, $P(X = x) = 0$ for all x. This is a surprising and counterintuitive result: **none of the values that X takes on have a nonzero probability.**

This is hard to understand in the abstract, so let's consider a concrete example. For the uniform random variable on $[0, 1]$, the distribution function is

$$F_X(x) = \begin{cases} 0, & x < 0 \\ x, & 0 \leq x < 1 \\ 1, & x \geq 1 \end{cases}.$$

Consider the probability of a symmetric interval centered around 0.5, $P(X \in (0.5 - \epsilon, 0.5 + \epsilon])$. Using Property 4 of the CDF, we can calculate this probability as a function of $0 < \epsilon \leq 0.5$ as follows

$$\begin{aligned} P(X \in (0.5 - \epsilon, 0.5 + \epsilon]) &= P(0.5 - \epsilon < X \leq 0.5 + \epsilon) \\ &= F_X(0.5 + \epsilon) - F_X(0.5 - \epsilon) \\ &= (0.5 + \epsilon) - (0.5 - \epsilon) \\ &= 2\epsilon \end{aligned}$$

In fact, it is not hard to see that $P(X \in (x - \epsilon, x + \epsilon]) = 2\epsilon$ for any $x \in (0, 1)$ provided ϵ is small enough that $(x - \epsilon, x + \epsilon] \subseteq [0, 1]$.

Note that $P(X = x) \leq P(X \in (x - \epsilon, x + \epsilon])$ for all $\epsilon > 0$. Then

$$\begin{aligned} P(X = x) &\leq \lim_{\epsilon \to 0} P(X \in (x - \epsilon, x + \epsilon]) \\ &= \lim_{\epsilon \to 0} 2\epsilon = 0. \end{aligned}$$

Since $P(X = x) \geq 0$, it must be that $P(X = x) = 0$ for all x.

FIGURE 8.23
Image of a rectangular block of wood weighing approximately 100g.

If you have not encountered this before, it may seem to make no sense – how can X have no probabilities that are nonzero? Let's start with an analogy. Consider a rectangular wood block, such as the one shown in Fig. 8.23. If the wood block weighs 100 g and the mass is evenly distributed, how much mass is in the horizontal range $[0, 50]$? That range contains half the block, so the answer is 50 g. How much mass is in the horizontal range $[0, x]$, where $0 \leq x \leq 100$? The range contains $x/100$ of the mass, so the answer is x grams. But if we ask how much mass is at any point, then we are back to the same issue. For instance, the mass at horizontal location 0 is the mass in the range $[0, 0]$, which is 0 grams.

The way to resolve this issue is that we do not specify the mass at any particular location on the horizontal axis – instead, we specify the mass density. So, if the horizontal axis is in millimeters, the mass density is 100 g/ 100 mm = 1 g/mm. Since the mass is assumed to be evenly distributed, the mass in any horizontal range $[x_0, x_1]$, where $0 \leq x_0 < x_1 \leq 100$ is then $(x_1 - x_0)\text{mm} \cdot (1\text{g/mm}) = (x_1 - x_0)$ g. If our mass is nonuniformly distributed, then we can create a function $\rho(x)$ that gives the mass density at position x. Then the mass over a horizontal region $[x_0, x_1]$ is

$$\int_{x_0}^{x_1} \rho(x)\, dx.$$

8.5.3 Density Functions

We can apply the same idea as mass density in physical objects to probability distributions. The CDF for a random variable X is the total probability in the range $(-\infty, x)$. If we want to create a probability density, $f_X(x)$, then the CDF should be the *integral* of the probability density on this range:

$$F_X(x) = \int_{-\infty}^{x} f(u)\, du.$$

WARNING

There are several important things to note about the equation above:

- First, it is essential to distinguish uppercase and lowercase letters: $F_X(x)$ is a distribution function (CDF), whereas $f_X(x)$ is a density function.
- The distribution function at the point x is defined as the integral of the density function from $-\infty$ to x. Because x is in the region of integration, we **cannot use x as the variable of integration**. You can use any unused variable as the variable of integration (here, I chose u) to represent a variable that is changing values as we integrate from $u = -\infty$ to $u = x$. **If you try to use x both in the region of integration and as the variable of integration, the answer will almost surely come out wrong!**

Then we have the following definition of the probability density function (from the Fundamental Theorem of Calculus):

Definition

probability density function (pdf)

If X is a continuous random variable with distribution function $F_X(x)$, then the probability density function $f_X(x)$ is

$$f_X(x) = \frac{d}{dx} F_X(x),$$

wherever the derivative is defined.

Note:

First, note the notation:

- We will use lowercase *pdf* to denote the probability density function, in the same way that we use lowercase f to represent such a function.
- We will use uppercase *CDF* to denote the cumulative distribution function, in the same way that we use uppercase F to represent such a function.

This consistency in capitalization will make remembering the differences between these functions easier. Although most books follow this convention, not all do, and you may encounter PDF to mean the probability density function or probability distribution function in some texts.

SymPy for Derivatives

SymPy (pronounced "Sim Pie") is a powerful library for performing symbolic mathematics in Python. Symbolic mathematics operates on symbols (often called variables in mathematics), such as x and y, and functions, such as $x^2 + y^2$, instead of numbers or specific values.

I will briefly demonstrate how to use this library to find density functions from distribution functions and vice versa. As usual, the first step is to import the library. Here, we will use the short name `sp` to refer to SymPy:

```
import sympy as sp
```

The next step is to create a variable that maps to a symbol name. We can use the SymPy function `sp.symbols()` to create one or more symbols. The function `sp.symbols()` can be called with multiple different types of arguments. However, we will always pass an argument that is a string that consists of either a single variable name or a comma-separated list of variable names. The output will be a tuple that will automatically unpack into a set of comma-separated variables. We can create a SymPy symbol object for the variable x as follows:

```
x = sp.symbols('x')
```

If we want to create SymPy objects for variables y and z, we can use this:

```
y, z = sp.symbols( 'y, z' )
```

Now we can create a function of any of these variables. For instance, if we want to create the function $g(x) = x^2 + 5x + 6$, we can just do the following:

```
g = x**2 + 5*x + 6
```

We can take derivatives using `sp.diff()`. It takes two arguments: the function to take the derivative of and the variable to differentiate. So we can get

$$\frac{d}{dx}g(x)$$

using SymPy as follows

```
sp.diff(g, x)
```

$$2x + 5$$

The SymPy function `sp.integrate()` performs symbolic integration. It has a similar call pattern to `sp.diff()`: the first argument is the function to integrate, and the second argument is the variable of integration:

```
sp.integrate(2*x+5, x)
```

$$x^2 + 5x$$

(Note that the integral of the derivative is only equal to the original expression up to a constant.)

As previously mentioned, many CDFs (and pdfs) are *piecewise functions*. For instance, consider again the distribution of the Uniform random variable T with range $[0, 365]$. We saw that if $t \in [0, 365]$, then

$$F_T(t) = \frac{t}{365}.$$

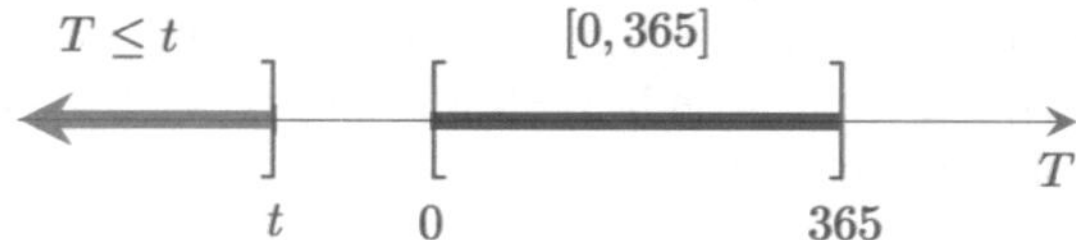

FIGURE 8.24
Region for $T \leq t$ when $t < 0$ when T is a Uniform random variable on $[0, 365]$.

However, that does not generalize to values of t outside of this interval. For instance, since we know that the range of T is $[0, 365]$, then there are no values of T for which $T \leq t$ if $t < 0$. For example, see Fig. 8.24 for an illustration of the relation between these regions. Thus,

$$F_T(t) = P(T \leq t) = 0, \text{ if } t < 0.$$

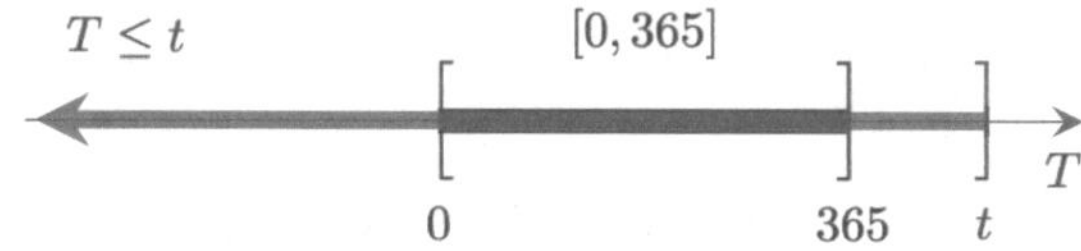

FIGURE 8.25
Region for $T \leq t$ when $t > 365$ when T is a Uniform random variable on $[0, 365]$.

Similarly, if $t > 365$, then the region $T \leq t$ contains the region $T \leq 365$, as shown in Fig. 8.25. Thus, all of the probability of T is contained in the region $T \leq t$, and

$$F_T(t) = P(T \leq t) = 1, \text{ if } t \geq 365.$$

(We could also have considered the region $t \geq 365$, but we already have a functional relationship for $0 \leq t \leq 365$, which contains the point $t = 365$.

Putting all this together, $F_T(t)$ is a piecewise function given by

$$F_T(t) = \begin{cases} 0, & t < 0 \\ \frac{t}{365}, & 0 \leq t \leq 365 \\ 1, & t > 365 \end{cases}$$

We can use the `sp.Piecewise()` function to implement this in SymPy. The arguments of `sp.Piecewise()` are a sequence of tuples, where the first element of the tuple is the function, and the second element is the condition that specifies the region. Here is an implementation of the function $F_T(t)$ above:

```
F=sp.Piecewise( (0, x<0), (x / 365, (x >=0) & (x <=365)), (1, x >365) )
F
```

$$\begin{cases} 0 & \text{for } x < 0 \\ \frac{x}{365} & \text{for } x \leq 365 \\ 1 & \text{otherwise} \end{cases}$$

Note that the region for the second "piece" does not match the one we specified; that is because `sp.Piecewise()` evaluates the conditions in order, so it is not necessary to specify

$x \geq 0$ for the second "piece". Similarly, the last piece corresponds to every condition not covered by the first two. We could have taken advantage of this in creating SymPy version as follows:

```
F = sp.Piecewise( (0, x<0), (x / 365, x <=365), (1, True) )
F
```

$$\begin{cases} 0 & \text{for } x < 0 \\ \frac{x}{365} & \text{for } x \leq 365 \\ 1 & \text{otherwise} \end{cases}$$

SymPy can differentiate piecewise functions, but the derivative is also piecewise. This works fine if the function being differentiated is continuous, but derivatives of functions with step discontinuities (like CDFs of discrete random variables) require using generalized functions (Dirac delta functions).

Here is the density function $f_T(t)$:

```
sp.diff(F, x)
```

$$\begin{cases} 0 & \text{for } x < 0 \\ \frac{1}{365} & \text{for } x \leq 365 \\ 0 & \text{otherwise} \end{cases}$$

This is a good time to emphasize that **probability densities are not probabilities**. Like probabilities, they are nonnegative (because a density function is the derivative of a monotonically nondecreasing function). However, unlike probabilities, they are not upper-bounded by 1. For example, suppose that U is a Uniform random variable with range $[0, 0.1]$. Then the CDF is

$$F_U(u) = \begin{cases} 0, & u < 0 \\ \frac{u}{0.1}, & 0 \leq u \leq 0.1 \\ 1, & u > 1 \end{cases},$$

and the density function is

$$f_u(u) = \frac{d}{du} F_U(u) = \begin{cases} 10, & 0 \leq u \leq 0.1 \\ 0, & \text{otherwise} \end{cases}.$$

Now, consider the time to the first Poisson event, T_0, which has the CDF

$$F_{T_0}(t) = \begin{cases} 1 - P(t) = 1 - e^{\lambda t}, & t \geq 0 \\ 0, & t < 0 \end{cases}.$$

Then a SymPy implementation is:

```
t, l = sp.symbols('t,lambda')
FT0 = sp.Piecewise( (0, t<0) , (1 - sp.exp(-l * t), t>=0))
FT0
```

$$\begin{cases} 0 & \text{for } t < 0 \\ 1 - e^{-\lambda t} & \text{otherwise} \end{cases}$$

The density function is given by

$$f_{T_0} = \frac{d}{dt} F_{T_0}(t),$$

which is

```
sp.diff(FT0, t)
```

$$\begin{cases} 0 & \text{for } t < 0 \\ \lambda e^{-\lambda t} & \text{otherwise} \end{cases}$$

8.5.4 Properties of Density Functions

Some of the most important properties of density functions are:

1. The probability density function is nonnegative: $f_X(x) \geq 0, -\infty < x < \infty$

Proof: Since the CDF, $F_X(x)$, is a nondecreasing function, its derivative is nonnegative.

2. The density function integrates to 1:

Proof: From the Fundamental Theorem of Calculus and the fact that $P(-\infty) = 0$,

$$F_X(x) = \int_{-\infty}^{x} f_X(x) dx.$$

Then

$$\int_{-\infty}^{\infty} f_X(x) dx = F_X(\infty) = 1.$$

3. Calculate the probability of an interval by integrating the density over it. I.e.,

$$P(a < X \leq b) = \int_{a}^{b} f_X(x) dx, a \leq b$$

Proof:

$$\begin{aligned} P(a < X \leq b) &= F_X(b) - F_X(a) \\ &= \int_{-\infty}^{b} f_X(x)\, dx - \int_{-\infty}^{a} f_X(x)\, dx \\ &= \int_{a}^{b} f_X(x)\, dx. \end{aligned}$$

4. Almost any nonnegative function that integrates to 1 is a valid density function for some random variable.

If $g(x)$ is a nonnegative piecewise continuous function with finite integral

$$\int_{-\infty}^{+\infty} g(x)dx = c, -\infty < c < +\infty,$$

then $f_X(x) = \frac{g(x)}{c}$ is a valid pdf for a random variable.

Proof: The proof is beyond the scope of this book.

Now that continuous random variables have been introduced, along with pdfs and their properties, we are ready to learn more about important continuous random variables and how to work with them using SciPy.stats in the next section.

Terminology review and self-assessment questions

Interactive flashcards to review the terminology introduced in this section and self-assessment questions are available at fdsp.net/8-5, which can also be accessed using this QR code:

8.6 Important Continuous Random Variables

There are numerous types of continuous random variables; in fact, Scipy.stats has over 100 (as of release 1.10.1). A list of the continuous random variables in Scipy.stats can be found here:

https://docs.scipy.org/doc/scipy/reference/stats.html#continuous-distributions

In this section, we will present five continuous random variables that are used in the remainder of this textbook. We have already encountered two of these: the Continuous Uniform random variable and the Exponential random variable. We introduce three new continuous random variables: the Normal (or Gaussian), Chi-squared, and Student's t random variables.

8.6.1 Continuous Uniform Random Variable

We introduced the continuous Uniform random variable in Section 8.5, where we used it to draw random arrival times for Poisson events observed on a fixed interval. We found the CDF to be linear over the random variable's range, and the pdf to be constant over that range. The range can be any interval (closed or open). If the range is $[a, b]$ for a Uniform random variable U, then we write $U \sim \text{Uniform}[a, b]$ (read: "U is distributed Uniform on $[A, B]$"). Here is a formal definition of a *Uniform random variable*:

Definition

Uniform random variable

If U is a Uniform random variable on the interval $[a, b]$, then the pdf $f_U(u)$ is

$$f_U(u) = \begin{cases} \frac{1}{b-a}, & u \in [a, b] \\ 0, & \text{otherwise,} \end{cases}$$

and the CDF $F_U(u)$ is

$$F_U(u) = \begin{cases} 0, & u < a \\ \frac{u}{b-a}, & u \in [a, b] \\ 1, & u > b. \end{cases}$$

Note that either or both endpoints may be excluded from the range of U, in which case the regions in the piecewise definition change slightly (for instance, by dropping or adding an equality sign), but the functions are unchanged.

We can use SciPy.stats to create a Uniform distribution by using `stats.uniform(start, length)`.

WARNING

Note carefully the form of `stats.uniform(start, length)` – the range is thus [start, start+length]. To create a variable U representing a Uniform distribution on $[a, b]$, we need to do `U = stats.uniform(a, b-a)`.

Three different Uniform random variables are generated in the code below. Their pdfs and CDFs are shown in Fig. 8.26. The code to generate these plots is available online at fdsp.net/8-6.

```
import scipy.stats as stats

U1 = stats.uniform(0, 1)
U2 = stats.uniform(2, 0.2)
U3 = stats.uniform(3, 5)
```

Note that the probability density is concentrated wherever the CDF increases quickly. Because the pdf is the derivative of the CDF, a steeper slope in the CDF translates into a higher density in the pdf.

Applications: Uniform random variables are used for any situation in which some random occurrence is equally likely to occur over a continuous range. The following are a few examples. A Uniform random variable may be used to model the random phase of a carrier waveform in a wireless communication system. Most computer random number systems generate floating-point approximations of Uniform random variables. Numerical errors in quantization can often be modeled as Uniform.

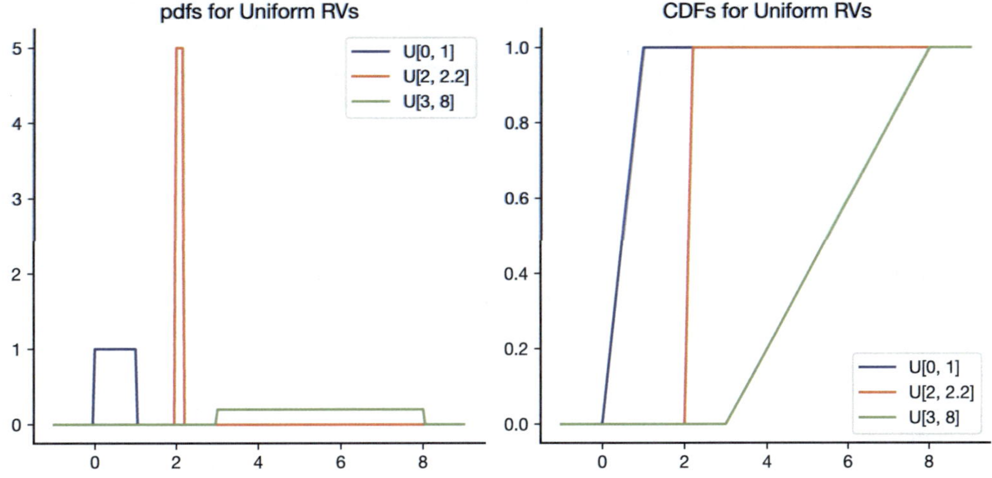

FIGURE 8.26
The pdfs and CDFs for three different Uniform random variables.

The Uniform distribution can also be shown to be the "most random" in a certain sense (maximizing a measure of randomness called entropy) over all distributions with finite range.

8.6.2 Exponential Random Variable

We found that the Exponential random variable models the time until the first Poisson arrival in an observation period. The Exponential random variable is an example of a continuous random variable with an infinite range, since it can take on any value from $[0, \infty]$. A formal definition of the *Exponential random variable* is:

> Definition
>
> **Exponential random variable**
>
> If X is an Exponential random variable with real parameter $\lambda > 0$, then X is a continuous random variable with pdf
>
> $$f_X(x) = \begin{cases} 0, & x < 0 \\ \lambda e^{-\lambda x}, & x \geq 0, \end{cases}$$
>
> and CDF
>
> $$F_X(x) = \begin{cases} 0, & x < 0 \\ 1 - e^{-\lambda x}, & x \geq 0. \end{cases}$$

> **Note:**
>
> Some books specify the exponential distribution in terms of a parameter μ such that $\mu = 1/\lambda$. The two definitions are equivalent, but the forms of the pdf and CDF are simpler and easier to remember when specified in terms of λ.

We can use SciPy.stats to create an exponential distribution with parameter `lam` using `stats.expon(scale= 1/lam)`. The following code generates three exponential random variables with different values of λ. Plots of the pdfs and CDFs of these Exponential random variables are shown in Fig. 8.27. The code to generate these plots is available online at fdsp.net/8-6. As we demonstrated before with the Uniform random variable, it is possible for $f_X(x)$ to be greater than 1 because probability densities **are not probabilities**. In particular, at $x = 0$, $f_X(0) = \lambda$, so $f_X(0) > 1$ whenever $\lambda > 1$.

```
X0 = stats.expon(scale=1 / 1)
X1 = stats.expon(scale=1 / 5)
X2 = stats.expon(scale=1 / 0.5)
```

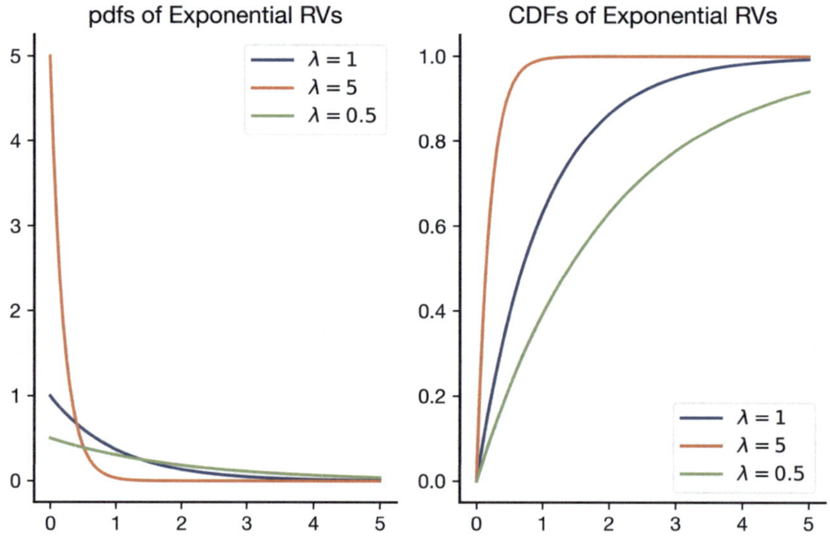

FIGURE 8.27
The pdfs and CDFs for three exponential random variables with different values of the parameter λ.

Again, note that the region in which the CDF is increasing from near 0 to near 1 represents the region where most of the probability is concentrated. This is particularly easy to see for exponential random variables: these regions all start at 0. We can then find a region of the form $[0, x]$ that contains 99% of the probability in two ways:

- From the graph, we can look on the y-axis at 0.99 and then look right to see what value of x intersects gives that value of $F_X(x)$. From the graph, the regions that contain approximately 99% of the probability are: $[0, 1]$ for $\lambda = 5$ and $[0, 4.5]$ for $\lambda = 1$.
- We can also find these from the *inverse CDF*:

Definition

inverse CDF

If X is a random variable with cumulative distribution function (CDF) $F_X(x) = P(X \leq x)$, then the *inverse CDF*, which is also called the *quantile function* or *percent-point function* given a value p, gives a value of x such that $F_X(x) = p$. It is the inverse of the CDF, $F^{-1}(p)$.

SciPy.stats calls the inverse CDF the percent-point function and provides a `ppf()` method for continuous random variables. Thus, the regions that contain 99% of the probability for the values of λ considered are:

```
lambdas = [5, 1, 1 / 2]
for lam in lambdas:
    X = stats.expon(scale=1 / lam)
    upper = X.ppf(0.99)
    print(f"lambda = {lam:>3}: [0, {upper :.2f}]")
```

```
lambda =   5: [0, 0.92]
lambda =   1: [0, 4.61]
lambda = 0.5: [0, 9.21]
```

Example 8.14: Lifetime of LED Headlights

An automotive company finds that 95% of its new LED headlights last more than five years. If the life of these headlights follows an Exponential distribution, what is the probability that one of these headlights lasts more than ten years?

Solution: The challenge in this problem is that we know that the distribution is Exponential, but we do not know the parameter λ that characterizes that distribution. However, we are given enough information to determine λ.

Let the life of one of these headlights in years be X, where $X \sim \text{Exponential}(\lambda)$. If 95% of the headlights last more than five years, then the probability that a particular headlight lasts more than five years is 0.95. In math, $P(X > 5) = 0.95$. Thus,

$$\begin{aligned}
P(X > 5) &= 1 - P(X \leq 5) \\
&= 1 - F_X(5) \\
&= 1 - \left(1 - e^{-\lambda(5)}\right) \\
&= e^{-5\lambda} = 0.95
\end{aligned}$$

Now, take the natural log of both sides. We need to know the natural log of 0.95:

```
np.log(0.95)
```

```
-0.05129329438755058
```

So,

$$\begin{aligned} -5\lambda &\approx -0.051 \\ \lambda &\approx 0.0102. \end{aligned}$$

The probability that a headlight lasts more than ten years is then

$$\begin{aligned} P(X > 10) &= 1 - P(X \leq 10) \\ &= 1 - F_X(10) \\ &= e^{-0.0102(10)}, \end{aligned}$$

which is approximately

```
np.exp(-0.0102 * 10)
```

```
0.9030295516688768
```

Thus, under these assumptions, about 90% of headlights will last over ten years.

Note:

There is something interesting in these results that might not be immediately obvious. The probability that a headlight still functions decreases with the number of years, as expected. However, consider those 95% of bulbs that are still functioning after five years. If we apply the same 95% rule to those bulbs to see how many are still functioning after another five years, we get $(0.95)^2 = 0.902$, which is essentially the same as the result we got above. This would mean that **after five years of use, those bulbs have the same probability of lasting another five years as a brand new bulb has of lasting for five years.**
We do not yet have the mathematical tools to analyze this, but it turns out that this is a special property, called the *memoryless property*, that is unique to the exponential distribution (among all continuous distributions). After finishing this chapter, you will have the necessary knowledge to understand this property, which is investigated more on the website for this book at fdsp.net/8-8.

Applications: As previously noted, the Exponential random variable models the time until the first Poisson arrival. It also models the time between two consecutive Poisson arrivals. It is also used in a related application called *survival analysis* when there is a constant *hazard rate*. For example, it can be used to model the lifetime of electrical devices.

8.6.3 Normal (Gaussian) Random Variable

The Normal, or Gaussian, random variable is one of the most commonly used random variables in statistical analyses. The term *Normal random variable* is more often used in the field of Statistics, whereas *Gaussian random variable* is more often used in electrical engineering. Either one is easily understood by the other community. Because this book is meant to be accessible to readers across engineering and the sciences and because SciPy.stats uses the term Normal, I will also use that terminology.

We have already seen that the PMFs of the Binomial and Poisson random variables take on a bell shape for large numbers of trials or large values of α, respectively. It turns out that the average of almost any type of random variable converges to this bell shape under a set of very mild conditions. The shape is the shape of the Normal pdf:

Definition

Normal random variable

If X is a Normal, or Gaussian, random variable with real parameters μ and $\sigma > 0$, the pdf of X is

$$f_X(x) = \frac{1}{\sigma\sqrt{2\pi}} \exp\left\{-\frac{1}{2}\left[\frac{x-\mu}{\sigma}\right]^2\right\}, \quad -\infty < x < \infty$$

The parameter μ is called the *mean*, and the parameter σ is called the *standard deviation.*

Note:

The Normal density is sometimes specified in terms of σ^2 (the variance) instead of σ (the standard deviation) because it can be argued that variance is a more fundamental property of random variables.
I have chosen to use σ for several reasons:

1. The units of σ are the same as the units of μ. We will see that either σ or σ^2 control the spread of the probability density, but σ has units that more directly translate to values we can understand and interpret on graphs.
2. We will see that the term $(x - \mu)/\sigma$ arises in other aspects of working with Normal random variables and in other applications.
3. SciPy.stats specifies Normal densities in terms of μ and σ, not σ^2.

In general, it is best to specify whether a parameter of a Normal random variable indicates standard deviation or variance. For instance, we might write $X \sim$ Normal ($\mu = 0$, $\sigma = 5$).

From the form of the pdf, we can make several observations:

- The density is symmetric, and it is centered at μ.
- The density rolls off exponentially fast in the square of the distance from μ.
- The spread of the density away from the mean is controlled by the parameter σ.

In SciPy.stats, we can create a Normal distribution using `stats.norm(mu, sigma)`. Note that we have to pass σ (**NOT** σ^2) to `stats.norm`.

Below are plots of the pdf for different values of μ and σ.

```
x = np.linspace(-8, 18, 1000)

all_params = [(-5, 0.3), (0, 1), (10, 3)]

for params in all_params:
    N = stats.norm(params[0], params[1])
    mylabel = f"Normal({N.mean()}, $\sigma=${N.std() :.1f})"
    plt.plot(x, N.pdf(x), label=mylabel)

plt.legend()
plt.title("pdfs for Normal densities");
```

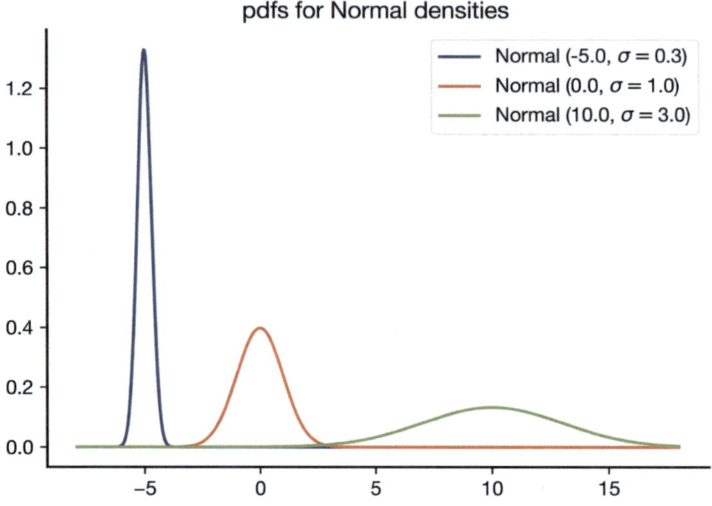

The pdf is centered on μ and symmetric about μ. Larger values of σ result in the density being spread out wider around μ but also having a lower peak value. It should be somewhat intuitive that this would happen, as we know the pdf must integrate to 1.

Now you may be wondering – why isn't the CDF included in this definition? We usually don't specify the CDF for the Normal random variable because **the CDF can only be written as an integral equation**:

$$F_X(x) = \int_{-\infty}^{x} \frac{1}{\sigma\sqrt{2\pi}} \exp\left\{-\frac{1}{2}\left[\frac{t-\mu}{\sigma}\right]^2\right\} dt.$$

The CDF is the area under the pdf from $-\infty$ to x, as shown by the shaded area in Fig. 8.28.

We will refer to a normal distribution with $\mu = 0$ and $\sigma = 1$ as the *standard Normal* distribution. The CDF for the standard Normal random variable is often denoted by $\Phi(x)$.

We will often need to calculate the survival function, $1 - F_X(x)$ for a Normal random variable. For a standard Normal random variable, the survival function is often denoted $Q(x)$ and is typically written as

$$Q(x) = \frac{1}{\sqrt{2\pi}} \int_{x}^{\infty} e^{-\frac{t^2}{2}}\ dt.$$

Note that we almost always evaluate $Q(x)$ for $x > 0$. The Q-function is the area from x to ∞ under the pdf of the standard Normal random variable, as shown in Fig. 8.29.

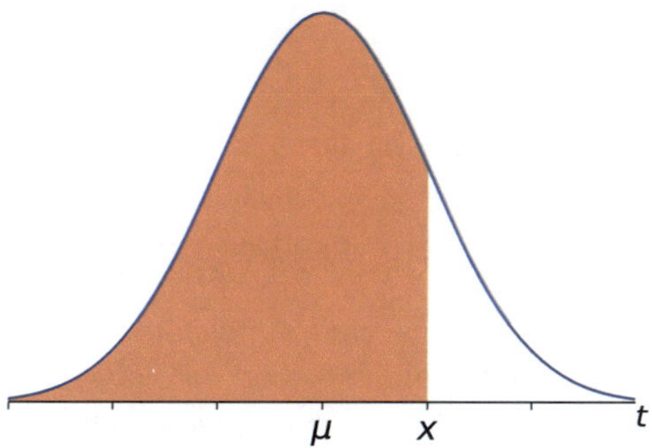

FIGURE 8.28
Area under pdf corresponding to CDF value at x for Normal distribution.

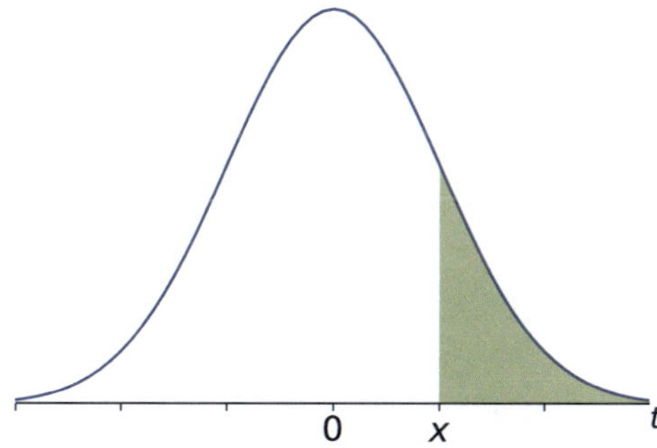

FIGURE 8.29
Area under standard Normal density corresponding to $Q(x)$.

The CDF and survival function for a Normal (μ, σ) random variable are

$$\Phi\left(\frac{x-\mu}{\sigma}\right)$$

and

$$\mathrm{Q}\left(\frac{x-\mu}{\sigma}\right), \tag{8.3}$$

respectively.

WARNING

Note that the argument of these functions has the same form that we saw in the Normal pdf. The denominator is σ, not σ^2.

Since the $Q()$ function is the survival function of the standard Normal random variable, we can create a very simple function to evaluate it:

```
def q(x):
  return stats.norm.sf(x)
```

For $x > 0$, another convenient form for $Q(x)$ is

$$Q(x) = \frac{1}{\pi} \int_0^{\pi/2} \exp\left(-\frac{x^2}{2\sin^2\theta}\right) \, d\theta.$$

There are many good approximations and bounds for the Q function that are appropriate for programming into an electronic calculator or for using in analysis when the integral forms prevent further simplification of expressions. The Wikipedia page on the Q-function: https://en.wikipedia.org/wiki/Q-function#Bounds_and_approximations has an excellent list. One that is worth pointing out because of its simplicity and accuracy for large values of its argument is the improved Chernoff bound, which is given by

$$Q(x) \leq \frac{1}{2} e^{-x^2/2}.$$

In this book, I will assume that you have access to a computer and SciPy.stats, so I am not going to spend additional time on how to evaluate the Q function. The inverse Q function is simply the inverse survival function, which in SciPy.stats is `norm.isf(p)`. For clarity of notation, we can define a `qinv()` function as

```
def qinv(p):
    return stats.norm.isf(p)
```

The CDFs for the same set of parameters as used for the pdfs are shown in Fig. 8.30. The code to generate these plots is available online at fdsp.net/8-6. The dotted lines show that the point that achieves $F_X(x) = P(X \leq x) = 0.5$ is μ for each of these distributions. This should be expected from our previous observations: since the distribution is symmetric around μ, it must be that half of the probability is on each side of μ.

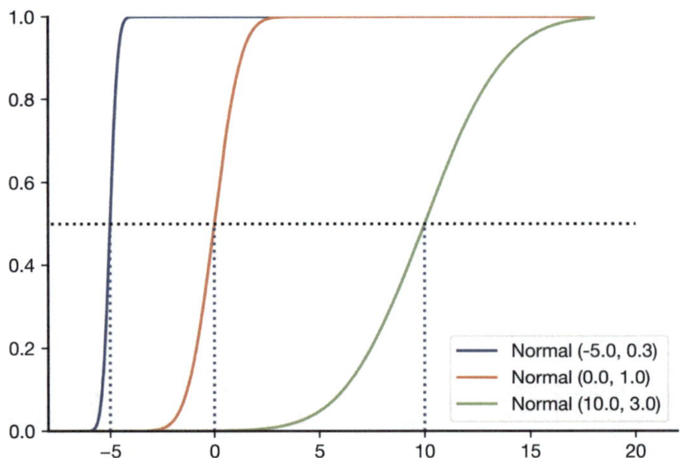

FIGURE 8.30
CDFs for several Normal random variables.

Note that the steepness of the CDF curves depends on the value of σ (and not on the value of μ). Larger σ results in the CDF curve being less steep because the probability is spread out over a wider range.

Evaluating Normal Probabilities

For a Normal random variable, the probability of any interval can be written as a simple function involving the Q-function with positive arguments. Consider $X \sim$Normal (μ, σ), where $\sigma > 0$. We consider several cases below.

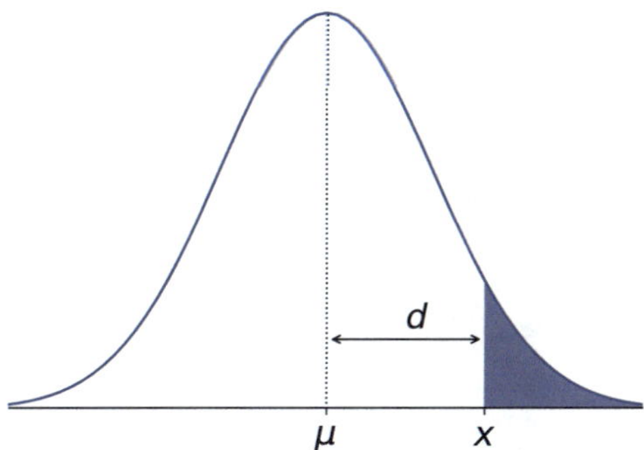

FIGURE 8.31
Right tail of a Normal random variable.

1. The easiest case is a "right tail", $P(X > x)$, where $x > \mu$. This scenario is illustrated in Fig. 8.31. From (8.3),

$$P(X > x) = Q\left(\frac{x - \mu}{\sigma}\right).$$

Note that the distance between the threshold x and the mean μ is $d = x - \mu$. Then we can write

$$P(X > x) = Q\left(\frac{d}{\sigma}\right).$$

Although introducing d may seem like an unnecessary step at this point, we will soon show why this is useful.

2. Now, consider a "left tail", $P(X < y)$, where $y < \mu$. This scenario is illustrated in Fig. 8.32.

Note that

$$P(Y \leq y) = F_Y(y) = \Phi\left(\frac{y - \mu}{\sigma}\right).$$

However, we are going to use the standard that every Normal probability should be expressed in terms of $Q()$ instead of Φ. There are several ways to proceed. One is to use the mathematical relation $\Phi(z) = 1 - Q(z)$. An alternative is to take advantage of the symmetry of the Normal density around the mean, μ. If we reflect the region $Y \leq y$ around an axis located at μ, then the probability of the reflected region will be equal to the $P(Y \leq y)$. To preserve the symmetry of the region, the reflected region must be at the same distance, $d = \mu - y$, from μ. Thus, the reflected region is $X \geq \mu + d = 2\mu - y$. The result is shown in the dark-shaded region shown in Fig. 8.33.

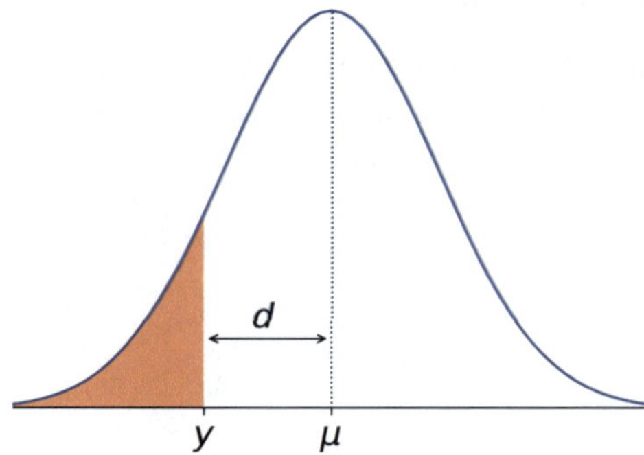

FIGURE 8.32
Left tail of a Normal random variable.

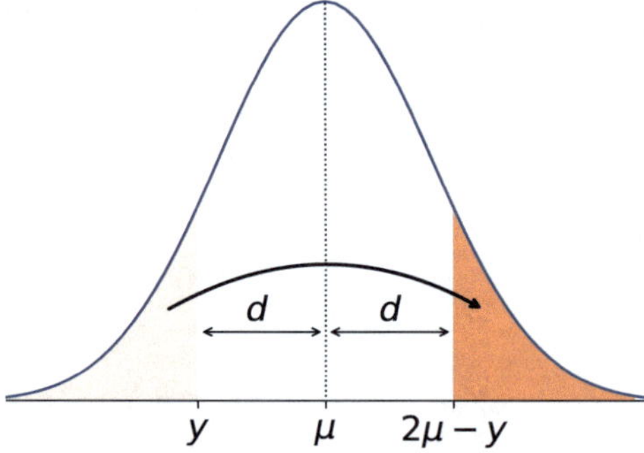

FIGURE 8.33
Reflecting a left tail of a Normal random variable around the mean.

Then using (8.3), the probability of the reflected region is

$$\begin{aligned} P(X \geq 2\mu - y) &= Q\left(\frac{2\mu - y - \mu}{\sigma}\right) \\ &= Q\left(\frac{\mu - y}{\sigma}\right) \\ &= Q\left(\frac{d}{\sigma}\right). \end{aligned}$$

Probability of a Normal Tail

The probability in a tail of a Normal density that is at distance d from the mean is $Q(d/\sigma)$.

3. Consider next the probability of a region that does not include the mean. For example, $P(a < X \leq b)$, where $a > \mu$. The corresponding region of probability density is shown in Fig. 8.34.

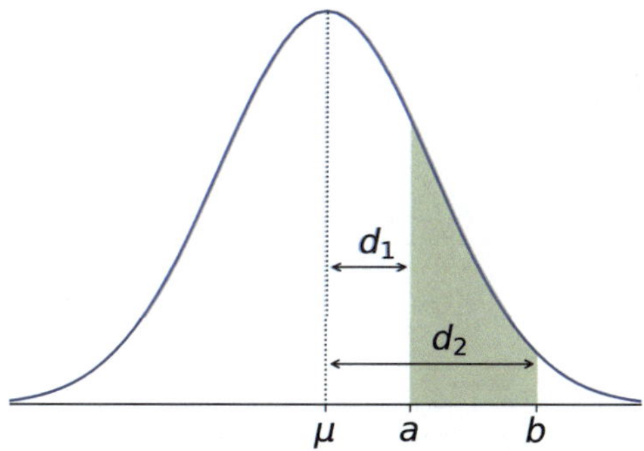

FIGURE 8.34
A typical non-tail region of a Normal random variable.

Note that the event $Y > a$ can be written as $a < Y \leq b \cup Y > b$, where the events $\{s \mid a < Y \leq b\}$ and $\{s \mid Y > b\}$ are mutually exclusive. Then

$$P(Y > a) = P(a < Y \leq b) + P(Y > b)$$
$$\Rightarrow P(a < Y \leq b) = P(Y > a) - P(Y > b).$$

Let d_1 and d_2 be the distances from the mean to a and b, respectively; i.e., $d_1 = a - \mu$ and $d_2 = b - \mu$. Then

$$P(a < Y \leq b) = Q\left(\frac{d_1}{\sigma}\right) - Q\left(\frac{d_2}{\sigma}\right).$$

4. For our final case, consider a region that includes the mean. For example, $P(g < X \leq h)$, where $g < \mu < h$. The corresponding region of probability density is shown in Fig. 8.35.

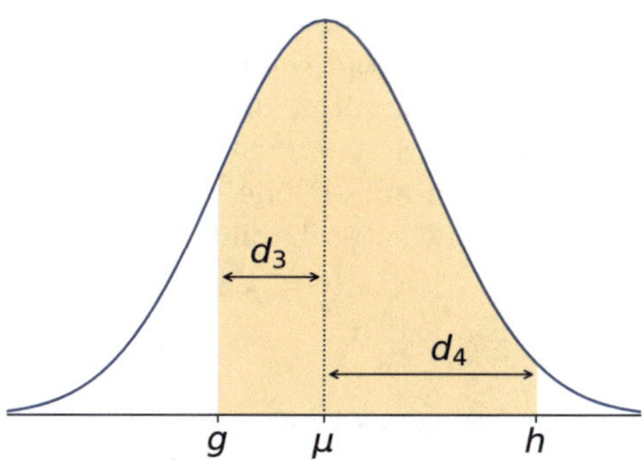

FIGURE 8.35
A region containing the mean for a Normal random variable.

For this type of probability, the easiest way to write it in terms of $Q()$ functions is to consider the complimentary event. I.e.,

$$P(g < X \leq h) = 1 - P\left(\overline{g < X \leq h}\right)$$
$$= 1 - \left[P\left(X \leq g \cup X > h\right)\right].$$

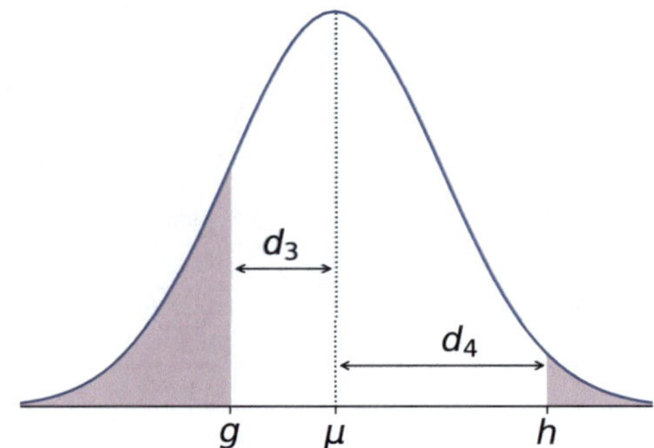

FIGURE 8.36
Complement of region containing the mean for a Normal random variable; the result is two disjoint tails.

The region of probability density for $X \leq g \cup X > h$ is shown in Fig. 8.36.

This probability can be calculated as the sum of two tail probabilities, which we already know how to calculate. The result is

$$\begin{aligned} P(g < X \leq h) &= 1 - Q\left(\frac{d_3}{\sigma}\right) - Q\left(\frac{d_4}{\sigma}\right) \\ &= 1 - Q\left(\frac{\mu - g}{\sigma}\right) - Q\left(\frac{h - \mu}{\sigma}\right). \end{aligned}$$

Example 8.15: Modeling Heights of Adults Using a Normal Distribution

The following code loads data for approximately 14,000 people in the US National Health and Nutrition Examination Survey (NHANES) for 2017–March 2020: https://wwwn.cdc.gov/Nchs/Nhanes/Search/DataPage.aspx?Component=Examination&Cycle=2017-2020. The data comes from the Body Measures data set. The code generates a histogram of the data, which is shown in Fig. 8.37. This data set includes both adults and children, and the combined data is clearly not Normal. However, there seems to be a strong Normal component in the higher height, which likely corresponds to the adults in the survey.

Using this data set, estimate the probability that an adult is over 7 feet tall.

```
import pandas as pd
df=pd.read_sas("https://wwwn.cdc.gov/Nchs/Nhanes/2017-2018/P_BMX.XPT")
counts, bins, plot = plt.hist(df['BMXHT']/2.54,bins=50);
plt.xlabel('Height (in)');
plt.ylabel('Counts');
```

This problem is very different from our previous problems because we cannot answer it directly from the data. Since this data set contains no heights greater than 7 feet, any resampling of the data will also have no heights greater than 7 feet. Thus, an estimate that uses only resampling will always estimate the probability of a person being over 7 feet tall as 0, which we know is not correct.

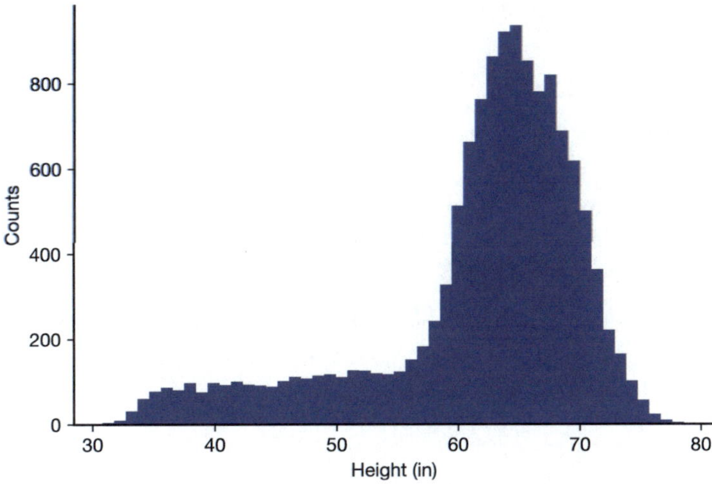

FIGURE 8.37
Heights of people in the NHANES 2017–2020 data set.

One way to deal with this problem is to use a *model-based* approach. That is, we assume that we have a model for the data and use this model to answer the question. Looking at the right-hand side of the histogram, we can choose to model the heights of adults in the US using a Normal distribution. We need to know two parameters to specify the Normal distribution: the mean μ and the variance σ^2. We don't yet know how to estimate these parameters directly, and even if we did, the fact that the adult data is mixed with juvenile data would make this challenging.

Interactive Exercise

For now, we will estimate these parameters by manually searching for a good match between the Normal density and the normalized histogram. Download the Jupyter notebook at

https://www.fdsp.net/notebooks/important-continuous-rvs-widget.ipynb.

Run the code and use the sliders to adjust the parameters of the Normal distribution until its density (the smooth curve) matches the data in the histogram as best as you can.

Your values may differ, but the values that I found to best match are:

- $\mu = 65.05$
- $\sigma = 4.9$
- scale $= 0.91$

Given these values, we can model the height of a random US adult as $H \sim$Normal($\mu = 65.1$, $\sigma = 4.9$). Then the probability that a randomly chosen adult is over 7 feet (i.e., 84 inches) is $P(H > 84)$. We can directly evaluate this using SciPy.stats as follows:

```
H = stats.norm(65.05, scale = 4.9 )
H.sf(12*7)
```

```
5.5012921622610294e-05
```

We can also easily express this using the Q function by noting that the distance from the mean to the threshold is $d = 84 - 65.1 = 18.9$, so

$$P(H > 84) = Q\left(\frac{d}{\sigma}\right) \approx Q\left(\frac{18.9}{4.9}\right) \approx Q(3.86),$$

which is

```
q(3.86)
```

```
5.669351253425653e-05
```

So the probability that a randomly chosen adult in the US is over 7 feet tall should be about 6×10^{-5}. This estimate may be low because the data we used to generate this model also includes adolescents, which may skew the data toward lower heights. Fortunately, we can check this result using another data set.

The Behavioral Risk Factor Surveillance System (BRFSS) Survey is a telephone survey of US residents that is sponsored by the US Centers for Disease Control and other federal agencies. Here is a link to the 2020 BRFSS survey data: https://www.cdc.gov/brfss/annual_data/annual_2020.html, which includes over 400,000 respondents. In this data set, the relative frequency of heights over 7 feet is 5.5×10^{-5}, which is extremely close to our estimate using the NHANES data set.

Example 8.16: Grades Following a Normal Distribution

A professor asks her TA how the students did on an exam. The TA provided the following information:

- The data looks like it follows a Normal distribution with mean $\mu = 83$.
- 10% of the students got a C or lower (less than 70).

If the threshold for an A is 90, what is the probability that a randomly chosen student got an A?

Solution

As in the last example, we have a *model* for our data. This time, we do not have to estimate the model's parameters from data, but we are also not directly given all the parameters. We know $\mu = 83$, but we will also need σ^2 before we can answer the question. To get σ^2, we can use the second piece of information we were given: 10% of students got a C or lower. If G is the grade on the exam, then $P(G \leq 70) = 0.1$.

Thus,

$$
\begin{aligned}
P(G \leq 70) &= 0.1 \\
Q\left(\frac{\mu - 70}{\sigma}\right) &= 0.1 \\
\frac{83 - 70}{\sigma} &= Q^{-1}(0.1) \\
\sigma &= \frac{13}{Q^{-1}(0.1)}
\end{aligned}
$$

Thus, the standard deviation, σ, is approximately

```
13/qinv(0.1)
```

```
10.143953898940929
```

Now that our model is fully specified, we can answer the main question: what is the probability that a randomly chosen student got an A ($G \geq 90$)?

$$
\begin{aligned}
P(G \geq 90) &= Q\left(\frac{90 - \mu}{\sigma}\right) \\
&= Q\left(\frac{90 - 83}{10.1}\right),
\end{aligned}
$$

which is approximately

```
q( (90-83)/10.1 )
```

```
0.24413302914341584
```

Thus, the probability that a randomly chosen student got an A is 24.4%.

Central Limit Theorem

The Central Limit Theorem (CLT) says that the CDF function for the average of M of almost any type of random variables will converge to a Normal CDF as M goes to infinity. This is a remarkable result that is useful for two different reasons:

1. Many phenomena can be modeled using the Normal distribution if they come from aggregating or averaging other phenomena.
2. For many random models, such as Binomial with a large number of trials or Poisson with a large average value, it becomes difficult to calculate the probabilities. In such cases, we can approximate the probabilities using a Normal approximation and compute the probabilities using the Q-function.

In reality, there are many different Central Limit Theorems, and the details of all of them are outside of the scope of this book. If you are interested in learning more, the Wikipedia page on the Central limit theorem: https://en.wikipedia.org/wiki/Central_limit_theorem is a good place to start.

We will demonstrate how to apply the CLT to solve problems relating to the means of data in Chapter 9 once we introduce means and variances for random variables.

Applications of Normal Random Variables: As mentioned above, many phenomena that come from aggregate effects or averaging can be modeled using the Normal distribution.

- In statistics, an average of a large number of data points can be assumed to be Normal.
- The motion of electrons inside the receivers of mobile phones and other radio equipment causes thermal noise, which has a distribution that is approximately Normal.
- Many properties in populations of people or animals, such as height or weight, tend to follow a Normal distribution; this may be because there are many different physiological factors that affect these properties, and the resulting values are caused by the aggregate effects of these many factors.

8.6.4 Chi-Squared Random Variable

The Chi-squared (sometimes written χ^2) random variable provides a connection between two distributions we have already been working with. Let's start with a definition, but note that the form of the definition is a bit different than for the previous random variables:

Definition

Chi-squared random variable

Let N_i, $i = 0, 1, \ldots, M-1$ be independent, standard Normal random variables with mean $\mu = 0$ and variance $\sigma^2 = 1$. Then

$$X = \sum_{i=0}^{M-1} N_i^2$$

is a (central) Chi-squared random variable with M degrees of freedom (dof).

I am not giving a formula for the pdf because it is more complicated than the other pdfs considered in this book, and we will not use the pdf later in this book. The CDF is not generally in closed form (meaning that it contains an integral that cannot be simplified), and we will use SciPy.stats to compute the CDF. The Chi-squared random variable with $M = 2$ is an Exponential random variable with $\lambda = 1/2$.

If X is Chi-squared with M degrees of freedom, we write $X \sim$ Chi-squared(M) or $X \sim \chi^2(M)$.

We can use SciPy.stats to create a Chi-squared distribution by using `stats.chi2(dof)`:

```
X = stats.chi2(dof)
```

Fig. 8.38 shows the pdfs of several Chi-squared random variables with different degrees of freedom. Since a random variable that is $\chi^2(M)$ can be created by summing the squares of M standard normal random variables, it should be no surprise that the probability density shifts away from 0 as M increases.

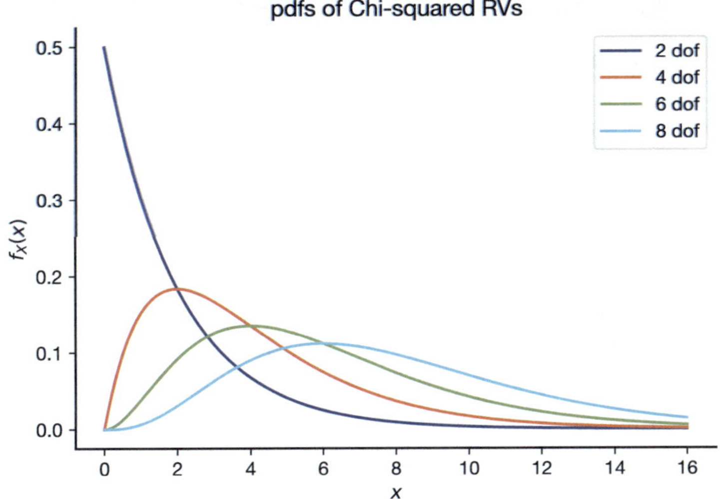

FIGURE 8.38
The pdfs of several Chi-squared random variables with different degrees of freedom.

Applications Squaring is often used in engineering and science applications to compute power, energy, or squared error. When the items being squared have a standard Normal distribution, a sum of squares of those items has a Chi-squared distribution. The Chi-squared distribution also arises in contingency tests, which we introduce in Chapter 11.

8.6.5 Student's t Random Variable

The Student's t random variable arises when estimating the mean μ of data from a Normal distribution for which the variance σ^2 is not known. The distribution takes its name from a paper by William Sealy Gosset that was published under the pen name *Student*. The paper, "The Probable Error of a Mean" is available at https://www.york.ac.uk/depts/maths/histstat/student.pdf

Gosset was working at the Guinness Brewery in Dublin, Ireland when he published this paper. He developed the distribution to perform statistical tests on small samples in the brewery. Thus, his derivation of the t distribution was motivated by practical consideration.

Note:

Gosset also coined the term *Pearson's correlation* for a measure of statistical dependence of random variables, which will be introduced in Section 12.1. A good discussion of Gosset's work is in the Biometrika article " 'Student' as Statistician": https://www.jstor.org/stable/2332648?seq=1.

Like the Chi-squared random variable, the Student's t random variable depends on the sum of multiple other random variables, and it also has a degrees of freedom parameter, ν. A brief definition of Student's t distribution follows. More insight on Student's t distribution and its application will be covered in Section 9.5.

Definition

Student's t random variable

If T is a Student's t random variable with ν degrees of freedom, the pdf is

$$f(t) = \frac{\Gamma\left(\frac{\nu+1}{2}\right)}{\sqrt{\nu\pi}\,\Gamma\left(\frac{\nu}{2}\right)} \left(1 + \frac{t^2}{\nu}\right)^{-(\nu+1)/2}.$$

Like the Chi-squared and Normal random variables, the CDF for the Student's t random variable can only be written in terms of standard integral functions. When $\nu \to \infty$, the Student's t random variable converges to a normal random variable.

We can use SciPy.stats to create a Student's t distribution by using `stats.t(dof)`:

```
T = stats.t(dof)
```

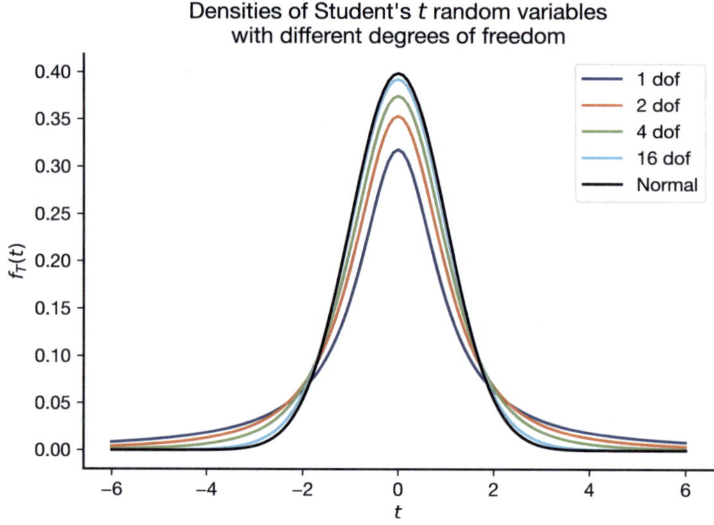

FIGURE 8.39
The pdfs for several Student's t random variables with different degrees of freedom.

Fig. 8.39 shows the pdfs of several Student's t random variables with different degrees of freedom. Also shown is a standard Normal random variable for comparison purposes. The densities for the Student's t random variables are very similar to those for the standard Normal random variable. In fact, for $\nu = 16$, the density is almost identical to that of the standard Normal random variable. The difference is that smaller dofs result in a greater spread of the density away from the mean. Let's check this by printing out the probability $P(|T| > 2)$ for different values of ν. Note that by symmetry, $P(|T| > 2) = 2 * P(T > 2)$.

```
print(f'{"DOFs": >6} | {"P(|T|>2)" : >11}')
print('-'*19)
for nu in [1, 2, 4, 16]:
    T = stats.t(nu)
    print(f'{nu :6} | {"":4}{2*T.sf(2) : .3f}')
```

```
    DOFs |     P(|T|>2)
--------------------
       1 |       0.295
       2 |       0.184
       4 |       0.116
      16 |       0.063
```

Thus, the Student's t random variable is similar to a standard normal random variable except that more of its probability density is spread out away from 0.

8.6.6 Discussion

We introduced several important continuous random variables, with an emphasis on random variables that are common in statistical data or statistical tests. As we noted, SciPy.stats has classes for over 100 types of continuous distributions, which can be used to model a wide variety of random phenomena.

One of the first tools to use in determining whether data can be modeled using a particular distribution is visualizing the distribution of the data. That is the topic of the next section.

Terminology review and self-assessment questions

Interactive flashcards to review the terminology introduced in this section and self-assessment questions are available at fdsp.net/8-6, which can also be accessed using this QR code:

8.7 Histograms of Continuous Random Variables and Kernel Density Estimation

In the example of using the Poisson random variable to motivate the continuous exponential random variable and in the example of modeling adult heights by a Normal random variable, we saw that the histogram of data has the same shape as the pdf. This will generally be true, provided there are a sufficient number of independent values from the random distribution. A detailed discussion is provided online at fdsp.net/8-7.

The values returned by `plt.hist()` with the `density=True` keyword parameter are equal to the bin counts divided by the total number of observed values and divided by the bin width. The resulting value in each bin is an estimate of the pdf over that bin called an *empirical pdf*:

> Definition
>
> **empirical pdf**
>
> A pdf created from data, for instance, by dividing the range of the data into bins and assigning probability density to a bin that is proportional to the relative frequency of points in the bin and inversely proportional to the bin's width.

If the bin widths are small and the number of observed values is large, then the empirical pdf will generally be a good approximation of the pdf of the random variable.

8.7.1 Kernel Density Estimation

Histograms can be considered to be built out of rectangular blocks, where each block represents one point in a bin. For continuous densities, we can get a better approximation of the density by replacing the rectangular blocks with a smooth shape (called a *kernel*) and centering each kernel on the data. This approach is called *kernel density estimation (KDE)*:

> Definition
>
> **kernel density estimation**
>
> A technique for estimating the pdf of a data set under the assumption that the pdf is smooth. The estimated pdf can be constructed by taking a smooth shape, called a kernel, centering a copy of the kernel at each data point, adding all of the copies, and normalizing the result.

A detailed explanation of KDE with animations is available online at fdsp.net/8-7. When the Normal pdf shape is used for the kernel, it is typically called a **Gaussian kernel**. SciPy.stats has a `stats.gaussian_kde()` function for performing KDE using a Gaussian kernel. It uses a heuristic to choose the variance of the Gaussian (called the kernel bandwidth). The function `stats.gaussian_kde()` returns a *function* that is an empirical pdf and that can be called just like the standard pdf method of SciPy.stats continuous random variables:

```
Nvals3=N.rvs(size=1000)

Nkde=stats.gaussian_kde(Nvals3)
```

```
x=np.linspace(-5,5,100)
plt.plot(x,Nkde(x),  label='Empirical pdf')
plt.plot(x,N.pdf(x), label='True pdf')
plt.legend();
plt.xlabel('$n$');
plt.ylabel('$f_N(n)$ or $\hat{f}_N(n)$');
plt.title('Comparison of true pdf and empirical pdf from stats.gaussian_kde()');
```

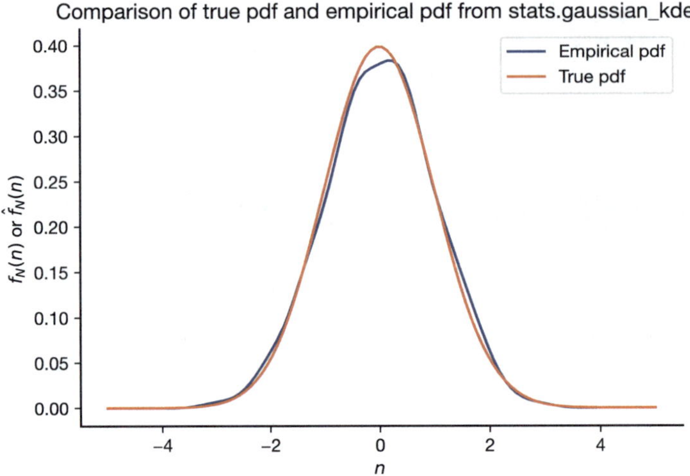

Terminology review and self-assessment questions

Interactive flashcards to review the terminology introduced in this section and self-assessment questions are available at fdsp.net/8-7, which can also be accessed using this QR code:

8.8 Conditioning with Random Variables

Random variables can depend on events or on other random variables. Let's start with the simplest case. We are given a random variable for which the distribution of the random variable depends on some other event. This will be most clear through an example:

Example 8.17: Motivating Example: Binary Communications

In a binary communication system, a transmitter attempts to communicate a stream of bits to a receiver. The received signal is a noisy version of the transmitted signal. After some signal processing, the received signal is converted into a *decision statistic*, X, which is a random variable that can be used to decide whether the transmitted signal was a 0 or a 1. For the decision statistic to be useful in deciding which bit was transmitted, its distribution must change depending on which bit was transmitted. In many cases, X is a Normal random variable, and the standard deviation of X does not depend on which bit was transmitted. The mean will change depending on which bit was transmitted. Let's use the following model:

$$\begin{cases} X \sim \text{Normal}(+1, \sigma), & 0 \text{ transmitted} \\ X \sim \text{Normal}(-1, \sigma), & 1 \text{ transmitted.} \end{cases}$$

Let T_i denote the event that i is transmitted. Then we can write a conditional distribution function for X given that i was transmitted as

$$F_X\left(x \mid T_i\right) = P\left(X \leq x \mid T_i\right).$$

We can create a conditional pdf using our usual approach of taking the derivative of the corresponding (conditional) CDF,

$$f_X\left(x \mid T_i\right) = \frac{d}{dx} F_X\left(x \mid T_i\right).$$

An example of the conditional densities for this binary communication system example is shown below:

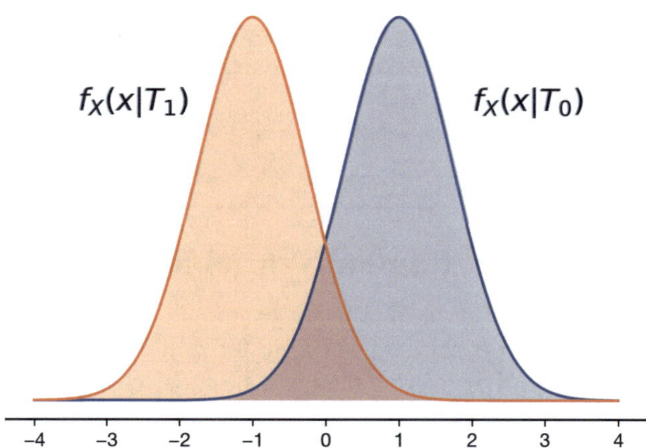

Note that we only have conditional information about the output of the system given the input of the system in the form of the conditional densities $f_X(x \mid T_0)$ and $f_X(x \mid T_1)$. These are called *likelihoods*:

Definition

likelihood (discrete-input, continuous-output stochastic system)

Consider a stochastic system with a discrete set of possible input events $\{A_0, A_i, \ldots\}$, where the output is a continuous random variable, X, whose distribution depends on the input event. Then the *likelihoods* are the conditional pdfs for X given A_i, which are denoted by $f_X(x|A_i)$.

To calculate probabilities on X, the likelihoods are not sufficient because they are conditioned on either T_0 or T_1. We need $P(T_0)$ and $P(T_1)$ so we can use techniques like the Law of Total Probability or Bayes' rule. As in Section 7.1, these are *a priori* probabilities. Even though this system has a continuous output, it has a discrete input, so our former definition of *a priori* probabilities holds without modification.

Now let's see how we can use conditional densities and distributions:

(a) Calculate the probability that $X > 2$ if $P(T_0) = 0.4$ and $P(T_1) = 0.6$ if $\sigma = 2$.

We can consider T_i to be the hidden state in this problem. To find $P(X > 2)$, we can apply the Law of Total Probability, where we condition on the hidden state:

$$P(X > 2) = P(X > 2 \mid T_0)\, P(T_0) + P(X > 2 \mid T_1)\, P(T_1).$$

We are given $P(T_0)$ and $P(T_1)$ in the problem statement, and we know that conditioned on a particular T_i, X is Normal with some mean $\mu \in \{+1, -1\}$ and standard deviation $\sigma = 2$. Here are two different approaches to solve this problem using the conditional distributions:

1. We can use SciPy.stats to create each of the conditional distributions and then use the built-in methods to evaluate the probability. Start by creating the two conditional distributions:

```
XgivenT0 = stats.norm(1, 2)
XgivenT1 = stats.norm(-1, 2)
```

Then $P(X > 2 \mid T_i)$ is simply the survival function of X using the conditional distribution given T_i. So, $P(X > 2 \mid T_0)$ and $P(X > 2 \mid T_1)$ can be evaluated as shown below:

```
XgivenT0.sf(2), XgivenT1.sf(2)
```

```
0.3085375387259869, 0.06680720126885807
```

Putting these values into the Total Probability expression gives the value for $P(X > 2)$,

```
XgivenT0.sf(2) * 0.4 + XgivenT1.sf(2) * 0.6
```

```
0.1634993362517096
```

2. An alternative approach that gives an answer in terms of the Q-function is available online at fdsp.net/8-8.

(b) Calculate the probability of the events T_0 and T_1 given that $X > 2$ if $P(T_0) = 0.4$ and $P(T_1) = 0.6$ if $\sigma = 2$.

Let's try to write one of these probabilities using our usual definition for evaluating a conditional probability,

$$P(T_0 \mid X > 2) = \frac{P(T_0 \cap X > 2)}{P(X > 2)}.$$

Since we do not directly know how to compute $P(T_0 \cap X > 2)$, we need to rewrite it using the Chain Rule in terms of probabilities that we do know: $P(T_0 \cap X > 2) = P(X > 2 \mid T_0)\, P(T_0)$. We have already shown how to calculate $P(X > 2)$ using the Law of Total Probability. Putting these together, we get

$$\begin{aligned} P(T_0 \mid X > 2) &= \frac{P(X > 2 \mid T_0)\, P(T_0)}{P(X > 2)} \\ &= \frac{Q\left(\frac{2-1}{\sigma}\right) P(T_0)}{P(X > 2)}. \end{aligned}$$

Note that the top line of this equation is just a different form of Bayes' Rule. Then the error probability given the specified parameters is

```
q(1/2)*0.4/(q(1/2)*0.4+q(3/2)*0.6)
```

```
0.754834963368876
```

The probability of T_1 given $X > 2$ is

$$P(T_1 \mid X > 2) = \frac{P(X > 2 \mid T_1)\, P(T_1)}{P(X > 2)}$$
$$= \frac{Q\left(\frac{2-(-1)}{\sigma}\right) P(T_0)}{P(X > 2)}.$$

```
q(3/2)*0.6/(q(1/2)*0.4+q(3/2)*0.6)
```

```
0.24516503663112402
```

Note that $P(T_0 \mid X > 2) + P(T_1 \mid X > 2) = 1$. This is expected because T_0 and T_1 are complementary events.

(c) Suppose now that we want to calculate the probability of the events T_0 and T_1 given that $X = 2$ if $P(T_0) = 0.4$, $P(T_1) = 0.6$, and $\sigma = 2$.

This looks a lot like the previous problem. If we try to apply the same technique as in the previous example, we get the following expression for $P(T_0 \mid X = 2)$:

$$P(T_0 \mid X = 2) = \frac{P(X = 2 \mid T_0)\, P(T_0)}{P(X = 2)}$$

This might look okay, except that $P(X = 2 \mid T_0) = 0$. In addition, $P(X = 2) = P(X = 2 \mid T_0)\, P(T_0) + P(X = 2 \mid T_1)\, P(T_1) = 0$. So the result is $0/0$. The problem is that we are conditioning on an event that has zero probability. But keep in mind that every time that X is received, it takes on *some* value with zero probability. Being able to answer this type of question is important, but we don't have the math to deal with it yet. This type of conditional probability is called *point conditioning*, which we discuss further in Chapter 10.

The Memoryless Property of the Exponential Distribution

In Example 8.14, the time that had passed since a headlight was first in service did not seem to affect its future lifetime. This is an example of the memoryless property of the exponential distribution, which is explored more on this book's website at fdsp.net/8-8.

Terminology review and self-assessment questions

Interactive flashcards to review the terminology introduced in this section and self-assessment questions are available at fdsp.net/8-8, which can also be accessed using this QR code:

8.9 Chapter Summary

We introduced *random variables* as tools to model numerical random phenomena. We discussed two classes of random variables and the tools (functions) to work with them:

- *Discrete random variables* have finite or countably infinite ranges. The values in the range take on nonzero probabilities. We can work with these probabilities through the probability mass function (PMF) or cumulative distribution function (CDF). Common examples include Discrete Uniform, Bernoulli, Binomial, Geometric, and Poisson random variables.

- *Continuous random variables* have ranges that are uncountably infinite, such as intervals of the real line. We showed that each value of a continuous random variable has zero probability, but intervals of values can have nonzero probability. The values have *probability density*, and we introduced the probability density function (pdf) as one way to calculate probabilities for continuous random variables; the other way is to use the CDF. Common examples include Continuous Uniform, Exponential, Normal, Chi-Squared, and Student's t random variables.

We gave an example using a statistical data set on adult heights in the US that showed how random variables can be used to model statistical data and that models can be used to answer questions that cannot be answered through resampling approaches.

We showed that normalized histograms and normalized cumulative histograms approximate the pdfs and CDFs, respectively, of random variables, as the number of values from the random variables becomes large and the size of the bins becomes small. We also introduced kernel density estimation (KDE) as an alternative to histograms that generally provides a better estimate of the shape of the pdf for continuous random variables.

Finally, we introduced conditional distribution functions and showed how to use them to solve problems relating to a communication system.

Access a list of key take-aways for this chapter, along with interactive flashcards and quizzes at fdsp.net/8-9, which can also be accessed using this QR code:

9

Expected Value, Parameter Estimation, and Hypothesis Tests on Sample Means

Summary statistics for data, such as the average/sample mean, were introduced in Section 3.4. Similar measures, called moments, can be created for random variables, where the moments are computed from the random variables' distributions instead of observations of those random variables. In this section, I show the connections between the average of random values from a distribution and the moments of that distribution. Moments are often used to characterize the distribution of random variables, and we investigate techniques to estimate moments and characterize these estimates. Finally, knowledge of moments is used to revisit the discussion of sample distributions, bootstrap distributions, confidence intervals, and power.

9.1 Expected Value

Let's start by showing how we can use the idea of the average of a data set to build a similar concept for a random variable. Let X be a discrete random variable with a finite range denoted by $\text{Range}(X) = \{a_0, a_1, \ldots, a_{K-1}\}$. Let the PMF of X be denoted by $p_X(x)$.

Now suppose we have n random values sample from this distribution, $x_0, x_1, \ldots, x_{n-1}$. Then the average of the data is

$$\overline{x} = \frac{1}{n} \sum_{i=1}^{n} x_i. \tag{9.1}$$

We would like to find a similar average for X without having to sample values from the distribution of X. We will call this statistic for X an *ensemble average* because it is computed over the ensemble of potential values that X takes on and is computed from the distribution of X.

We can use *relative frequency* to connect the average of the sample values to the ensemble average. Note that in (9.1), some of the sample values x_i may actually be the same number. For instance, the range of X may consist of only 10 values, but we draw 100 sample values. This means that some of those 100 sample values must be a repeat of a value from the range of X. For each possible value a_k, let n_k be the number of time a_k appears in the sample $x_0, x_1, \ldots, x_{n-1}$. The total contribution of all the terms with value a_k to the sum in (9.1) is then $n_k \cdot a_k$. Then we can rewrite (9.1) as

$$\overline{x} = \frac{1}{n} \sum_{k=0}^{K-1} n_k \cdot a_k. \tag{9.2}$$

DOI: 10.1201/9781003324997-9

Let's move the factor $1/n$ inside the summation in (9.2) to yield

$$\overline{x} = \sum_{k=0}^{K-1} a_k \left(\frac{n_k}{n}\right). \tag{9.3}$$

Note that n_k/n is the relative frequency of outcome k. If the experiment possesses statistical regularity, then as $n \to \infty$,

$$\lim_{n\to\infty} \frac{n_k}{n} = p_X(k),$$

where $p_X(k)$ is the probability of outcome k. Applying this to (9.3) and moving the limit inside the summation yields

$$\lim_{n\to\infty} \overline{x} = \sum_{i=0}^{K-1} a_k p_X(k).$$

The average converges to a value that does not depend on the data samples from the distribution of X but instead depends directly on the distribution of X through $p_X(x)$.

We use this approach to define the *expected value* or *mean* of X:

Definition

expected value (discrete random variable)
mean (discrete random variable)

The expected value, or ensemble mean, is denoted by $E[X]$ or by μ_X and is given by

$$\mu_X = E[X] = \sum_x x p_X(x).$$

Continuous random variables do not have PMFs, and our arguments regarding convergence of the sample average do not apply in the same way. If X is a continuous random variable, then $\mu_X = E[X]$ is defined as follows:

Definition

expected value (continuous random variable)
mean (continuous random variable)

The expected value, or ensemble mean, is denoted by $E[X]$ or by μ_X and is given by

$$\mu_X = E[X] = \int_{-\infty}^{\infty} x f_X(x)\, dx.$$

Special Cases

There are some special cases where $E[X]$ may not be defined. Such cases are outside the scope of this book. In some cases, $E[X]$ may be defined and still be infinite.

The concept of expected value is broader than just the mean. For a random variable X, the mean is defined above and is $\mu_X = E[X]$. But we compute expected values for functions of X, like $E[X^2]$ or $E[(X - \mu_X)^2]$.

Why do we care about the mean?

There are several reasons we care about the mean.

1. As we already saw, the limit of the average value is the mean for most experiments.
2. If we wish to use a constant value to estimate a random variable, then the mean is the value that minimizes the mean-square error.
3. The mean is commonly used as a parameterization of distributions.

Examples

Example 9.1: Rolling a fair 6-sided die

Let D be a random variable whose value is the top face when a fair 6-sided die is rolled. The PMF of D is

$$p_D(d) = \begin{cases} \frac{1}{6}, & d = 1, 2, 3, 4, 5, 6 \\ 0, & \text{o.w.} \end{cases}$$

Then the mean of D is

$$\begin{aligned} E[D] &= \sum_{d=1}^{6} d \cdot p_D(d) \\ &= \sum_{d=1}^{6} d \cdot \frac{1}{6}, \end{aligned}$$

which is

```
mu_d = 0

## Be careful! To include 6, we need to set the upper limit of the range to 7
for d in range(1, 7):
  mu_d += d* (1/6)

print(f'E[D] = {mu_d}')
```

```
E[D] = 3.5
```

The plot in Fig. 9.1 illustrates the PMF of D, and the value of $E[D]$ is labeled.

Example 9.2: Bernoulli Random Variable

Calculating the mean of a Bernoulli random variable may seem trivial, but it will be used to demonstrate an important property of expected values. From Section 8.4,

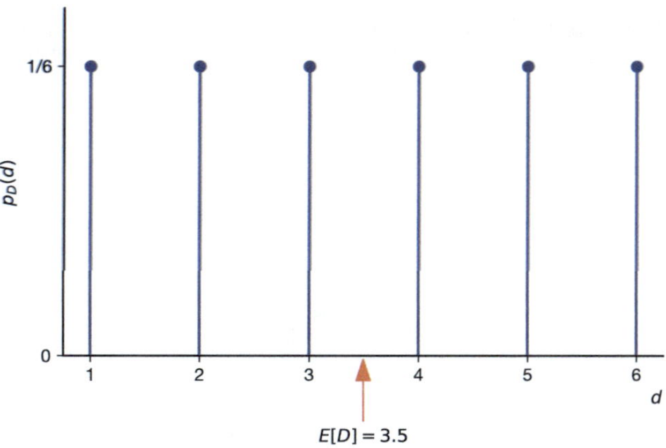

FIGURE 9.1
PMF and expected value for a random variable D that is the outcome of rolling a fair die.

the PMF of a Bernoulli random variable B with probability of success p is

$$p_B(b) = \begin{cases} 1-p, & b=0 \\ p, & b=1 \\ 0, & \text{o.w.} \end{cases}$$

Then

$$\begin{aligned} E[B] &= \sum_{b=0}^{1} b p_B(b) \\ &= 0 \cdot (1-p) + 1 \cdot (p) = p \end{aligned}$$

The PMFs of Bernoulli (p) random variables are shown in Fig. 9.2 for two different values of p. In each plot, the expected value $E[B]$ is labeled. In each example, $E[B]$ is not a value in the range of B. $E[B]$ can be visualized as the value that would make the PMF balance if the values of the PMF were masses on the x-axis.

Although this example is very simple, it can help us find the expected value of the Binomial random variable, which has a much more complicated PMF. To do that, we need to know more about the properties of expected value.

Properties of Expected Value

1. Expected value of a constant is that constant.

A constant c can be treated as a discrete random variable with all of its probability mass at c:

$$p_C(x) = \begin{cases} 1, & x = c \\ 0, & x \neq c \end{cases}.$$

Then we can find the expected value of the constant as

$$E[c] = \sum_{x=c} x p_X(x) = c(1) = c.$$

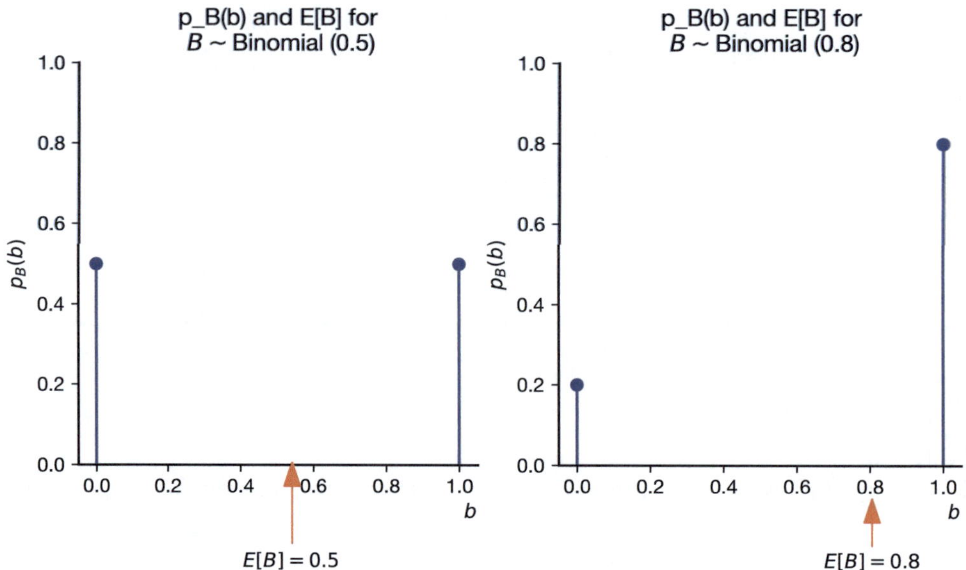

FIGURE 9.2
PMF and expected value for two Bernoulli (p) random variables.

2. Expected value is a linear operator.

If X and Y are random variables, and a and b are arbitrary constants, then

$$E[aX + bY] = aE[X] + bE[Y].$$

Note that this result holds regardless of whether X and Y are independent.

This result generalizes easily, so if X_i, $i = 0, 1, \ldots, N-1$ are random variables and a_i, $i = 0, 1, \ldots, N-1$ are arbitrary constants, then

$$E\left[\sum_{i=0}^{N-1} a_i X_i\right] = \sum_{i=0}^{N-1} a_i E\left[X_i\right].$$

Example 9.3: Example: Expected Value of Binomial RV

Suppose we want to find the formula for the mean of a general Binomial random variable with N trials with probability of success p. Let X denote this random variable. We now know two ways to find $E[X]$ analytically.

1. We can write an equation for the mean using the values and the PMF, where the PMF is

$$E[X] = \sum_{x=0}^{N} x \cdot \binom{N}{x} p^x (1-p)^{N-x}.$$

This can be manipulated into a very simple final result by expanding the binomial coefficient and then canceling factors, or we could solve this using Python for specific values of N and p. However, there is a simpler way.

2. Recall from Section 8.4 that we can think of a Binomial (N, p) random variable as the sum of N independent Bernoulli (p) random variables. Then we can use the fact that expected value is a linear operator to find the mean quickly.

Let $B_i, \quad i = 1, 2, \ldots, N$ be the Bernoulli(p) random variables. Then

$$
\begin{aligned}
E[X] &= E\left[\sum_{x=0}^{N} B_i\right] \\
&= \sum_{x=0}^{N} E\left[B_i\right] && \text{(by linearity)} \\
&= \sum_{x=0}^{N} (p) && \text{(Using mean of Bernoulli RV)} \\
&= Np.
\end{aligned}
$$

SciPy.stats distributions have a `mean()` method. If we have a Binomial (10, 0.25) random variable, we can find its mean using SciPy.stats as follows:

```
import scipy.stats as stats
X = stats.binom(10, 0.25)
print(f'E[X] = {X.mean()}')
```

```
E[X] = 2.5
```

The results match our formula, $E[X] = Np = (10)(0.25) = 2.5$. The PMF for this random variable is shown in Fig. 9.3 with the expected value labeled.

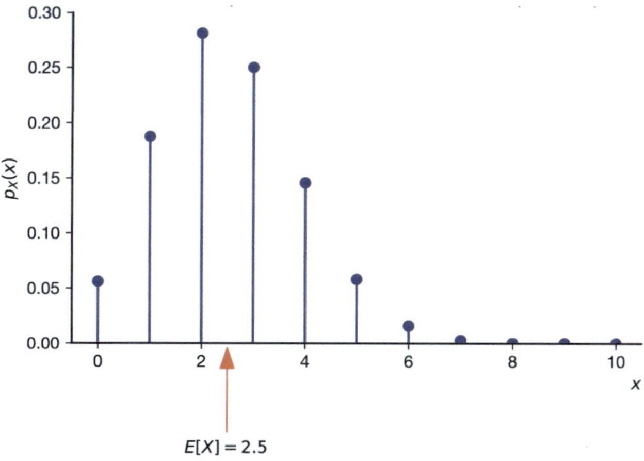

FIGURE 9.3
PMF and expected value for a random variable $X \sim$ Binomial(10, 0.25).

Continuous random variables require integration to find the mean, which can sometimes be complicated and introduce errors in calculation. To ease the burden of doing calculus, in the next section I show how to use the SymPy module to evaluate the expected value of a continuous random variable.

Terminology review and self-assessment questions

Interactive flashcards to review the terminology introduced in this section and self-assessment questions are available at fdsp.net/9-1, which can also be accessed using this QR code:

9.2 Expected Value of a Continuous Random Variable with SymPy

Consider a random variable X with the simple density function shown in Fig. 9.4. Note that I left a parameter m in the density, where m is the slope of the line. We should be able to figure out m from what we know about the pdf. In particular, the pdf must integrate to 1. Rather than carry out that integration by hand, let's use SymPy. We start by defining the variables we will need:

```
import sympy as sp

x, m = sp.symbols("x,m")
```

Now we can define our function. For this first version, let's ignore the fact that the function really needs to be defined piecewise – we will just do all our computations on $[0, 3]$. Since the function is linear in x with slope m, it is just $f_X(x) = mx$ on $[0, 3]$. In SymPy, we can write:

```
fX1 = m*x
```

Let's integrate the density function on $[0, 3]$ to find the total probability as a function

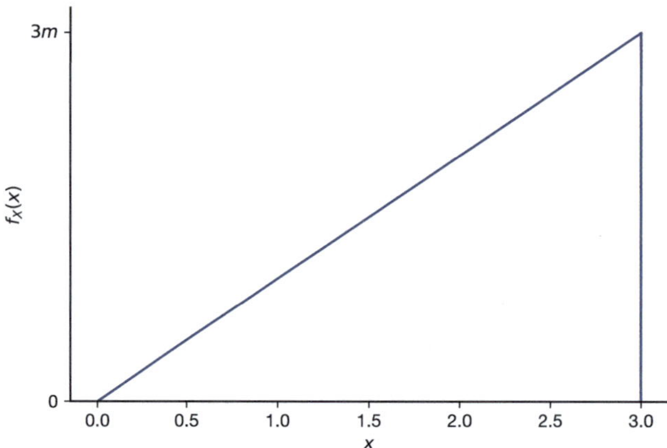

FIGURE 9.4
The pdf $f_X(x)$ for an example random variable X.

of m. Mathematically, we would write

$$\int_0^3 f_X(x)\, dx = \int_0^3 mx\, dx = 1.$$

In SymPy, we can use the `integrate()` function. The first argument of `integrate()` is the function to be integrated. The second argument is either:

- *For indefinite integrals:* the variable to be integrated
- *For definite integrals:* a tuple containing the variable to be integrated, the lower limit of integration, and the upper limit of integration

To find the value of m, we need to calculate the definite integral as x goes from 0 to 3, so the integral can be calculated using SymPy as:

```
sp.integrate(fX1, (x, 0, 3) )
```

$$\frac{9m}{2}$$

The result is $9m/2$. This is not hard to verify. We could calculate the integral by hand, or we could simply observe that we are finding the area of a triangle with base $b = 3$ and height $h = 3m$. Since the area of a triangle is $bh/2$, the result is $9m/2$.

For $f_X(x)$ to be a valid pdf, it must integrate to 1, so $9m/2 = 1$, or $m = 2/9$. Let's make a new version of $f_X(x)$ in SymPy with this substitution. Rather than manually substituting it, I will show you how to let SymPy do the work for you. We can put in `2/9` for the upper bound, but the result will be converted to a decimal. SymPy can work with fractions if instead of 2/9, we pass `sp.Rational(2,9)` to tell SymPy to use the rational form of 2/9:

```
fX = fX1.subs(m, sp.Rational(2,9) )
fX
```

$$\frac{2x}{9}$$

Let's confirm that `fX` now integrates to 1:

```
sp.integrate(fX, (x, 0, 3) )
```

$$1$$

Now we are ready to find the expected value $E[X]$. From Section 9.1, the expected value of a continuous random variable X is

$$E[X] = \int_{-\infty}^{\infty} x f_X(x)\, dx.$$

Again, we will ignore the parts of the integral where $f_X(x) = 0$. Then this is easy to carry out in SymPy:

```
sp.integrate(x * fX, (x, 0, 3) )
```

$$2$$

So the mean (i.e., expected value) of X is 2. When calculating results, it is helpful to perform some basic checks to determine if the result is reasonable. The value we calculated is reasonable because the mean is toward the middle of the variable's range, but it is to the right of the center of the range, where there is more density.

What expected value does not mean

(Or what the *mean* doesn't mean!)

1. The mean is not the most likely value to occur. The value with the highest density is 3. For a continuous random variable, all the values occur with probability 0, but the probability of getting values close to 3 is higher than the probability of getting values close to 2.

The value 3 is the *mode* of X:

> Definition
>
> **mode(s) (of a random variable)**
>
> The value(s) with the highest probability (for a discrete random variable) or the highest probability density (for a continuous random variable).

2. The mean is not the value in the middle of the distribution. Since the range of X is $[0, 3]$, the value in the middle of the distribution is 1.5.

3. The mean is not the value that splits the probability of the distribution equally. We can solve for that value using SymPy:

First, note that we are looking for the point c such that $P(X \leq c) = P(X > c)$. Since those two probabilities add to 1, we can simply find c such that $P(X \leq c) = 1/2$.

Second, note that $P(X \leq c)$ is in the form of the CDF, $F_X(x) = P(X \leq x)$. Let's find the CDF of X using indefinite integration on the density function:

```
FX = sp.integrate(fX, x)
FX
```

$$\frac{x^2}{9}$$

The CDF $F_X(x)$ is a quadratic function. Thus, we can use the quadratic equation to find the values where $F_X(x) = 1/2$, or we can use SymPy's `sp.nonlinsolve()` function. The arguments of `sp.nonlinsolve()` are a list of functions that must evaluate to 0 and a list of arguments to be found. First, we need to rewrite the function we are going to solve to get it in the right form:

$$F_X(c) = \frac{1}{2}$$
$$F_X(c) - \frac{1}{2} = 0$$

Now we can use `sp.nonlinsolve()`:

```
c = sp.symbols('c')
solns = sp.nonlinsolve([FX.subs(x, c) - sp.Rational(1,2) ], [c])
solns
```

$$\left\{\left(-\frac{3\sqrt{2}}{2},\right),\left(\frac{3\sqrt{2}}{2},\right)\right\}$$

Since this is a quadratic equation, there are two answers, but only one is in $[0,3]$. We can get the floating point values by iterating over `solns`, getting the first (and only) element of each returned tuple, and using `sp.N()` to convert the result to a float:

```
for x in solns:
  print(sp.N(x[0]))
```

```
-2.12132034355964
2.12132034355964
```

Since we are looking for a value in $[0,3]$, the value that divides the probability into equal halves is $c \approx 2.12$. This is not the mean; this is the *median*:

Definition

median (of a random variable)

For a random variable X with distribution function $F_X(x)$, the median is a value $\tilde{X}$ such that $P\left(X \leq \tilde{X}\right) = P\left(X > \tilde{X}\right)$. An equivalent condition is $F_X\left(\tilde{X}\right) = 1/2$. The median is not necessarily unique.

So, the mean is not necessarily equal to the median.

The tools we developed in this section allow us to easily calculate the expected values of continuous random variables. In the next section, we apply these tools as we learn about moments, which are expected values of powers of a random variable.

Terminology review and self-assessment questions

Interactive flashcards to review the terminology introduced in this section and self-assessment questions are available at fdsp.net/9-2, which can also be accessed using this QR code:

9.3 Moments

The distribution of a random variable is often characterized not only in terms of the mean $\mu_X = E[X]$, but also in terms of other expected values called *moments*. We are going to consider the two most common types of moments. The first of these are just called *moments* or nth *moments*:

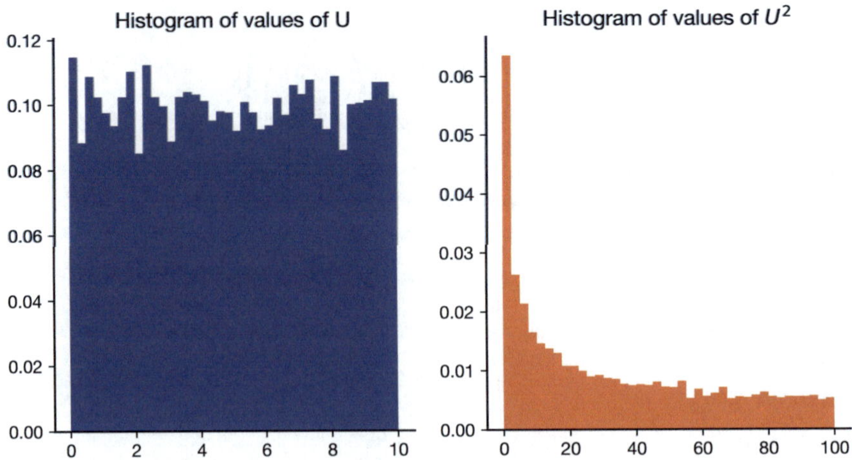

FIGURE 9.5
Empirical densities for values from U and U^2.

> **Definition**
>
> ***n*th moment**
>
> The nth moment of a random variable X is $E[X^n]$ for $n = 1, 2, \ldots$.

Note that the mean is the first moment.

Example 9.4: Estimating a Second Moment

Let's consider a second moment $E[U^2]$ of a random variable U. If we let $g(u) = u^2$, then $g(U)$ is a function of a random variable. A function of a random variable is also a random variable. Determining the density of Y is beyond the scope of this text, but I want to give you some insight into that density. Let's make a random variable $U \sim$ Uniform(0,10) using Scipy.stats, draw 10,000 random values, and compute the squares of those 10,000 random values:

```
import scipy.stats as stats

U=stats.uniform(0, 10)
Uvals = U.rvs(10_000)
Uvals2 = Uvals ** 2
```

We can approximate the pdfs of U and U^2 using their empirical pdfs, which are shown in Fig. 9.5. The two densities look quite different. We can use the averages of the values and the squares of the values to estimate the expected values. Thus, the expected values $E[U]$ and $E[U^2]$ are approximately:

```
print(f'Avg of U values: {Uvals.mean(): .2f}')
print(f'Avg of U**2 values: {(Uvals2).mean(): .2f}')
```

```
Avg of U values:  4.99
Avg of U**2 values:  33.40
```

The true mean of U is

$$\int_0^{10} u\left(\frac{1}{10}\right)\, du = \left.\frac{u^2}{20}\right|_0^{10} = 5.$$

The average of values from the random variable U is very close to the true mean. Note that $E[U^2] \approx 33.40 \neq E[U]^2 = 25$. This implies that $E[U^2] \neq (E[U])^2$. This seems like bad news because:

1. We can't get $E[U^2]$ from $E[U]$.
2. We don't know how to get the distribution of $Y = U^2$ from the distribution of U.

Fortunately, we can calculate moments without having to find the distribution of a function of a random variable using the LOTUS rule:

Definition

Law of the Unconscious Statistician (LOTUS)

Let $g(x)$ be a real function. If X is a discrete random variable, then

$$E\left[g(X)\right] = \sum_x g(x) p_X(x).$$

If X is a continuous random variable, then

$$E\left[g(X)\right] = \int_{-\infty}^{\infty} g(x) f_X(x)\, dx.$$

It is called the "Law of the Unconscious Statistician" because it is what you might try if you did not realize that $g(X)$ is itself a random variable with its own distribution that is different than that of X. Yet it can be proven to be correct.

Example 9.4 (continued)

Let's see what this means for our example of $E[U^2]$. We are evaluating $E[f(U)]$ for the function $g(u) = u^2$. Thus, by the LOTUS rule

$$\begin{aligned} E[U^2] &= \int_{-\infty}^{\infty} g(u) f_U(u)\, du \\ &= \int_{-\infty}^{\infty} u^2 f_U(u)\, du, \end{aligned}$$

which I will show how to calculate with SymPy:

```
import sympy as sp
u = sp.symbols('u')
sp.integrate(u**2 * (1/10), (u, 0, 10) )
```

33.3333333333333

9.3.1 Interpretation of Moments

You might be wondering about the purpose of defining different moments. Let's try to get some insight by looking at the equation to calculate the nth moment for a continuous random variable,

$$E[X^n] = \int_{-\infty}^{\infty} x^n f_X(x)\, dx.$$

The nth moment is the integral of the product of two functions. Let's plot these two functions along with their product for two Normal density functions – don't worry about the parameters I chose for the distributions, focus on building intuition through the plots.

We start with the 1st moment (the mean),

$$E[X] = \int_{-\infty}^{\infty} x f_X(x)\, dx.$$

We will use the following interpretation. For this integral, each probability density value $f_X(x)$ is weighted by the position x of that density value. In Fig. 9.6, I illustrate the integrand for two different Normal densities. The left-hand sides of the graphs show the pdf and the linear weighting function, and the right-hand side of the graphs show their product.

Normal distribution 1 has a pdf that is symmetric around 0, and the weighting function applies equal positive and negative components. The resulting mean is 0: since $x f_X(x)$ is an odd symmetric function around 0, the integral is 0. The pdf for Normal distribution 2 is symmetric around 1 and has more spread away from 1. The product of the pdf and the weighting function may at first glance seem symmetric, but further inspection should reveal that it is not. However, the mean is still 1. The parameter μ of the Normal distribution that we have been calling the mean is, in fact, always the mean of that distribution. A proof is available online at fdsp.net/9-3.

Now consider the 2nd moment,

$$E[X^2] = \int_{-\infty}^{\infty} x^2 f_X(x)\, dx.$$

The weight function is now a quadratic that is always positive and increases with the *square of the distance from the point* $x = 0$. The second moment will be large when the probability mass or density is distant from the origin. This is shown in Fig. 9.7 for the two example Normal densities. Note the extreme difference in the values of the second moment for these distributions. The second distribution has most of its probability density away from $x = 0$, whereas the first distribution has its probability density highly concentrated around 0.

If we have different distributions that are centered at 0, then the second moment measures the spread of the distribution, as shown in Fig. 9.8. However, if the mean is nonzero, the second moment will also depend on the mean. If we want to measure the spread around

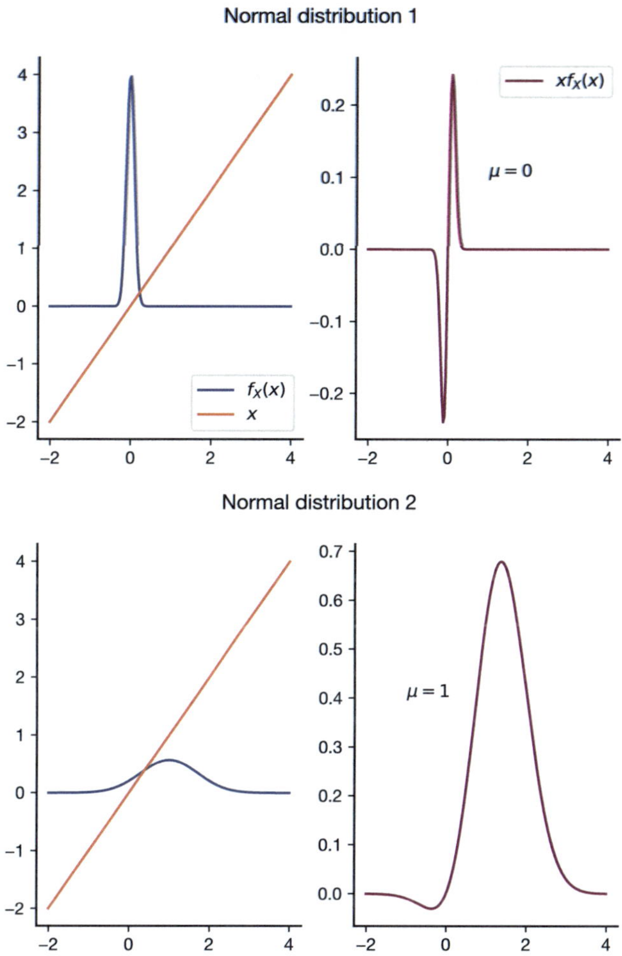

FIGURE 9.6
Integrand for computing mean for two different Normal densities.

the mean, we should subtract the mean from the random variable before evaluating the second moment. This creates a *central moment*:

> Definition
>
> **nth central moment**
>
> Let X be a random variable with mean μ_X. Then the nth *central moment* of X is $E[(X - \mu_X)^n]$ for $n = 2, 3, \ldots$.

The most comment moment after the mean is the *variance*:

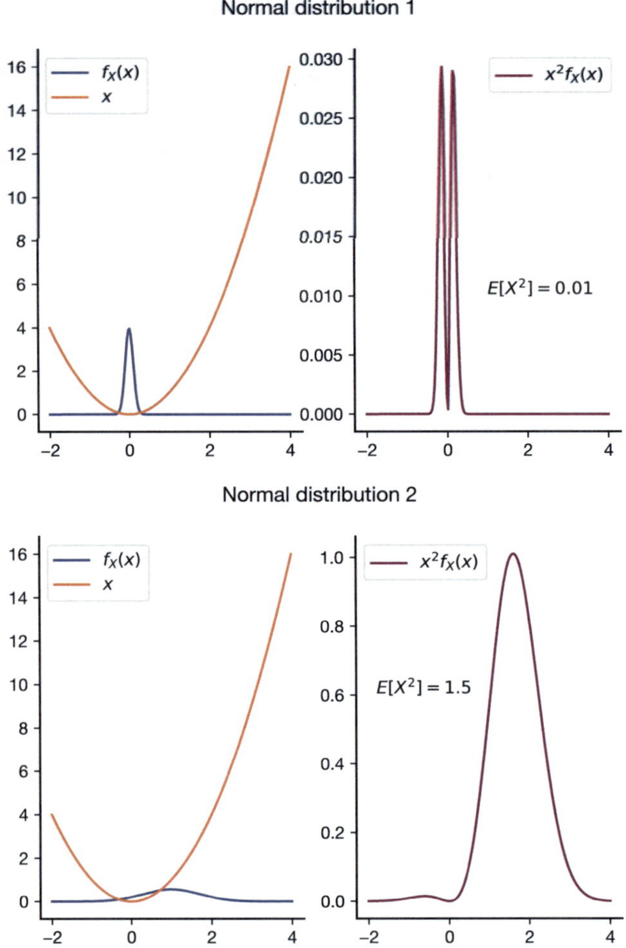

FIGURE 9.7
Integrand for computing second moment for two different Normal densities.

> **Definition**
>
> **variance (random variable)**
>
> Let X be a random variable with mean μ_X. **The variance is the second central moment** and is denoted by $\text{Var}(X)$ or σ_X^2. I.e., $\sigma_X^2 = E\left[(X - \mu_X)^2\right]$.

The variance of a random variable may be infinite, even if the mean is finite.

We can interpret central moments using a similar visualization as we did for non-central moments. For example, the formula for the variance of a continuous random variable is

$$\text{Var}(X) = \int_{-\infty}^{\infty} (x - \mu)^2 f_X(x)\, dx.$$

The example in Fig. 9.9 shows the density and weighting factors for two Normal distributions with the same σ but different means. Because the quadratic weighting function is shifted to align with the mean of the distribution, the product of the two functions is

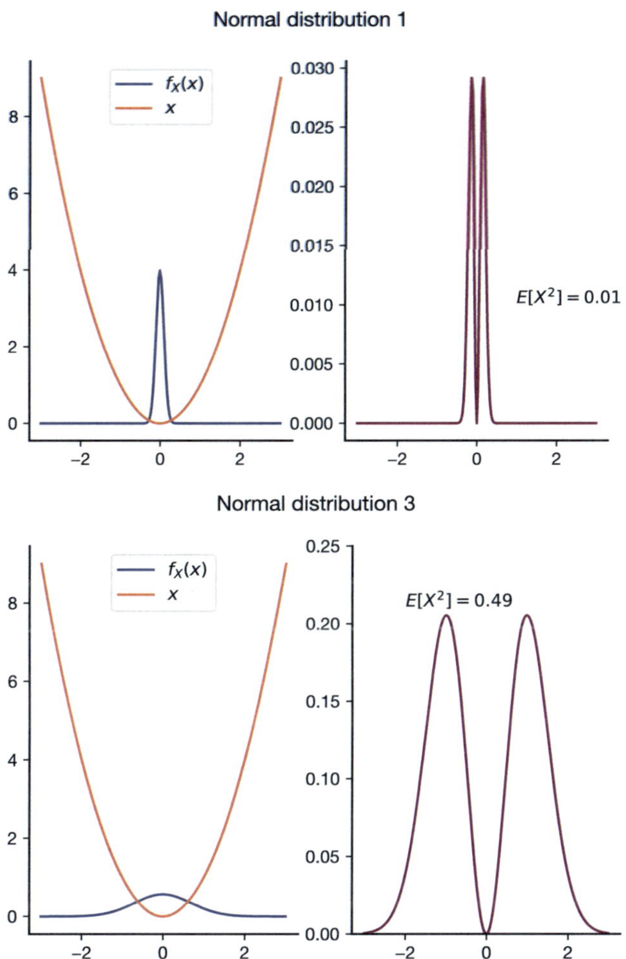

FIGURE 9.8
Integrand for second moment of two Normal distributions with zero means.

even around the mean. The variance of the Normal random variable is more challenging to derive than the mean, so the derivation is omitted. For a Normal random variable with parameter σ, the variance is σ^2. The value σ is called the *standard deviation*:

Definition

standard deviation (random variable)

Let X be a random variable with finite variance $\sigma_X^2 = E[(X - \mu_X)^2]$. Then the *standard deviation* of X is denoted by σ_X and can be computed as

$$\sigma_X = \sqrt{E\left[(X - \mu_X)^2\right]}.$$

For a given value x, $(x - \mu_x)^2$ is the distance squared of x from the mean of the random variable X. If instead of a particular x, we have random variable X, then $(X - \mu_X)^2$ is a

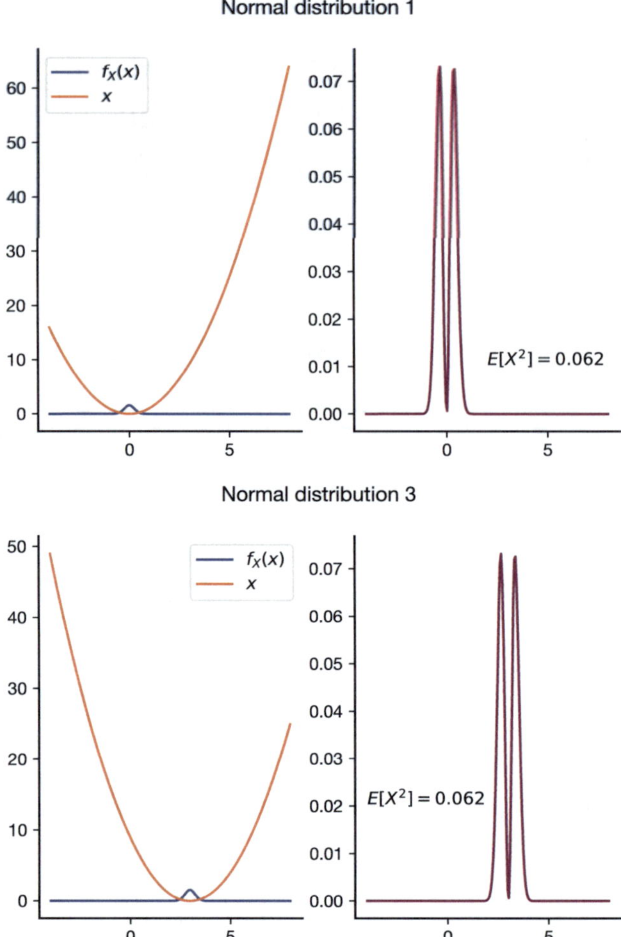

FIGURE 9.9
Integrands for second moment of Normal random variables with same values of σ^2 but different means.

random distance squared. Thus, $\sigma_X^2 = E\left[(X - \mu_X)^2\right]$ is the expected squared distance of the random variable from its mean. The more probability assigned to values far away from the mean of a random variable, the higher the value of σ_X^2.

To illustrate how different values of μ and σ affect the Normal pdf, let's create three different random variables: $Y_0 \sim \text{Normal}(\mu = -4, \sigma = 3)$, $Y_1 \sim \text{Normal}(\mu = 0, \sigma = 1)$, $Y_0 \sim \text{Normal}(\mu = 2, \sigma = 1/2)$. We can create these distributions in SciPy.stats as follows:

```
Y0 = stats.norm(loc = -4, scale =3 )
Y1 = stats.norm(loc = 0, scale = 1)
Y2 = stats.norm(loc = 2, scale = 1/2)
```

The resulting pdfs are shown in Fig. 9.10. (Code to generate this figure is available online at fdsp.net/9-3.) Means change the center of the pdf but do not affect the spread. Larger standard deviations σ_i (and hence larger variances) cause the pdf to spread out more, in which case the magnitude of the pdf must be smaller, since the integral of the pdf must be 1.

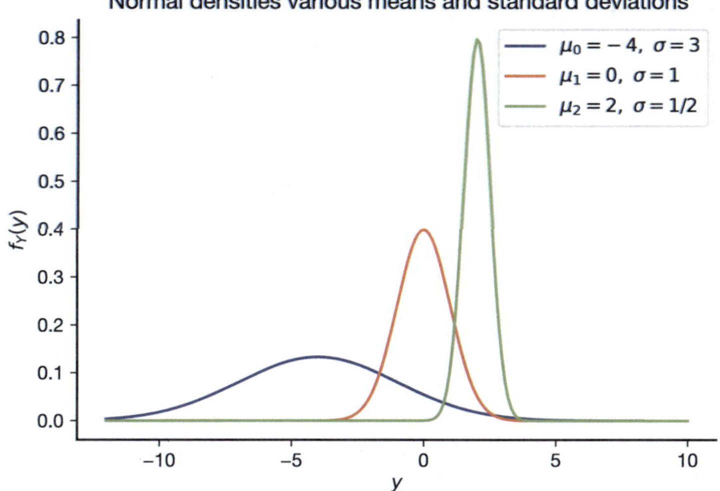

FIGURE 9.10
PDFs for several Normal random variables.

Example 9.5: Mean and variance of an exponential random variable

Recall that the density of the exponential random variable is

$$f_X(x) = \begin{cases} 0, & x < 0 \\ \lambda e^{-\lambda x}, & x \geq 0. \end{cases}$$

Remember that **`lambda`** is a keyword, so we will use the variable `L` in place of **`lambda`**; however, we can use 'lambda' for the string that SymPy associates with the variable `L`, and SymPy will know to typeset it as λ. Let's set up our variables and density function:

```
x, L = sp.symbols('x, lambda')
fX = L * sp.exp(- L * x)
```

WARNING

Note that we had to use SymPy's `sp.exp()` function, not NumPy's `np.exp()` function, so that SymPy will know how to integrate it. This also demonstrates the importance of importing libraries into their own namespaces because otherwise we have a potential confusion between these two functions.

We can first check that our pdf integrates to 1. To do that, we will need to learn that to express ∞ in SymPy, we can use `sp.oo` (because `oo` looks like ∞).

```
sp.integrate(fX, (x, 0, sp.oo) )
```

$$\begin{cases} 1 & \text{for } |\arg(\lambda)| < \frac{\pi}{2} \\ \int\limits_0^\infty \lambda e^{-\lambda x}\, dx & \text{otherwise} \end{cases}$$

We see that `sp.integrate()` can actually handle complex integration, but we only have real integration, so the first result applies. The integral of the pdf is 1.

Now let's compute the mean:

```
sp.integrate(x*fX, (x, 0, sp.oo) )
```

$$\begin{cases} \frac{1}{\lambda} & \text{for } |\arg(\lambda)| < \frac{\pi}{2} \\ \int\limits_0^\infty \lambda x e^{-\lambda x}\, dx & \text{otherwise} \end{cases}$$

The mean is $1/\lambda$. This integral comes up often enough that it may be worth memorizing; otherwise, you may need to look it up in a table or perform integration by parts if you don't use SymPy.

Let's calculate the variance:

```
muX = 1/L
sp.integrate( (x- muX)**2 * fX, (x, 0, sp.oo) )
```

$$\begin{cases} \frac{1}{\lambda^2} & \text{for } |\arg(\lambda)| < \frac{\pi}{2} \\ \int\limits_0^\infty \lambda \left(x - \frac{1}{\lambda}\right)^2 e^{-\lambda x}\, dx & \text{otherwise} \end{cases}$$

The variance of the exponential random variable is $\sigma_X^2 = 1/\lambda^2$.

We often compute the variance using a simpler formula. Consider the definition of the variance and expand the quadratic term:

$$E\left[(X - \mu_X)^2\right] = E\left[X^2 - 2\mu_X X + \mu_X^2\right].$$

Note that μ_X is some constant. Thus, we have a linear combination of terms, and expected value is a linear operator. We can thus rewrite this expression as

$$E\left[(X - \mu_X)^2\right] = E\left[X^2\right] - 2\mu_X E\left[X\right] + \mu_X^2.$$

In the second term, $E[X] = \mu_X$. Then combining the second and third terms yields

$$E\left[(X - \mu_X)^2\right] = E\left[X^2\right] - \mu_X^2.$$

Let's test this for the exponential random variable. The 2nd moment is

```
sp.integrate( x**2 * fX, (x, 0, sp.oo) )
```

$$\begin{cases} \frac{2}{\lambda^2} & \text{for } |\arg(\lambda)| < \frac{\pi}{2} \\ \int\limits_0^\infty \lambda x^2 e^{-\lambda x}\, dx & \text{otherwise} \end{cases}$$

The result is $E[X^2] = 2/\lambda^2$. Then the variance is

$$\sigma_X^2 = E[X^2] - \mu_X^2 = \frac{2}{\lambda^2} - \left(\frac{1}{\lambda}\right)^2 = \frac{1}{\lambda^2},$$

which agrees with our previous result.

Properties of Variance

Variance has several important properties that we will use later in this book. To see these, let X be a random variable and b and c constant values. Then the following hold:

1. The variance of a constant is 0, $\text{Var}[c] = 0$. This is because the mean of a constant is that constant, so

$$E\left[(c - E[c])^2\right] = E\left[(c-c)^2\right] = 0.$$

2. Adding a constant to a random variable does not change its variance. This is because adding a constant shifts the mean by that constant: $E[X+c] = \mu_X + c$, so

$$\begin{aligned} \text{Var}[X+c] &= E\left[(\{X+c\} - E\{X+c\})^2\right] \\ &= E\left[(X + c - \mu_X - c)^2\right] \\ &= E\left[(X - \mu_X)^2\right] = \text{Var}[X]. \end{aligned}$$

3. Multiplying by a constant scales the variance by the **square** of that constant, $\text{Var}[cX] = c^2\,\text{Var}[X]$. The proof uses a similar algebraic approach as for adding a constant and is omitted.

One important implication of Property 3 and the linearity of expectation relates to a form that we have encountered several times when working with the Normal random variable, $(X - \mu_X)/\sigma_X$:

$$\begin{aligned} E\left[\frac{X-\mu_X}{\sigma_X}\right] &= \frac{E[X - \mu_X]}{\sigma_X} && \text{by linearity} \\ &= \frac{E[X] - \mu_X}{\sigma_X} && \text{by linearity} \\ &= \frac{\mu_X - \mu_X}{\sigma_X} = 0, \end{aligned}$$

and

$$\begin{aligned} \text{Var}\left[\frac{X-\mu_X}{\sigma_X}\right] &= \frac{\text{Var}[X - \mu_X]}{\sigma_X^2} && \text{by Property 3} \\ &= \frac{\sigma_X^2}{\sigma_X^2} = 1. \end{aligned}$$

So for any random variable X with finite mean μ_X and finite standard deviation σ_X, $(X - \mu_X)/\sigma_X$ has mean 0 and variance 1.

Any linear operation on a Normal random variable gives a Normal random variable:

$$\text{If } X \sim \text{Normal}(\mu_X, \sigma_X), \text{ then } \frac{X - \mu_X}{\sigma_X} \sim \text{Normal}(0, 1).$$

4. For independent random variables (where knowing one does not change the distribution of the others), **the variance of the sum is the sum of the variances,**

$$\text{Var}\left[\sum_{i=0}^{N-1} X_i\right] = \sum_{i=0}^{N-1} \text{Var}\left[X_i\right].$$

Terminology review and self-assessment questions

Interactive flashcards to review the terminology introduced in this section and self-assessment questions are available at fdsp.net/9-3, which can also be accessed using this QR code:

9.4 Parameter Estimation

We often have the following problem: we observe data points $x_0, x_1, \ldots, x_{N-1}$ that come from some common underlying distribution. However, the details of the underlying distribution are not completely known. In this section, we consider the problem of estimating parameters of the distribution, such as the mean and variance. Let θ denote some parameter of the distribution that we wish to estimate. Then there are two different philosophies for how to treat θ, and these different philosophies result in different mathematical approaches:

- θ can be treated as constant but unknown. The parameter θ is considered deterministic (not random). This is considered *classical parameter estimation* and is the approach taken in this section.
- θ can be considered to be random. This typically results in a Bayesian approach to parameter estimation in which the *a posteriori* distribution for θ is determined from the data and some *a priori* distribution on θ.

In this section, the discussion and examples use continuous random variables, so the mathematics will use density functions $f_X(x)$. In most cases, the results generalize to discrete random variables if the pdfs are replaced by PMFs. It is helpful to make explicit the effect on the density function of the parameter(s) being estimated; i.e., we want to show that the density at some point x depends on θ. We will use $f_X(x;\theta)$ to make this effect clear. We do not use a conditioning bar ($|$) here because θ is not a value of a random variable.

The correct way to interpret $f_X(x;\theta)$ is that it represents a family of density functions for different values of θ. For instance, if θ is the parameter of an exponential random variable, then some of the possible density functions are shown in Fig. 9.11:

Now suppose we have samples from some exponential density $f_X(x;\lambda)$. How should we estimate λ? More generally, how should we estimate θ from the observations $x_0, x_1, \ldots, x_{N-1}$?

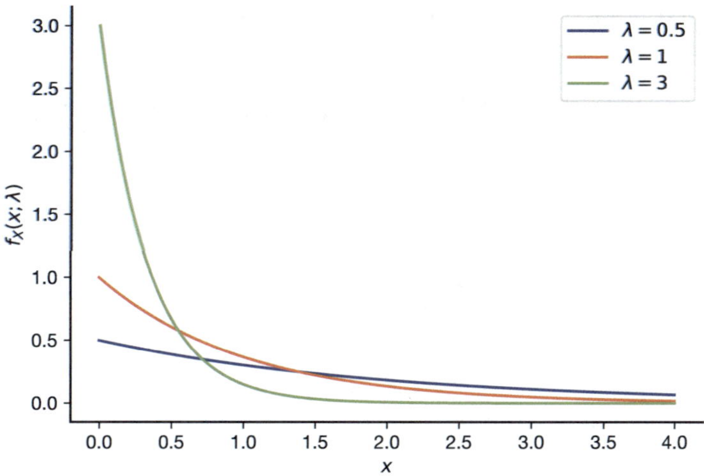

FIGURE 9.11
Samples from family of exponential densities parameterized by λ.

First, let's simplify the notation for the observations by writing them as a *vector*,

$$\mathbf{x} = [x_0, x_1, \ldots, x_{n-1}] .$$

We wish to estimate θ from $\mathbf{x}$. We will use "hats" on an item (like $\hat{\theta}$) to indicate an estimate of a parameter. If the estimate $\hat{\theta}$ is a single value, then we call that a *point estimate*:

Definition

point estimate

Given a vector of observed values $\mathbf{x}$ from a common distribution, a *point estimate*, or simply *estimate*, for a parameter θ is a numerical value $\hat{\theta}$ that is a function of the observed data.

Given the data $\mathbf{x}$, let the (point) estimate for θ be the value of a function evaluated on the data, $\hat{\theta} = g(\mathbf{x})$. Then the main problem of parameter estimation is how to choose a **good** function. Rather than choose a function for one specific set of data $\mathbf{x}$, we want to choose a function that will have good properties for any data. In this case, the data itself becomes a vector of random variables, which we will denote by a bold upper-case letter. The random data is then $\mathbf{X} = [X_0, X_1, \ldots, X_{n-1}]$. If we apply a function to the random variables, its output will itself be a random variable, which we call an *estimator*:

Definition

estimator

Given a vector of random variables $\mathbf{X}$ from a common distribution, a *point estimator*, or simply *estimator*, for a parameter θ of the distribution is a random variable $\hat{\Theta}$ that is a function of $\mathbf{X}$.

If our estimator is $\hat{\Theta} = g(\mathbf{X})$, then we wish to choose the function $g()$ such that the estimator has good properties. However, we need to define what we mean by "good". Our first step is to introduce some additional terminology:

Definitions

estimator error

The difference between the estimator for a parameter and the true value, $\hat{\Theta} - \theta$.

estimator bias

The difference between the mean of an estimator and the true value, $E[\hat{\Theta}] - \theta$.

unbiased estimator

An estimator is unbiased if its mean is the true value, $E[\hat{\Theta}] = \theta$, i.e., if the estimator bias is zero.

9.4.1 Estimating Mean and Variance

Suppose that $X = \{X_1, X_2, \cdots, X_N\}$ are independent observations that come from a common distribution. For our purposes, we do not provide a technical definition for independence of multiple random variables. Informally, independence generally means that knowing the value of one or more of the random variables does not change the distribution of any of the remaining random variables.

Estimating the Mean

Let μ_X denote the mean of the random variables. Then μ_X is usually estimated via the **sample-mean estimator**,

$$\hat{\mu}_X = \frac{1}{N} \sum_{i=0}^{N-1} X_i$$

The expected value of the sample-mean estimator is

$$\begin{aligned} E\left[\hat{\mu}_X\right] &= E\left[\frac{1}{N} \sum_{i=0}^{N-1} X_i\right] = \frac{1}{N} \sum_{i=0}^{N-1} E\left[X_i\right] \quad \text{(by linearity)} \\ &= \frac{1}{N} \sum_{i=0}^{N-1} \mu_X = \mu_X. \end{aligned}$$

Thus, this estimator is **unbiased**.

The sample-mean estimator can be shown to be the estimator that minimizes the sum of the mean-square errors to the random variables,

$$\min_c E\left[\sum_{i=0}^{N-1} (X_i - c)^2\right] = \widehat{\mu}_X.$$

(The derivation is similar to finding the summary statistic that minimizes the MSE to data.)

The sample-mean estimator converges to the true mean if the variance of the distribution is finite. Fig. 9.12 shows the sample-mean estimate as a function of the number of random

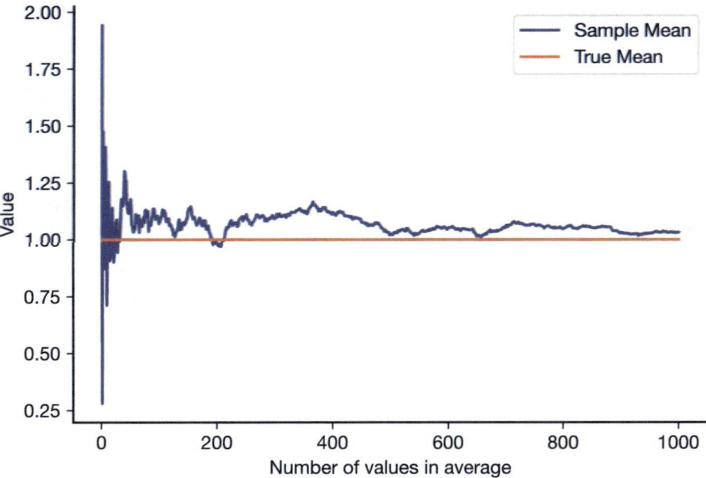

FIGURE 9.12
Comparison of the sample mean to the true mean as a function of the number of values used to compute the average. Data drawn from Normal (1,2) distribution.

values used to create the estimate, when the sample-mean estimate is computed on sets of random values from a Normal (1, 2) distribution. (Code to generate this graph is available online at fdsp.net/9-4.)

Evaluating bias via simulation

Consider again what it means for the average (sample mean) to be an unbiased estimator. **It does not mean that the average converges to the true mean for large** N **– unbiased has nothing to do with the number of samples in the data.** Instead, it means that for any N, the expected value of the average is equal to the true mean.

We can estimate the expected value of an estimator through simulation by calculating the value of the estimator for many different random samples. Then the expected value of the estimator is the average over all of those samples. Below is a simulation that estimates the expected value of the sample-mean estimator for $N = 10$:

```
num_sims = 100_000
data_size = 10 # This is our N
X = stats.norm(1, 2)

total_avg = 0
for sim in range(num_sims):
  Xvals = X.rvs(data_size)
  avg = Xvals.mean()
  total_avg += avg

print('The estimated mean of the sample-mean estimator is'
      +f'{total_avg/num_sims: 0.2f}')
```

```
The estimated mean of the sample-mean estimator is 1.00
```

The estimated mean of the sample-mean estimator is equal to the true mean, so this matches our theoretical result: the sample-mean estimator is unbiased.

> **Important!**
>
> The mean estimator is special because of the Central Limit Theorem (see Section 8.6) – regardless of the type of distribution of the underlying data[1], the distribution of the mean estimator will be approximately Normal if there are at least tens of data points.

Estimating the Variance

Let σ_X^2 denote the variance of the random variables. Then there are two cases that should be considered for estimating the variance.

Known Mean: First, consider the case where the mean of the random variables, μ_X, is known. Let the sample-variance estimator for this case be defined by

$$\hat{\sigma_X^2} = \frac{1}{N}\sum_{i=1}^{N}(X_i - \mu_X)^2. \tag{9.4}$$

The expected value of the sample-variance estimator is

$$\begin{aligned} E\left[\hat{\sigma_X^2}\right] &= E\left[\frac{1}{N}\sum_{i=1}^{N}(X_i - \mu_X)^2\right] \\ &= \frac{1}{N}\sum_{i=1}^{N} E\left[(X_i - \mu_X)^2\right] \quad \text{(by linearity)} \\ &= \frac{1}{N}\sum_{i=1}^{N}\sigma_X^2 \\ &= \sigma_X^2. \end{aligned}$$

So if the mean is known, the sample-variance estimator in (9.4) is an unbiased estimator and is commonly used for this purpose.

As with the sample-mean estimator, let's conduct a simulation to confirm that the sample-variance estimator is unbiased.

```
num_sims = 100_000
data_size = 10 # This is our N
X = stats.norm(1, 2)

total_var = 0
for sim in range(num_sims):
  Xvals = X.rvs(data_size)
  var = ((Xvals-1)**2).mean()
```

(continues on next page)

[1]Actually the distribution will have to satisfy some mild restrictions, such as having finite mean and variance.

(continued from previous page)

```
    total_var += var

print('The estimated mean of the sample-variance estimator is' +
      f'{total_var/num_sims: 0.2f}')
```

```
The estimated mean of the sample-variance estimator is 4.00
```

The estimated mean of the sample-variance estimator matches the true variance (4), confirming the theoretical result that the sample-variance estimator is unbiased.

Unknown Mean:

Now consider the case where the variance of the data is being estimated when the mean is unknown. If we want to apply the sample-variance estimator of the form in (9.4), we will have to replace the true value of μ_X with its estimator, giving

$$\hat{\sigma}_X^2 = \frac{1}{N}\sum_{i=1}^{N}(X_i - \hat{\mu}_X)^2. \tag{9.5}$$

Let's run a simulation to check if this is an unbiased estimator:

```
num_sims = 100_000

data_size = 10 # This is our N
X = stats.norm(1, 2)

total_var = 0
for sim in range(num_sims):
  Xvals = X.rvs(data_size)
  avg = Xvals.mean()
  var = ((Xvals-avg)**2).mean()
  total_var += var

print('The estimated mean of the sample-variance estimator is' +
      f'{total_var/num_sims: 0.2f}')
```

```
The estimated mean of the sample-variance estimator is 3.61
```

The estimated mean of the sample-variance estimator is now not close to the true variance! If you rerun this block, you may see different values, but each value should be close to 3.6. This may seem surprising, but to see if this estimator is unbiased, we would need to calculate

$$E\left[\frac{1}{N}\sum_{i=0}^{N-1}\left(X_i - \sum_{k=0}^{N-1} X_k\right)^2\right]. \tag{9.6}$$

Evaluating this expectation is somewhat tricky, but it can be evaluated in closed form, and the value of (9.6) is $\sigma^2(N-1)/N$. That is, this estimator produces a value that is biased too low by a factor of $(N-1)/N$. This matches our simulation example, which for a sample size of 10 was biased too low by a factor $(10-1)/10 = 0.9$.

NumPy's built-in variance estimator `np.var()` also uses the biased estimator. Fortunately, this is easy to fix if we want an unbiased estimator. We just multiply the previous estimator by a factor of $N/(N-1)$, which yields the following unbiased estimator when the true mean is unknown:

$$\hat{\sigma_X^2} = \frac{1}{N-1}\sum_{i=1}^{N}(X_i - \hat{\mu}_X)^2. \tag{9.7}$$

The change in the denominator is often referred to as a *degrees-of-freedom (dof) correction.* A detailed discussion of dof is beyond the scope of this text, but a simple interpretation is that we started with the number of dof equal to the size of our data (N), and we used one dof in calculating the sample mean. Thus, there are $N-1$ dofs remaining.

Because a dof correction is so common in calculating the sample variance, NumPy supports passing a `ddof` parameter to `np.var()`, where `ddof` is the "Delta Degrees of Freedom". In other words, we don't need to pass $N-1$, we only need to pass the difference from N, which is `ddof=1`.

Note, however, that there is a price to pay for having an unbiased estimator of the variance. For each of the biased and unbiased estimators, the following code estimates the mean of the estimator and the mean-square error of the estimator:

```
num_sims = 100_000

data_size = 10 # This is our N
X = stats.norm(1, 2)

total_var_biased = 0
total_var_unbiased = 0
total_error_biased = 0
total_error_unbiased = 0

for sim in range(num_sims):
  Xvals = X.rvs(data_size)

  var_biased = Xvals.var()
  total_var_biased += var_biased
  total_error_biased += (var_biased - X.var())**2

  var_unbiased = Xvals.var(ddof = 1)
  total_var_unbiased += var_unbiased
  total_error_unbiased += (var_unbiased - X.var())**2


print('Means: ',
      f'biased estimator: {total_var_biased/num_sims: 0.2f}, ',
      f'unbiased estimator: {total_var_unbiased/num_sims: 0.2f}\n')

print('MSEs: ',
      f'biased estimator: {total_error_biased/num_sims: 0.2f}, ',
      f'unbiased estimator: {total_error_unbiased/num_sims: 0.2f}')
```

```
Means:  biased estimator:  3.61,  unbiased estimator:  4.01

MSEs:  biased estimator:  3.03,  unbiased estimator:  3.56
```

The dof correction results in the estimator being unbiased. However, the unbiased estimator has a higher MSE than the biased estimator. This is an example of a property called the bias-variance tradeoff that plays an important role in statistics and machine learning problems. Even though the unbiased estimator produces a higher MSE, we will still use the unbiased variance estimator in (9.7) whenever we need to estimate the variance of data.

Let's introduce additional notation to help distinguish between unbiased and biased variance estimators and variance estimates. This additional notation is also commonly used in other statistics literature. Consider data $\mathbf{x} = [x_0, x_1, \ldots, x_{n-1}]$ with average $\overline{\mathbf{x}}$. We will use an uppercase S to indicate an estimator and a lowercase s to indicate an estimate. A subscript of n indicates division by n and is thus a biased estimator/estimate. A subscript of $n-1$ indicates division by $n-1$ and is thus an unbiased estimator/estimate. Thus, the estimators are

$$S_n^2 = \frac{1}{n}\sum_{i=0}^{n-1}\left(x_i - \hat{\mu}_X\right)^2, \qquad \text{(biased)}$$

$$S_{n-1}^2 = \frac{1}{n-1}\sum_{i=0}^{n-1}\left(x_i - \hat{\mu}_X\right)^2 \qquad \text{(unbiased)},$$

and the corresponding estimates are

$$s_n^2 = \frac{1}{n}\sum_{i=0}^{n-1}\left(x_i - \overline{\mathbf{x}}\right)^2, \qquad \text{(biased)}$$

$$s_{n-1}^2 = \frac{1}{n-1}\sum_{i=0}^{n-1}\left(x_i - \overline{\mathbf{x}}\right)^2 \qquad \text{(unbiased)}.$$

The variance of a data vector is defined as either s_n^2 or s_{n-1}^2.

Terminology review and self-assessment questions

Interactive flashcards to review the terminology introduced in this section and self-assessment questions are available at fdsp.net/9-4, which can also be accessed using this QR code:

9.5 Confidence Intervals for Estimates

The parameter estimates we have considered up to this point are point estimates: they are single numerical values computed from the data. Point estimates are commonly used, but they inherently do not provide any quantification of the ambiguity or reliability of the estimate. An *interval estimate* is an alternative that provides more information:

Definition

interval estimate

Given a vector of observed values $\mathbf{x}$ from a common distribution, an *interval estimate* for a parameter θ is an interval $[a, b]$ that is likely to contain the true value of θ according to some criterion.

The most common interval estimates are confidence intervals for the mean estimate.

9.5.1 Bootstrap Confidence Intervals for Estimates

Confidence intervals are easily generated for any parameter estimates by using bootstrap sampling. To generate a $c\%$ confidence interval, the bootstrap distribution of the estimator is generated, and then a region containing the center $c\%$ of the probability is determined. If you have been following along in the text, you are probably capable by now of generating a $c\%$ bootstrap confidence interval based on that brief description. If you feel less confident, you can see a more detailed description of the procedure on the website https://www.fdsp.net/9-5.html.

Let's illustrate this with an example. The code below takes data and a confidence interval and provides a point estimate and a confidence interval for the mean of the distribution of the data:

```
import numpy as np
import numpy.random as npr

def find_mean_CI(x, C):

  n = len(x)

  alpha = (100 - C)/2

  avg = x.mean()

  # Now do bootstrap sampling
  # Start by specifying the number of bootstrap values to simulate
  num_bs_sims = 1000

  # Create an array of zeros to store the bootstrap values
  bs_averages = np.zeros(num_bs_sims)

  for bs_sim in range(num_bs_sims):
    bs_sample = npr.choice(x, n)
    bs_avg = bs_sample.mean()
    bs_averages[bs_sim] = bs_avg

  # Now find the values that define the C%
  percentiles = np.percentile(bs_averages, [alpha, 100-alpha])

  print(f'The mean estimate is {avg:.2f}')
```

(continues on next page)

(continued from previous page)

```
print(f'A {C}% confidence interval for the average is [{percentiles[0]: .2f},
↪{percentiles[1]:.2f}]')
```

Here are the 90%, 95%, and 99% confidence intervals for the mean estimate of a small example data set:

```
x = np.array([ 1, 1.2, 1.4, 1.2, 1, 1.2, 1.5, 1.45, 1.05, 1.65, 1.8])
find_mean_CI(x, 90)
```

```
The mean estimate is 1.31
A 90% confidence interval for the average is [ 1.20, 1.44]
```

```
find_mean_CI(x, 95)
```

```
The mean estimate is 1.31
A 95% confidence interval for the average is [ 1.17, 1.46]
```

```
find_mean_CI(x, 99)
```

```
The mean estimate is 1.31
A 99% confidence interval for the average is [ 1.13, 1.49]
```

Note that the larger the specified confidence c, the wider the confidence interval must be to ensure that $c\%$ of the bootstrap distribution is within the interval.

It is important to remind ourselves of what a $c\%$ confidence interval means. In particular, it does not mean that there is a $c\%$ probability that a particular confidence interval contains the true mean. Rather, it means that over the long run, the true mean would lie in $c\%$ of the confidence intervals that are generated according to this procedure.

For data sample sizes of less than approximately 100, the $c\%$ confidence intervals generated via bootstrap sampling are generally not wide enough to achieve the specified confidence. Let's illustrate this with an experiment: I drew 25 data samples of length 10 from a Normal (1, 2) distribution. For each sample, I computed the sample mean, and bootstrap sampling was used to generate a 95% confidence interval for the sample mean. The results are shown in Fig. 9.13. For each of the 25 samples, the star marker indicates the average (sample mean), and a horizontal line shows the confidence interval. A vertical line through all the results shows the true mean (1). Confidence intervals that contain the true mean are indicated as solid, green lines; confidence intervals that do not contain the true mean are indicated by dashed, red/orange lines.

For the seed I arbitrarily chose, 2 out of 25 confidence intervals did not contain the true mean. Thus 8% of the CIs did not contain the true mean. In the long run, we expect at least 5% of confidence intervals to not contain the true mean. However, the bootstrap estimator for CIs is not conservative enough for small data, so 8% may be accurate. A simulation of 1000 such samples resulted in a 10% chance of the bootstrap confidence interval not containing the true mean. (The simulation code is available online at fdsp.net/9-5.) Below is a table that shows the probability[2] that the confidence interval contains the true mean for several different values of the data size, N. For 10 samples from the distribution, the

[2]These probabilities are approximations; they were estimated via simulations of 100,000 random samples of the data for each data size. The simulation code is available online at fdsp.net/9-5.

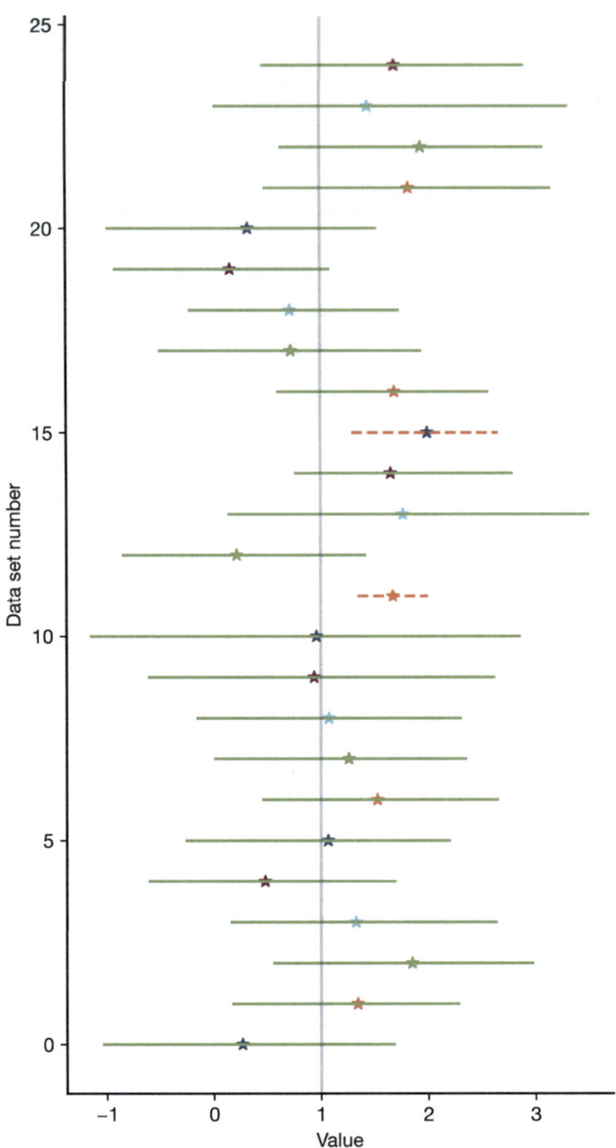

FIGURE 9.13
Confidence intervals for 25 random samples of Normal (1, 2) data. The mean estimate is indicated by a star and the true mean is indicated by a vertical line.

95% confidence interval generated by bootstrapping does not contain the true mean 10% of the time, when the value should be 5% of the time.

Data Size (N)	Probability CI does not contain true mean
5	16.3%
10	10.0%
25	6.9%
50	5.9%
75	5.6%
100	5.5%

Based on this distribution, bootstrap confidence intervals can only be considered accurate if they are generated from more than 100 data samples. However, better confidence intervals can be generated through bootstrap sampling using algorithms such as the BC_α method or ABC method[3], which are available in the Python bootstrap-stat library.

We will continue to use percentile-based bootstrap confidence intervals in this book for simplicity, but readers who need more accurate results should use one of the two algorithms mentioned above.

9.5.2 Analytical Confidence Intervals for Mean Estimates

Recall that the mean estimator is special because the CLT implies that its distribution will be Normal. We can use the Normal model to apply analytical approaches as an alternative to bootstrap estimators. The first step is to characterize the parameters of the distribution of the sample-mean estimator.

Linear Combinations of Independent Normal RVs

Suppose we have a data sample $\mathbf{X} = [X_0, X_1, \cdots, X_{n_X-1}]$, where different values in the sample are assumed to be independent. We know from above that the sample-mean estimator is unbiased. Thus, the expected value of the sample-mean estimator is the true mean. We will also need the variance of the sample-mean estimator in our analysis. Let σ_X^2 be the variance of the X_i. Then the variance of the sample-mean estimator is

$$\begin{aligned} \operatorname{Var}[\hat{\mu}_X] &= \operatorname{Var}\left[\frac{1}{n_X}\sum_{i=0}^{n_X-1} X_i\right] \\ &= \frac{1}{n_X^2}\operatorname{Var}\left[\sum_{i=0}^{n_X-1} X_i\right], \end{aligned}$$

where the second line of the equation follows from Property 3 of variance (see Section 9.3). Since the random variables in $\mathbf{X}$ are independent, we can apply Property 4 of variance to

[3]See Chapter 14 of *An Introduction to the Bootstrap* by Bradley Efron and Robert J. Tibshirani, Chapman & Hall/CRC, 1993.

get

$$\begin{aligned}
\text{Var}\left[\hat{\mu}_X\right] &= \frac{1}{n_X^2}\sum_{i=0}^{n_X-1}\text{Var}\left[X_i\right] \\
&= \frac{1}{n_X^2}\sum_{i=0}^{n_X-1}\sigma_X^2 \\
&= \frac{1}{n_X^2} n_X \sigma_X^2 \\
&= \frac{\sigma_X^2}{n_X}.
\end{aligned}$$

The variance of the sample-mean estimator decreases linearly with the number of samples. (This can be used to show that the sample-mean estimator converges to the true mean as $n_X \to \infty$ if σ_X^2 is finite.)

The standard deviation of the mean estimator is called the *standard error of the mean*:

Definition

standard error of the mean (exact)

For n samples from a random variable with known standard deviation σ_X, the *standard error of the mean (SEM)* is the standard deviation of the mean estimator,

$$\sigma_{\hat{X}} = \frac{\sigma_X}{\sqrt{n_X}}.$$

The analysis above is sufficient to characterize $\hat{\mu}_X$:

$$\hat{\mu}_X \sim \text{Normal}\left(\mu_X, \frac{\sigma_X}{\sqrt{n_X}}\right).$$

Note, however, that μ_X is unknown (it is the parameter we are trying to estimate), and σ_X may be known but is more often also unknown and must be estimated from the data. The analysis of the confidence intervals is different depending on whether the standard deviation of the data is known or unknown.

Confidence Intervals for Mean Estimate with Known Data Variance

Recall that given a specified confidence $c\%$, we wish to create a procedure that for each data sample $\mathbf{x}$ will determine an interval I_c such that $P(\mu_X \in I_c) = c\%$, where μ_X is the true mean, which is unknown. Then for a particular data sample $\mathbf{x}$ with average $\overline{\mathbf{x}}$, we can calculate the probability that the mean belongs to a confidence interval of the form $(\hat{\mu}_X - d, \hat{\mu}_X + d]$ as

$$\begin{gathered}
P\left(\mu \in (\hat{\mu}_X - d, \hat{\mu}_X + d]\right) \\
P\left(\hat{\mu}_X - d \leq \mu \cap \hat{\mu}_X + d > \mu\right) \\
P\left(\hat{\mu}_X - \mu \leq d \cap \hat{\mu}_X - \mu > -d\right).
\end{gathered}$$

Given a particular confidence level $c\%$, we would like to calculate the corresponding value of d. Observing that $\hat{\mu}_X - \mu$ has mean zero, we start by rewriting this expression in terms of tail probabilities:

$$\begin{aligned} P\left(\hat{\mu}_X - \mu \leq d \cap \hat{\mu}_X - \mu > -d\right) &= 1 - P\left(\hat{\mu}_X - \mu > d \cup \hat{\mu}_X - \mu \leq -d\right) \\ &= 1 - P\left(\hat{\mu}_X - \mu > d\right) + P\left(\hat{\mu}_X - \mu \leq -d\right), \end{aligned}$$

where the second line comes from the fact that we are asking about the probability that $\hat{\mu}_X - \mu$ is in mutually exclusive regions of the real line, which results in mutually exclusive events. Under the assumption that $\hat{\mu}_X - \mu$ is Normal with mean 0, the pdf of $\hat{\mu}_X - \mu$ is symmetric around $x = 0$. We can use this symmetry to write

$$1 - P\left(\hat{\mu}_X - \mu > d\right) + P\left(\hat{\mu}_X - \mu \leq -d\right) = 1 - 2P\left(\hat{\mu}_X - \mu > d\right).$$

Then for a given confidence c, we have

$$\begin{aligned} 1 - 2P\left(\hat{\mu}_X - \mu > d\right) &= c \\ P\left(\hat{\mu}_X - \mu > d\right) &= \frac{1-c}{2}. \end{aligned}$$

For future convenience, let $\gamma = (1-c)/2$.

We have several ways to proceed from here. One approach is to create a SciPy distribution for the random variable $\hat{\mu}_X - \mu$, which we know is Normal with mean 0. The variance of $\hat{\mu}_X$ is $\sigma_X^2/\sqrt{n}$, and this is also the variance of $\hat{\mu}_X - \mu$ by Property 2 of variance. Note that the probability is in the form of a survival function, which is the Q function for a Normal random variable. Since we have SciPy available, the simplest and most direct way is to make a Normal distribution object with the appropriate mean and variance and use the inverse survival function method `isf()` to solve for the value of d. The code below shows how to find the value of d for a 95% confidence interval when $\sigma = 2$:

```
sigma_X = 2
n = 10
gamma = (1-0.95)/2

N = stats.norm(loc = 0, scale = sigma_X / np.sqrt(n) )
N.isf(gamma)
```

```
1.2395900646091231
```

SciPy.stats provides an even simpler way to get the CI once the distribution object has been set up. The `interval() method` takes as a parameter the confidence level (as a probability, not a percentage), and returns the corresponding CI range, centered around 0:

```
N.interval(0.95)
```

```
(-1.2395900646091231, 1.2395900646091231)
```

Just as the distance d is added to the observed value of $\hat{\mu}_X$ (the average of the data), this confidence interval must also be shifted by adding the average of the data. I will delay presenting results on the performance of this approach to estimating CIs until we can also compare with results for the more typical case when the variance is unknown.

Confidence Intervals for Mean Estimate with Unknown Data Variance

If the standard deviation or variance of the distribution is not known, then we must estimate it from the data. We will use the unbiased variance estimator,

$$S_{n-1}^2 = \frac{1}{n-1} \sum_{i=0}^{n-1} (X_i - \hat{\mu}_X)^2 .$$

If $\hat{\mu}_X$ is Normal(μ_X, σ_X) , then

$$\frac{X - \mu_X}{\sigma_X} \sim \text{Normal}(0, 1). \tag{9.8}$$

If we have to replace the true variance by its estimate, then the distribution changes. Using the unbiased variance estimator instead of the true variance, the distribution of

$$\frac{\hat{\mu} - \mu_X}{S_{n-1}/\sqrt{n}} \tag{9.9}$$

has a **Student's t-distribution with $\nu = n - 1$ degrees of freedom (dof)**. For convenience, we will denote this distribution by t_ν. Recall from Section 8.6 that the pdf of the Student's t random variable is very similar to that of the Normal random variable, except that the density is spread further from the mean, especially for small ν.

Note:

Normalizing by the true mean and variance, as in (9.8), results in a random variable with mean 0 and variance 1.

The normalization in (9.9) results in a random variable with mean 0 and variance equal to $\nu/(\nu - 2)$ for $\nu > 2$, where $\nu = n - 1$ is the number of degrees of freedom of the Student's t distribution. Note that the variance of the Student's t distribution is always greater than 1 but converges to 1 as ν goes to ∞.

We can generate a SciPy distribution for a Student's t variable of the form in (9.9), by setting the following parameters:

- The `df` parameter is set to the number of degrees of freedom, $\nu = n - 1$.
- The `loc` parameter is set to the true mean, μ_X.
- The `scale` parameter is set to the **sample** standard error of the mean (SSEM), which is obtained from the SEM equation by replacing the standard deviation with its unbiased estimate, yielding $s_{n-1}^2 = s_{n-1}^2/\sqrt{N}$.

The SSEM can be computed from the data using the `stats.sem()` function:

```
# Some sample data
x = [-9, -4, 1, 4, 9, 16]

print(f'SSEM using np.std(x, ddof=1)/np.sqrt(len(x)): { np.std(x, ddof=1)/np.
→sqrt(len(x))}')
print(f'SSEM using stats.sem(x): {stats.sem(x)}')
```

```
SSEM using np.std(x, ddof=1)/np.sqrt(len(x)): 3.66439323459939
SSEM using stats.sem(x): 3.66439323459939
```

We will often be working with a t variable with mean 0. If we have data in a variable `x`, then we can generate a SciPy t-distribution as follows:

```
# Some sample data
x = [-9, -4, 1, 4, 9, 16]

n = len(x)
sigma_t = stats.sem(x)
T = stats.t(df = n-1, scale = sigma_t)
```

The average value is

```
print(f'{np.mean(x):.2f}')
```

```
2.83
```

If the true mean were 0, the probability of seeing such a large value of the mean could be computed using the `T` object as:

```
print(f'{T.sf(np.mean(x)):.3f}')
```

```
0.237
```

To create analytical confidence intervals with the standard deviation estimated from the data, we can again find a region $(\hat{\mu}_X - d, \hat{\mu}_X + d]$ that satisfies

$$P\left(\hat{\mu}_X - \mu > d\right) = \alpha,$$

where $\alpha = (1-c)/2$. However, $\hat{\mu}_X - \mu$ is no longer Normal. It will have a Student's t distribution if we normalize it by dividing by $S_{n-1}/\sqrt{n}$. In practice, we perform this normalization using the *observed* standard deviation estimate, s_{n-1} in place of the random estimator. Then we can use SciPy to create a Student's t distribution with the appropriate degrees of freedom and scaling, and then use the `interval()` method to find the confidence interval.

This will be most clear using an example. The data below is the first data sample from the confidence intervals above:

```
# The data
x0= np.array([-0.41957876,  0.96561764,  1.63882274, -3.53066214, -1.75490733,
        4.89996147, -0.12762015, -0.68747518,  1.44907716,  0.21724457])
n0=len(x0)

# Set C and calculate alpha
C = 95/100
alpha = (1-C)/2

# Calculate the SSEM
sigma_t0 = stats.sem(x0)
```

(continues on next page)

(continued from previous page)

```
# Now create the scaled Student's t distribution object using SciPy
T0 = stats.t(df = n0 - 1, scale=sigma_t0)

# And now find the confidence interval
T0.interval(0.95)
```

```
(-1.604600189983367, 1.604600189983367)
```

This compares to a confidence interval of approximately $[-1.24, 1.24]$ when the variance is known. The use of the sample variance estimator results in the CIs generally being wider than when the variance is known. Fig. 9.14 shows the confidence intervals for all three methods: bootstrap, analytical with known variance, and analytical with variance estimated from the data.

The confidence intervals produced using the Student's t distribution are very similar to those produced by bootstrap sampling because the width of the confidence interval is now determined from the data. Data samples 11 and 15 do not include the true mean for either the bootstrap CIs or the analytical model with estimated variance. Just like the bootstrap confidence intervals, the Student's t model does not provide confidence intervals that are conservative enough when the number of samples is small. Various correction factors are available in the literature to provide better confidence intervals for small data samples.

Recall that the data comes from a distribution with a standard deviation of 2, corresponding to a variance of 4. Fig. 9.15 shows the sample variances for each of the data samples in this example, with the values for samples 11 and 15 highlighted in orange. Note that having a low sample variance does not necessarily imply that the bootstrap confidence interval will be small. The sample variance for data sample 7 is similar to that of data sample 11, and yet the confidence interval for data sample 7 is not that smaller than the analytical result.

This example shows that analysis can produce better estimates of confidence intervals when the standard deviation of the distribution is known. Analytical confidence intervals with unknown variance offered similar performance as bootstrap sampling. Both require correction factors to provide accurate CIs when the number of samples is small.

Terminology review and self-assessment questions

Interactive flashcards to review the terminology introduced in this section and self-assessment questions are available at fdsp.net/9-5, which can also be accessed using this QR code:

9.6 Testing a Difference of Means

One of the most common statistical tests is whether two data sets come from distributions with the same means. The mean is special because of the Central Limit Theorem (see Section 8.6) – regardless of the type of distribution of the underlying data, the distribution of the sample mean will be approximately Normal if there are at least tens of data points. This allows two approaches to conducting tests involving sample means: we can use bootstrap resampling, or we can use analysis by applying the assumption that sample means are

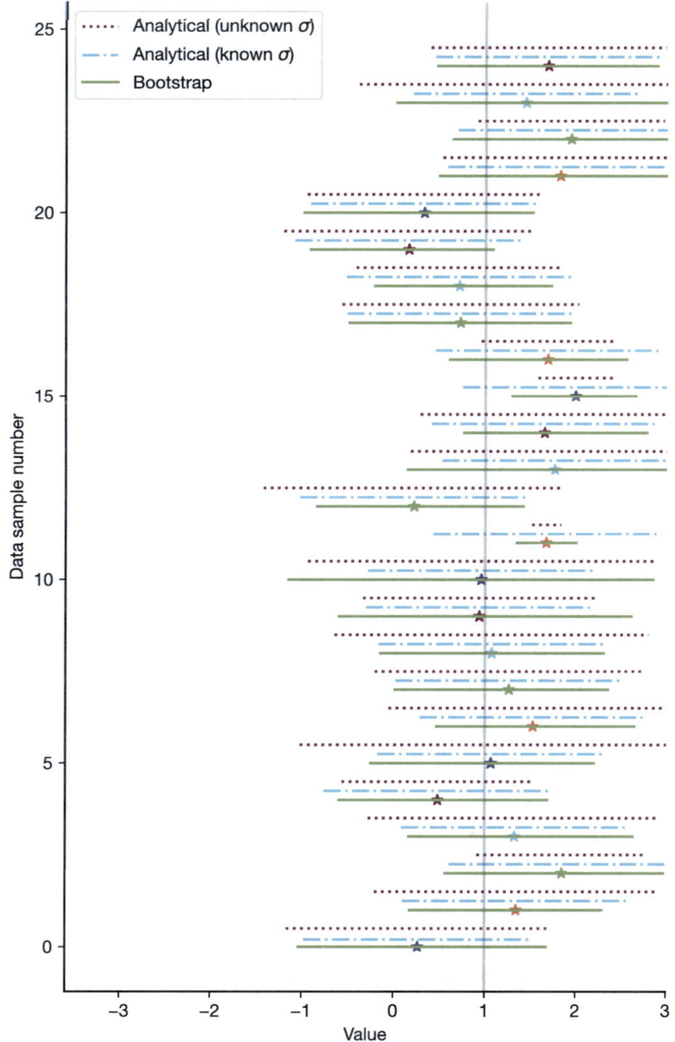

FIGURE 9.14
Comparison of confidence intervals generated in three different ways: bootstrapping, analytical with known variance, and analytical with variance estimated.

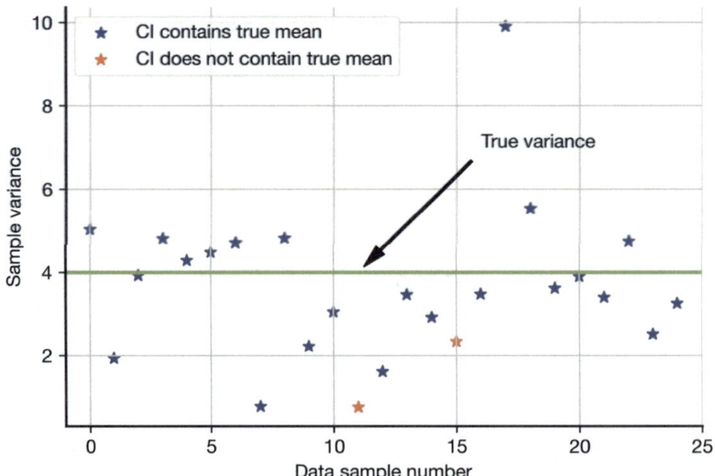

FIGURE 9.15
Sample variance versus true variance for 25 random samples of Normal data.

approximately Normal. To understand how to apply the analytical approaches, we need to characterize the sample mean estimators and the test statistic, which is the difference between the sample mean estimators.

9.6.1 Characterizing the Difference of Sample-Mean Estimators

Suppose we have two data samples $\mathbf{X} = [X_0, X_1, \cdots, X_{n_X-1}]$ and $\mathbf{Y} = [Y_0, Y_1, \cdots, Y_{n_Y-1}]$, where the data samples are assumed to be independent.[4] From the results in Section 9.5, the sample-mean estimators for these data samples have the following distributions:

$$\hat{\mu}_X \sim \text{Normal}\left(\mu_X, \frac{\sigma_X}{\sqrt{n_X}}\right), \text{ and}$$
$$\hat{\mu}_Y \sim \text{Normal}\left(\mu_Y, \frac{\sigma_Y}{\sqrt{n_Y}}\right).$$

In addition, $\hat{\mu}_X$ and $\hat{\mu}_Y$ are independent.

To assess a difference in means, we will create a test statistic that is the difference between sample-mean estimators. For convenience, we will consider the case $T = \hat{\mu}_X - \hat{\mu}_Y$, but the result for $T = \hat{\mu}_Y - \hat{\mu}_X$ is almost identical. The mean of the test statistic is

$$\begin{aligned} E[T] &= E\left[\hat{\mu}_X - \hat{\mu}_Y\right] \\ &= E\left[\hat{\mu}_X\right] - E\left[\hat{\mu}_Y\right] && \text{(by linearity)} \\ &= \mu_X - \mu_Y && \text{(the estimators are unbiased).} \end{aligned}$$

To find the variance of T, we first rewrite the formula for the test statistic slightly, as $T = \hat{\mu}_X + (-1)\hat{\mu}_Y$. Since $\mathbf{X}$ and $\mathbf{Y}$ are independent, so are functions computed from them.

[4] Actually, the data within each group are only *conditionally independent* given the group it belongs to

Thus,

$$\begin{aligned}
\sigma_T^2 &= \text{Var}[T] \\
&= \text{Var}\left[\hat{\mu}_X + (-1)\hat{\mu}_Y\right] \\
&= \text{Var}\left[\hat{\mu}_X\right] + \text{Var}\left[(-1)\hat{\mu}_Y\right] && \text{(by Property 4 of variance)} \\
&= \frac{\sigma_X^2}{n_X} + (-1)^2\frac{\sigma_Y^2}{n_Y} && \text{(from above and Property 3)} \\
&= \frac{\sigma_X^2}{n_X} + \frac{\sigma_Y^2}{n_Y}
\end{aligned}$$

Importantly, a linear combination of independent Normal random variables is also a Normal random variable. Thus, we have a characterization of the test statistic:

$$T \sim \text{Normal}\left(\mu_X - \mu_Y, \sqrt{\frac{\sigma_X^2}{n_X} + \frac{\sigma_Y^2}{n_Y}}\right).$$

9.6.2 Statistical Inference for a Difference of Sample Means *with Known and Equal Variances*

Consider the case where we know that the data come from distributions with the same variance, σ^2, and that we know the value of σ^2. (The case of known but unequal variances requires only a minor modification but is rare enough that I omit it here.) Suppose we have samples from these distributions,

$$\mathbf{x} = \left[x_0, x_1, \ldots, x_{n_X-1}\right],$$

and

$$\mathbf{y} = \left[y_0, y_1, \ldots, y_{n_Y-1}\right].$$

Let the averages of the data be denoted by $\overline{\mathbf{x}}$ and $\overline{\mathbf{y}}$. Denote the true (but unknown) means of the distributions as μ_X and μ_Y.

We can now easily conduct an NHST. The null hypothesis is that the data have the same mean and so $E[T] = 0$. Let the observed value of the test statistic be $t = \overline{\mathbf{x}} - \overline{\mathbf{y}}$. Below I assume that $t > 0$; if not, then interchange the role of x and y.

Under the assumption that variances of the X_i and Y_k are both equal to σ^2, the variance of T simplifies to

$$\begin{aligned}
\sigma_T^2 &= \frac{\sigma^2}{n_X} + \frac{\sigma^2}{n_Y} \\
&= \sigma^2\left(\frac{1}{n_X} + \frac{1}{n_Y}\right).
\end{aligned}$$

Once we know the mean and variance of T, then the NHST is a straightforward application of the results in Section 8.6:

- For a one-sided hypothesis test, the p-value is

$$P(T \geq t|H_0) = Q\left(\frac{t}{\sigma_T}\right) = Q\left(\frac{t}{\sigma\sqrt{\frac{1}{n_X} + \frac{1}{n_Y}}}\right).$$

- For a two-tailed NHST, the p-value is

$$P(|T| \geq t|H_0) = 2Q\left(\frac{t}{\sigma\sqrt{\frac{1}{n_X} + \frac{1}{n_Y}}}\right).$$

In both these cases, the only information that is needed is the observed value of the test statistic, the standard deviation of the data, and the number of samples in each group (n_X and n_Y).

9.6.3 Binary Hypothesis Tests with *Unknown Variance*

In most cases, the underlying distributions' variances are not known and must be estimated from the data. As in the previous section, scaling the test statistic based on the sample variance will result in a statistic that is no longer Normal. The scaled test statistic depends on what we can assume about the relationship between the variances of the two samples. If the samples can be assumed to have the same variance, then the test statistic can be scaled so that the resulting variable has a Student's t distribution. If the variances of the data samples are unequal, then the test statistic can be scaled so that the resulting distribution is well-approximated by a Student's t distribution. Before considering each of these cases, we start with some preliminaries:

We wish to determine the probability of observing a difference in averages with magnitude as large as $t = \overline{x} - \overline{y}$ under the null hypothesis, under which the mean of the test statistic is zero. To calculate this probability, we will need the variance estimators for each sample, which in turn depend on the mean estimators. The mean estimators are given by

$$\hat{\mu}_X = \frac{1}{n_x}\sum_{i=0}^{n_x-1} X_i, \text{ and}$$

$$\hat{\mu}_Y = \frac{1}{n_y}\sum_{i=0}^{n_y-1} Y_i.$$

The unbiased variance estimator for each sample is

$$S_x^2 = \frac{1}{n_x - 1}\sum_{i=0}^{n_x-1} \left(X_i - \hat{\mu}_X\right)^2, \text{ and}$$

$$S_y^2 = \frac{1}{n_y - 1}\sum_{i=0}^{n_y-1} \left(Y_i - \hat{\mu}_Y\right)^2.$$

Now we consider the two cases:

1. Test for equal means with *unknown, equal variances*

If equal variances can be assumed, then we only need a single estimator of the variance, called the pooled variance estimator. Since we do not know whether the means are equal, the pooled variance estimator must use the mean estimator for each data sample when centralizing the data for that sample. Because two mean estimators are used, the degrees of freedom adjustment requires us to divide by $n_x + n_y - 2$ when calculating the unbiased sample variance estimator for the pooled data. The resulting pooled variance can be written in either of the following forms:

$$S_p^2 = \frac{1}{n_x + n_y - 2}\left[\sum_{i=0}^{n_x-1}(X_i - \hat{\mu}_X)^2 + \sum_{i=0}^{n_y-1}(Y_i - \hat{\mu}_Y)^2\right]$$
$$= \frac{(n_x - 1)S_X^2 + (n_y - 1)S_Y^2}{n_x + n_y - 2}$$

Then

$$\frac{T}{S_p\sqrt{\frac{1}{n_x} + \frac{1}{n_Y}}} \sim t_\nu$$

for $\nu = n_x + n_y - 2$ degrees of freedom.

2. Test for equal means with *unknown, unequal variances*

If we cannot assume that the variances are equal, then the distribution of the following normalized form will be approximately equal to a Student's t distribution:

$$\frac{T}{S_d} \sim t_\nu,$$

where

$$S_d = \sqrt{\frac{S_x^2}{n_x} + \frac{S_y^2}{n_y}}. \tag{9.10}$$

Here, the value of ν must be determined from the sample standard errors of the means, which we denote by

$$s_{\overline{x}}^2 = s_x^2/n_x, \text{ and}$$
$$s_{\overline{y}}^2 = s_y^2/n_x.$$

Then ν is the largest integer that satisfies

$$\nu \leq \frac{\left(s_{\overline{x}}^2 + s_{\overline{y}}^2\right)^2}{s_{\overline{x}}^2/(n_x - 1) + s_{\overline{y}}^2/(n_y - 1)}.$$

WARNING

The following example concerns deaths caused by firearms, which may be a sensitive topic for some readers. In addition, the example is particularly about assessing the effect of gun legislation on firearms deaths, which is a politically sensitive topic. However, I have chosen to include this example because of its relevance to the national discussion around these topics. Readers should know that a simple analysis like the one below can suggest a relationship between factors that may actually be attributable to other underlying causes.

Please read this example with an open mind and the desire to see what the data indicates.

Example 9.6: Permitless Carry and Firearms Mortality

In this example, we will consider the effect of "permitless carry" on different types of firearms deaths. Here "permitless carry" (variants of which are also known as "constitutional carry") allows gun owners in a state to carry a loaded firearm without having to apply to the government for a gun permit. There are actually many variations on permitless carry, but I consider a state to allow permitless carry if a state's citizens can generally carry (either open or concealed) a loaded handgun without a gun permit.

The two research hypotheses I will consider are:

1. There is a difference in firearm homicide rates between states with permitless carry and those without.
2. There is a difference in firearm suicide rates between states with permitless carry and those without.

Because this example is quite long, I am not going to use the blue highlight bar for most of the example. Below, I use the blue highlight bar to indicate specific statistical tests.

To answer these questions, we need data from two separate sources. We can use firearms mortality data from CDC WONDER, which is the Wide-ranging ONline Data for Epidemiologic Research. In particular, I have used WONDER SEARCH to access data from Underlying Cause of Death 1999-2020: https://wonder.cdc.gov/ucd-icd10.html. I have downloaded data by state for the following Causes of Death, shown using IC-10 codes:

X93 Assault by handgun discharge
X94 Assault by rifle, shotgun, and larger firearm discharge
X95 Assault by other and unspecified firearm discharge

I selected to download the total deaths across these categories, the population, and the crude rate. Here, the crude rate is defined as the deaths per population times 100,000. The resulting download is a tab-separate value (TSV) file, which is very similar to the CSV files that we saw previously, except the field separators are tabs instead of commas. The resulting file is available at https://www.fdsp.net/data/wonder-homicides-2020.tsv.

Data for suicides was retrieved from the same WONDER database using the following IC-10 codes:

X72 Intentional self-harm by handgun discharge
X73 Intentional self-harm by rifle, shotgun, and larger firearm discharge
X74 Intentional self-harm by other and unspecified firearm discharge

The resulting TSV is available at https://www.fdsp.net/data/wonder-suicides-2020.tsv.

To load these data into Pandas dataframes, we can use `pd.read_csv()`, but we need to tell that function that the data is separated by tabs instead of commas. To do this, we will pass the keyword argument `sep = '\t'`, where `\t` is a special code that translates to the tab character. The following Python code loads the two data sets:

```
import pandas as pd
```

(continues on next page)

(continued from previous page)

```
suicides = pd.read_csv('https://www.fdsp.net/data/wonder-suicides-2020.tsv',
                       sep='\t')
homicides= pd.read_csv('https://www.fdsp.net/data/wonder-homicides-2020.tsv',
                       sep='\t')
```

Let's merge these two dataframes into a single dataframe. To do that, I am first going to do some data cleaning:

1. We will drop the Notes, Crude Rate, and State Code columns from the `suicides` dataframe.
2. We will drop all of the above plus the Population column from the `homicides` dataframe (because we will merge this with the `suicides` dataframe that already has the same information).
3. We will relabel the `Deaths` column of each dataframe to match the type of death.

Here are these first steps:

```
#1.
suicides.drop(columns=['Notes', 'Crude Rate', 'State Code'], inplace=True)

#2.
homicides.drop(columns=['Notes', 'Population', 'Crude Rate', 'State Code'],
               inplace=True)

#3.
suicides.rename({'Deaths': 'Suicides'}, axis=1, inplace=True)
homicides.rename({'Deaths': 'Homicides'}, axis=1, inplace=True)
```

Let's check each dataframe now:

```
suicides.head(3)
```

	State	Suicides	Population
0	Alabama	542	4921532
1	Alaska	133	731158
2	Arizona	830	7421401

```
homicides.head(3)
```

	State	Homicides
0	Alabama	564
1	Alaska	27
2	Arizona	382

A few notes:

1. We are interested in mortality **rates**, and we could have preserved the `Crude Rate` column instead of the `Deaths` column. However, the `Crude Rate` is computed from `Deaths` and `Population`, and preserving these separately will allow us to analyze the data in different ways. Moreover, the `Crude Rate` for suicide in some states is

listed as `unreliable`, corresponding to fewer than 20 suicide deaths in that state. Since we never use these rates for a single state in isolation, the values are useful to our analysis, and we will use all the rate data.

2. These dataframes are of different sizes:

```
len(suicides), len(homicides)
```

```
(50, 49)
```

When we merge them, we need to decide how to handle any discrepancies regarding which states are included. I will use the approach that the combined data set contains only states with entries in both the suicides and homicides dataframes. This is called an *inner join.* We will use the `merge()` method of the `suicides` Pandas dataframe, which takes as argument the dataframe to be merged. We will specify the keyword argument `on = 'State'` to specify that we are matching up the rows from the different dataframes based on the entry in the `State` column. We will pass the keyword argument `how = 'inner'`, to do an inner join and preserve only those entries that appear in *both* of the dataframes being merged.

```
all_deaths=suicides.merge(homicides, on='State', how='inner')
all_deaths.head(3)
```

	State	Suicides	Population	Homicides
0	Alabama	542	4921532	564
1	Alaska	133	731158	27
2	Arizona	830	7421401	382

Let's check the length of the merged dataframe:

```
len(all_deaths)
```

```
48
```

Even though the smaller of the dataframes had 49 rows, the inner join produced only 48 rows. The `homicides` dataframe has entries for all 50 states, but the `suicides` dataframe has entries for only 48 states and the District of Columbia.

Now, let's go ahead and compute the homicide and suicide rates (scaled up by 100,000):

```
all_deaths['Homicide Rate'] = \
    all_deaths['Homicides'] / all_deaths['Population'] * 100_000

all_deaths['Suicide Rate'] = \
    all_deaths['Suicides'] / all_deaths['Population'] * 100_000

all_deaths.head(3)
```

	State	Suicides	Population	Homicides	Homicide Rate	Suicide Rate
0	Alabama	542	4921532	564	11.459846	11.012831
1	Alaska	133	731158	27	3.692772	18.190323
2	Arizona	830	7421401	382	5.147276	11.183872

The second data source we need is on permitless carry of firearms. I used the table at Wikipedia: Constitutional Carry - Ages to carry without a permit:

https://en.wikipedia.org/wiki/Constitutional_carry#Ages_to_carry_without_a_permit

(with consultation of the source documents) to create a CSV file with each row containing data on a state that allows a *state resident* to carry a *handgun* without a permit as of 2020. For each such state, the age at which a handgun can be carried without a permit is listed for both *open* and *concealed* carry. If permitless carry of a handgun by a state resident is not allowed for one of these categories, the entry is "N/A". States not in this CSV file do not allow permitless carry as of 2020. The resulting CSV file is available on the book's website at https://www.fdsp.net/data/permitless-carry-2020.csv.

Let's load this data into another Pandas dataframe:

```
permitless_df = pd.read_csv('https://www.fdsp.net/data/permitless-carry-2020.csv')
permitless_df.head(3)
```

	State	Permitless_open	Permitless_concealed
0	Alabama	18.0	NaN
1	Alaska	16.0	21.0
2	Arizona	18.0	21.0

We will again use the `merge()` method of the Pandas dataframe, but this time we will use it on the `all_deaths` dataframe. We will specify the keyword argument `on = 'State'` to specify that we are matching up the rows from the different dataframes based on the entry in the `State` column. For this merge operation, we do not want to do an inner join because that would drop all of the death data for states that do not allow permitless carry. Instead, we will perform a *left join*, which means we will preserve all of the keys in the `all_deaths` dataframe, which appears to the left of the `permitless` dataframe.

Here is that left join:

```
df = all_deaths.merge(permitless_df, on = 'State', how = 'left')
```

It will be convenient to index this dataframe by `State` and remap the order of the columns:

```
df.set_index('State', inplace=True)
df2 = df[ ['Population', 'Homicides', 'Homicide Rate', 'Suicides',
        'Suicide Rate', 'Permitless_open', 'Permitless_concealed'] ]
df2.head()
```

	Population	Homicides	Homicide Rate	Suicides	Suicide Rate	Permitless_open	Permitless_concealed
State							
Alabama	4921532	564	11.459846	542	11.012831	18.0	NaN
Alaska	731158	27	3.692772	133	18.190323	16.0	21.0
Arizona	7421401	382	5.147276	830	11.183872	18.0	21.0
Arkansas	3030522	282	9.305328	364	12.011132	18.0	18.0
California	39368078	1731	4.396963	1552	3.942280	NaN	NaN

We can see that the left join worked as expected by noting the entry for California. California is not one of the states in the `permitless` dataframe, yet its firearms mortality data is preserved in the merged dataframe.

Finally, let's split `df` back into two separate dataframes based on whether they allow permitless carry (open or concealed, at any age). When using `df.query()`, we can combine logical conditions using "|" to represent logical **or** or "&" to represent logical **and**. Thus, the following queries can be used to partition `df`:

```
permitless = df2.query('Permitless_concealed >0 | Permitless_open >0')
permit = df2.query('Permitless_concealed.isnull() & Permitless_open.isnull()')
```

As a check, we can see that the sizes of `permitless` and `permit` equal the size of `df`:

```
len(permitless), len(permit), len(df)
```

```
(17, 31, 48)
```

Example 9.6: Effect of Permitless Carry on Homicide Rates

Let's conduct a test to see whether the average homicide rate differs between permitless carry states and states that require a permit to carry a gun. In this analysis, we will use the simplest approach, which is to directly compute the average of the homicide rates. We discuss an alternative approach in the exercises further below.

Computing the average homicide rate for each class of states is easy:

```
permitless['Homicide Rate'].mean(), permit['Homicide Rate'].mean()
```

```
6.066280811037099, 5.391510100924644
```

The difference in means is

```
Hdiff = permitless['Homicide Rate'].mean() - permit['Homicide Rate'].mean()
print(f'{Hdiff:.2f}')
```

```
0.67
```

The homicide rate for permitless carry states is higher than for states that do not allow permitless carry. We can perform bootstrap resampling to determine if the observed difference is statistically significant. Since our initial research hypothesis is that there is a difference in homicide rates between permitless carry states and those without, it makes sense to carry out a two-tailed test. Our null hypothesis is that there is no difference in homicide rate between these two classes of states, so we will pool the homicide rate data for all states and draw random bootstrap samples representing each class of states. We then determine the relative frequency of a difference in averages as high as the one observed in the data. Because we are interested in conducting statistical tests on both homicide rates and suicide rates and most of the programming code is identical, I am providing a function that can carry out a bootstrapping test on any column of the data:

```
import numpy.random as npr

def bootstrap_permit (df, permitless, permit, column_name, num_sims = 10_000):
  pooled=df[column_name]
  permitless_len = len(permitless)
  permit_len = len(permit)

  diff = permitless[column_name].mean() - permit[column_name].mean()
  print(f'Observed difference in means was {diff:.2f}')
  count = 0

  for sim in range(num_sims):
    # Draw the bootstrap samples
    bs_permitless = npr.choice(pooled, permitless_len)
    bs_permit =npr.choice(pooled, permit_len)

    # Now compute the statistic for the bootstrap samples
    bs_t =  bs_permitless.mean() - bs_permit.mean()

    # And conduct a two-sided test
    if abs(bs_t) >= diff:
      count+=1

      print('Prob. of observing absolute difference as large as data =~ ',
            f'{count/num_sims: .2g}')
```

Then we can carry out the two-tailed hypothesis test for a difference of means as follows:

```
bootstrap_permit(df2, permitless, permit, 'Homicide Rate')
```

```
Observed difference in means was 0.67
Prob. of observing absolute difference as large as data =~  0.53
```

Thus, under the null hypothesis that the means are identical, we see a difference in homicide rates this large more than 50% of the time. The observed difference in means is not statistically significant. An analytical T-test yields $p = 0.57$. I am omitting the details here, as I will show the analytical T-test for the next research question in detail. However, the discussion and code for this analytical T-test are available online at fdsp.net/9-6.

Example 9.6: Effect of Permitless Carry on Suicide Rates

Now let's consider whether permitless carry affects suicide rates. As before, we start by computing the mean suicide rates for both the permitless and the permit group, as well as the difference in means:

```
permitless['Suicide Rate'].mean(), permit['Suicide Rate'].mean()
```

```
12.55598040437242, 7.238123660784345
```

The difference in sample means is

```
Sdiff = permitless['Suicide Rate'].mean() - permit['Suicide Rate'].mean()
Sdiff
```

```
5.317856743588075
```

The difference in average suicide rates (5.32) is *much* larger than the difference in average homicide rates (0.67). The larger difference is much less likely to be attributed to randomness in sampling, but we should conduct a statistical test to confirm whether the result should be considered statistically significant. We can carry out a two-tailed hypothesis test on the difference of means using the function we created previously:

```
bootstrap_permit(df2, permitless, permit, 'Suicide Rate')
```

```
Observed difference in means was 5.32
Prob. of observing absolute difference as large as data =~  0
```

The probability of observing such a large difference in means under the null hypothesis is so small that most runs of the bootstrap simulation will not generate any events where a difference this large occurs under the null hypothesis, unless we use a much larger number of simulation points. If we want to accurately estimate the probability of seeing such a large difference under the null hypothesis, the analytical approach can estimate that probability without requiring a huge number of bootstrap samples. As for the case of homicide rates, we start by finding the sample SEM for each group:

```
sem_x = stats.sem(permitless['Suicide Rate'])
sem_y = stats.sem(permit['Suicide Rate'])
sem_x, sem_y
```

```
(0.8939434974895945, 0.588546275254346)
```

```
nx = len (permitless)
ny = len(permit)
nx, ny
```

```
(17, 31)
```

Then the standard error for the decision statistic is

```
Sd = np.sqrt(sem_x**2 + sem_y**2)
Sd
```

```
1.0702904721708462
```

Next, we need to estimate the number of degrees of freedom to use in the Student's t distribution:

```
nu = (sem_x**2 +sem_y**2) ** 2 / ( sem_x**4/(nx-1) + sem_y**4/(ny-1) )
nu
```

```
29.882264069086958
```

Putting this all together, we can create the SciPy distribution object as

```
HT = stats.t(df = nu, scale = Sd)
```

and the probability of such a large difference under H_0 is approximately

```
2*HT.sf(Sdiff)
```

```
2.5702508897423874e-05
```

For comparison, I ran the bootstrap simulation for 10 million points, and the resulting probability estimate was approximately 4×10^{-5}. Thus, either method produces a similar result, but the analytical solution is much faster. This test for a difference of means is common enough that SciPy.stats has a built-in method for carrying out the analytical T-test, `stats.ttest_ind()`. Here, the "ind" refers to the fact that the samples must be *independent*. The call signature for this function is shown below:

```
scipy.stats.ttest_ind(a, b, axis=0, equal_var=True, nan_policy='propagate',␣
↪permutations=None, random_state=None, alternative='two-sided', trim=0)
```

Here `a` and `b` are the two data samples. The main thing that we will need to be aware of here is that the default is to assume that the variances are **equal**, which we cannot assume for our data set; so we will need to pass the keyword argument `equal_var=False`:

```
stats.ttest_ind(permitless['Suicide Rate'], permit['Suicide Rate'],
                equal_var=False)
```

```
Ttest_indResult(statistic=4.96861074807289, pvalue=2.5702508897423874e-05)
```

The result is identical that we found by creating the appropriate Student's t variable ourselves.

Exercises

1. Use the Student's t random variable to determine 95% confidence intervals for the mean difference under the null hypothesis for each of the statistics above. Are the resulting confidence intervals compatible with the observed differences of means?
2. The tests above are based on averaging over states. But the states involved have very different populations. Compute the average rate over populations by summing up the number of homicides for each group and then dividing by the corresponding total population for the group. Conduct a bootstrapping NHST based on the observed rates. Repeat for suicide rates.

Terminology review and self-assessment questions

Interactive flashcards to review the terminology introduced in this section and self-assessment questions are available at fdsp.net/9-6, which can also be accessed using this QR code:

9.7 Sampling and Bootstrap Distributions of Parameters

Now that we understand how moments are used to parameterize distributions, we can discuss different types of distributions that come up in working with random data and parameter estimates. In particular, we have introduced the general idea of the bootstrap distribution in Section 5.6.

9.7.1 Sampling Distribution of an Estimator

Consider a scenario in which we have independent random variables $\mathbf{X} = [X_0, X_1, \ldots, X_{n-1}]$ from a distribution that is characterized by some parameter θ. Let $\hat{\Theta}$ be an estimator for θ; that is, $\hat{\Theta}$ is some function $\hat{\Theta} = g(\mathbf{X})$, where $g()$ is chosen to make $\hat{\Theta}$ be a "good" (e.g., unbiased, low MSE) estimator for θ. Then $\hat{\Theta}$ is itself a random variable and hence has some distribution that is generally different than the distribution of the X_i. The distribution of $\hat{\Theta}$ is called the *sampling distribution*:

> Definition
>
> **sampling distribution (estimator)**
>
> Given a vector of independent random variables $\mathbf{X}$ and a parameter estimator $\hat{\Theta} = g(\mathbf{X})$, the *sampling distribution* is the probability distribution of $\hat{\Theta}$.

This will be made much more clear with an example. Suppose we have 25 random values from some random distribution, and we want to estimate the sampling distribution for the mean estimator. An empirical estimate of the sampling distribution is created by creating multiple sample mean values from independent draws of length-25 samples. You can see how the empirical estimate of the sampling distribution changes with the number of sample means available using the visualization labeled "Sampling Distribution for Uniform [0, 1] Random Variable on the website https://fdsp.net/9-7. Some examples of the different empirical sampling distributions from the interactive visualization are shown in Fig. 9.16.

Results for samples drawn from several other distributions are also available online at https://fdsp.net/9-7. All show that the sample mean roughly converges to a bell shape for the average of 25 samples, **regardless of the initial distribution**.

As the number of simulated averages goes to infinity, the empirical distribution of the estimator will converge to the true sampling distribution. Note that in each of the simulation iterations, the data is **drawn from the original distribution**.

9.7.2 Bootstrap Distribution of an Estimator

Next, we consider the bootstrap distribution of an estimator. For a given parameter, we can create bootstrap samples from observed data by resampling with replacement and then computing the parameter from the bootstrap samples, as we did to create confidence intervals in Section 9.5. Since the bootstrapped value of the parameter will depend on the bootstrap sample, which is randomly chosen, the bootstrap sample will be a random variable, and thus its distribution can be characterized. Before we illustrate this with an example, let's consider some observations about the bootstrap distribution:

1. The bootstrap distribution is inherently a discrete distribution. If there are n values in the original sample, then the number of ways to create a bootstrap sample is n^n. Note,

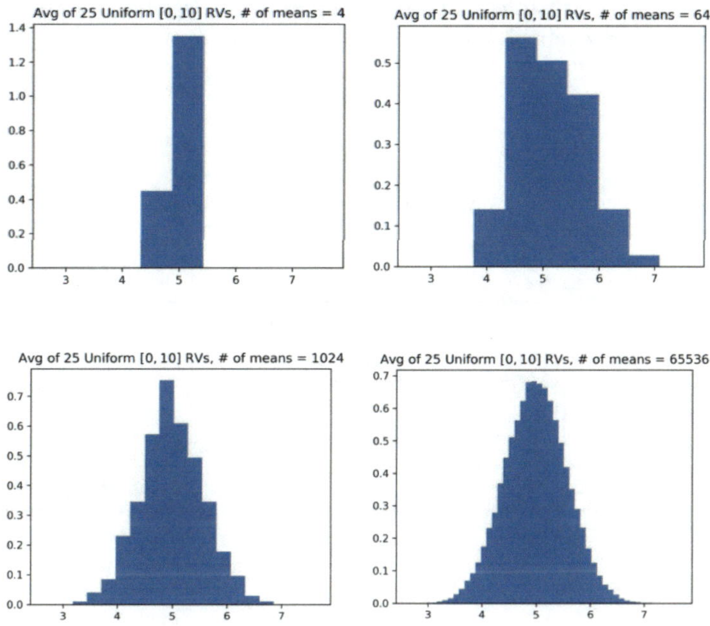

FIGURE 9.16
Examples of empirical estimates of the sampling distribution of the average (sample mean) based on different numbers of simulated averages.

however, that the number of possible values grows quickly with the number of values in the original sample. For instance, if there were 25 values in the original sample, then the number of possible values of the parameter estimator is:

```
25 ** 25
```

```
88817841970012523233890533447265625
```

2. The range of a bootstrap distribution is always finite. For instance, if the mean estimator is used, then the minimum possible value will occur when every value in the bootstrap sample is equal to the minimum value in the original data sample. Thus, the minimum possible value of the mean estimator is equal to the minimum value in the original data sample. We can make a similar conclusion for the maximum.

An implication of both these observations is that if the data comes from a continuous distribution, the bootstrap distribution will never converge to the sampling distribution as the number of bootstrap samples goes to infinity. Even with these limitations, the bootstrap distribution is usually used as an approximation of the sampling distribution. However, the bootstrap distribution is usually only a good approximation of the sampling distribution if the sample size is sufficiently large (at least 10s of data points in the sample).

Fig. 9.17 shows bootstrap samples and estimates from a random sample of data from a Normal ($\mu = 3$, $\sigma = 2$) distribution. The topmost plot shows a set of ten data points drawn from this distribution. Below that are eight plots, each of which shows a different bootstrap sample from the data. Unique markers and colors are used to indicate the different points in the original data, and multiple values of the same point in a bootstrap sample are stacked vertically. The plot for each bootstrap sample also includes a labeled arrow marking the sample mean.

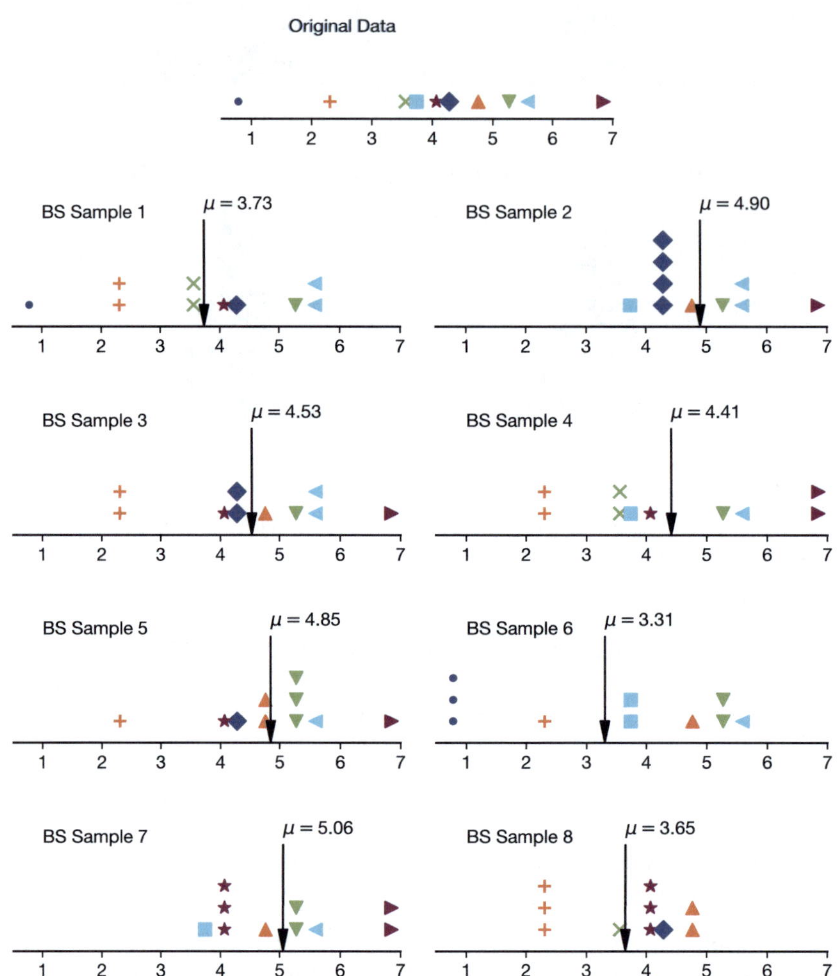

FIGURE 9.17
An example of eight different bootstrap samples and bootstrap estimates of the mean for small data set.

If we repeat the bootstrap process many times and compute the histogram of the data, we will get an empirical **bootstrap distribution** for this data sample. The function below plots the empirical histogram of the bootstrap distribution as a function of the number of bootstrap samples varies from 100 to 1 million, along with the sampling distribution for the mean estimator (the solid curve).

```
import numpy as np
import numpy.random as npr
```

(continues on next page)

(continued from previous page)

```
import scipy.stats as stats

def bootstrap_vs_sampling(num_data_samples = 10,
                          num_bs_samples = 1_000_000, seed=21490):

  np.random.seed(seed)
  N=stats.norm(3,2)
  nvals = N.rvs(num_data_samples)

  avgs = np.zeros(num_bs_samples)

  for sample in range(num_bs_samples):
    bs = npr.choice(nvals, len(nvals))
    avg =bs.mean()
    avgs[sample] = avg

  fig = plt.figure()
  fig.set_dpi(100)
  fig.set_size_inches(8, 10)
  gs = gridspec.GridSpec(5,1)

  # Set up sampling distribution for mean estimator
  mu_hat = stats.norm(3, 2/np.sqrt(num_data_samples))
  x = np.linspace(1.5, 5)

  for i in range(5):
    ax = plt.subplot(gs[i,0])

    ax.hist(avgs[:10**(i+2)], density=True, bins=10*(i+1), label='BS Dist')
    ax.plot(x, mu_hat.pdf(x) , label = 'Sampling Dist')
    ax.legend()
    ax.set_title(f'Bootstrap Distribution with {10**(i+2)} samples')

  plt.tight_layout()
```

Fig. 9.18 shows empirical bootstrap distributions for the mean estimator when there are 10 values in the original data sample. The different distributions shown depend on the number of bootstrap samples drawn from the data. The solid, red curve shows the sampling distribution for the mean estimator. As the number of bootstrap samples increases, the bootstrap distribution looks more like a Normal distribution. More importantly, the bootstrap distribution for the mean estimator differs significantly from the sampling distribution for the mean estimator, in terms of both the location of the mode of the distribution and the spread of the distribution. That is because the original data sample is small (10 samples), and the bootstrap distribution is limited to resampling from these 10 samples.

The results in Fig. 9.19 show the same experiment with 100 samples from the original distribution. Although the bootstrap and sampling distributions still differ, the shapes of the two distributions are much more similar. There is an offset in the location of the modes/means of the distributions, but this is to be expected from the limited size of the data. These results suggest that the bootstrap distribution may provide a reasonable ap-

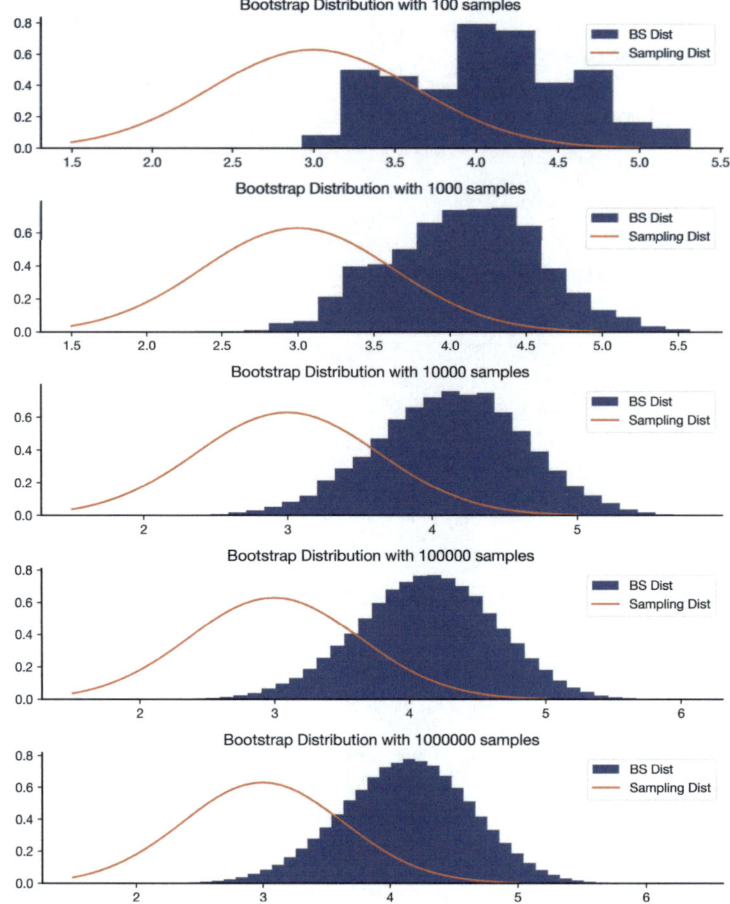

FIGURE 9.18
Empirical bootstrap distributions for the mean estimator based on 10 values from a Normal ($\mu = 3$, $\sigma = 2$) distribution, as a function of the number of bootstrap samples.

proximation of the sampling distribution if the number of data samples is at least 100, and that it is a poor approximation if the number of data samples is small (less than 20).

Terminology review and self-assessment questions

Interactive flashcards to review the terminology introduced in this section and self-assessment questions are available at fdsp.net/9-7, which can also be accessed using this QR code:

9.8 Effect Size, Power, and Sample Size Selection

Up to now, we have considered experiments for which the data has already been collected. In this section, I consider one aspect of experimental design: how to select the sample size for an experiment. That value will depend on the characteristics of the underlying

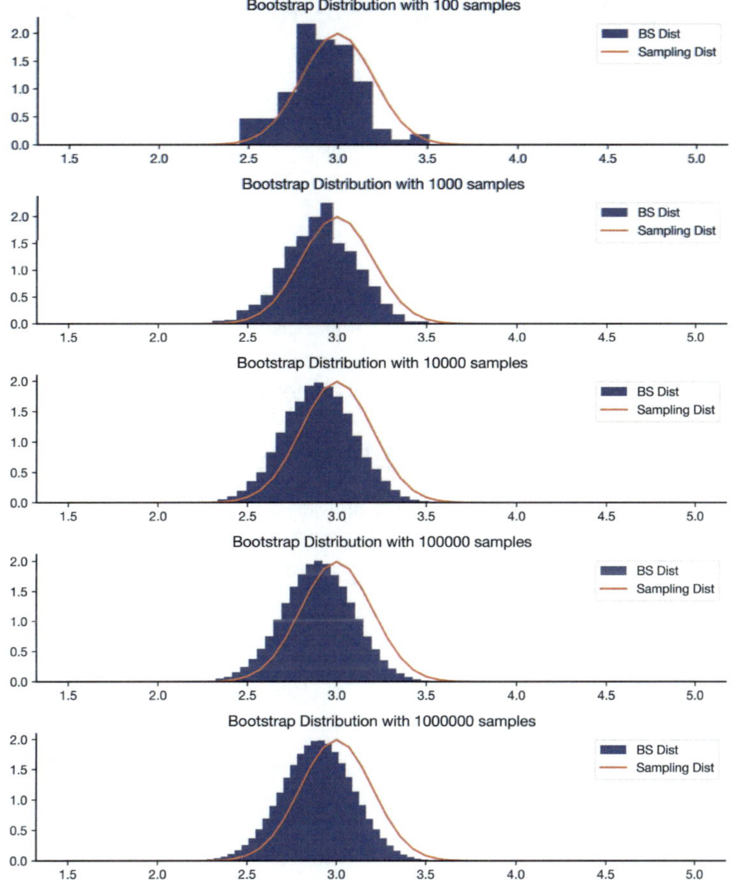

FIGURE 9.19
Empirical bootstrap distributions for the mean estimator based on 100 values from a Normal ($\mu = 3$, $\sigma = 2$) distribution, as a function of the number of bootstrap samples.

data (which are at least partially unknown) and on the performance criteria, including the acceptable probabilities of Type I and Type II errors. A full discussion and analysis of sample size selection is on the book's website at fdsp.net/9-8. Here, I just provide some brief highlights.

Suppose we observe n_X random values $\mathbf{X} = [X_0, X_1, \ldots, X_{n_X-1}]$ and n_Y random values $\mathbf{Y} = [Y_0, Y_1, \ldots, Y_{n_Y-1}]$ from these two distributions. Let $\hat{\mu}_X$ and $\hat{\mu}_Y$ be the mean estimators that are computed from $\mathbf{X}$ and $\mathbf{Y}$, which are assumed to have a common standard deviation, σ.

Sample size selection requires making some assumptions about the alternative hypothesis, H_a. In this case, we need to specify how much the means differ. This is usually done through a normalized difference,

$$d = \frac{\mu_X - \mu_Y}{\sigma}$$

called *Cohen's d*, which is a type of *effect size*:

Definition

effect size

One of many measures of separation between distributions. For a difference of means, Cohen's d is standard:

$$d = \frac{\mu_X - \mu_Y}{\sigma}.$$

In practice, the effect size is not known before the experiment, but it is often practical to make some assumption about the effect size. Effect sizes are also often specified by descriptors, which are adjectives that indicate the relative effect size and which have been specified in the statistics literature according to the table below.

Cohen's d	Effect size descriptor
0.01	Very small
0.2	Small
0.5	Medium
0.8	Large
1.2	Very large
2.0	Huge

Let α and β be the required probability of Type I and Type II error, respectively. If we want equal group sizes ($n_X = n_Y$), then the group sizes for a one-tailed test are given by

$$n_X = \frac{2\left[Q^{-1}(\alpha) + Q^{-1}(\beta)\right]^2}{d^2}.$$

(A function to implement this equation and a general equation for unequal group sizes are provided online.)

For example, for a statistical significance of 0.05, power of 0.8, and effect size of 0.8, the size of each group should be at least 20, for 40 total participants. The smaller the effect size, the larger the groups must be. For instance, if the effect size is only $d = 0.2$, the required group sizes for $\alpha = 0.05$ and power $= 0.8$ are 310.

Additional examples and formulas for two-sided tests are available on the book's website at fdsp.net/9-8.

Terminology review and self-assessment questions

Interactive flashcards to review the terminology introduced in this section and self-assessment questions are available at fdsp.net/9-8, which can also be accessed using this QR code:

9.9 Chapter Summary

Expected values, and especially moments, are commonly used to characterize distributions. In this chapter, I introduced expected value and moments and showed how to calculate them. Useful properties of expected values, means, and variances were introduced. Then we explored parameter estimation, focusing on estimation of the mean and variance of a random distribution based on samples from that distribution. I showed how to find confidence intervals using bootstrap resampling and by using analytical methods for the special case of the mean estimator. I also showed how to conduct an analytical NHST (commonly called a T-test when the underlying variances are not known) for a difference of means between two groups.

We applied our new knowledge to study the sampling distribution for an estimator, and revisited the concept of a bootstrap distribution in the context of an estimator. These two distributions are closely related, and we showed by example how the bootstrap distribution can converge to the sampling distribution as the number of data samples becomes large (≥ 100). Finally, I introduced the concept of effect size and showed how to calculate the number of data samples needed for a given combination of significance level (probability of Type-I error), power, and effect size.

Access a list of key take-aways for this chapter, along with interactive flash-cards and quizzes at fdsp.net/9-9, which can also be accessed using this QR code:

10

Decision-Making with Observations from Continuous Distributions

In this chapter, we consider different approaches to decision-making for stochastic systems in which the input is from a discrete set but the output is a continuous random variable. There are many examples of such systems, including:

- detection of disease by a test that measures the amount of some indicator molecule
- detection of a vehicle by a radar system
- determination of which bit was sent over a noisy communication channel
- classification of RF signals from noisy measurements

This chapter particularly builds on the material on conditional probabilities in Chapter 6, Bayesian methods in Chapter 7, and conditional distributions in Section 8.8. Although this chapter focuses on decision-making, we also introduce some important general techniques for working with conditional probabilities involving continuous random variables. In the next section, we start with non-Bayesian approaches to binary decision-making when the observation is a continuous random variable.

10.1 Binary Decisions from Continuous Data: Non-Bayesian Approaches

In Section 7.3, we considered how to make optimal decisions in a discrete stochastic system (with discrete inputs and outputs) using a Bayesian framework. However, in many applications, the output of the system is continuous, even if the input is discrete. For example, the input of a binary communication system consists of bits, but the received noisy waveform is converted inside the receiver into a continuous random variable. Similarly, many medical tests for disease are based on chemical tests whose outputs can be modeled as continuous random variables.

In this section, we consider scenarios in which the system input or hidden state is binary and the output is a continuous random variable. More specifically, we consider a binary hypothesis test in which the data comes from one of two continuous densities, $f_0(x|H_0)$ or $f_1(x|H_1)$, and we wish to decide between H_0 and H_1 based on an observed value x. In many binary hypothesis tests, H_1 corresponds to the event that some phenomenon is present (such as a vehicle is present in a RADAR system or a disease is present in a medical test), and H_0 corresponds to the event that the phenomenon is not present.

Any deterministic decision rule can then be written as follows. Let $\{R_0, R_1\}$ be a partition of the real line. Then the decision rule is:

DOI: 10.1201/9781003324997-10

- If $x \in R_0$, decide H_0.
- If $x \in R_1$, decide H_1.

The regions R_0 and R_1 can be chosen to optimize some criterion that measures costs or rewards for making correct or erroneous decisions.

Consider a scenario where the *a priori* probabilities $P(H_0)$ and $P(H_1)$ are not known, and thus we cannot apply a Bayesian test. Then we might instead focus on determining R_0 and R_1 based on performance criteria that do not depend on these *a prioris*. Let $\hat{H}_i$ indicate that the decision was H_i. We will use the following metrics, where the terminology is especially used when the test is used to detect some phenomenon:

- The probability of false alarm, which is $\alpha = P(\hat{H}_1|H_0)$.
- The probability of miss, which is $\beta = P(\hat{H}_0|H_1)$.

Let's introduce a concrete example to motivate our work:

Example 10.1: Prostate-Specific Antigen Test

The PSA (Prostate-Specific Antigen) values for men in their 60s without cancer are approximately[1] Normal $(2, \sigma = 1)$. The PSA values for men in their 60s with cancer are approximately Normal $(4, \sigma = 1.5)$. *These distributions are not from actual medical data – they have been chosen to illustrate the ideas in context of this application.*

Let X denote the PSA value. Then the conditional densities for X given H_0 and given H_1 are shown below:

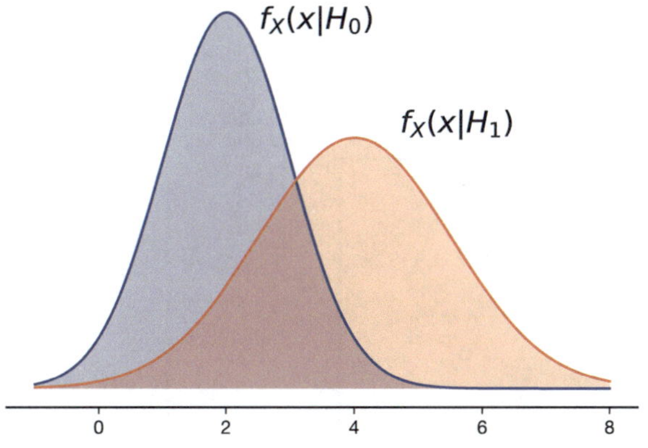

10.1.1 Maximum Likelihood Decision Rule

In binary hypothesis tests, we usually have to know the likelihoods for the system, which specify how the output depends on the input. For a discrete-input, continuous-output system, the *likelihoods* are the conditional pdfs of the outputs given the inputs. For example, if the input events are of the form A_i and the output is a continuous random variable X, then the likelihoods are of the form $f_X(x|A_i)$.

[1] Unlike the other examples in this book, the models in this problem are not based on actual data.

Recall from Section 7.3 that the maximum likelihood (ML) rule chooses an input that has the maximum likelihood among all the likelihoods. The only difference from our previous application is that the likelihoods are now conditional pdfs. For our binary hypothesis test, the ML rule is:

- If $f_X(x|H_0) > f_X(x|H_1)$, decide H_0.
- If $f_X(x|H_0) \leq f_X(x|H_1)$, decide H_1.

(Note that deciding H_1 when $f_X(x|H_0) = f_X(x|H_1)$ is arbitrary and basically meaningless – the probability of getting that exact value of X is zero because X is a continuous random variable.)

Example 10.1 (continued)

We can determine the ML decision rules by finding the values of x where $f_X(x|H_0) < f_X(x|H_1)$. The code below generates a plot as a function of the observed value x that has a value of 0 where the decision is $\hat{H}_0$ and a value of 1 where the decision is $\hat{H}_1$. The resulting plot is shown in Fig. 10.1.

```
import numpy as np
import scipy.stats as stats
import matplotlib.pyplot as plt

G0 = stats.norm(loc = 2, scale = 1)
G1 = stats.norm(loc = 4, scale = 1.5)
x=np.linspace(-10,10, 1001)
plt.plot(x, (G0.pdf(x) < G1.pdf(x)) )
plt.title('ML decision rule');
```

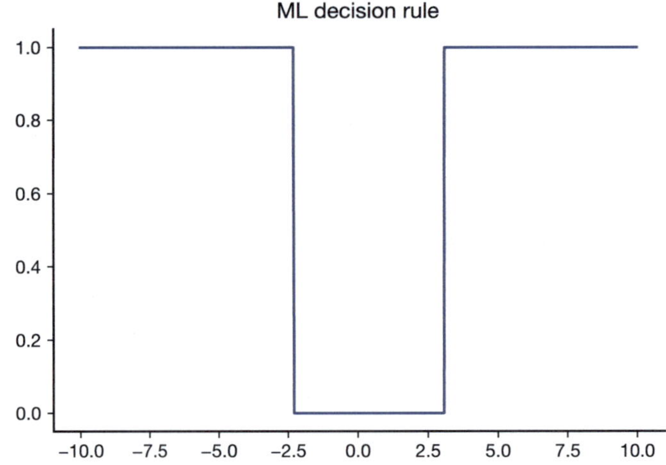

FIGURE 10.1
Plot of likelihoods for Example 10.1.

So the ML rule decides H_0 if $-2.28 \leq X \leq 3.08$ and decides H_1 otherwise. Given these decision regions, we can evaluate the probabilities of false alarm and miss. The

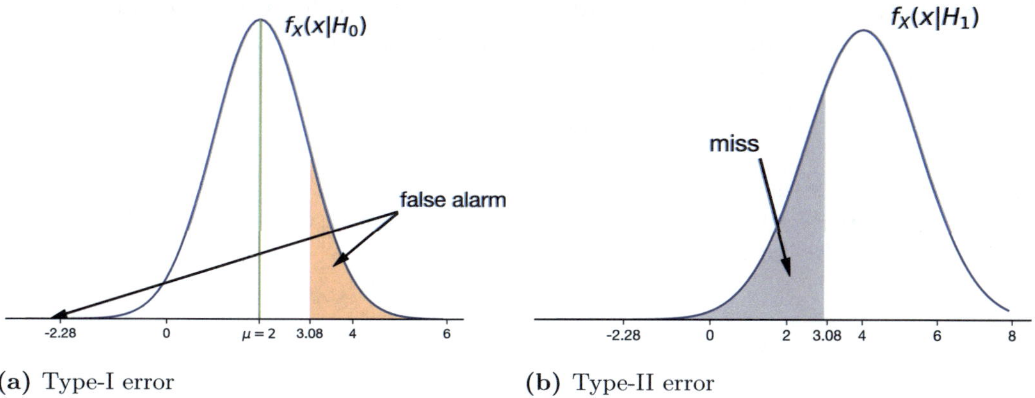

(a) Type-I error

(b) Type-II error

FIGURE 10.2
Regions for false alarm (Type-I error) event for Example 10.1.

probability of false alarm is evaluated under the condition H_0 and can be calculated as the probability in the two tails of $f_X(x|H_0)$ shown in Fig. 10.2a. (The probability in the left tail is very small, and the area is too small to be seen in the image.)

The probability of false alarm is

$$Q\left(\frac{2-(-2.28)}{1}\right) + Q\left(\frac{3.08-2}{1}\right),$$

where the numerators of the $Q()$ functions are the distances to the tails, and the denominators are the σ values under H_0. Let's create a function to calculate the probability of false alarm for decision regions of this form:

```
def q(x):
  return stats.norm.sf(x)

def prob_false_alarm(gamma0, gamma1, mean0=2, sigma0=1):

  # Check that the decision region boundary conditions are satisfied
  assert mean0 > gamma0
  assert gamma1 > mean0

  return q((mean0-gamma0) / sigma0) + q( (gamma1 - mean0) /sigma0)
```

```
print(f'The probability of false alarm is {prob_false_alarm(-2.28, 3.08) : .3f}')
```

```
The probability of false alarm is  0.140
```

Similarly, we can calculate the probability of miss by calculating the probability of observing a value in $-2.28 < X \leq 3.08$ under H_1. This region is shown in Fig. 10.2b. We can express this using the $Q()$ function as

$$Q\left(\frac{4-3.08}{1.5}\right) - Q\left(\frac{4--2.28}{1.5}\right).$$

Let's create a function to evaluate the probability of miss when the decision region is of this form:

```
def prob_miss(gamma0, gamma1, mean1=4, sigma1=1.5):

  # Check that the decision region boundary conditions are satisfied
  assert gamma0 < mean1
  assert gamma1 < mean1

  return q((mean1-gamma1) / sigma1) - q( (mean1-gamma0) /sigma1)
```

```
print(f'The probability of miss is {prob_miss(-2.28, 3.08) : .3f}')
```

```
The probability of miss is  0.270
```

10.1.2 Generalizing the ML Rule

Suppose we want either a lower probability of false alarm or a lower probability of miss. How should we go about selecting the decision regions? Let's start by rewriting the ML decision rule as shown:

$$f_X(x|H_0) \underset{H_1}{\overset{H_0}{\gtrless}} f_X(x|H_1)$$

$$\frac{f_X(x|H_0)}{f_X(x|H_1)} \underset{H_1}{\overset{H_0}{\gtrless}} 1.$$

The ratio on the left is a *likelihood ratio*:

Definition

likelihood ratio

A ratio of the likelihoods, typically denoted by $L(x)$. For instance,

$$L(x) = \frac{f_X(x|H_0)}{f_X(x|H_1)}.$$

We can change the probabilities of false alarm and miss if we compare $L(x)$ to a different threshold than 1. In other words, let's consider the following decision rule:

$$L(x) \underset{H_1}{\overset{H_0}{\gtrless}} c.$$

Example 10.2: Generalized Likelihood Decision Rule for PSA Test

Consider again the PSA test from Example 1, where the likelihoods are conditionally Normal. Below is a function to find the decision region as a function of c. I have called c by the variable name `threshold` to make its purpose more clear:

```
def likelihood_ratio_region(threshold, mean0=2, sigma0=1, mean1=4, sigma1=1.5,
                            lower=-10, upper=10):
  '''calculate the decision regions using a likelihood ratio comparison
  Also plots a function that is 1 when decision H1 and 0 when decision H0

  Parameters
  ----------
  threshold: threshold to compare likelihood ratio to
  mean0, sigma0, mean1, sigma1: parameters of conditional Normal distributions
  lower, upper: region to use when determining decision rule

  Returns
  -------
  float, float: edges of decision region for H0
  '''

  # Set up likelihoods
  G0 = stats.norm(loc = mean0, scale = sigma0)
  G1 = stats.norm(loc = mean1, scale = sigma1)

  # Calculate likelihood ratios over the specified region
  x=np.arange(lower, upper, 0.01)
  L = G0.pdf(x)/ G1.pdf(x)

  # Plot decision rule
  plt.plot(x, (L < threshold) )
  plt.title('Decision as a function of $x$');

  # Return the edges of the decision region
  # (this is a bit tricky)
  return np.round(x[np.where(L>threshold)[0][0]], 2), \
         np.round(x[np.where(L>threshold)[0][-1]], 2)
```

If $c = 1$, we get the ML rule. If we choose a value of c smaller than 1, we will increase the region over which we decide H_0, so the probability of false alarm should decrease. Here is the decision region for H_0 when $c = 1/3$:

```
likelihood_ratio_region(1/3)
```

```
(-2.94, 3.74)
```

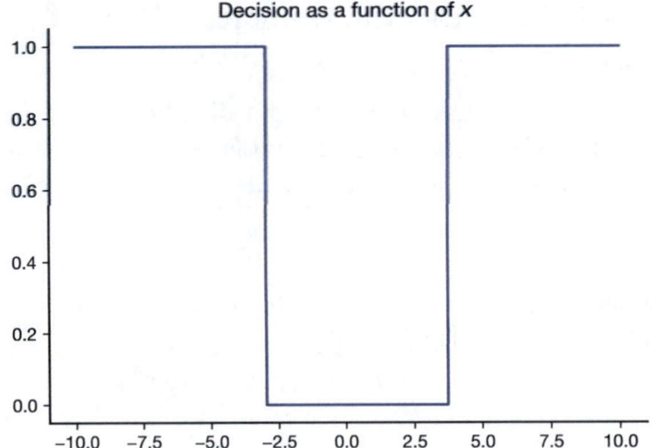


```
print(f'The probability of false alarm is {prob_false_alarm(-2.94, 3.74) : .3f}')
```

```
The probability of false alarm is  0.041
```

As expected, the probability of false alarm decreased from 0.14 for the ML rule to 0.041 when $c = 1/3$. There is a price to pay for this because the probability of miss must then increase:

```
print(f'The probability of miss is {prob_miss(-2.94, 3.74) : .3f}')
```

```
The probability of miss is  0.431
```

If we instead wanted to decrease the probability of miss, we could choose a value of $c > 1$. The decision region for H_0 when $c = 3$ is shown below:

```
likelihood_ratio_region(3)
```

```
(-1.4, 2.2)
```

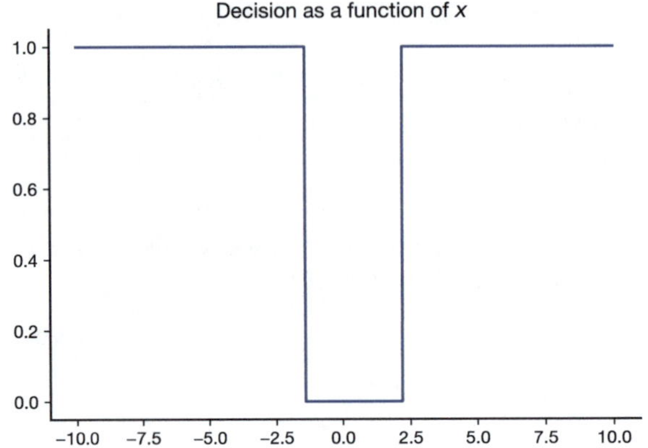

```
print(f'The probability of false alarm is {prob_false_alarm(-1.4, 2.2) : .3f}')
print(f'The probability of miss is {prob_miss(-1.4, 2.274) : .3f}')
```

```
The probability of false alarm is  0.421
The probability of miss is  0.125
```

Comparing the likelihood ratio to a threshold is optimal in the following sense. If we use this rule and choose a value of c that achieves a specified probability of false alarm, α, this rule achieves the minimum possible probability of miss over all rules that achieve the same α. This result is called the Neyman-Pearson Lemma, and this detection rule is the Neyman-Pearson detector.

10.1.3 Illustrating Performance Tradeoffs: ROC Curves

A common way to illustrate the tradeoff between the probability of false alarm and the probability of miss is through a ROC curve. ROC stands for *receiver operating characteristic*:

Definition

receiver operating characteristic (ROC)

For a binary hypothesis test, a plot of the probability of false alarm (sometimes called the False Positive Rate or FPR) versus the probability of detection (sometimes called the True Positive Rate or TPR).

The probability of detection is $1-\beta$. ROC curves were originally developed in the context of detecting objects in RADAR systems. For a Neyman-Pearson detector, a ROC curve can be generated by sweeping the likelihood-ratio threshold, c, over a region that essentially covers $0 \leq \alpha \leq 1$ and $0 \leq \beta \leq 1$, recording the pairs $\alpha, 1-\beta$, and then generating a line plot of the probability of detection as a function of the probability of false alarm.

Example 10.3: ROC for PSA Test with Simplified Decision Regions

Let's generate a ROC curve for an even simpler detector that uses a single decision threshold, γ. The decision rules are then:

- Decide H_0 if $x \leq \gamma$.
- Decide H_1 if $x > \gamma$.

Then the probability of false alarm and miss simplify to

$$\alpha = Q\left(\frac{\gamma - \mu_0}{\sigma_0}\right), \text{ and}$$
$$\beta = Q\left(\frac{\mu_1 - \gamma}{\sigma_1}\right).$$

The function below calculates and prints the probability of false alarm and miss. On the website for this book, there is a version of this function that plots the likelihoods and shows the regions corresponding to Type-I and Type-II errors.

```
def binary_hypothesis_perf (gamma, mu0=2, sigma0=1, mu1=4, sigma1=1.5):
  ''' Evaluate performance of binary hypothesis test for 2 Normal likelihoods
  '''

  # Calculate the probabilities
  alpha = q( (gamma - mu0) / sigma0)
  beta = q( (mu1 - gamma) / sigma1)

  print(f'The probability of false alarm is {alpha:.2g}')
  print(f'The probability of miss is {beta:.2g}')
```

For instance, for our example distributions, Fig. 10.3 shows the densities and the ML decision rule using a single threshold $\gamma = 3.08$.

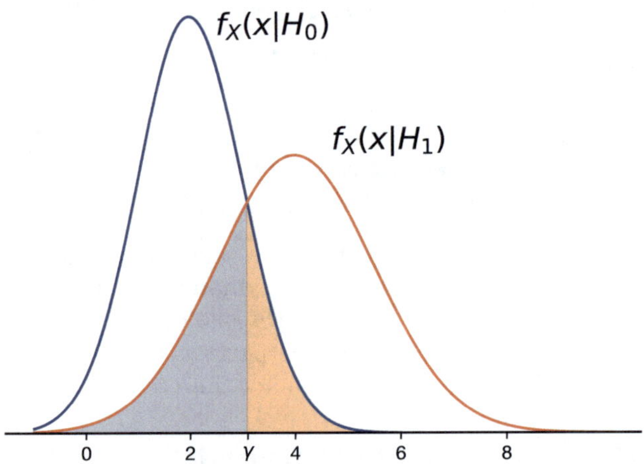

FIGURE 10.3
Illustration of two Normal densities with corresponding false alarm and miss regions shown using shading under the pdfs.

The performance metrics are computed to be:

```
binary_hypothesis_perf(3.08)
```

```
The probability of false alarm is 0.14
The probability of miss is 0.27
```

The following code generates a ROC curve for this simple detector. In addition to the ROC curve, it also includes a Reference, which is the performance if the output of the system, X, is not used in the decision. If H_1 is selected with probability p, then the probability of false alarm will be p (since when H_0 is true, we will still choose H_1 probability p), and the probability of detection will also be p (since, when H_1 is true, we also decide H_1 with probability p). Thus, without using the observed value, we get a linear relation between α and $1 - \beta$.

```
gammas = np.arange(-10, 10, 0.1)
# False alarm: Given H0, mu = 2 and sigma = 1
mu0 = 2
sigma0 =1

# Miss: Given H1, mu = 4 and sigma = 1.5
mu1 = 4
sigma1 = 1.5

alphas = np.zeros_like(gammas)
betas = np.zeros_like(gammas)

for i, gamma in enumerate(gammas):
  alphas[i] = q( (gamma - mu0) / sigma0)
  betas[i] = q( (mu1 - gamma) / sigma1)

plt.plot(alphas, 1-betas, label='ROC curve')
plt.xlim(0, 1)
plt.ylim(0, 1)

plt.xlabel('Probability of false alarm')
plt.ylabel('Probability of detection')
plt.title('ROC curve for example binary hypothesis test');
plt.grid()

p=np.arange(0, 1, 0.1)
plt.plot(p, p, label='Reference');
plt.legend();
```

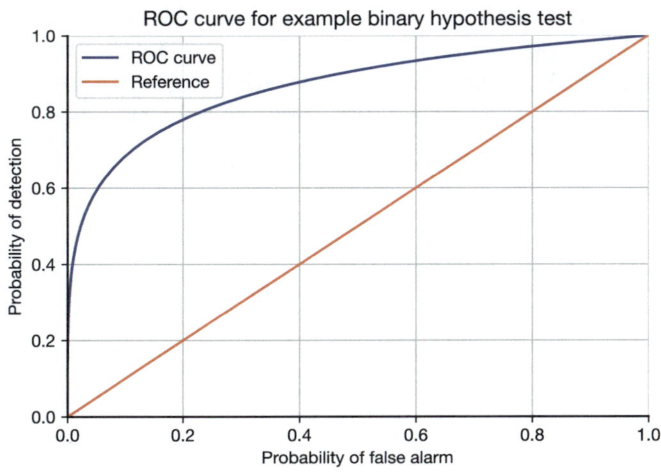

The further the ROC curve is from the reference (in terms of being more toward the upper left-hand corner), the better performance of the detector. I.e., it can achieve a higher $1-\beta$ for a given α, or vice versa.

10.1.4 Quantifying Overall Detector Performance: Area Under the Curve (AUC)

The individual points on the ROC curve tell the performance of a given detector for a *specific detection threshold. However, we can also quantify the overall performance of a* detector by providing a measure of how close the ROC curve is to the upper left-hand corner of the graph. Such a measure should depend on the entire ROC curve. We typically quantify this using the *area under curve*:

Definition

area under curve (AUC)

For a binary hypothesis test, a measure of the overall performance of the detector. AUC is computed by integrating the area under the ROC curve.

We have seen that random guessing yields the diagonal reference line, and the area under the reference line is 1/2. If a detector makes perfect decisions, then it can achieve a probability of detection of $1-\beta = 1$ with probability of false alarm $\alpha = 0$, so the ideal ROC curve rises instantly from 0 to 1 and then remains at 1. The area under the ideal ROC curve is thus 1. We can compute the area under the ROC for our example using NumPy's `np.trapz()` numerical integration routine. It is important to note that depending on the form of the detector, the α values may come out in reverse numerical order:

```
alphas[::10]
```

```
array([1.00000000e+00, 1.00000000e+00, 1.00000000e+00, 1.00000000e+00,
       1.00000000e+00, 1.00000000e+00, 9.99999999e-01, 9.99999713e-01,
       9.99968329e-01, 9.98650102e-01, 9.77249868e-01, 8.41344746e-01,
       5.00000000e-01, 1.58655254e-01, 2.27501319e-02, 1.34989803e-03,
       3.16712418e-05, 2.86651572e-07, 9.86587645e-10, 1.27981254e-12])
```

Whenever that is the case, we need to use `np.flip()` on both vectors to put them in the correct order. We pass the flipped `alphas` vector as the `x` keyword parameter and the flipped `1-betas` vector as the `y` keyword parameter of `np.trapz()`. (**Do NOT omit the keyword names because the default order is y, x, which will be confusing to most people.**) The final command is shown at the end of the code block below:

```
num_gammas = 101

gammas = np.linspace(-10, 10, num_gammas)
# False alarm: Given H0, mu = 2 and sigma = 1
mu0 = 2
sigma0 =1

# Miss: Given H1, mu = 4 and sigma = 1.5
mu1 = 4
sigma1 = 1.5
```

(continues on next page)

(continued from previous page)

```
alphas = np.zeros(num_gammas)
betas = np.zeros(num_gammas)

for i, gamma in enumerate(gammas):
  alphas[i] = q( (gamma - mu0) / sigma0)
  betas[i] = q( (mu1 - gamma) / sigma1)

auc = np.trapz(x = np.flip(alphas), y = np.flip(1-betas) )

print(f'The area under curve (AUC) for this experiment is {auc : .2g}')
```

```
The area under curve (AUC) for this experiment is  0.87
```

Exercises

1. Plot the performance if the variance of each PSA test is reduced by a factor of 4. What is the AUC?

2. If the variance of each PSA test is reduced by a factor of 2 (from the original values), what is the decision threshold for $\alpha = 0.1$? What value of β is achieved?

Terminology review and self-assessment questions

Interactive flashcards to review the terminology introduced in this section and self-assessment questions are available at fdsp.net/10-1, which can also be accessed using this QR code:

10.2 Point Conditioning

In this section, we build upon some of the concepts from Section 8.8, where we began to study conditioning with random variables. We often have cases where the observed output of a system is best modeled as a random variable, and we would like to make decisions based on the observed value of the random variable. However, if the output is a continuous random variable, our previous approaches break down because the probability of a continuous random variable taking on any particular value is zero.

Example 10.4: Optimal Decisions in a Binary Communication System with Continuous Outputs

Consider again the binary communication system introduced in Example 8.17 (from Section 8.8), for which the output is conditionally Normal given the input. In particular, the output random variable has conditional distributions given by

$$\begin{cases} X \sim \text{Normal}(+1, \sigma), & 0 \text{ transmitted} \\ X \sim \text{Normal}(-1, \sigma), & 1 \text{ transmitted.} \end{cases}$$

Let T_i denote the event that i is transmitted. As in Section 7.3, we want to determine the probabilities of the inputs given the observed value of the output. The difference is that in Section 7.3, the output was a discrete event, whereas now we observe a particular value of a continuous random variable: we observe some received value $X = x$.

We would like to calculate the probabilities of T_0 and T_1 given an observation $X = x$. The form of these probabilities may look familiar: we are asking about probabilities of the form $P(T_i|X = x)$. This is the probability of an input given the observation or output and hence is an *a posteriori* probability (APP). Given the APPs, we would also like to find the maximum *a posteriori* (MAP) decision rule.

Let's revisit the example from Section 8.8 to identify the issue:

Example 8.17(c) Revisited

Suppose that we want to calculate the probabilities of the events T_0 and T_1 given that $X = 2$ if $P(T_0) = 0.4$, $P(T_1) = 0.6$, and $\sigma = 2$. A direct application of our previous Bayes' rule approaches yields

$$P(T_0|X = 2) = \frac{P(X = 2|T_0)P(T_0)}{P(X = 2)}.$$

This is problematic because $P(X = 2|T_0) = 0$ and $P(X = 2) = 0$, so the fraction is $0/0$.

This problem is caused by conditioning on an event that has zero probability. But keep in mind that every time that X is received, it takes on *some* value, even though that value has zero probability. Being able to answer this type of question is important, but we don't have the math to deal with it yet. This type of conditional probability is called *point conditioning*:

Definition

point conditioning

A conditional probability in which the conditioning statement is (or includes) the event that a continuous random variable is equal to a particular value. An example is $P(A|X = x)$, where X is a continuous random variable.

We can evaluate a conditional probability with point conditioning by treating it as a limit and doing some careful manipulation:

$$\begin{aligned}
P(A|X = x) &= \lim_{\Delta x \to 0} P(A|x < X \leq x + \Delta x) \\
&= \lim_{\Delta x \to 0} \frac{F_X(x + \Delta x|A) - F_X(x|A)}{F_X(x + \Delta x) - F_X(x)} P(A) \\
&= \lim_{\Delta x \to 0} \frac{\frac{F_X(x+\Delta x|A) - F_X(x|A)}{\Delta x}}{\frac{F_X(x+\Delta x) - F_X(x)}{\Delta x}} P(A)
\end{aligned}$$

Taking the limit in the numerator and denominator yields

$$P(A|X=x) = \frac{f_X(x|A)}{f_X(x)} P(A), \tag{10.1}$$

provided $f_X(x|A)$ and $f_X(x)$ exist, and $f_X(x) \neq 0$. (The result looks like what you would do if you didn't know any better – treat the densities as if they were probabilities, and everything works out!)

10.2.1 Total Probability for Continuous Distributions

Note that the form above is almost a Bayes' rule form. If A is some input event and X is the observed output, then $f_X(x|A)$ is the likelihood of X given A. In this context, the conditional probability $P(A|X=x)$ is the *a posteriori* probability of A given that the output of the system is $X = x$. However, in Bayes' rule, the denominator usually needs to be found using total probability, but we do not yet have any Law of Total Probability for point conditioning.

In the binary communication system example, we can use the partitioning events $\{T_0, T_1\}$. We can easily create a general Total Probability rule for CDFs because CDFs are probability measures:

Total Probability for CDFs

If $\{A_i\}$ forms a partition of S, then from our previous work on the Law of Total Probability, we have

$$\begin{aligned} F_X(x) &= P(X \leq x) \\ &= \sum_i P(X \leq x|A_i)P(A_i) \\ &= \sum_i F_X(x|A_i)P(A_i) \end{aligned}$$

Total Probability for pdfs

To derive a similar rule for pdfs, we note that $f_X(x) = \frac{d}{dx}F_X(x)$ and $f_X(x|A_i) = \frac{d}{dx}F_X(x|A_i)$. Substituting the Total Probability for CDFs into the definition of the pdf $f_X(x)$ yields

$$\begin{aligned} f_X(x) &= \frac{d}{dx}F_X(x) \\ &= \frac{d}{dx}\sum_i F_X(x|A_i)P(A_i) \\ &= \sum_i \left[\frac{d}{dx}F_X(x|A_i)\right] P(A_i) \\ &= \sum_i f_X(x|A_i)P(A_i). \end{aligned}$$

The equation is almost identical to the equation for the Total Probability for CDFs, except CDFs (denoted by F) are replaced by pdfs (denoted by f) everywhere.

Total Probability for Events with Point Conditioning

We have one more case that often arises, which is when we wish to calculate the probability of some event but we only know conditional probabilities for that event given the value of some continuous random variable. Consider again the point-conditioning form for the probability of an event, which we found in (10.1),

$$P(A|X=x) = \frac{f_X(x|A)}{f_X(x)} P(A).$$

Multiplying both sides by $f_X(x)$ and integrating over x yields

$$\Rightarrow P(A|X=x) f_X(x) = f_X(x|A) P(A)$$
$$\Rightarrow \int_{-\infty}^{\infty} P(A|X=x) f_X(x)\, dx = \int_{-\infty}^{\infty} f_X(x|A)\, dx P(A),$$

where on the right side, we have pulled out $P(A)$ from the integral because it does not depend on x. The remaining integral on the right evaluates to 1 because it is the integral of a density from $-\infty$ to ∞. Swapping the two sides yields

$$P(A) = \int_{-\infty}^{\infty} P(A|X=x) f_X(x)\, dx. \tag{10.2}$$

Bayes' Rule for the Probability of an Event Under Point Conditioning:

We are now ready to derive a general formula for Bayes' rule for scenarios in which the observation is a continuous-valued random variable for which the distribution depends on some discrete underlying hidden state or input. Examples include the binary communication system described at the beginning of this section and many types of sensing systems, such as when the sensor output (e.g., sound, RADAR, or seismic) depends on whether some type of vehicle is present. A generic formulation of this type of problem is as follows:

Let $\{A_i,\ i = 0, 1, \ldots, n-1\}$ be a partition of S; for instance, if $\{A_i\}$ represents all of the different values of the hidden state, and no two values in $\{A_i\}$ can occur at the same time, then $\{A_i\}$ will be a partition of S. Let X be some random variable for which we know the conditional density given A_i occurred for each $i = 0, 1, \ldots, n-1$. Then the conditional probabilities of the form $P(A_i \mid X = x)$ are the *a posteriori* probabilities:

> Definition
>
> ***a posteriori* probability (discrete-input, continuous-output stochastic system)**
>
> Consider a stochastic system with a discrete set of possible input events $\{A_0, A_i, \ldots\}$ and a continuous output random variable X, where the dependence of X on A_i can be expressed in terms of the likelihoods $f_X(x|A_i)$. Then the *a posteriori probabilities* are the conditional probabilities of the input events given the observed outputs, $P(A_i \mid X = x)$.

We can use (10.2) in (10.1) to get Bayes' Rule for this case:

Definition

Bayes' Rule (discrete-input, continuous-output stochastic system)

Consider a stochastic system with a discrete set of possible input events $\{A_0, A_i, \ldots\}$ and a continuous output random variable X, where the dependence of X on A_i can be expressed in terms of the likelihoods $f_X(x|A_i)$. Then

$$P(A_i|X=x) = \frac{f_X(x|A_i)P(A_i)}{\sum_{i=0}^{n-1} f_X(x|A_i)P(A_i)}. \tag{10.3}$$

Let's use this to find the *a posteriori* probabilities in our communications example.

Example 8.17(c) Revisited

Consider again the binary communication system with $P(T_0) = 0.4$ and $P(T_1) = 0.6$ if $\sigma = 2$. If $X = 2$ is received, what are the conditional probabilities for T_0 and T_1?

We have to use point conditioning because $P(X = 2) = 0$. Thus, we will specialize (10.3) as

$$P(T_i|X = x) = \frac{f_X(x|T_i)P(T_i)}{f_X(x|T_0)P(T_0) + f_X(x|T_1)P(T_1)},$$

where we can evaluate the particular probabilities by plugging in the values specified. Before we do that, let's try to gain some intuition about this formula. First, note that the denominator is the same for both $P(T_0|X = x)$ and $P(T_1|X = x)$. It can be considered a normalization constant. So, let's start by comparing the unnormalized numerator values. Below, I have plotted $f_X(x|T_0)(0.4)$ and $f_X(x|T_1)(0.6)$ and indicated the values at $x = 2$ with markers:

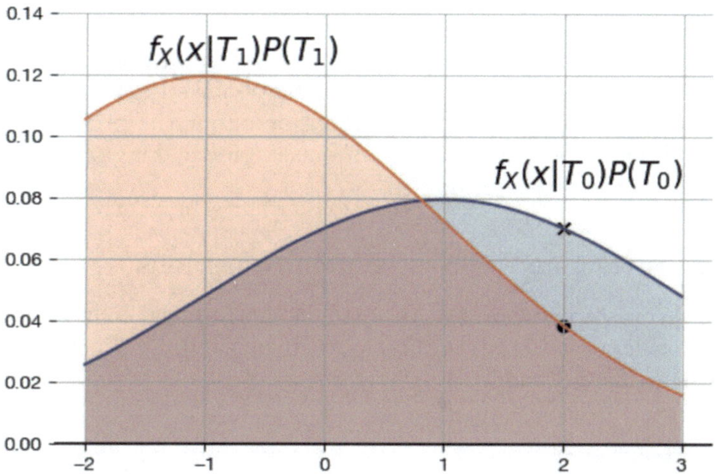

From the graph, $f_X(2|T_0)P(T_0) \approx 0.75$ and $f_X(2|T_1)P(T_1) \approx 0.4$, so we can conclude that $P(T_0|X = 2)$ is greater than $P(T_1|X = 2)$ by a factor of approximately $0.75/0.4 = 1.875$. Let's check out the exact values using (10.3). The denominator is

```
den=G0.pdf(2)*0.4 + G1.pdf(2)*0.6
den
```

```
0.10926834405262742
```

The conditional probability of T_0 given $X = 2$ is then

```
pT0_2 = G0.pdf(2)*0.4/den
pT0_2
```

```
0.6444049826448045
```

and the conditional probability of T_1 given $X = 2$ is

```
pT1_2 = G1.pdf(2)*0.6/den
pT1_2
```

```
0.3555950173551955
```

and the ratio is

```
pT0_2/pT1_2
```

```
1.8121878856393636
```

The analytical result matches our approximation from the graph.

As in the example of a binary communication system with discrete outputs, the *a posteriori* probabilities provide information about the probabilities of the inputs given the observed output, and these are useful for making optimal decisions, which we consider in the next section.

Terminology review and self-assessment questions

Interactive flashcards to review the terminology introduced in this section and self-assessment questions are available at fdsp.net/10-2, which can also be accessed using this QR code:

10.3 Optimal Bayesian Decision-Making with Continuous Random Variables

Consider a system in which the input can be characterized by the following:

- A discrete set of input events $A_i,\ i = 0, 1, \ldots, n-1$

- The output is a continuous random variable X
- The distribution of X depends on the input event A_i

In particular, the likelihoods are the conditional densities $f_X(x|A_i)$ of the density of X at x given the input A_i. Then the maximum *a posteriori* (MAP) decision rule chooses the input event A_i that maximizes the *a posteriori* probability $P(A_i|X = x)$. Let $\hat{A}_i$ denote the decision that input event A_i occurred. Then the MAP decision rule is

$$\begin{aligned}\hat{A}_i &= \arg\max_{A_i} P(A_i|X = x) \\ &= \arg\max_{A_i} \frac{f_X(x|A_i)P(A_i)}{f_X(x)}.\end{aligned}$$

Noting that $f_X(x)$ does not depend on the input event A_i, we can simplify that MAP rule as

$$\hat{A}_i = \arg\max_{A_i} f_X(x|A_i)P(A_i),$$

where $f_X(x)|A_i)$ is the likelihood of X under input event A_i, and $P(A_i)$ is the *a priori* probability of A_i.

For the case of equal *a priori* probabilities, we can eliminate the term $P(A_i)$ from the MAP rule (since it is the same for every A_i), yielding the rule

$$\hat{A}_i = \arg\max_{A_i} f_X(x|A_i),$$

which is a maximum likelihood (ML) rule.

Example 10.5: Binary Communication System with Continuous Outputs, Revisited

Consider again the binary communication system introduced in Section 8.8, for which the output X is conditionally Normal given the input. Here,

$$\begin{cases} X \sim \text{Normal}(+1, \sigma), & \text{0 transmitted} \\ X \sim \text{Normal}(-1, \sigma), & \text{1 transmitted,} \end{cases}$$

where A_i denotes the event that i is transmitted.

Since we have only two possible inputs, the MAP rule can be simplified to a single comparison. We will decide $\hat{A}_0$ if A_0 maximizes

$$f_X(x|A_0)P(A_0) \geq f_X(x|A_1)P(A_1)$$

and decide $\hat{A}_1$ otherwise.

The comparison is between weighted likelihoods, where the likelihoods are Normal pdfs with the same variance but different means, and the weights are the *a priori* probabilities of the inputs. Noting that $P(A_1) = 1 - P(A_0)$, we can create a function that draws the weighted likelihoods for a given value of $P(A_0)$ and a given value of σ. In addition, the function uses different colors to illustrate which weighted density is greater; in other words, the colors indicate the MAP decision regions. Although we have not shown it yet, the MAP decision region in this scenario can be characterized using a single decision threshold, and the value of that threshold is determined (approximately) by finding the smallest value of x such that the weighted density $f_X(x|A_0)P(A_0)$ is greater than $f_X(x|A_1)P(A_1)$.

```
def drawMAP(p0, sigma=1):
    ''' Draw the weighted densities for the binary communication system problem
    and shade under them according to the MAP decision rule.

    Inputs:
    p0= probability that 0 is transmitted
    sigma2= variance of the Normal noise (default is 1)'''

    # Set up random variables
    G0 = stats.norm(loc=1, scale=sigma)
    G1 = stats.norm(loc=-1, scale=sigma)
    x=np.linspace(-4, 4, 1001)
    p1=1-p0

    # plot the weighted densities:
    # these are proportional to the APPs
    plt.plot(x, p0*G0.pdf(x))
    plt.plot(x, p1*G1.pdf(x))

    # Add labels
    plt.annotate('$f_X(x|T_0)P(T_0)$', (2.5, 1.6*p0*G0.pdf(2.5)),
                 fontsize=12)
    plt.annotate('$f_X(x|T_1)P(T_1)$', (-4.2, 1.6*p1*G1.pdf(-2.5)),
                 fontsize=12);

    # Determine the regions where the APP for 0 is
    # bigger and the APP for 1 is bigger
    R0=x[np.where(p0*G0.pdf(x) >= p1*G1.pdf(x))]
    R1=x[np.where(p0*G0.pdf(x) < p1*G1.pdf(x))]

    # Fill under the regions found above
    plt.fill_between(R0,p0*G0.pdf(R0), alpha=0.3)
    plt.fill_between(R1,p1*G1.pdf(R1), alpha=0.3)

    # Print the MAP threshold
    print("MAP decision threshold is",round(R0[0],2))
```

The weighted densities, MAP decision regions, and the decision threshold for equal *a priori* probabilities are shown in the following figure. Note that this corresponds to the ML decision rule. Since the weighted densities are symmetric around 0, the ML decision threshold is 0. The decision rule is:

$$\hat{A}_0, \quad x \geq 0, \text{ and}$$
$$\hat{A}_1, \quad x < 0.$$

Here we have arbitrarily assigned the point $x = 0$ to $\hat{A}_0$, but it does not matter which decision region it is assigned to – any individual point has zero probability of occurring.

```
drawMAP(0.5)
```

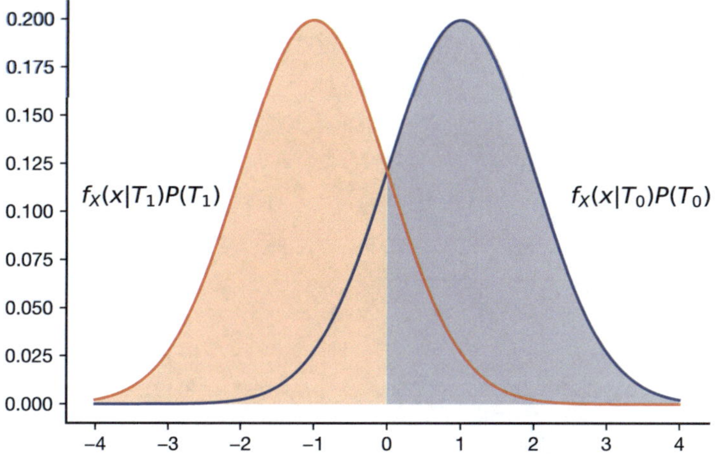

If we decrease $P(A_0)$ to 0.25, we get the weighted densities shown in the following figure. The decision threshold is now 0.55, and the decision rule is

$$\hat{A}_0, \quad x \geq 0.55, \text{ and}$$
$$\hat{A}_1, \quad x < 0.55.$$

Note that more of the real axis is assigned to the decision $\hat{A}_1$, which makes sense since A_1 is more likely to have been sent.

```
drawMAP(0.25)
```

```
MAP decision threshold is 0.55
```

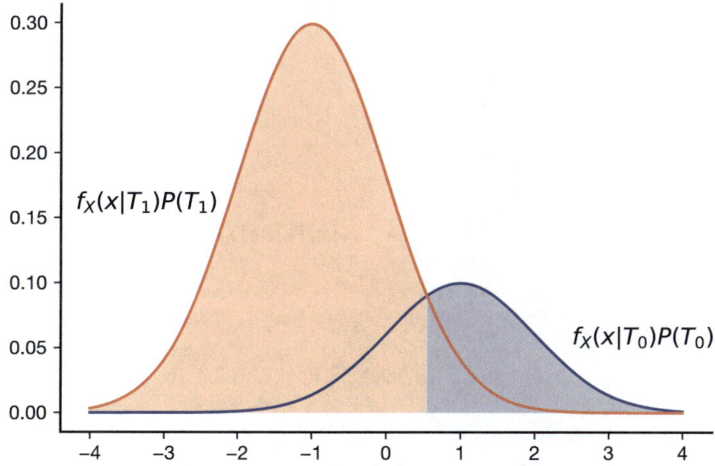

If we instead let $P(A_0) = 0.75$, the opposite effect occurs. The decision rule is

$$\begin{aligned} &\hat{A}_0, \ x \geq -0.55, \text{ and} \\ &\hat{A}_1, \ x < -0.55, \end{aligned}$$

and more of the real line is assigned to the decision $\hat{A}_0$.

```
drawMAP(0.75)
```

```
MAP decision threshold is -0.54
```

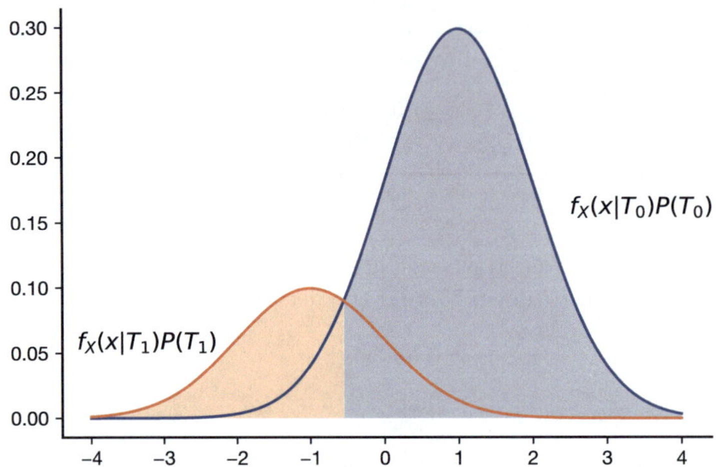

As we further increase $P(A_0)$, the decision region moves farther to the left:

```
drawMAP(0.9)
```

```
MAP decision threshold is -1.1
```

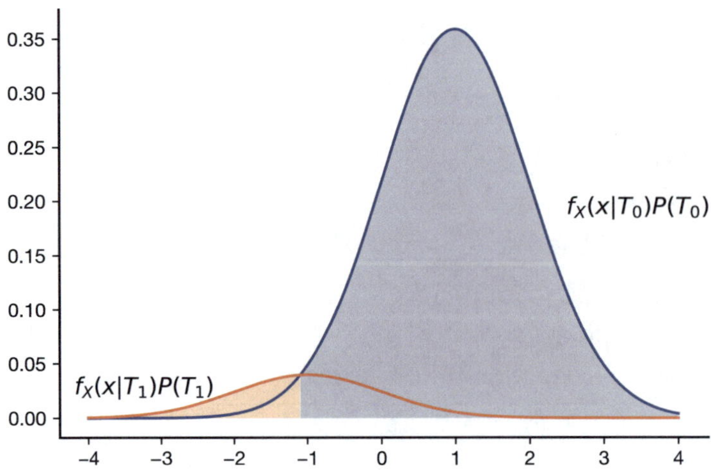

10.3.1 Analytical Value of the Decision Threshold

Note that in each of the figures illustrating the decision regions, the decision threshold is the value for which the weighted densities are equal. We can use analysis to find the value of that threshold, γ in terms of $P(A_0)$ and σ.

Consider first calculating the decision threshold for the ML decision rule when the underlying distributions are Normal random variables with equal variance but different means. When A_i is true, let the mean be μ_i. Then the ML threshold is the value of x such that the likelihoods are equal:

$$\begin{aligned}
f_X(x|A_0) &= f_X(x|A_1) \\
\frac{1}{\sigma\sqrt{2\pi}}\exp\left[-\left(\frac{\gamma-\mu_0}{\sigma}\right)^2\right] &= \frac{1}{\sigma\sqrt{2\pi}}\exp\left[-\left(\frac{\gamma-\mu_1}{\sigma}\right)^2\right] \\
-\left(\frac{x-\mu_0}{\sigma}\right)^2 &= -\left(\frac{\gamma-\mu_1}{\sigma}\right)^2 \\
\gamma^2 - 2\mu_0\gamma + \mu_0^2 &= \gamma^2 - 2\mu_1\gamma + \mu_1^2 \\
2\mu_0\gamma - 2\mu_1\gamma &= \mu_0^2 - \mu_1^2 \\
\gamma &= \frac{\mu_0^2 - \mu_1^2}{2(\mu_0 - \mu_1)} \\
\gamma &= \frac{\mu_0 + \mu_1}{2}.
\end{aligned}$$

Thus, the ML decision threshold is the average of the means.

The MAP decision threshold can be found by solving for equality between the weighted densities, where the weights are the *a priori* probabilities. The details of the algebraic manipulation are omitted, but the resulting expression is

$$\gamma_{MAP} = \frac{\mu_0 + \mu_1}{2} + \frac{\sigma^2}{\mu_0 - \mu_1}\log\left(\frac{P_1}{P_0}\right).$$

Let's interpret this threshold for the case that $\mu_0 > \mu_1$. If $P_1 > P_0$, then ratio $P_1/P_0 > 1$ and the logarithmic term is greater than 0. Since $\mu_0 - \mu_1 > 0$ under our assumption, the decision threshold moves to the right, and the MAP region for deciding $\hat{A}_1$ is bigger. Conversely, if $P_1 < P_0$, then the logarithm term will be negative, and the MAP decision threshold will move to the left, increasing the decision region for $\hat{A}_0$. The effect of the *a priori* information increases with σ^2. In other words, the more noisy the observation is, the more we should rely on the *a priori* information.

Terminology review and self-assessment questions

Interactive flashcards to review the terminology introduced in this section and self-assessment questions are available at fdsp.net/10-3, which can also be accessed using this QR code:

10.4 Chapter Summary

In this chapter, I introduced techniques for making decisions when the observation is a continuous random variable. I first introduced non-Bayesian approaches for binary hypothesis tests, including maximum-likelihood (ML) and general decision rules that tradeoff between the probability of false alarm and the probability of miss. ROC curves and area under the curve (AUC) were introduced as ways to visualize and quantify the overall performance of a detector.

I introduced point conditioning, in which we have conditional probabilities for which the condition is that a continuous random variable takes on a particular value. New forms of the Law of Total Probability and Bayes' Rule were developed for point conditioning. Then we applied Bayes' rule to develop the optimal MAP decision rule for systems with a discrete input but a continuous output. These rules were illustrated for a binary communication system in which the output is conditionally Normal with a mean that depends on the input to the system.

Access a list of key take-aways for this chapter, along with interactive flashcards and quizzes at fdsp.net/10-4, which can also be accessed using this QR code:

11

Categorical Data, Tests for Dependence, and Goodness of Fit for Discrete Distributions

In all of the examples that we have covered in this book up to this point, the data has been numerical in nature. However, another type of data that you may often encounter is *categorical data*:

> Definition
>
> **categorical data**
>
> Data that does not take on a specific numerical value but instead takes on one of several categories.

Examples of categorical data include:

- Handedness: whether a person is left-handed or right-handed
- Political affiliation: whether a person is Republican, Democrat, Independent, or other
- Assigned sex at birth: male or female
- Likert scale data, such as strongly disagree, disagree, neither agree nor disagree, agree, or strongly agree. Note that although Likert categories often have an associated number, those numbers should not be treated as numerical data.
- Country of citizenship
- Income range data: For instance, a survey may ask a respondent to choose whether their annual income is: a) less than $30,000; b) between $30,000 and $59,999; or c) $60,000 or more.

The examples above illustrate some sub-classes of categorical data. When the data has an inherent ordering, such as Likert scale data or income range data, that is considered *ordinal data*:

> Definition
>
> **ordinal data**
>
> Categorical data for which the categories have a natural ordering.

DOI: 10.1201/9781003324997-11

Categorical data that has no natural ordering is called *nominal data*:

> Definition
>
> **nominal data**
>
> Categorical data for which the categories have no natural ordering.

The examples above that are nominal data include handedness, political affiliation, assigned sex at birth, and country of citizenship.

Our previous statistical methods rely on treating numerical data as samples from a random distribution. These methods break down when the data is categorical. In this chapter, I introduce examples of categorical data and develop resampling and analytical techniques for determining whether different features of categorical data are dependent. Then I show that similar techniques can be applied to the discrete goodness-of-fit problem, in which we evaluate whether discrete numerical data could come from some model distribution.

Terminology review and self-assessment questions

Interactive flashcards to review the terminology introduced in this section and self-assessment questions are available at fdsp.net/11, which can also be accessed using this QR code:

11.1 Tabulating Categorical Data and Creating a Test Statistic

We will consider statistical tests for independence among different categorical features. The first steps are to tabulate the categorical data, compare the tabulated data to what would be expected if the variables were independent, and use the result to generate a test statistic. Let's illustrate this with two health-related questions for college students:

Example 11.1: Comparing Marijuana Use and Exercise with Self-Reported Sex for College Students

1. Many states now permit the licensed use of medical marijuana. Marijuana has often been used recreationally by college students. Is nonmedical use of marijuana by college students dependent on students' self-reported sex?
2. Many college students exercise to stay healthy. Is college students' exercise in the last 30 days dependent on students' self-reported sex?

We can answer these questions using data from the Behavioral Risk Factor Surveillance System (BRFSS) survey. For more information, see the BRFSS web page at https://www.cdc.gov/brfss/.

Because this example is quite long, I am not going to use the left-column blue highlighting for the entire example. Instead, I will break this example down into multiple parts and number and highlight the start of each new part.

We will limit the investigation in this section to people who reported that they were living in college housing and were answering the survey on a cell phone. (The number of respondents living in college housing and answering via a landline was much smaller and used some different questions.) The full data set is available here:

https://www.cdc.gov/brfss/annual_data/2021/files/LLCP2021XPT.zip

However, this file is over 1 GB in size and uses only numbers to encode the responses. I have pre-processed this file to limit it to only the college cellphone respondents and to reinterpret several variables of interest as strings:

- Respondent's sex is stored in the `SEXVAR` column of the data as a 1 or 2. These have been mapped to the strings `male` or `female` in a column called `sex`.
- The response to whether the respondent has exercised in the previous 30 days is stored in the `EXERANY2` column in the original data as a 1 or 2. These have been mapped to the corresponding values of `yes` or `no` and stored in a column called `exercised`. Rows with values of 9 in the original data (for people who refused to answer) were dropped.
- The response to the type of marijuana use (if any) in the last 30 days was encoded as a number in the `RSNMRJN2` column in the original data and mapped to a string in a new column `marijuana_use`. Values of 1 ("medical use only") were assigned a value of `medical`. Values of 2 ("nonmedical use only") were assigned a value of `nonmedical`. Values of 3 ("both medical and nonmedical use") were assigned a value of `both`.

Let's start by loading the dataframe:

```
college = pd.read_csv('https://www.fdsp.net/data/college.csv', index_col=0)
college.head()
```

	SEXVAR	EXERANY2	MARIJAN1	RSNMRJN2	sex	exercised	marijuana_use
3209	1.0	2.0	NaN	NaN	male	no	NaN
3320	1.0	1.0	NaN	NaN	male	yes	NaN
4163	1.0	1.0	NaN	NaN	male	yes	NaN
4172	2.0	2.0	NaN	NaN	female	no	NaN
4434	1.0	1.0	NaN	NaN	male	yes	NaN

There are many `NaN` entries in columns corresponding to marijuana use because many students did not report using marijuana in the last 30 days. Let's filter out those values for now so that we can start working with those students who did report using marijuana:

```
college2 = college.dropna()
college2.head()
```

	SEXVAR	EXERANY2	MARIJAN1	RSNMRJN2	sex	exercised	marijuana_use
8392	1.0	1.0	3.0	2.0	male	yes	nonmedical
44864	1.0	1.0	8.0	2.0	male	yes	nonmedical
49893	1.0	1.0	30.0	3.0	male	yes	both
50301	1.0	1.0	1.0	2.0	male	yes	nonmedical
51119	1.0	1.0	30.0	3.0	male	yes	both

Example 11.2: Assessing Dependence Between Marijuana Use and Student's Sex

We wish to determine whether the categorical data in the `sex` and `marijuana_use` columns are dependent.

The first step is to determine the absolute frequency of each combination of `sex` and `marijuana_use`. For instance, if we want to count the number of entries that are `male` and `medical`, we can use the `query()` method and then use `len()` to find the length of the resulting data frame:

```
len(college2.query('sex=="male" & marijuana_use=="medical"'))
```

```
3
```

We could write for loops to iterate over the various combinations, but fortunately, Pandas provides a *much* easier way with the `pd.crosstab()` function, which cross-tabulates data. For our purposes, we can pass `pd.crosstab()` two columns of the dataframe, and it will count the entries. The order in which the data is passed to `pd.crosstab()` matters in terms of how the table is constructed and, to a lesser extent, how it is interpreted:

- The first argument will determine the rows of the table. When one of the categorical features represents population subgroups, it is usually mapped to the rows. For instance, for our data, our `sex` variable creates population subgroups `male` and `female`.
- The second argument determines the columns of the table.

```
contingency = pd.crosstab(college2['sex'], college2['marijuana_use'])
contingency
```

marijuana_use	both	medical	nonmedical
sex			
female	6	2	14
male	12	3	10

Note that `NaN` entries are ignored by `pd.crosstab()`, so we would get the same result with the original data frame.

This type of table is called a *contingency table* or *cross tabulation*:

> Definition
>
> **contingency table,**
> **cross tabulation**
>
> For data consisting of two or more categorical features, a table that lists the number of occurrences (the counts) for each combination of feature outcomes. The totals across each category (rows and columns for a two-way table) are also usually computed and shown.

Contingency tables also often include information about the occurrences of individual features, which may be obtained by summing the counts in the contingency across the rows or columns. Such sums are called *marginal sums* or *marginal counts*. We can ask `pd.crosstab()` to append the marginal sums using the `margins = True` keyword argument:

```
contingency2 = pd.crosstab(college2['sex'], college2['marijuana_use'],
                           margins=True)
contingency2
```

marijuana_use	both	medical	nonmedical	All
sex				
female	6	2	14	22
male	12	3	10	25
All	18	5	24	47

Note that a larger number of male respondents reported using marijuana for both purposes, whereas a larger number of female respondents reported using it for only nonmedical use. Since the number of males and females differ, it is helpful to find the relative frequencies for each row by dividing by the row sums. For instance, the entry for `female`, `both` should be $6/22 \approx 0.273$.

We can find all of the relative frequencies by row using `pd.crosstab()` and passing the keyword argument `normalize = 'index'`:

```
pd.crosstab(college2['sex'], college2['marijuana_use'],
            normalize='index', margins=True)
```

marijuana_use	both	medical	nonmedical
sex			
female	0.272727	0.090909	0.636364
male	0.480000	0.120000	0.400000
All	0.382979	0.106383	0.510638

For males, the relative frequency of `both` is higher than `medical` or `nonmedical`, whereas for females, `nonmedical` is higher than `medical` or `both`. However, we only have 47 entries in the table, so is there sufficient data for this difference to be significant enough to ensure that sex and marijuana use are dependent? To answer this question, we first need to generate a summary statistic for the table. Then we can use either resampling or analysis to determine the probability of seeing such a large value of the decision statistic.

To create the decision statistic, let's determine how much this table differs from the expected number of entries for each table cell if there were no dependence. The `All` entries in the table are the probabilities for each category of marijuana use, neglecting respondent sex. These are equal to the total number of entries in the corresponding column divided by the total number of entries in the table. For instance, the entry for `both` is $18/47 \approx 0.383$.

If we want the expected number of entries in the `female`, `both` cell, then we can multiply the total number of females times the relative frequency of `both`, which yields $22 \cdot 18/47 \approx 8.43$. If we repeat this for each cell, we find something interesting – **if there is no dependence, the expected value for each cell is equal to the marginal row sum times the marginal column sum, divided by the total number of entries**.

The expected contingency table is shown below:

```
# Set these to variables to make the code below more concise
sex_sums = contingency2.loc[:, 'All']
use_sums = contingency2.loc['All', :]

# Get a dataframe of the right size,
expected_contingency = contingency2.copy()

# Loop over the rows and columns, but
# ignore the sums
for sex in sex_sums.index:
  if sex != 'All':
    for use in use_sums.index:
      if use != 'All':
        expected_contingency.loc[sex, use] = \
          sex_sums.loc[sex] * use_sums.loc[use] \
          / sex_sums.loc['All']

expected_contingency
```

marijuana_use	both	medical	nonmedical	All
sex				
female	8.425532	2.340426	11.234043	22
male	9.574468	2.659574	12.765957	25
All	18.000000	5.000000	24.000000	47

As usual, there are easier ways to do this using SciPy.stats. In fact, SciPy.stats has a whole submodule called `contingency` that is dedicated to working with contingency tables. To get the expected contingency table, we can pass the contingency table to `stats.contingency.expected_freq()`:

```
import scipy.stats as stats

expected_contingency2 = stats.contingency.expected_freq(contingency)
print(expected_contingency2)
```

```
[[ 8.42553191  2.34042553 11.23404255]
 [ 9.57446809  2.65957447 12.76595745]]
```

Now, we can determine the difference between the observed contingency table and the expected contingency table by subtracting the entries in each cell:

```
diffs = contingency2.copy()

for sex in sex_sums.index:
    for use in use_sums.index:
        diffs.loc[sex, use] = \
          contingency2.loc[sex, use] - \
```

(continues on next page)

(continued from previous page)

```
        expected_contingency.loc[sex, use]

diffs
```

marijuana_use	both	medical	nonmedical	All
sex				
female	-2.425532	-0.340426	2.765957	0
male	2.425532	0.340426	-2.765957	0
All	0.000000	0.000000	0.000000	0

We can also do this subtraction in NumPy using matrix subtraction as follows:

```
ndiffs = contingency.to_numpy() \
      - stats.contingency.expected_freq(contingency)
print(ndiffs)
```

```
[[-2.42553191 -0.34042553  2.76595745]
 [ 2.42553191  0.34042553 -2.76595745]]
```

Note that each row and column sums to zero, and this is always true for such a table of differences. These fixed row sums limit the number of different values that can appear in the table. The absolute values of the cells can only take on three distinct values, and if two of those values are known, the third is also fixed. Thus, the degrees of freedom (dofs) for this table is two. We give a definition and formula for the *degrees of freedom* below:

Definition

degrees of freedom (contingency table)

For a contingency table, the *degrees of freedom*, abbreviated dofs, is the number of values in the table that can be selected independently while satisfying the row and column totals. For a table with r rows and c columns, the number of dofs is

$$n_{dof} = (r-1)(c-1).$$

The next step is to convert the table of differences to a summary statistic. The summary statistic should increase the more the observed table differs from the expected table under the assumption of no dependence. We can't use the sum of the differences in the table because it is always zero. We could use the sum of the squares of the differences, but in general tables, some cells may have much larger expected values than others, and bigger differences are likely to occur in such cells. Let I be the set of indices of all cells, and let O_i and E_i be the observed value and expected value, respectively, for cell i. Then the standard test statistic for the contingency table is

$$C = \sum_{i \in I} \frac{\left(O_i - E_i\right)^2}{E_i}.$$

The statistic C is called the chi-squared statistic, also written χ^2-statistic, because its distribution can be modeled as a chi-squared random variable when the number of entries in each cell is sufficiently large.

For this example, the normalized squares of the differences are shown in the following matrix:

```
norm_sq_diffs = ndiffs**2 / expected_contingency2
print(norm_sq_diffs)
```

```
[[0.69825919 0.04951644 0.68101225]
 [0.61446809 0.04357447 0.59929078]]
```

The chi-squared statistic is the sum of these entries:

```
C = norm_sq_diffs.sum()
print(f'C = {C:.2f}')
```

```
C = 2.69
```

In the next section, we will conduct a null hypothesis test to determine if the observed difference is statistically significant.

Terminology review and self-assessment questions

Interactive flashcards to review the terminology introduced in this section and self-assessment questions are available at fdsp.net/11-1, which can also be accessed using this QR code:

11.2 Null Hypothesis Significance Testing for Dependence in Contingency Tables

When the relative frequencies in a contingency table vary by row or column, that suggests that the categorical data in the rows may depend on the categorical data in the columns, and vice versa. However, the observed dependence may also be caused by randomness in the data sample. To determine whether we can be confident that any observed dependence cannot reasonably be attributed to randomness, we can conduct an NHST. The null hypothesis, H_0, will be that the data in the rows and columns are not dependent. We have already defined the summary statistic that we will use, C, so we just need to determine the probability of seeing such a large value of C under H_0. We can do this using resampling or analysis, each of which we cover below.

11.2.1 Resampling NHST Contingency Test

Example 11.2 (continued)

Let's use resampling to test whether the observed dependence between marijuana use and reported sex for college students is statistically significant.

To perform resampling under H_0, we wish to break the dependence between `sex` and `marijuana_use`. The dependence will be broken if we randomly assign values of `sex` to values of `marijuana_use`. Since we will reuse the exact same data but reassign the mapping between the variables, this is a type of permutation test.

We can choose to permute the data for either feature; the result will be the same. Let's permute the data in the `marijuana_use` feature. We can call the dataframe's `sample()` method with `frac=1` to get a sample that is the same size as the original data. The `sample()` method samples *without replacement* by default, so with `frac=1`, it randomly permutes the data. If we wanted to perform bootstrap resampling, we could pass the keyword argument `replace=True` to `sample()` to specify sampling with replacement.

An example using the permutation test is shown below (I am only including the first 10 rows to conserve space). Try running this code a few times to see different random permutations of the data in the `marijuana_use` feature:

```
print(college2['marijuana_use'].sample(frac=1))
```

```
140980    nonmedical
155844          both
337471       medical
125117    nonmedical
79778     nonmedical
391169          both
49893           both
389980    nonmedical
138159          both
141105    nonmedical
        ...
```

To carry out the NHST, we recompute the chi-squared statistic for each permuted vector and determine the relative frequency of exceeding the chi-squared statistic of the original data. That relative frequency is an estimate of the p-value, the probability of observing such a large test statistic under H_0. A Python function to carry out this type of NHST on two features is below:

```
def nhst_contingency(feature1, feature2,
                     observed_C, num_sims = 1_000, seed = 92375):

  ''' Estimate the probability of seeing such a large
  value of the chi-squared statistic under the null
  hypothesis by using a permutation test
  '''

  npr.seed(seed)

  # Calculate the expected contingency once, outside the loop
  contingency = pd.crosstab(feature1, feature2)
  expected_contingency = stats.contingency.expected_freq(contingency)

  # Set up the counter for how many times we see a chi-squared
  # statistic as large as the original data
  count = 0
```

(continues on next page)

(continued from previous page)

```
for sim in range(num_sims):

    # Calculate the contingency table with one feature permuted
    sample_contingency  = pd.crosstab(feature1.to_numpy(),
                                      feature2.sample(frac=1).to_numpy())

    # Caclulate the test statistic using the new sample
    sample_C = ((sample_contingency.to_numpy()  - expected_contingency)**2 \
                / expected_contingency).sum()

    if sample_C >= observed_C:
      count +=1

  print('Prob. of observing chi-squared value as large as original')
  print(f'data under H0  =~ {count/num_sims: .2g}')

  return count/num_sims
```

```
nhst_contingency(college2['sex'],
                 college2['marijuana_use'],
                 C);
```

```
Prob. of observing chi-squared value as large as original
data under H0  =~  0.27
```

Before interpreting the results, I want to discuss one fine point about the simulation that may confuse people building simulations like this for the first time, which is why I converted each Pandas Series into a NumPy vector. If we use the Pandas Series objects directly, `pd.crosstab()` will pair up the variables with the same index, and the effect of shuffling the data will be negated. When using NumPy vectors, the variables are paired based on position, and the shuffling effect is preserved.

The p-value is approximately 0.27, which is above our threshold of $\alpha = 0.05$, so we fail to reject the null hypothesis. The data is too small to be able to conclude that the observed dependence between sex and type of marijuana could not be caused by randomness in the sample.

Example 11.3: Assessing Dependence Between Student's Sex and Exercise in Last 30 Days

Let's try a contingency test with more data. We will compare reported sex with whether the respondent reported exercising in the last 30 days.

Let's jump straight to creating the contingency table:

```
pd.crosstab(college['sex'], college['exercised'], margins=True)
```

exercised	no	yes	All
sex			
female	99	480	579
male	80	594	674
All	179	1074	1253

To calculate the expected contingency using `stats.contingency.expected_freq()`, we need to pass it the contingency table without the marginal totals:

```
ex_contingency=pd.crosstab(college['sex'], college['exercised'])
exp_ex_contingency = stats.contingency.expected_freq(ex_contingency)
print(exp_ex_contingency)
```

```
[[ 82.71428571 496.28571429]
 [ 96.28571429 577.71428571]]
```

The difference table is:

```
ex_contingency - exp_ex_contingency
```

exercised	no	yes
sex		
female	16.285714	-16.285714
male	-16.285714	16.285714

The raw data suggests that men may be more likely to have exercised in the last 30 days than women, and this time we have 1253 data points. Let's calculate the chi-squared statistic:

```
ex_C = ((exp_ex_contingency- ex_contingency.to_numpy() )**2
        / exp_ex_contingency).sum()
print(f'For sex/exercise table, C = {ex_C:.2f}')
```

```
For sex/exercise table, C = 6.95
```

This is a much larger value of the chi-squared statistic. Let's determine the p-value using a permutation test:

```
nhst_contingency(college['sex'], college['exercised'],
                 ex_C)
```

```
Prob. of observing chi-squared value as large as original
data under H0  =~  0.01


0.01
```

Since the p-value of 0.01 is less than our significance threshold of $\alpha = 0.05$, the result is statistically significant at the $p < 0.05$ level. We can reject the null hypothesis and conclude that the `sex` and `exercised` features are dependent. Note, however, that this test does not allow us to say anything else about how these features depend on each other.

11.2.2 Analytical NHST

Under H_0, the chi-squared statistic can be shown to be well approximated by a standard chi-squared random variable, provided that the entries in the table are sufficiently large. The typical criterion for being able to model the test statistic as a chi-squared random variable is that the expected values in each cell are all at least 4. Thus, the analytical model is likely to be less accurate if applied to the marijuana-use example and more accurate for the example about exercise.

The standard chi-squared random variable requires a single parameter: the degrees of freedom (dof). Recall that the number of degrees of freedom for a contingency table with r rows and c columns is $(r-1)(c-1)$. Then under H_0, the chi-squared statistic is a chi-squared random variable with $(r-1)(c-1)$ degrees of freedom. From Section 8.6, the probability density becomes more concentrated toward higher values as the degrees of freedom increase.

Example 11.4: Assessing Dependence Between Marijuana Use and Student's Sex – Analytical Test

Let's apply this model to calculate the p-value for our first example comparing `sex` to `marijuana_use`. The contingency table has two rows and three columns, and hence $(2-1)(3-1) = 2$ degrees of freedom. We can create a chi-squared distribution with 2 dof in SciPy.stats using `stats.chi2(2)`, where the argument is the dofs.

WARNING

Be careful to use `stats.chi2()` because `stats.chi()` is a different type of random variable!

```
dof = 2
chi2_rv1=stats.chi2(2)
```

The probability of getting a chi-squared statistic as large as the 2.69 we observed can be calculated using the survival function as follows:

```
chi2_rv1.sf(2.69)
```

```
0.2605397078599757
```

For this example, the value from the analysis is approximately equal to the value found through resampling, even though some of the cell values are smaller than our usual threshold for applying the chi-squared model.

As usual, there are also methods that can reduce our effort in determining the p-value. The function `stats.chi2_contingency()` takes as input a contingency table and returns:

- the chi-squared statistic,
- the analytical p-value,
- the number of degrees of freedom, and
- the expected contingency table.

```
stats.chi2_contingency(contingency)
```

```
Chi2ContingencyResult(statistic=2.686121212121211, pvalue=0.26104548728461613,
↪ dof=2, expected_freq=array([[ 8.42553191,  2.34042553, 11.23404255],
       [ 9.57446809,  2.65957447, 12.76595745]]))
```

The results match our analysis and confirm that the data is not sufficient to reject the null hypothesis.

Example 11.5: Assessing Dependence Between Student's Sex and Exercise in Last 30 Days – Analytical Test

Now let's use `stats.chi2_contingency()` to calculate the p-value for the data features `sex` and `exercised`. Recall that the contingency table is

```
ex_contingency
```

```
exercised  no  yes
sex
female     99  480
male       80  594
```

There is $(2-1)(2-1) = 1$ degree of freedom, so we can calculate the probability of seeing such a large chi-squared statistic as:

```
chi2_rv2 = stats.chi2(1)
chi2_rv2.sf(ex_C)
```

```
0.008360480726183703
```

We can also calculate the p-value from `stats.chi2_contingency()`:

```
stats.chi2_contingency(ex_contingency)
```

```
Chi2ContingencyResult(statistic=6.534102100555383, pvalue=0.
↪010582562809445315, dof=1, expected_freq=array([[ 82.71428571, 496.
↪28571429],
       [ 96.28571429, 577.71428571]]))
```

The p-value slightly differs from our previous analysis. The reason is that the statistic is not really a chi-squared random variable under H_0 because we are processing a set of discrete values and representing the result by a continuous distribution. To compensate for this when there is one degree of freedom, `stats.chi2_contingency()` applies an adjustment called *Yate's continuity correction* to produce a more accurate estimate of the true p-value.

To find the p-value without the continuity correction, pass the `correction=False` argument:

```
stats.chi2_contingency(ex_contingency, correction=False)
```

```
Chi2ContingencyResult(statistic=6.954582494119101, pvalue=0.
↪008360480726183703, dof=1, expected_freq=array([[ 82.71428571, 496.
↪28571429],
       [ 96.28571429, 577.71428571]]))
```

Now the result matches our analysis exactly. In practice, it is better to use the continuity correction. The analytical p-value with the continuity correction is very close to our result via resampling. Again, since $p \approx 0.011$ is less than our significance threshold of $\alpha = 0.05$, we reject the null hypothesis and conclude that these data features are dependent.

11.2.3 Fisher's Exact Test

For small tables, instead of using random permutations of the data, we can use every possible permutation of the data. The resulting test is called Fisher's Exact Test. If you wish to implement this test, you can use the `itertools` library to generate all permutations of the data for one of the features.

Terminology review and self-assessment questions

Interactive flashcards to review the terminology introduced in this section and self-assessment questions are available at fdsp.net/11-2, which can also be accessed using this QR code:

11.3 Chi-Square Goodness-of-Fit Test

We can use different forms of contingency tables and chi-squared tests to determine whether some discrete numerical data come from a given distribution. In this section, we will show how to calculate the chi-squared test statistic given a model distribution and use this to calculate a p-value for the observed data given the model distribution. These types of tests are called *chi-squared goodness-of-fit tests* or *one-way chi-square tests* and exist in two basic varieties:

- The first variety is the standard NHST model. We observe some characteristic in the data that makes us think that it may not have come from some default distribution. The null hypothesis, H_0, is that it came from the default distribution, and we estimate the probability of seeing such a large value of the chi-squared statistic under H_0.
- In the second variety, we have a model distribution for the data , and we wish to determine whether the model is reasonable. We can still conduct an NHST, but now H_0 is the

hypothesis that the data came from the model distribution, and we conclude that the model is consistent with the data if the probability of seeing such a large value of the chi-squared statistic is large (typically much more than the usual threshold of 0.05).

11.3.1 Standard NHST

We start with an interesting observation: it appears that Major League Baseball Players are much more likely to have birthdays in August and the following months than in July and the months immediately previous to that. This is not a new observation – some examples of articles and books discussing this follow.

- I first encountered this phenomenon while reading Malcolm Gladwell's book *Outliers: The Story of Success*, published by Little, Brown, and Company, 2008. Gladwell writes, "more major league [baseball] players are born in August than in any other month".
- In a 2008 *Slate* article by Greg Spira entitled "The Boys of Late Summer," the author states, "Since 1950, a baby born in the United States in August has had a 50 percent to 60 percent better chance of making the big leagues than a baby born in July". In addition, Spira indicates that the book *The Baseball Astrologer*, published by Total/Sports Illustrated in 2000, makes the observation that "the sign under which an individual was born played a significant role in whether he made it in pro ball".

Let's look at the data and try to analyze it using what we know about contingency tables and chi-squared tests. We start by importing the data from the Baseball Databank, maintained by the Chadwick Baseball Bureau and used under the Creative Commons Attribution-ShareAlike 3.0 Unported License:

```
# The data through 2022 can be loaded using the following:
df = pd.read_csv('https://www.fdsp.net/data/baseball.csv')
```

The dataframe contains information for essentially every Major League Baseball (MLB) player, with data going back to the first MLB game on May 4, 1871. Take a look at the column names to see what features are included in this dataset:

```
df.columns
```

```
Index(['playerID', 'birthYear', 'birthMonth', 'birthDay', 'birthCountry',
       'birthState', 'birthCity', 'deathYear', 'deathMonth', 'deathDay',
       'deathCountry', 'deathState', 'deathCity', 'nameFirst', 'nameLast',
       'nameGiven', 'weight', 'height', 'bats', 'throws', 'debut',
       'finalGame', 'retroID', 'bbrefID'],
      dtype='object')
```

Let's look at the data in the 'birthMonth' column:

```
df['birthMonth'].head()
```

```
0    12.0
1     2.0
2     8.0
3     9.0
4     8.0
Name: birthMonth, dtype: float64
```

```
df['birthMonth'].min(), df['birthMonth'].max()
```

```
(1.0, 12.0)
```

The data in `birthMonth` is the standard numerical encoding (January=1, February=2, etc.) of the player's month of birth. Since the data is numerically encoded, we can start by plotting a histogram of the birth months to check whether we can see the effects described by these authors:

```
import numpy as np
import matplotlib.pyplot as plt

mybins = np.arange(0.5, 13.5, 1)
plt.hist(df['birthMonth'], bins = mybins);
```

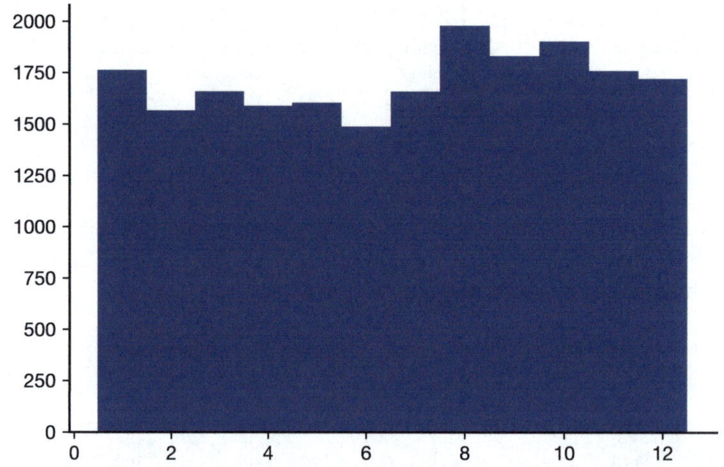

There is certainly a large discrepancy that occurs starting in month 8 (August). This graph is not very attractive, so in future graphs, I will make a more attractive version by using `np.unique()` to tabulate the data and `plt.bar()` to make the plot.

If we show the data in the form of a table, it is called a *one-way table*:

Definition

one-way frequency table,
one-way contingency table

A table that shows the relative frequencies for a single categorical variable.

We now wish to determine whether the observed variations in the numbers of players born in different months are caused by some real effect or if it could just be explained by randomness in the data. We formulate this as an NHST with H_0 representing the scenario in which birth month does not influence whether a person plays MLB. I purposefully left the description of H_0 vague because we are going to explore two different models that are consistent with this description of H_0.

Example 11.6: MLB Player Birth Months – Simple Null-Hypothesis Model

In the simple model, we assume that an MLB player's birthday is equally likely to be any month, so we model this using a 1/12th probability of being born in each month. As of late June 2023, there were 20536 players with known birth months in the data set (a few others are unknown and omitted), but that number may change when using the GitHub link.

```
df.dropna(subset='birthMonth', inplace=True)
len(df)
```

```
20536
```

The expected number of players born in each month can be computed as follows. The resulting expected value is plotted for comparison with the data in Fig. 11.1.

```
len(df)/12
```

```
1711.3333333333333
```

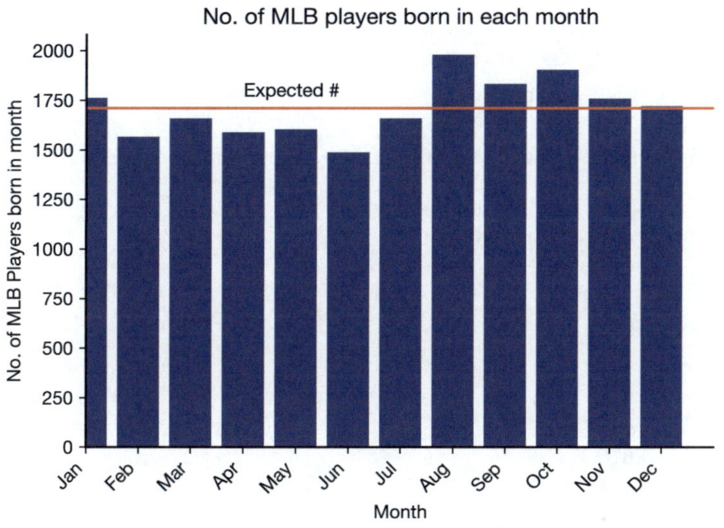

FIGURE 11.1
Comparison of actual players born in each month to the expected number of players born in each month, assuming a player is equally likely to be born in any of the 12 months.

We can tabulate the differences between the actual and expected values as shown in the following code. The resulting differences are plotted in Fig. 11.2.

```
expected = len(df)/12
errors=counts - expected
```

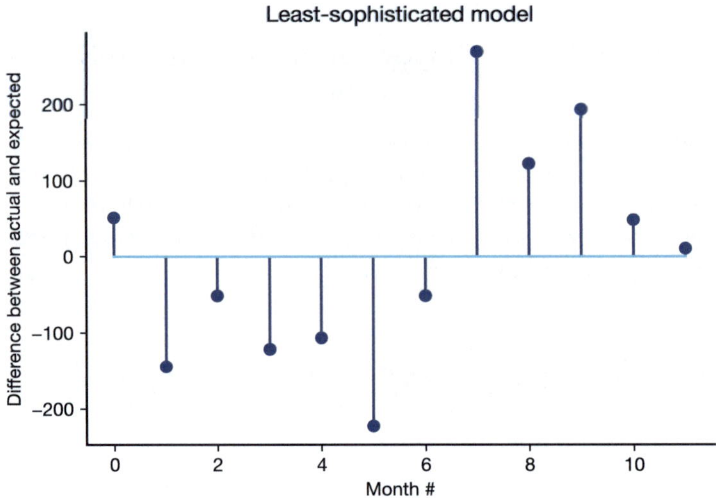

FIGURE 11.2
Errors between actual and expected births per month, simple model.

We can convert these to normalized squared errors by squaring and dividing by the expected value:

```
expected =len(df)/12
print('Month | Normalized squared error:')
print('-'*33)
for i, count in enumerate(counts):
  month = months[i]
  print(f'{month:^5} | {(count - expected)**2/expected:>10.1f}')
```

```
Month | Normalized squared error:
---------------------------------
 Jan  |        1.6
 Feb  |       12.2
 Mar  |        1.5
 Apr  |        8.6
 May  |        6.6
 Jun  |       28.9
 Jul  |        1.5
 Aug  |       42.5
 Sep  |        8.8
 Oct  |       21.9
 Nov  |        1.4
 Dec  |        0.1
```

Note the size of the August and October entries. The sum of all the normalized squared errors is the test statistic:

```
normalized_errors = (counts-expected)**2/expected
test_stat = normalized_errors.sum()
print(f'The chi-squared test statistic value is {test_stat}')
```

```
The chi-squared test statistic value is 135.56018698870275
```

Under H_0, the test statistic has a distribution that is approximately chi-squared. The number of degrees of freedom will be one less than the number of cells in the table because given the total number of entries in the table and 11 of the cells, the 12th cell's value can be determined. To determine if the observed value of the test statistic is statistically significant, we will use the survival function of the chi-squared random variable with 11 degrees of freedom:

```
import scipy.stats as stats
chi2 = stats.chi2(11)

print('Probability of seeing such a test statistic value')
print(f'as large as {test_stat:.2f} is approximately {chi2.sf(test_stat):.3g}')
```

```
Probability of seeing such a test statistic value
as large as 135.56 is approximately 1.3e-23
```

The p-value is incredibly small, but our model can easily be criticized. The months have different numbers of days, and even accounting for that, the number of births varies by month. This motivates us to use a more sophisticated model:

Example 11.7: MLB Player Birth Months – More-Sophisticated Null-Hypothesis Model

The assumption that people's birthdays are evenly distributed throughout the year is unjustified. Most MLB players are from the US, so to perform a better test, our null hypothesis model should be based on the actual distribution of birthdays in the US by month.

The following code loads a data file that was extracted from the National Center for Health Statistics (http://www.cdc.gov/nchs/data_access/vitalstatsonline.htm) by Andrew Collier (https://github.com/datawookie) and made publicly available as an R package called the Lifespan Package. The file contains the number of births from January 1994 to December 2014, broken down by month, day of the week, and sex. I have translated the data on births to a CSV file that can be loaded as follows:

```
births=pd.read_csv('https://www.fdsp.net/data/births.csv')
births.head()
```

	year	month	dow	sex	count
0	1994	Jan	Sun	F	19980
1	1994	Jan	Sun	M	20268
2	1994	Jan	Mon	F	26015
3	1994	Jan	Mon	M	27132
4	1994	Jan	Tue	F	22615

Then we can loop over the months and add up the counts for each month:

```
births_by_month = np.zeros(12)
for i, month in enumerate(months):
  births_by_month[i] = \
    births.query('month=="' + month + '"')['count'].sum()
print(births_by_month)
```

```
[6906798. 6448725. 7080880. 6788266. 7112239. 7059986. 7461489. 7552007.
 7365904. 7220646. 6813037. 7079453.]
```

(A more elegant approach to doing this using the dataframe's `groupby()` method is shown online at fdsp.net/11-3.)

Note that August is the month with the most births. It has significantly more births than the winter months:

```
plt.bar(months, births_by_month/1e6);
plt.title('US births by month, 1994\u20142014');
plt.ylabel('Total births (millions)');
plt.xlabel('Month');
```

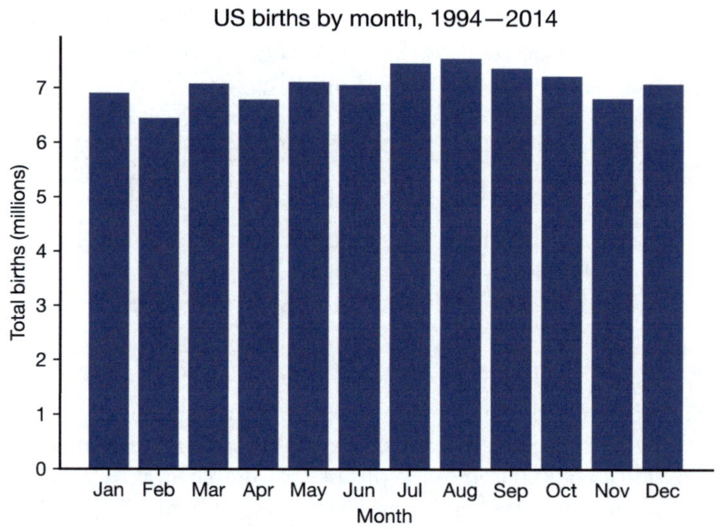

We can convert this to a PMF by dividing by the total number of births over these months. Then the expected number of MLB player births for each month is the value of the PMF times the total number of MLB players with known birth months:

```
pmf3 = births_by_month / births_by_month.sum()
expected3 = pmf3 * len(df)
```

Fig. 11.3 compares the data to the expected values under this more-sophisticated model for H_0. The results in the figure suggest that this more-sophisticated model is a better match for the data. This can be seen in the closer correspondence between the expected and observed values in the left plot, as well as smaller differences in the right plot.

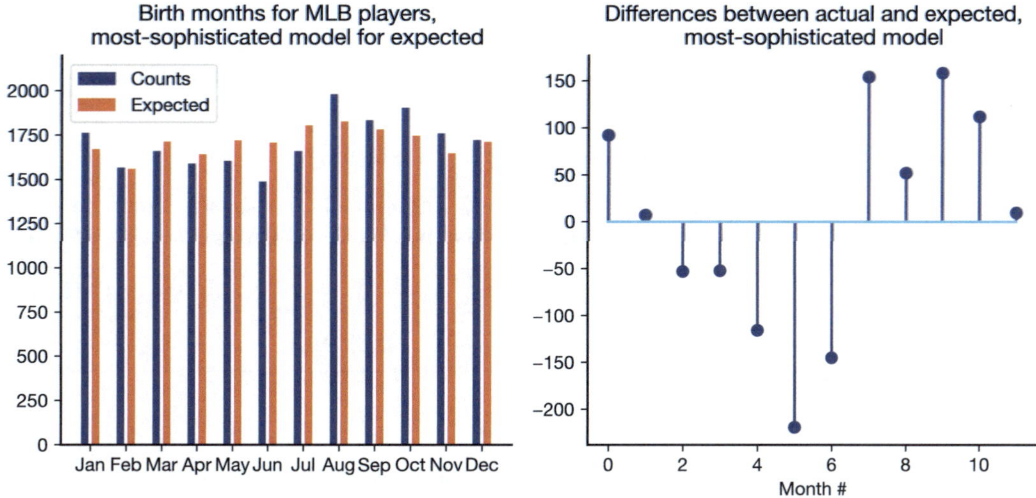

FIGURE 11.3
Comparison of observed and expected players born in each month (left graph) and differences (right graph).

The test statistic value under this H_0 model is:

```
errors3 = (counts-expected3)**2 / expected3
test_stat3= errors3.sum()
print(test_stat3)
```

```
92.36709879680593
```

The test statistic for this more-sophisticated model is much smaller than the for the previous model. The resulting probability of seeing such a large value of the test statistic under H_0 is approximately:

```
chi2.sf(test_stat3)
```

```
5.7217627082031914e-15
```

The resulting p-value is much larger than for the simple model, but it is still very small. We can reject H_0 — the distribution of birthdays for MLB players is different than that of the general public in the US. The dramatic change in errors between July and August might indicate that there is some issue related to that boundary that causes the observed differences. Gladwell points out in *Outliers* that "The cutoff date for almost all nonschool baseball leagues in the United States is July 31". This means that players born in August are the oldest in their league, which results in them being selected for more elite teams that get more practice and better coaching. These differences may compound and eventually influence which players go on to become professional baseball players.

11.3.2 Testing Fit of a Model Distribution

In the standard NHST test, we formulate a test to determine whether the data is sufficiently different from some baseline distribution that we can reject the null hypothesis, H_0. We can conduct a similar test to determine whether discrete data fits a given model distribution. I have included an example of checking whether the distribution of named storms in the North Atlantic fits a Poisson distribution on the book's website at fdsp.net/11-3.

Terminology review and self-assessment questions

Interactive flashcards to review the terminology introduced in this section and self-assessment questions are available at fdsp.net/11-3, which can also be accessed using this QR code:

11.4 Chapter Summary

In this section, we introduced techniques for working with categorical data and for performing goodness-of-fit tests for discrete data. We showed how categorical data can be summarized into contingency tables. Then the contingency tables can be used to calculate a summary statistic called the chi-squared statistic. The chi-squared statistic can be used to determine statistical significance through a NHST using either resampling with a permutation test or through analysis. One-way contingency tables can be used to compare discrete data to a reference distribution. As in the case of two-way contingency tables, the resulting differences can be converted to a chi-squared statistic that can be used to determine the probability of observing such a large value of the chi-squared statistic under a reference distribution. This can be used to carry out an NHST or to check whether a proposed distribution is a reasonable model for observed data.

Access a list of key take-aways for this chapter, along with interactive flashcards and quizzes at fdsp.net/11-4, which can also be accessed using this QR code:

12

Multidimensional Data: Vector Moments and Linear Regression

All of the tests we have done so far have worked on one-dimensional data; i.e., we look at some statistics from samples of a single feature, but we tried to determine whether that feature was influenced by some other binary event. For example, in Chapter 3, we considered COVID-19 rates across states and tried to determine whether they were affected by two different types of state classifications: low GDP versus high GDP or urban versus rural.

Here is another example that seems to show an even stronger relation. Fig 12.1 shows the average annual temperature in Miami-Dade County (Florida) from 1895 to 2022. There seems to be a strong, approximately linear relation between year and temperature. We are ready to apply more sophisticated approaches to study such relationships, but we need some new tools, which we begin to introduce in the next section.

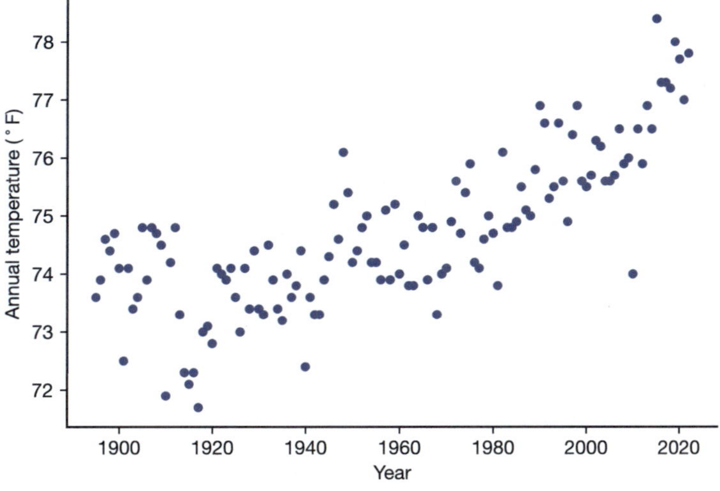

FIGURE 12.1
Average annual temperature in Miami-Dade County, Florida, from 1895 to 2022.

> Note:
>
> This and the following chapter assume that the reader has previous knowledge of vectors and matrices and their operations. If you do not have this expertise, there are many good books on the topic, but *Introduction to Linear Algebra for Data Science with Python* was written alongside this book to cover these topics.

DOI: 10.1201/9781003324997-12

12.1 Summary Statistics for Vector Data

Using vectors to represent data allows us to efficiently analyze that data. In this section, we review techniques for computing summary statistics with vector data, and we introduce a new summary statistic that provides a measure of dependence between variables. We start by introducing the use of matrices to facilitate working with multiple vectors of data in NumPy. In this section, we will revisit the Covid data set from Chapter 3 and show how to use both Pandas and NumPy techniques to compute summary statistics on the data. Let's start by loading the data into a dataframe, setting the index to the `state` column, and computing normalized COVID rates and GDPs per capita:

```
import pandas as pd
df = pd.read_csv( 'https://www.fdsp.net/data/covid-merged.csv' )

df.set_index('state', inplace=True)
df["gdp_norm"] = df["gdp"] / (df["population"] / 1000)
df["cases_norm"] = df["cases"] / (df["population"] / 1000)
```

We can consider each column of this dataframe to be a vector of data. In fact, it is easy to convert any column to a vector using the dataframes `to_numpy()` method. For example, we can create a vector of the number of cases like this:

```
import numpy as np

cases = df['cases'].to_numpy()
print(cases)
```

```
[ 7068    353   7648   3281  50470  15207  27700   4734  33683  25431 ... ]
```

This offers us flexibility in working with data because it makes it easy to work with all of the tools that NumPy offers. Note that `cases` is a *view* into the dataframe, and changes to `cases` affect the original dataframe. If you need a separate copy, pass the keyword argument `copy=True`.

12.1.1 From Dataframes to Matrices

Recall from Section 1.6.7 that a matrix is a two-dimensional table of numbers. When the values in our dataframe are all numeric, we can convert our dataframe directly into a two-dimension NumPy array:

```
covid_array = df.to_numpy()
```

This array has a lot of rows, so let's print the first five. We can do this using indexing. By using the index range `:5`, we will get the first five rows (I have reduced the number of digits of precision to make the output more concise):

```
print(covid_array[:5])
```

```
[[7.06800e+03 4.90318e+06 2.30750e+05 5.90400e+01 4.70612e+01 1.44151e+00]
 [3.53000e+02 7.31545e+05 5.46747e+04 6.60200e+01 7.47386e+01 4.82540e-01]
 [7.64800e+03 7.27871e+06 3.79018e+05 8.98100e+01 5.20721e+01 1.05073e+00]
 [3.28100e+03 3.01780e+06 1.32596e+05 5.61600e+01 4.39380e+01 1.08721e+00]
 [5.04700e+04 3.95122e+07 3.20500e+06 9.49500e+01 8.11141e+01 1.27732e+00]]
```

Compare the values in `covid_array` with the values in the dataframe. Each column in `df` *has been converted into a column of the* NumPy *array* `covid_array`. *Note that all of* the variables have been converted to floating-point values because a NumPy array can only have one data type, and the percent urban data requires a floating-point representation.

Each data feature (i.e., number of cases, population, GDP, percent urban) occupies one of the columns of the NumPy array. The entries in this matrix can be indexed by row and then column. For instance, since row 4 corresponds to California and column 1 corresponds to population, we can retrieve the population of California as follows:

```
covid_array[4,1]
```

```
39512223.0
```

We can get all the data for California using two different indexing approaches. If we omit the column, we will get the whole row:

```
print(covid_array[4])
```

```
[5.04700e+04 3.95122e+07 3.20500e+06 9.49500e+01 8.11141e+01 1.27732e+00]
```

As an alternative, we can pass just a colon as the range to indicate to retrieve all the values in that dimension: `covid_array[4, :]`. Using `:` for the column index is not particularly helpful, as we could have just omitted it. But if we want to retrieve a column, then it becomes very useful. To get all the population data, we can use `:` as the index of the rows and 1 as the column index:

```
populations = covid_array[:, 1]
print(populations)
```

```
[ 4903185.   731545.  7278717.  3017804. 39512223.  5758736.  3565287. ... ]
```

Note that some of the libraries that we will use expect each data feature to be in a different row, while the Pandas dataframe `to_numpy()` method puts each data feature into a different column. We can *transpose* the matrix to interchange the rows and columns. We can get the transpose of a NumPy matrix by appending `.T`. Because `covid_array.T` has 50 columns, I only print the first five below:

```
print(covid_array.T[:,:5])
```

```
[[7.06800000e+03 3.53000000e+02 7.64800000e+03 3.28100000e+03 5.04700000e+04]
 [4.90318500e+06 7.31545000e+05 7.27871700e+06 3.01780400e+06 3.95122230e+07]
 [2.30750100e+05 5.46747000e+04 3.79018800e+05 1.32596400e+05 3.20500010e+06]
```

(continues on next page)

(continued from previous page)

```
[5.90400000e+01 6.60200000e+01 8.98100000e+01 5.61600000e+01 9.49500000e+01]
[4.70612673e+01 7.47386695e+01 5.20721990e+01 4.39380424e+01 8.11141428e+01]
[1.44151200e+00 4.82540377e-01 1.05073463e+00 1.08721441e+00 1.27732626e+00]]
```

12.1.2 Averages, Medians, and Variances

We often use the average (sample mean), median, and variance as summary statistics for individual data features. Both Pandas and NumPy have methods for calculating these different features, but they occasionally have some differences in their behavior. In this section, I show how to use each method and point out any behaviors that users need to be aware of.

Let's start with Pandas. The Pandas dataframe object has methods called `mean()`, `median()`, and `var()`, which calculate the sample mean, median, and variance, respectively. When called on the whole dataframe, it will calculate these values for each column of numerical data. Examples for the COVID dataframe are shown below:

```
df.mean()
```

```
cases          2.137406e+04
population     6.550675e+06
gdp            4.296362e+05
urban          7.358180e+01
gdp_norm       6.201696e+01
cases_norm     2.718524e+00
dtype: float64
```

```
df.median()
```

```
cases          7.020500e+03
population     4.558234e+06
gdp            2.551785e+05
urban          7.373500e+01
gdp_norm       6.104737e+01
cases_norm     1.540303e+00
dtype: float64
```

```
df.var()
```

```
cases          2.189225e+09
population     5.460149e+13
gdp            3.134142e+11
urban          2.121263e+02
gdp_norm       1.402868e+02
cases_norm     9.898749e+00
dtype: float64
```

In Section 9.4 we introduced two different estimates for the variance:

$$s_n^2 = \frac{1}{n}\sum_{i=0}^{n-1}(x_i - \bar{\mathbf{x}})^2, \qquad \text{(biased)}$$

$$s_{n-1}^2 = \frac{1}{n-1}\sum_{i=0}^{n-1}(x_i - \bar{\mathbf{x}})^2 \qquad \text{(unbiased)}.$$

As a reminder, the corresponding estimator, S_{n-1}^2 is an unbiased estimator for the variance, but it does have a higher mean-squared error than the biased estimator. A quick check of the help for `df.var()` will indicate that it is s_{n-1}^2, the unbiased estimate.

Now, let's see how to perform these operations on a NumPy array, where each column represents a different data feature. As with Pandas, NumPy arrays have `mean()` and `var()` methods, but they do not have a `median()` method, and they might not give us the results we expect. Let's start with the `mean()` method:

```
covid_array.mean()
```

```
1166970.6795473746
```

By default, the `mean()` method finds the average of all the values in the array. To calculate the `mean()` for each column, we need to specify that the mean should be computed across the rows, which is axis 0. If we pass the `axis=0` keyword argument, we get back a vector of the feature means that has the same values that we found with Pandas.

Although NumPy arrays do not have a `median()` method, NumPy does offer a `np.median()` function to get the medians:

```
print(np.median(covid_array, axis=0))
```

```
[7.02050e+03 4.55823e+06 2.55178e+05 7.37350e+01 6.10473e+01 1.54030e+00]
```

Finally, we can use the NumPy array's `var()` method to find the variances:

```
print(covid_array.var(axis=0))
```

```
[2.14544e+09 5.35094e+13 3.07145e+11 2.07883e+02 1.37481e+02 9.70077e+00]
```

These variances do not match the ones from Pandas. As discussed in Section 9.4, NumPy returns the s_n^2 estimate by default. Since the mean is being estimated, we need to subtract one degree of freedom. By using the keyword argument `ddof=1`, we can get the s_{n-1}^2 estimate:

```
print(covid_array.var(ddof=1, axis=0))
```

```
[2.18922e+09 5.46014e+13 3.13414e+11 2.12126e+02 1.40286e+02 9.89874e+00]
```

The fact that NumPy returns the means as an array is not just a convenience. If we consider each data point to be a vector of values for the different features, then the mean of those vectors is the vector of the means.

Example 12.1: Two-dimensional Vector Mean with COVID Data

To illustrate this concept, let's plot two features against each other along with the mean vector. The figure below shows the COVID-19 rate per 1000 people as a function of the percent of the population that lives in an urban area. The average value of these two features is shown by an x.

```
import matplotlib.pyplot as plt

plt.scatter(df["urban"], df["cases_norm"])
plt.xlabel("Percent of population in urban area")
plt.ylabel("Covid Rate per 1000 People");

plt.scatter(df['urban'].mean(), df['cases_norm'].mean(), color='C1', marker='X');
```

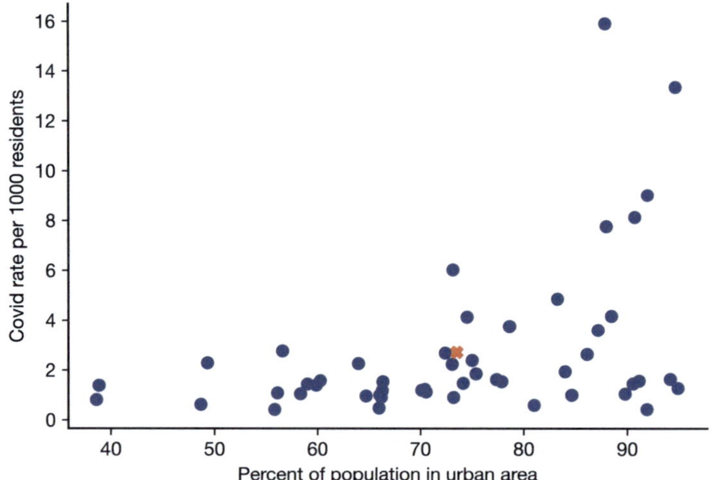

Note from the scatter plot that there seems to be a general trend that the COVID rate increases with the percentage of the population that lives in an urban area. The statistics we have studied in previous chapters cannot measure this dependence because they operate on only one variable or feature at a time. Let's start by introducing new summary statistics that can measure dependence among the data features.

12.1.3 Measuring Dependence through Moments: Covariances, and Correlations

To measure dependence between two features, we generalize the concept of variance. The math will be easier if we start with random variables. Let X and Y be random variables. Then the variances of these random variables are

$$\operatorname{Var}[X] = E\left[(X - E[X])^2\right], \text{ and } \operatorname{Var}[Y] = E\left[(Y - E[Y])^2\right].$$

We create a new *joint moment* called *covariance* that combines these two:

> Definition
>
> **covariance (random variables)**
>
> For random variables X and Y, the *covariance* is the joint moment given by
>
> $$\mathrm{Cov}(X, Y) = E\Big[(X - E[X])\,(Y - E[Y])\Big].$$

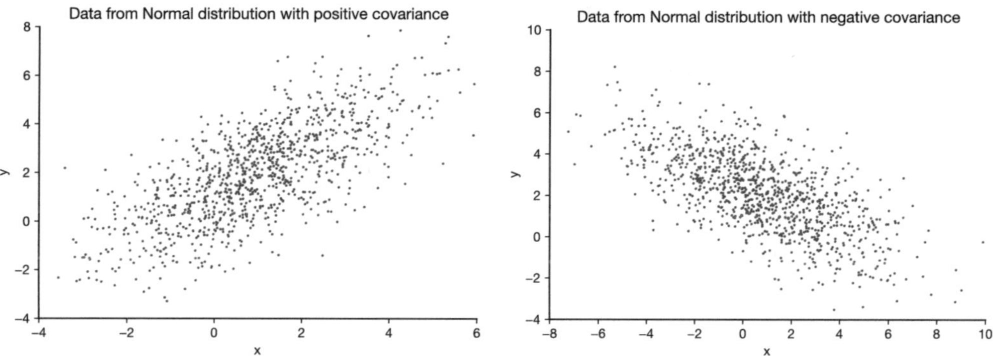

FIGURE 12.2
Scatter plots of data from two different Normal distributions.

It can be shown that if X and Y are independent random variables, then $\mathrm{Cov}(X, Y) = 0$. However, the converse is not true: $\mathrm{Cov}(X, Y) = 0$ does not mean that X and Y are independent; instead, we say they are *uncorrelated*:

> Definition
>
> **uncorrelated**
>
> Jointly distributed random variables X and Y are *uncorrelated* if and only if $\mathrm{Cov}(X, Y) = 0$.

- Roughly speaking, a positive covariance indicates that the values of $X - E[X]$ and $Y - E[Y]$ obtained in a single experiment "tend" to have the same sign. For example, the data in the left-hand plot in Fig. 12.2 are drawn from a Normal distribution with positive covariance.
- Similarly, a negative covariance indicates that the values of $X - E[X]$ and $Y - E[Y]$ obtained in a single experiment "tend" to have the *opposite* sign. The data in the right-hand plot of Fig. 12.2 are drawn from a Normal distribution with negative covariance.

Computing covariance for random variables requires understanding *joint probability distributions* – although I introduce the very basics in Section 13.1, using joint distributions for computing covariance is outside the scope of this book. However, we will compute the covariance for vectors of data. If $\mathbf{x}$ and $\mathbf{y}$ are equal-length samples from some random variables X and Y, then the unbiased (sample) covariance is:

Definition

covariance (data vectors)

For n-vectors $\mathbf{x}$ and $\mathbf{y}$, the unbiased sample *covariance* is given by

$$\text{Cov}(\mathbf{x}, \mathbf{y}) = \frac{1}{n-1} \sum_{i=0}^{n-1} (x_i - \overline{x})(y_i - \overline{y}).$$

Note that the covariance of a feature with itself is just the variance of that feature.

Pandas dataframes have a `cov()` method that returns all the pairwise covariances:

```
df.cov()
```

	cases	population	gdp	urban	gdp_norm	cases_norm
cases	2.189225e+09	1.566015e+11	1.410528e+10	2.428692e+05	2.696033e+05	1.195509e+05
population	1.566015e+11	5.460149e+13	4.044665e+12	4.833276e+07	2.386045e+07	3.638665e+06
gdp	1.410528e+10	4.044665e+12	3.134142e+11	3.787249e+06	2.641812e+06	4.303970e+05
urban	2.428692e+05	4.833276e+07	3.787249e+06	2.121263e+02	9.481075e+01	1.945257e+01
gdp_norm	2.696033e+05	2.386045e+07	2.641812e+06	9.481075e+01	1.402868e+02	1.934676e+01
cases_norm	1.195509e+05	3.638665e+06	4.303970e+05	1.945257e+01	1.934676e+01	9.898749e+00

NumPy arrays do not have a covariance method. However, NumPy does have a `np.cov()` function for computing the pairwise covariances. It expects each feature to be in a row, so we need to transpose the data before calling `np.cov()`. Fortunately (but somewhat inconsistently), NumPy uses the unbiased estimator for covariance by default:

```
print(np.cov(covid_array.T))
```

```
[[2.18922e+09 1.56601e+11 1.41052e+10 2.42869e+05 2.69603e+05 1.19550e+05]
 [1.56601e+11 5.46014e+13 4.04466e+12 4.83327e+07 2.38604e+07 3.63866e+06]
 [1.41052e+10 4.04466e+12 3.13414e+11 3.78724e+06 2.64181e+06 4.30397e+05]
 [2.42869e+05 4.83327e+07 3.78724e+06 2.12126e+02 9.48107e+01 1.94525e+01]
 [2.69603e+05 2.38604e+07 2.64181e+06 9.48107e+01 1.40286e+02 1.93467e+01]
 [1.19550e+05 3.63866e+06 4.30397e+05 1.94525e+01 1.93467e+01 9.89874e+00]]
```

The result is a matrix whose i, jth entry is the covariance between feature i and feature j. The i, ith entries are the variances. This type of matrix is called a *covariance matrix*.

NumPy's `np.cov()` can also calculate the covariance for two separate vectors. For instance, we can get the covariance between the features 3 and 4 as follows:

```
np.cov(covid_array[:,3], covid_array[:,4])
```

```
array([[212.12630894,  94.81074707],
       [ 94.81074707, 140.28677723]])
```

One problem with covariances is that they are hard to interpret because they can take on very large or very small values, depending on the variances of the features. To get around this, we often use a normalized version of the covariance called the *correlation coefficient*. As with covariance, we start by defining it in terms of random variables:

Definition

correlation coefficient (random variables)

For random variables X and Y, the *correlation coefficient* is

$$\rho = \frac{\text{Cov}(X, Y)}{\sigma_X \sigma_Y},$$

where σ_X and σ_Y are the standard deviations of X and Y, respectively.

It can be shown that $|\rho| \leq 1$. Correlation coefficients with magnitudes closer to 1 generally indicate greater dependence among the variables.

The correlation coefficient for data vectors is usually denoted by r or R and is given by:

Definition

correlation coefficient (data vectors)

For n-vectors $\mathbf{x}$ and $\mathbf{y}$, the *correlation coefficient* or *Pearson's correlation coefficient* is given by

$$r = \frac{\text{Cov}(\mathbf{x}, \mathbf{y})}{\sigma_x \sigma_y},$$

where σ_x and σ_y are the standard deviations of $\mathbf{x}$ and $\mathbf{y}$, respectively.

Pandas dataframes have a `corr()` method for computing the pairwise correlation coefficients:

```
df.corr()
```

	cases	population	gdp	urban	gdp_norm	cases_norm
cases	1.000000	0.452948	0.538489	0.356394	0.486487	0.812115
population	0.452948	1.000000	0.977734	0.449099	0.272626	0.156513
gdp	0.538489	0.977734	1.000000	0.464480	0.398414	0.244354
urban	0.356394	0.449099	0.464480	1.000000	0.549607	0.424512
gdp_norm	0.486487	0.272626	0.398414	0.549607	1.000000	0.519170
cases_norm	0.812115	0.156513	0.244354	0.424512	0.519170	1.000000

Note that the correlation coefficient is much easier to interpret than the covariance. If we look at the normalized cases, we can see that it is most correlated with the non-normalized number of cases, followed by the normalized GDP. The correlations between COVID rates and either urban index or population are lower. The correlation coefficients give us an easy way to look for dependence during exploratory data analysis.

The equivalent function in NumPy is `np.corrcoef()`. Here, `corrcoef` is short for correlation coefficient. As with `np.cov()`, the data features are expected to be in the rows of the array, so we have to transpose the array before passing it to `np.corrcoef()`:

```
np.round(np.corrcoef(covid_array.T), 3)
```

```
array([[1.   , 0.453, 0.538, 0.356, 0.486, 0.812],
       [0.453, 1.   , 0.978, 0.449, 0.273, 0.157],
       [0.538, 0.978, 1.   , 0.464, 0.398, 0.244],
       [0.356, 0.449, 0.464, 1.   , 0.55 , 0.425],
       [0.486, 0.273, 0.398, 0.55 , 1.   , 0.519],
       [0.812, 0.157, 0.244, 0.425, 0.519, 1.   ]])
```

As with `np.cov()`, we can use `np.corrcoef()` to calculate the correlation coefficient between two vectors, like

```
np.corrcoef( covid_array[:,4], covid_array[:,5])
```

```
array([[1.        , 0.51917023],
       [0.51917023, 1.        ]])
```

Example 12.2: Covariance and Correlation Between Weight and Height in US Adults

"The Behavioral Risk Factor Surveillance System (BRFSS) is the nation's premier system of health-related telephone surveys that collect state data about U.S. residents regarding their health-related risk behaviors, chronic health conditions, and use of preventive services". https://www.cdc.gov/brfss/index.html.

The BRFSS 2021 survey contains over 400,000 records and over 300 variables. It takes a long time to load and work with the full set of survey results, so I have extracted data for two variables to analyze. The variables are as follows, and I performed data cleaning for each variable as described:

- **HTIN4**: A computed variable that lists height in inches. Invalid responses ("Don't know/Not sure", "Refused", or "Not asked or Missing") have been dropped.
- **WEIGHT2**: The reported weight in pounds. Again, I have dropped invalid responses, as above.

Let's load the resulting data:

```
brfss = pd.read_csv('https://www.fdsp.net/data/brfss21-hw.csv')
brfss.head()
```

	HTIN4	WEIGHT2
0	59.0	72.0
1	65.0	170.0
2	64.0	195.0
3	71.0	206.0
4	75.0	195.0

Let's start by plotting our data. For this data, we are interested in the dependence between height and weight, so I plot the weights as a function of the heights using a scatter plot. Because the data is so huge (almost 400,000 points), in all of the plots involving this data set, I only plot every tenth point. The scatter plot is shown in Fig. 12.3.

```
plt.scatter(brfss['HTIN4'][::10], brfss['WEIGHT2'][::10], 2, alpha=0.6)
plt.xlabel('Height (in)')
plt.ylabel('Weight (lbs)')
```

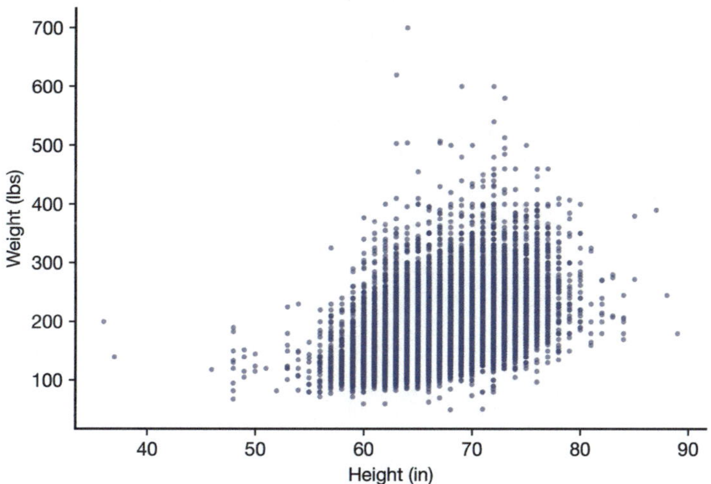

FIGURE 12.3
Scatter plot of height and weight data from BRFSS 2021 survey.

Before we determine the numerical values of the covariance and correlation coefficient for this pair of features, try to answer the following questions:

- What should the sign of the covariance between the Height and Weight features be?
- What should the magnitude of the correlation coefficient be?

Let's start by calculating the covariance matrix:

```
brfss.cov()
```

	HTIN4	WEIGHT2
HTIN4	17.632831	93.412371
WEIGHT2	93.412371	2236.288131

Hopefully, you guessed that the covariance is positive. Taller people tend to weight more than shorter people, so these features tend to move together.

Now let's check the correlation coefficient:

```
brfss.corr()
```

	HTIN4	WEIGHT2
HTIN4	1.000000	0.470413
WEIGHT2	0.470413	1.000000

The correlation coefficient is approximately 0.47, which indicates a fairly strong dependence between height and weight.

Example 12.3: Covariance and Correlation for Independent Data

Finally, let's look at what happens for some independent data. Below I generate completely separate samples of Normal random variables with different variances and show a scatter plot of the data:

```
import scipy.stats as stats

Y=stats.norm()
Z=stats.norm(scale=4)
y=Y.rvs(size=10_000)
z=Z.rvs(size=10_000)
plt.scatter(y,z, 1);
```

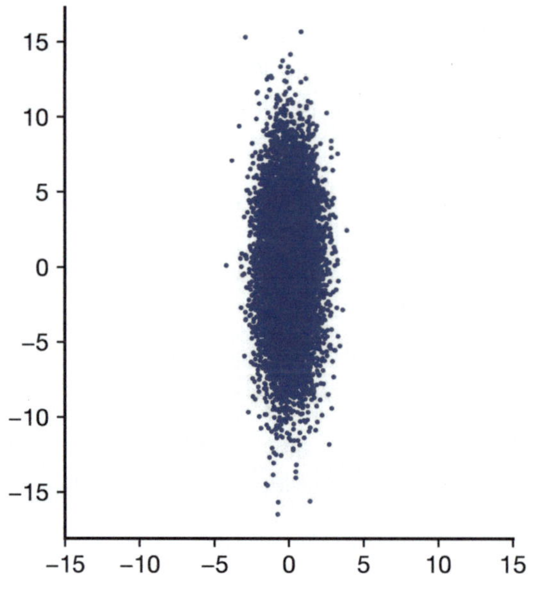

Since the shape of the scatter plot is an ellipse that is aligned with the x- and y-axes, there is no clear direction of dependence. Let's check the numerical value of the covariance and correlation coefficient:

```
np.cov(y,z), np.corrcoef(y,z)
```

```
(array([[1.01900446e+00, 2.66734173e-04],
        [2.66734173e-04, 1.57437054e+01]]),
 array([[1.00000000e+00, 6.65943125e-05],
        [6.65943125e-05, 1.00000000e+00]]))
```

Note the small sample covariance and very small correlation coefficient. When random variables are independent, their covariance is zero; however, the sample covariance will generally not be exactly zero.

Example 12.4: Data Sets with Different Correlations

More examples of data sets with different correlations are shown in this image from the Wikipedia page for correlation (https://en.wikipedia.org/wiki/Correlation):

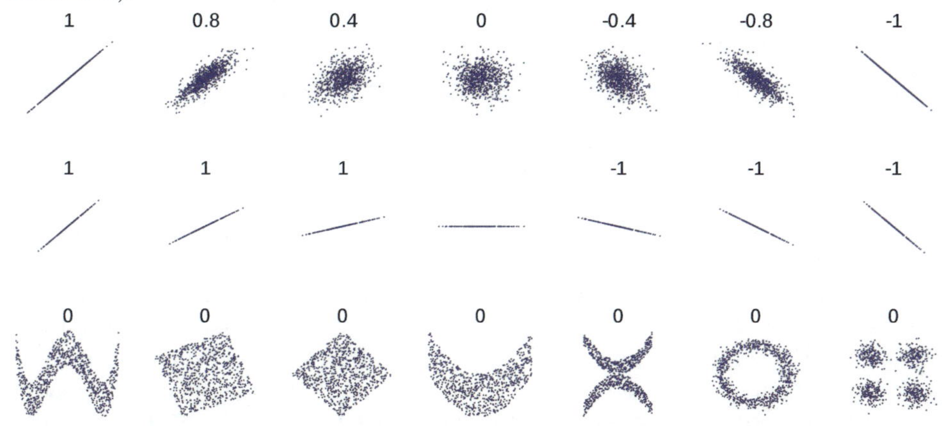

In particular, note that data can be uncorrelated and have a distribution that looks nothing like the circular distribution of data from independent Normal random variables. In fact, data can be uncorrelated and still be highly dependent.

In the next section, we use the concepts of covariance and correlation to help find the best functional relationships between data features to minimize an error metric.

Terminology review and self-assessment questions

Interactive flashcards to review the terminology introduced in this section and self-assessment questions are available at fdsp.net/12-1, which can also be accessed using this QR code:

12.2 Linear Regression

In analyzing data, we do not want to be limited to determining whether observed effects are statistically significant. Our analysis would be much more powerful if it allowed us to make predictions from data. In this section, we consider the problem of finding a good linear predictor from a data set. Let's use the BRFSS data set on heights and weights to illustrate this. Take a second to review the scatter plot of the data in Fig. 12.3.

In making this scatter plot, I have already made some choices about how to interpret this data. By putting the height on the x-axis, I am implicitly treating it as an independent variable, and by putting the weight on the y-axis, I am treating it as a dependent variable. However, this language is somewhat problematic because if weight is dependent on height, then height is also dependent on weight. Thus, we may instead ask how the weight *responds* to the height. The height is called the *explanatory variable*, and the weight is called the *response variable*:

Definitions

explanatory variable

A variable or feature used to predict or explain differences in another variable. Sometimes called an independent variable, especially if this variable is under an experimenter's control.

response variable

A variable or feature that is to be predicted or explained using another variable. Sometimes called a dependent variable, especially if this variable is measured as the result of an experiment.

When we consider prediction, we can ask questions about points for which we do not have data, such as "What is the predicted weight for a US adult who is 53 inches tall?". Or we can ask to give a single prediction for a point for which we do have data but for which the data does not give us a single answer. For instance, if we ask, "What is the predicted weight for a US adult who is 70 inches tall?", the data spans a wide range. Let's load the BRFSS height and weight data and use the Pandas dataframe `query()` method to get a view of the dataframe containing only those entries where the `HEIGHT` feature is equal to 70:

```
brfss = pd.read_csv('https://www.fdsp.net/data/brfss21-hw.csv')
height70 = brfss.query('HTIN4==70')
```

A convenient way to get the minimum, maximum, median, and average, along with other useful information, is to use the `describe()` method:

```
height70['WEIGHT2'].describe()
```

```
count    30995.000000
mean       198.180094
std         40.808096
min         70.000000
25%        170.000000
```

(continues on next page)

(continued from previous page)

```
50%          190.000000
75%          220.000000
max          568.000000
Name: WEIGHT2, dtype: float64
```

We can treat these as estimates of *conditional* statistics given that the height is 70 inches. We see that for people in the sample who are 70 inches tall, the weight spans from 70 to 568 pounds, with a (conditional) sample mean of 198.2 pounds. The median is the point 50% of the way through the data, and it is 190 pounds. Given this wide range, how should we choose a single value to predict the weight of someone who is 70 inches tall?

First, let's consider using only the values at a specific height to predict the weight. In particular, let's create two predictors: one that uses the conditional mean given the height and one that uses the conditional median given the height. To find these predictors, we can iterate over all of the heights in the data set and calculate the conditional mean and conditional median for data points with that specific height:

```
heights = np.sort(brfss['HTIN4'].unique())

mean_predictor = []
median_predictor = []

for height in heights:
  mean_predictor +=  [brfss.query('HTIN4 == ' +str(height))['WEIGHT2'].mean()]
  median_predictor +=  [brfss.query('HTIN4 == ' +str(height))['WEIGHT2'].median()]
```

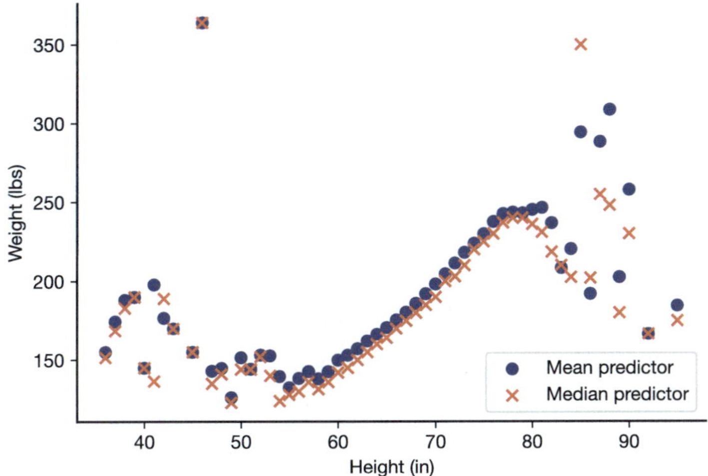

FIGURE 12.4
Conditional mean and conditional median weights for each height (in inches) for respondents in the BRFSS survey.

Fig. 12.4 shows the values of these predictors as a function of a person's height. Either of these approaches looks reasonable, with some limitations:

- The overall relation looks roughly linear in the middle of the range, but there are significant variations away from a line for high or low heights, where the data gets sparse.

- This can still only predict the weight for people of a specific height if the data includes people of that height.

We can resolve both these problems by using all of the data to find a single line that represents the relation between height and weight. This is called *linear regression*:

Definition

linear regression

A technique for determining a linear relationship between one or more explanatory variables and a response variable, where the parameters of the linear relationship are estimated from the data.

In our example, we only have one explanatory variable, and this is called *simple linear regression*:

Definition

simple linear regression

Linear regression with a single explanatory variable.

Given our data, how can we find such a linear relationship? Let's start simple and just look at the data and guess a line to fit it. Such an approach is *ad hoc*, meaning it does not use a standard, systematic technique. Consider these two observations:

- For a height of 60 inches, the weight data is approximately centered around 140 pounds.
- For a height of 75 inches, the weight data is approximately centered around 220 pounds.

The slope of the line going through those two points is:

```
m1 = (210-140)/(75-60)
print(f'{m1:.2f}')
```

```
4.67
```

An equation that goes through those points is

$$
\begin{aligned}
(y-140) &\approx 4.67(x-60) \\
y &\approx 4.67x - 140.
\end{aligned}
$$

```
b1=-140
```

Here is the scatter plot of the data with this line overlaid, along with an interpolated version of the mean predictor we previously created:

```
plt.scatter(brfss['HTIN4'][::10], brfss['WEIGHT2'][::10], 2, alpha=0.6)
plt.xlabel('Height (in)')
plt.ylabel('Weight (lbs)')
plt.title('BRFSS 2021 Data on US Residents\' Heights and Weights');
```

(continues on next page)

(continued from previous page)

```
x= np.arange(45,81,1)
plt.plot(x, m1 * x + b1, color='C1', linewidth=2)
plt.plot(heights, mean_predictor, color='C2', linewidth=2)
plt.xlim(45,80);
```

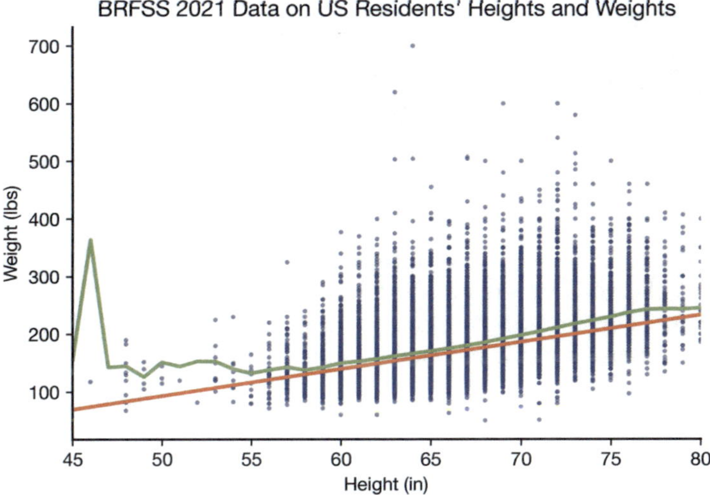

This simple approach to finding a line of fit produced a very good match for this data set, but this approach still suffers from several problems:

1. This approach required human judgment as to which two points to use – it is not easily automated.
2. For many data sets, it will not be as easy to choose two points to define the line.
3. This line is not necessarily optimal in any sense.

Let's consider measuring how good or bad this line of fit is. Letting **h** and **w** be vectors of the heights and weights, respectively, we can find the errors between the true and predicted weights as

$$\mathbf{w} - \hat{\mathbf{w}} = \mathbf{w} - (4.67\mathbf{h} - 140)\,.$$

The following code generates a graph that shows these errors:

```
w = brfss['WEIGHT2'].to_numpy()
h = brfss['HTIN4'].to_numpy()
errors = w-(m1*h +b1)
plt.scatter(np.arange(len(errors[::10])), errors[::10], 3, alpha=0.4);
```

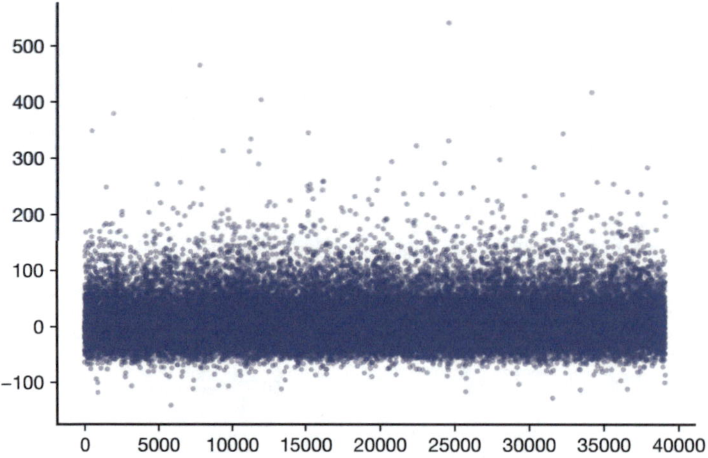

There is no way to make all these errors go to zero: if we move the line up, the positive errors will get smaller, but the negative errors will grow in magnitude. If we move the line down, the reverse will occur. The best we can do is to minimize some metric that combines all these errors. A common choice for this metric is the mean squared error (MSE),

$$MSE = \frac{1}{n} \sum_{i=0}^{n-1} (w_i - \hat{w}_i)^2 .$$

```
errors = w-(m1*h + b1)
print(f'The MSE for our linear predictor is ' +
      f'{np.sum(errors**2)/len(errors) : .1f}')
```

```
The MSE for our linear predictor is  1856.1
```

12.2.1 Linear Regression: Linear Prediction to Minimize the Mean-Square Error

We can use the MSE to find the *best* line to fit the data in the sense that it minimizes the MSE between the line and the data. The line that we find in this way is called the least-squares line of fit, and the approach we use to find this line of fit is called *ordinary least squares (OLS)*. A formal definition of ordinary least squares and a very general approach to solving least-squares problems are covered in the companion book, *Introduction to Linear Algebra for Data Science with Python*. However, for simple linear regression, we can solve for the coefficients of the line that minimizes the MSE using basic calculus.

Consider a simple linear regression problem with explanatory variable x and response variable y. Let the equation for our line of fit be $\hat{y}_i = mx_i + b$, where m and b are constants that we wish to find to minimize the MSE:

$$MSE = \sum_{i=0}^{n-1} [y_i - \hat{y}_i]^2 = \sum_{i=0}^{n-1} [y_i - (mx_i + b)]^2 .$$

We can find the values of m and b that minimize the MSE by taking derivatives with respect to m and b and setting the result equal to 0. Let's start with the derivative with respect to

b:

$$\frac{d}{db}\left\{\frac{1}{n}\sum_{i=0}^{n-1}\left[y_i - (mx_i + b)\right]^2\right\} = 0$$
$$\sum_{i=0}^{n-1}(2)\left[y_i - (mx_i + b)\right](-1) = 0$$
$$\sum_{i=0}^{n-1} y_i - m\sum_{i=0}^{n-1} x_i - \sum_{i=0}^{n-1} b = 0.$$

Rearranging, we have

$$b = \frac{1}{n}\sum_{i=0}^{n-1} y_i - (m)\frac{1}{n}\sum_{i=0}^{n-1} x_i$$
$$= \overline{y} - m\overline{x}.$$

Note that the solution for b depends on m, but we can substitute this value for b into the equation for the MSE and then take the derivative with respect to m:

$$\frac{d}{dm}\left\{\frac{1}{n}\sum_{i=0}^{n-1}\left[y_i - (mx_i + \overline{y} - m\overline{x})\right]^2\right\} = 0$$
$$\sum_{i=0}^{n-1}(2)\left[(y_i - \overline{y}) - m\,(x_i - \overline{x})\right](x_i - \overline{x}) = 0.$$

Rearranging and dividing both sides by $n-1$, we have

$$\frac{1}{n-1}\sum_{i=0}^{n-1}(y_i - \overline{y})(x_i - \overline{x}) = (m)\frac{1}{n-1}\sum_{i=0}^{n-1}(x_i - \overline{x})^2.$$
$$\Rightarrow m = \frac{\operatorname{Cov}(\underline{x}, \underline{y})}{\operatorname{Var}(\underline{x})}.$$

Let's use these equations to calculate the regression line for our height and weight data. As a reminder, we are taking h as the explanatory variable and w as the response variable. We can get the value of m from the covariances between **h** and **w**, which we can get from NumPy as follows:

```
K=np.cov(h,w)
print(K)
```

```
[[  17.63283104   93.41237144]
 [  93.41237144 2236.28813105]]
```

Then the slope m is

```
m = K[1,0] / K[0,0]
print(f'm = {m:.2f}')
```

```
m = 5.30
```

The value of b is then easily calculated from m and the means of $\mathbf{h}$ and $\mathbf{w}$:

```
b = w.mean() - m * h.mean()
print(f'b = {b}')
```

```
b = -171.97804815595876
```

In practice, we do not need to use these formulas. The `stats.linregress()` function from SciPy.Stats can find the parameters of this line. The arguments of `stats.linregress()` are two data vectors for which we wish to determine the least-squares line of fit. Let's run this on our height and weight data, and then we will discuss the output:

```
regress1 = stats.linregress(h,w)
print(regress1)
```

```
LinregressResult(slope=5.297638888874249, intercept=-171.97804815595,
                 rvalue=0.47041319501296, pvalue=0.0,
                 stderr=0.01589123944005, intercept_stderr=1.0688139979789)
```

The output is a `LinregressResult` object that has six attributes. The `slope` and `intercept` are equivalent to the values m and b that we found using the formulas derived from calculus. The slope and intercept for the line of fit from linear regression are also very close to the values using our *ad hoc* approach. The linear least-squares line of fit and the ad hoc linear predictor are shown with the BRFSS data in Fig. 12.5.

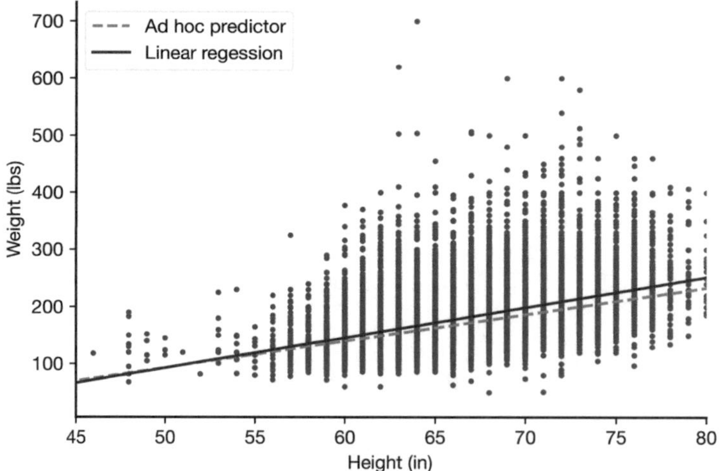

FIGURE 12.5
BRFSS height and weight data with ad hoc and linear least-squares predictors.

Now let's check the MSE to see whether this line is actually better than the MSE for the ad hoc predictor:

```
errors2 = w - (m2*h + b2)
print(f'The MSE for the least-squares linear predictor is' +
      f'{np.sum(errors2**2)/len(errors) : .1f}')
```

```
The MSE for the least-squares linear predictor is 1741.4
```

As expected, the least-squares linear predictor is better than our *ad hoc* approach, which achieved an MSE of approximately 1856.

12.2.2 Variance Reduction of Linear Prediction

Consider the `rvalue` element returned by `stats.linregress()`. Let's compare it with the correlation coefficient between the heights and weights:

```
brfss.corr()
```

	HTIN4	WEIGHT2
HTIN4	1.000000	0.470413
WEIGHT2	0.470413	1.000000

```
print(regress1.rvalue)
```

```
0.47041319501296763
```

We see that the `rvalue` is the correlation coefficient, for which we have used the notation r. We previously saw that the value of r indicates how close the data is to a linear relation. In fact, we can use r to calculate the MSE as

$$MSE = s_{w,n}^2 \left(1 - r^2\right), \tag{12.1}$$

where $s_{w,n}^2$ is the sample variance calculated by dividing by the number of data points.

Let's check this for our height and weight data:

```
r=regress1.rvalue
print(f'Analytical MSE = {w.var(ddof=0)*(1-r**2):.1f}')
```

```
Analytical MSE = 1741.4
```

Note:

In the computation above, I have explicitly passed the keyword argument `ddof=0` to ensure that variance is computed using the denominator equal to the length of `w`. In this particular case, it is not necessary to pass `ddof=0` because `w` is a NumPy array, and the default for NumPy's `var()` method is `ddof=0`. However, the default for the `var()` method of a Pandas Series is `ddof=1`. It is safest to pass `ddof=0` to be sure.

We see that the analytical MSE matches the MSE achieved from regression. For linear regression involving only two variables, the value r^2 is called the *coefficient of determination*:

Definition

coefficient of determination (simple linear regression)

In simple linear regression between two data vectors $\mathbf{x}$ and $\mathbf{y}$ with Pearson's correlation coefficient r, the *coefficient of determination* is the value r^2, which is also denoted R^2.

The MSE achieved by linear regression can be described in terms of *total variance* and *explained variance*:

Definitions

total variance (simple linear regression)

In simple linear regression between explanatory vector $\mathbf{x}$ and response vector $\mathbf{y}$, the total variance refers to the variance of the response vector, σ_y^2, which is the variance without using the explanatory vector to predict the values in $\mathbf{y}$.

explained variance (simple linear regression)

In simple linear regression between explanatory vector $\mathbf{x}$ and response vector $\mathbf{y}$, the explained variance is the reduction in the variance of the response data after subtracting off the values predicted from the explanatory data. The explained variance is $r^2\sigma_y^2$, where σ_y^2 is the variance of $\mathbf{y}$ and r^2 is the coefficient of determination.

In the context of our example, the idea is this: if we know a person's height, we should be able to make a better guess of their weight than if we did not know their height. If we use the linear predictor that minimizes the MSE, then r^2 is the proportion of the total variance in the person's weight that can be "explained" – i.e., predicted – using the person's height. In this case, the variance is reduced on average by:

```
print(f'The r^2 value is {100*regress1.rvalue**2:.1f}%')
```

```
The r^2 value is 22.1%
```

In general, we can compute r^2 given a response variable with sample variance s_n^2 and an achieved value of the MSE by rearranging (12.1). The resulting equation is

$$r^2 = 1 - \frac{\text{MSE}}{s_n^2}.$$

Note that our conclusion about height and weight is that they are *associated* – meaning that we can use one to predict the other. We generally cannot conclude from linear regression that there is a *causal* relationship, meaning that the value of one of the variables causes the value of the other variable to take on a particular distribution. For this example, being taller does not cause a person to weigh more, or vice versa. The most likely explanation for why they are associated is that height and weight are jointly affected by other factors, such as genetics, nutrition, and other environmental factors.

Example 12.5: Climate Change

Is the climate changing? That is a complicated question that is outside of the scope of this book, but we can try to answer it on a small scale. Since I live in Florida, let's consider the research question, "Is the annual temperature changing over time in Miami-Dade County, Florida?"

Let's start by loading the annual temperature data for Miami-Dade County from the National Oceanic and Atmospheric Administration:

```
df=pd.read_csv('https://www.ncei.noaa.gov/access/monitoring/climate-at-a-glance/'
               +'county/time-series/FL-086/tavg/ann/5/'
               +'1895-2022.csv?base_prd=true&begbaseyear=1895&endbaseyear=2022',
               skiprows=4)
# Alternate site for accessing data:
# df = pd.read_csv('https://www.fdsp.net/data/miami-weather.csv', skiprows=4)
df.head()
```

```
     Date  Value  Anomaly
0  189512   73.6     -1.1
1  189612   73.9     -0.8
2  189712   74.6     -0.1
3  189812   74.4     -0.3
4  189912   74.7      0.0
```

The `Value` column contains the annual temperature. The `Date` column contains the year followed by a two-digit month code, which can be ignored because this is annual data. Let's create a separate column for the year:

```
df['Year'] = df['Date'] //100
df.head()
```

```
Date  Value  Anomaly  Year
0  189512   73.6     -1.1  1895
1  189612   73.9     -0.8  1896
2  189712   74.6     -0.1  1897
3  189812   74.4     -0.3  1898
4  189912   74.7      0.0  1899
```

This is an example of *time-series data*:

> Definition
>
> **time-series data**
>
> Data that is collected over time, usually at regular intervals. Each data point is associated with a timestamp indicating when the data was collected.

Let's find the best linear regression curve for this data using `stats.linregress()`:

```
regress2 = stats.linregress(df['Year'], df['Value'])
regress2
```

```
LinregressResult(slope=0.027612559512909, intercept=20.630958443966,
                 rvalue=0.76691699973022, pvalue=4.91569850153399e-26,
                 stderr=0.00205843141937, intercept_stderr=4.0321553265906)
```

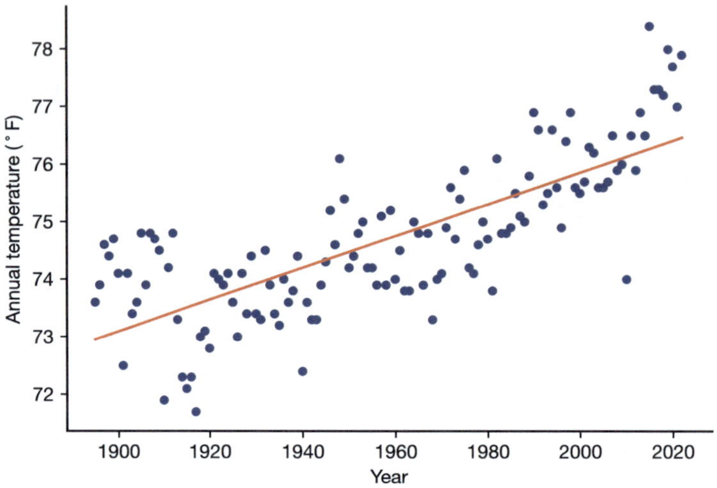

FIGURE 12.6
Annual temperature data for Miama-Dade County (Florida, USA) and linear least-squares line of fit.

The regression line is plotted with the data in Fig. 12.6. The line seems to be a good match for the data visually, and this is confirmed by checking r^2:

```
print(f'The r^2 value is {100*regress2.rvalue**2:.1f}%')
```

```
The r^2 value is 58.8%
```

Note that the temperature may be increasing over time even faster than linear, at least for the last decade or two. Thus, nonlinear relations should also be considered.

Example 12.6: COVID Rate versus GDP

In Section 12.1, we saw that the normalized COVID case rate and the normalized GDP had a correlation of approximately 0.52. The following code performs linear regression for these variables. The linear predictor is shown with the data in Fig. 12.7.

```
regress3 = stats.linregress(covid['gdp_norm'], covid['cases_norm'])
regress3
```

```
LinregressResult(slope=0.1379086689536938, intercept=-5.8341524970008,
                 rvalue=0.519170230371747, pvalue=0.0001119524186598497,
                 stderr=0.032768764050575, intercept_stderr=2.06822161723187)
```

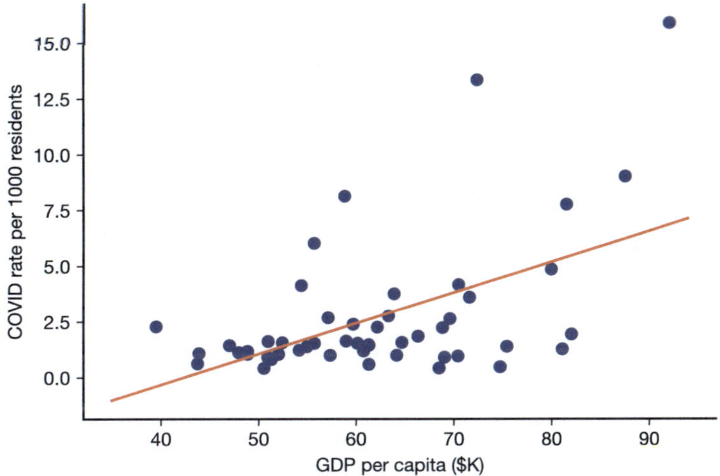

FIGURE 12.7
COVID-19 rates versus GDP per capita ($K) and linear least-squares line of fit.

```
print(f'The r^2 value is {100*regress3.rvalue**2:.1f}%')
```

```
The r^2 value is 27.0%
```

So, the linear regression curve can explain approximately 27% of the variance in the normalized COVID rates. Looking at the line versus the data, a linear equation may not be the best predictor. In the next section, we consider how to use linear regression to determine nonlinear relationships.

12.2.3 Correlation is not causation!

"Correlation is not causation" is a common expression used by statisticians. It means that just because two variables are correlated, that does not mean that one of the variables caused the other. Given features x and y that are correlated, any of the following can be true:

- x causes y.
- y causes x.
- x and y both "cause" each other. For instance, Wikipedia gives the example that people who cycle have a lower body mass index (BMI), but people with a lower BMI may be more likely to cycle.
- One or more other factors cause both x and y. For instance, in the height and weight data analysis, these third factors could include genetics and nutrition.

- The correlation is "spurious," meaning that it is a random effect.

Spurious correlations are easy to find, given enough different data sources. In fact, there is an entire website dedicated to spurious correlations at https://www.tylervigen.com/spurious-correlations. Two examples of spurious correlations from that website (CC BY) are shown in Fig. 12.8 and Fig. 12.9.

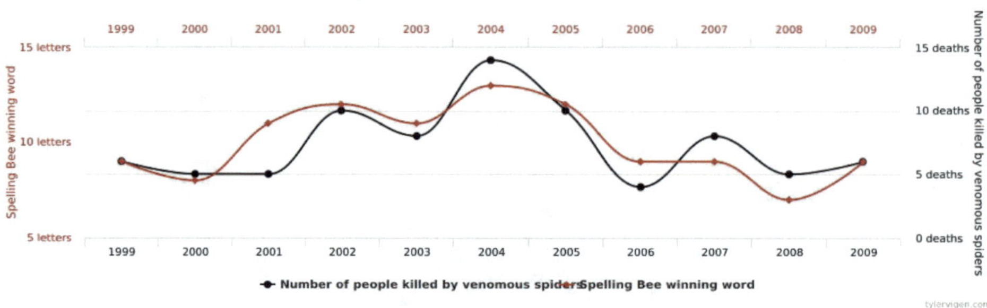

FIGURE 12.8
Example of spurious correlation between letters in winning word of Scripps National Spelling Bee and number of people killed by venomous spiders. Source: https://tylervigen.com/view_correlation?id=2941

To give you an idea of how strong some of the spurious correlations are, I have extracted the data from this last graph:

```
pets=[39.7, 41.9, 44.6, 46.8, 49.8, 53.1, 56.9, 61.8, 65.7, 67.1]
lawyers=[128553, 131139, 132452, 134468, 136571, 139371, 141030,
         145355, 148399, 149982]
```

Now we can run our own linear regression:

```
regress2 = stats.linregress(pets,lawyers)
print(regress2)
```

```
LinregressResult(slope=751.0366939629846, intercept=99122.3247603921,
                 rvalue=0.99838620404485, pvalue=2.961633157474e-11,
                 stderr=15.1036396068426, intercept_stderr=808.947609797420)
```

We see that the value of r^2 is very close to 1:

```
print(f'r^2 = {100*regress2.rvalue ** 2 : .1f}%')
```

```
r^2 =  99.7%
```

Most people would realize that these two features are both increasing over time. The increases may be attributable to related factors, such as increases in population or consumption (spending) over time.

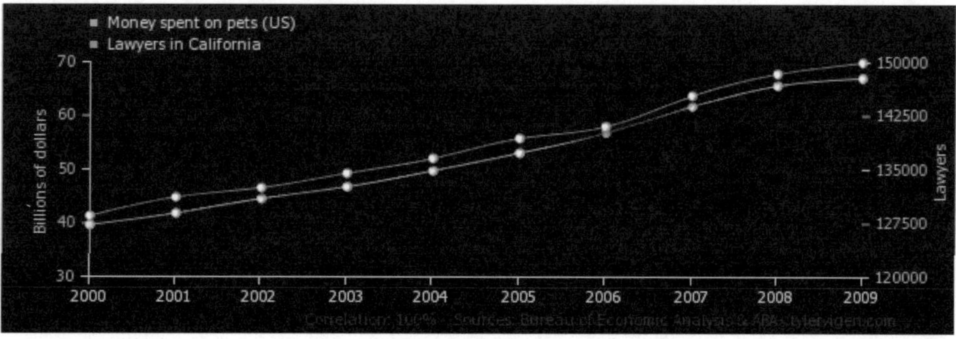

FIGURE 12.9
Example of spurious correlation between money spent on pets in the US and the number of lawyers in California. Source: https://www.tylervigen.com/view_correlation?id=2956

Terminology review and self-assessment questions

Interactive flashcards to review the terminology introduced in this section and self-assessment questions are available at fdsp.net/12-2, which can also be accessed using this QR code:

12.3 Null Hypothesis Tests for Correlation

If a correlation is observed, how can we know it is not just caused by random variations in the data? In the case of small data, the correlation coefficient between two vectors is likely to not be close to zero, even if the vectors come from independent variables. We can perform a null hypothesis significance test (NHST) to determine whether an observed correlation is statistically significant. As usual, the first step in NHST is to establish a null hypothesis, H_0. Since we are trying to assess whether the data is correlated, the null hypothesis is that the data is not correlated. We will consider two approaches: resampling and analytical techniques.

12.3.1 Resampling

Consider data vectors for two features $\mathbf{x}$ and $\mathbf{y}$. Under resampling, we wish to generate new samples from the same distributions as $\mathbf{x}$ and $\mathbf{y}$ but without the data being correlated. We can achieve this by drawing separate random samples from the $\mathbf{x}$ and $\mathbf{y}$ data. Two ways to think about how this will eliminate the correlation are:

1. Any correlation in the data is because the samples (x_i, y_i) vary together. Separate bootstrap samples from $\mathbf{x}$ and $\mathbf{y}$ inherently do not vary together.
2. Drawing bootstrap samples separately from $\mathbf{x}$ and $\mathbf{y}$ creates independent samples, and independent random variables are uncorrelated.

So to apply a resampling NHST:

1. Determine the number of simulation iterations to be run, the significance threshold (p-value) α, and whether the test will be one-sided or two-sided.
2. Initialize a counter to zero.

3. In each simulation iteration:
   i. Use resampling (such as bootstrapping) to separately draw sample vectors for each feature. The length of the vectors should be the same as the length of the original vectors **x** and **y**.
   ii. Calculate the correlation coefficient for the vectors.
   iii. If the sample correlation coefficient is larger than the observed correlation, increment the counter. Here, larger can either refer to one particular direction (for a one-sided test) or magnitude (for a two-sided test).
4. When the simulation iterations are completed, divide the counter by the number of iterations to calculate the p-value.
5. Compare the p-value to α. If $p \leq \alpha$, the result is statistically significant, and the null hypothesis is rejected. If $p > \alpha$, the data is not sufficient to reject the null hypothesis.

Let's demonstrate this using the COVID-19 data set, for which correlation coefficient between GDP per capita and COVID-19 rate is 0.52:

```
covid = pd.read_csv( 'https://www.fdsp.net/data/covid-merged.csv' )
covid['gdp_norm'] = covid['gdp'] / covid['population'] * 1000
covid['cases_norm'] = covid['cases'] / covid['population'] * 1000
rho = np.corrcoef(covid['gdp_norm'], covid['cases_norm'])[0,1]
```

Before we carry out our test, we need to determine our significance threshold and whether to perform a one-sided or two-sided test. We will use a significance threshold of $\alpha = 0.05$, and we will carry out a two-sided test, which is more conservative than a one-sided test. The code to implement the two-sided NHST is shown below:

```
num_sims = 20_000

num_large_correlation = 0
for sim in range(num_sims):

  # Bootstrap resampling
  gdp_norm_sample = covid['gdp_norm'].sample(frac=1, replace=True)
  cases_norm_sample = covid['cases_norm'].sample(frac=1, replace=True)

  # Two-sided test
  if abs(np.corrcoef(gdp_norm_sample, cases_norm_sample)[0,1]) > rho:
    num_large_correlation += 1

print(f'The probability of seeing a correlation as large as {rho: .2g}')
print(f'is approximately {num_large_correlation / num_sims : .2g}')
```

```
The probability of seeing a correlation as large as  0.52
is approximately  0.00015
```

Since the p-value is much less than 0.05, the observed correlation is statistically significant.

12.3.2 Analytical Tests

The mathematics behind the analytical NHST are outside the scope of this book, but it can be shown that the p-value can be approximated by the tail of a Student's t distribution with $n-2$ degrees of freedom. Rather than go into the details of that, we will take advantage of the `pvalue` returned by `stats.linregress()`. This is the p-value for a two-sided NHST in which the null hypothesis is that the features are uncorrelated.

For our example, the p-value is

```
regress = stats.linregress(covid['gdp_norm'], covid['cases_norm'])

print(f'The p-value for a correlation NHST is {regress.pvalue : .2g}')
```

```
The p-value for a correlation NHST is  0.00011
```

As expected, the p-values generated by bootstrap simulation and analysis are approximately equal.

Terminology review and self-assessment questions

Interactive flashcards to review the terminology introduced in this section and self-assessment questions are available at fdsp.net/12-3, which can also be accessed using this QR code:

12.4 Nonlinear Regression Tests

Consider again the linear regression result for COVID rates as a function of GDP per capita shown in Fig. 12.7. As shown in the figure, this data is better fit by a line with a positive slope (corresponding to a positive correlation) than a line with zero slope (corresponding to zero correlation) and that the correlation is statistically significant at the $p < 0.05$ level. However, this does not mean that a line is the best function to fit this data. The fact that the proportion of data points above the line varies as a function of normalized GDP, suggests that a nonlinear relation might achieve a lower MSE.

Doesn't simple linear regression and OLS require a linear relationship? The answer is both yes and no. Suppose we consider nonlinear regression, where the relationship between the explanatory variable and the response variable is a nonlinear function. We can use OLS to find the coefficients of the curve that minimizes the MSE provided the equation is linear in the *coefficients*. Let's clarify this using an example.

Suppose we want to model the relationship using the quadratic equation $\hat{y}_i = ax_i^2 + b$. It doesn't matter that we have x_i^2 instead of x_i in the equation. The a and b still act on x_i^2 in a linear fashion. In fact, we can still use `stats.linregress()`, but for our example, instead of passing arguments representing normalized GDPs and normalized COVID-19 rates, we pass the square of the normalized GDPs along with the normalized COVID-19 rates:

```
covid = pd.read_csv('https://www.fdsp.net/data/covid-merged.csv' )
covid['gdp_norm'] = covid['gdp'] / covid['population'] * 1000
covid['cases_norm'] = covid['cases'] / covid['population'] * 1000
```

(continues on next page)

(continued from previous page)

```
qregress = stats.linregress(covid['gdp_norm']**2, covid['cases_norm'])
print(qregress)
```

```
LinregressResult(slope=0.00113039489285474, intercept=-1.784499517877938,
                 rvalue=0.5547343058282337, pvalue=2.909396850664566e-05,
                 stderr=0.0002447159363981, intercept_stderr=1.044142342479318)
```

The resulting relationship is approximately $\hat{y}_i = 0.00113x_i^2 - 1.78$ and is shown in Fig. 12.10. (Code to generate this figure is available online at fdsp.net/12-4.) The value of r^2 achieved by linear and quadratic regression are:

```
regress1 = stats.linregress(covid['gdp_norm'], covid['cases_norm'])

print(f'Linear: r^2 = {regress1.rvalue**2: .2g}')
print(f'Quadratic: r^2 = {qregress.rvalue**2: .2g}')
```

```
Linear: r^2 =  0.27
Quadratic: r^2 =  0.31
```

So, quadratic regression achieves a slightly higher proportion of explained variance.

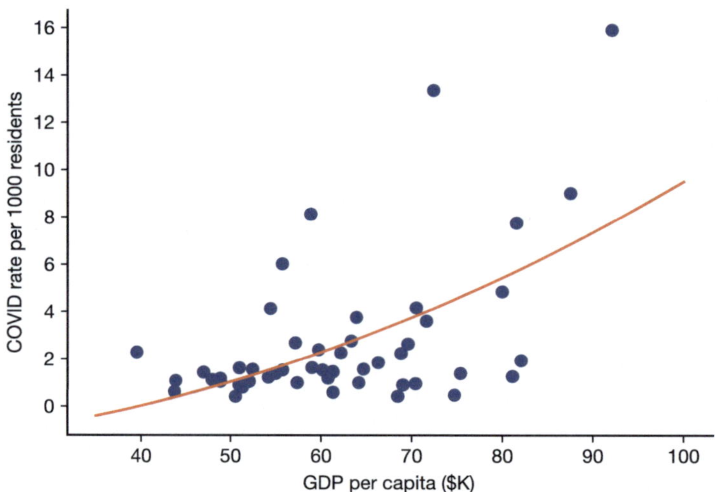

FIGURE 12.10
A least-squares quadratic fit for COVID-19 rates as a function of GDP per capita.

Exercise

Experiment with using other powers of `covid{[gdp_norm{]}}` in the linear regression. What power achieves the best r^2?

12.4.1 Non-polynomial Regression

We are not limited to polynomial functions of the explanatory variable. We may use any function that we think explains the relationship. To demonstrate this, let's introduce another COVID-19 data set that the students in my data-science class used to analyze the spread of COVID-19 in early 2020.

We will use data from the COVID-19 Data Repository by the Center for Systems Science and Engineering (CSSE) at Johns Hopkins University (JHU CSSE COVID-19 Data). The repository is available here: https://github.com/CSSEGISandData/COVID-19. This data is licensed under the Creative Commons Attribution 4.0 International license (CC BY 4.0).

We will use the archived time-series data for confirmed cases, which gives the number of confirmed cases by location (rows) and date (columns). The following code loads the data directly from the JHU CSSE COVID-19 Data GitHub repository:

```
csse_github = 'https://raw.githubusercontent.com/CSSEGISandData/COVID-19/master/'
data_path = 'archived_data/archived_time_series/'
file_name = 'time_series_19-covid-Confirmed_archived_0325.csv'
covid_ts = pd.read_csv(csse_github + data_path + file_name)

# Alternative if  original source data is not available:
# covid_ts = pd.read_csv('https://www.fdsp.net/data/covid-time-series.csv')

covid_ts.head()
```

	Province/State	Country/Region	Lat	Long	1/22/20	1/23/20	1/24/20	1/25/20	1/26/
0	NaN	Thailand	15.0000	101.0000	2	3	5	7	
1	NaN	Japan	36.0000	138.0000	2	1	2	2	
2	NaN	Singapore	1.2833	103.8333	0	1	3	3	
3	NaN	Nepal	28.1667	84.2500	0	0	0	1	
4	NaN	Malaysia	2.5000	112.5000	0	0	0	3	

5 rows × 66 columns

This file includes data for countries and for US states and cities. We can limit this to just the entries for the United States using the `query()` method:

```
us = covid_ts.query ('`Country/Region` == "US"')
us.head()
```

	Province/State	Country/Region	Lat	Long	1/22/20	1/23/20	1/24/20	1/25/20	1/:
98	Washington	US	47.4009	-121.4905	0	0	0	0	
99	New York	US	42.1657	-74.9481	0	0	0	0	
100	California	US	36.1162	-119.6816	0	0	0	0	
101	Massachusetts	US	42.2302	-71.5301	0	0	0	0	
102	Diamond Princess	US	35.4437	139.6380	0	0	0	0	

5 rows × 66 columns

We can select the date columns (4 to the end) using the `iloc` member and apply the `sum()` method to the numeric columns, which will give us a Pandas Series containing the sums by date:

```
us_sums = us.iloc[:, 4:].sum()
us_sums
```

```
1/22/20        1.0
1/23/20        1.0
1/24/20        2.0
1/25/20        2.0
1/26/20        5.0
             ...
3/19/20    13677.0
3/20/20    19100.0
3/21/20    25489.0
3/22/20    33272.0
3/23/20    33276.0
Length: 62, dtype: float64
```

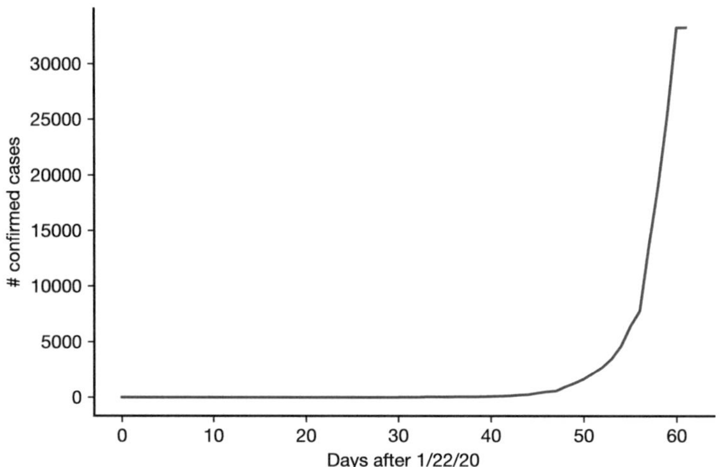

FIGURE 12.11
Confirmed COVID-19 cases in the United States by date.

Now let's see what type of trend the data might follow. Fig. 12.11 shows a plot of total confirmed COVID-19 cases versus time. Before day 40, there were fewer than 100 cases, so that was before there was a significant spread of COVID-19 in the US. We will exclude this part of the data in our analysis. The data for the last day (3/23/20) is approximately equal to the data for 3/22/20. This is completely unreasonable given the other data, and this inconsistency does not occur in other data sources, so we will also exclude this last data point.

Let's create a new variable with this restricted range of dates:

```
us_sums2 = us_sums[40:-1]
```

If we apply simple linear regression to fit a line to this data, we get the following:

```
days = range(len(us_sums2))
linregress = stats.linregress( days, us_sums2)
print('Linear regression results:')
print(linregress)
```

```
Linear regression results:
LinregressResult(slope=1182.812987012987, intercept=-5874.367965367965,
                 rvalue=0.79183101268279, pvalue=1.8962366136262583e-05,
                 stderr=209.297150620038, intercept_stderr=2446.778281444334)
```

The correlation coefficient is 0.79, which is not that low, but consider the regression line versus the data:

```
plt.plot(days, us_sums2);
plt.xlabel('Days after ' + us_sums2.index[0]);
plt.ylabel('# confirmed cases');

# Linear regression line of fit:
plt.plot(days, linregress.slope * days + linregress.intercept,
         color = 'C1');
```

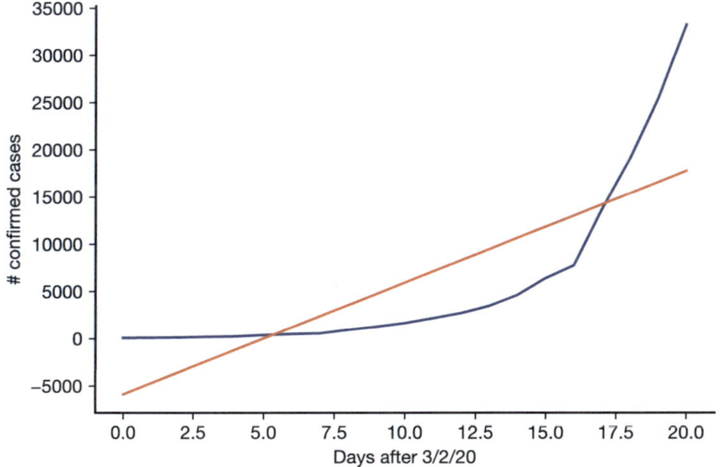

Clearly, a line is not a good match to this data. The rate of increase is so fast that it is likely that the trend is not polynomial but instead is exponential in the number of days. We can easily check this by plotting the data on a logarithmic y-axis, which we can do by calling `plt.semilogy()` instead of `plt.plot()`:

```
plt.semilogy(range(len(us_sums2)), us_sums2);
plt.xlabel('Days after ' + us_sums2.index[0]);
plt.ylabel('# confirmed cases');
```

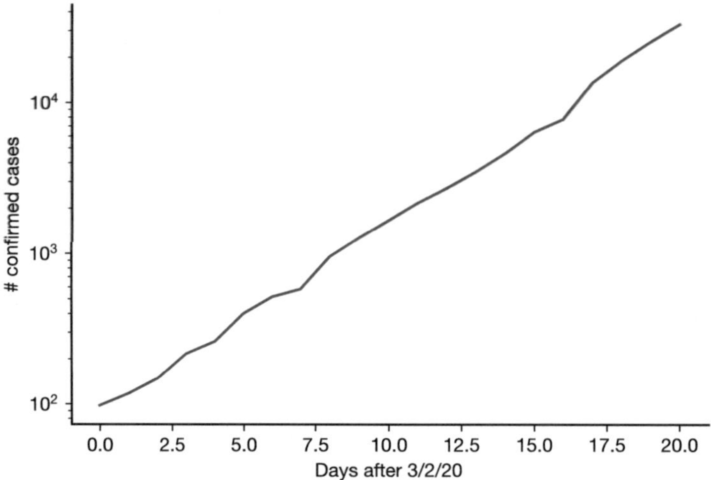

The relationship on a semi-log plot looks very linear! This implies that the number of cases is approximately exponential in the number of days since March 2, 2020.

To fit an exponential to this data, the best approach is to take the logarithm of the response variable (`us_sums2`) and then use linear regression with the original explanatory variable (`days`):

```
lsums2 = np.log(us_sums2)
log_regress = stats.linregress(days, lsums2)
print('The results of linear regression with the log')
print('of the response variable are given below')
print(log_regress)
```

```
The results of linear regression with the log
of the response variable are given below
LinregressResult(slope=0.292790859349689, intercept=4.4637623776277,
                 rvalue=0.99866601829409, pvalue=6.3581668291162e-26,
                 stderr=0.00347300567271, intercept_stderr=0.0406010059199033)
```

From these results we can get the following estimate:

$$\log \hat{y}_i \approx 0.293x_i + 4.46.$$

Taking e to the power of both sides gives

$$\begin{aligned}\hat{y}_i &\approx e^{0.293x_i + 4.46} \\ &\approx e^{0.293x_i} e^{4.46}\end{aligned}$$

The value of the second term is approximately:

```
c = np.exp(log_regress.intercept)
print(c)
```

```
86.81352067039107
```

So, the exponential relationship can be written as

$$\approx 86.8e^{0.293x_i}.$$

The value of r^2 (using the logarithmic data) is now:

```
log_regress.rvalue ** 2
```

```
0.9973338160953739
```

which is exceptionally high! Fig. 12.12 shows a plot of the estimate versus the original data. Comparing the two plots, it is easy to see how great the exponential fit is.

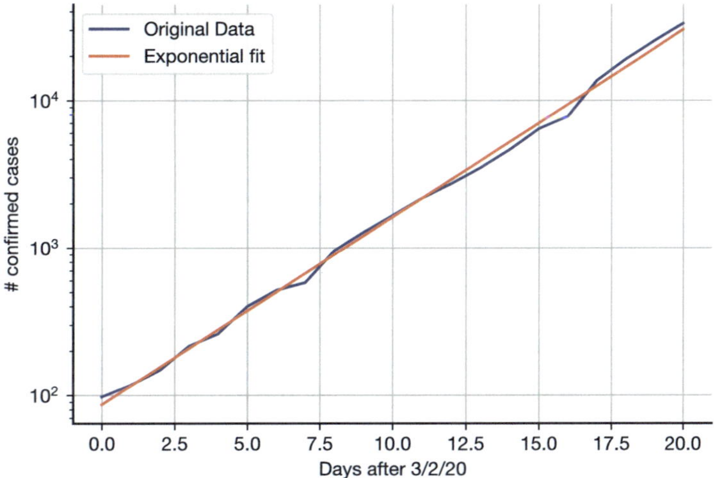

FIGURE 12.12
COVID time-series data versus least-squares exponential fit.

We have found that the data is growing exponentially and have the equation for the exponential curve, based on applying linear regression to the logarithm of the response variable. However, the fact that the number of confirmed cases grows as $\exp(0.293x)$, where x is the number of days, is not very easy to interpret by the public. To make this more clear, we often calculate the number of days that it takes cases to double, which we can solve as

$$\begin{aligned} e^{0.293x} &= 2 \\ 0.293x &= \log 2 \\ x &= \frac{\log 2}{0.293}, \end{aligned}$$

which is approximately

```
np.log(2) / log_regress.slope
```

```
2.367379849560459
```

The number of cases is doubling every 2.37 days. Similarly, the number of days to increase by a factor of 10 is

```
np.log(10) / log_regress.slope
```

```
7.864265633525107
```

This value is easier to interpret on the semi-log graph of the data than the doubling rate. Based on this rate, we can see that the number of cases should increase by a factor of 100 about every 15.7 days. Looking at day 0 of the graph, there are approximately 100 cases, and on day 16, there are about 80,000 cases, so the relation holds approximately.

We can also solve for the number of days when the US would hit 100,000 cases with no mitigation strategies:

$$\begin{aligned}
86.8e^{0.293x_i} &= 10^5 \\
e^{0.293x_i} &= \frac{10^5}{86.8} \\
0.293x_i &= \log\left(\frac{10^5}{86.8}\right) \\
x_i &= \frac{1}{0.293}\log\left(\frac{10^5}{86.8}\right)
\end{aligned}$$

The number of days to reach 100,000 cases is approximately:

```
np.log(10 ** 5/86.8)/0.293
```

```
24.059108680218174
```

Since day 0 for our calculations is March 2, 2020, we could have expected to reach 100,000 cases in the US by approximately March 26 or March 27 of 2020. According to the cumulative case data on Statista at https://www.statista.com/statistics/1103185/cumulative-coronavirus-covid19-cases-number-us-by-day/, 100,000 confirmed cases was reached sometime between March 27, 2020 and April 3, 2020, which is consistent with our prediction.

In this section, we have seen that the least-squares techniques we developed for finding linear regression curves can actually be applied to a variety of different relationships between data through appropriate transforms of the explanatory or response variables.

Terminology review and self-assessment questions

Interactive flashcards to review the terminology introduced in this section and self-assessment questions are available at fdsp.net/12-4, which can also be accessed using this QR code:

12.5 Chapter Summary

In this chapter, we introduced a new summary statistic called *covariance* that measures how two data features vary together. We showed how to find the best line of fit for data that consists of pairs of features, and we showed how that line of fit connects to the concept of correlation. Finding a line of fit for a pair of features is called *simple linear regression*, and ordinary least squares (OLS) finds the line of fit that minimizes the mean-squared error. Finally, we showed that we can perform nonlinear regression by applying functions to the explanatory variable and then finding the linear regression fit for the transformed data. Unlike our previous statistical tests, the functional fits found by regression allow us to create a predictor for data that is outside of our sample set.

Access a list of key take-aways for this chapter, along with interactive flash-cards and quizzes at fdsp.net/12-5, which can also be accessed using this QR code:

13

Working with Dependent Data in Multiple Dimensions

In Section 12.2, we used linear regression to analyze the dependence between features in a data set. In this chapter, we take a deeper dive into working with dependent data. We start by exploring techniques for modeling random phenomena as dependent random variables, with a particular focus on dependent Normal random variables. Then we explore the effects of linear transformations on the moments of vector random variables. Finally, we will explore how eigendecomposition can be used for dimensionality reduction, with applications to data visualization and decision-making.

13.1 Jointly Distributed Pairs of Random Variables

Let's start by considering how we can model pairs of random variables that may depend on each other. This will help us build models that we can use to interpret the techniques introduced later in this chapter.

A pair of random variables can be created by making them depend on a single outcome from a shared sample space. Consider the following simple example:

Example 13.1: Pair of Random Variables from Two Coin Flips

A fair coin is flipped two times, and the top faces are observed. Two random variables are created based on the observed values of the coin faces:

$$X = \begin{cases} 1, & \text{at least one heads observed} \\ 0, & \text{no heads observed,} \end{cases}$$

$$Y = \begin{cases} 1, & \text{at least one tails observed} \\ 0, & \text{no tails observed.} \end{cases}$$

Formally, we can define the sample space

$$S = \{\text{HH}, \text{HT}, \text{TH}, \text{TT}\}.$$

We can define the random variables X and Y as functions of which outcome is chosen from this sample space. Again, note that the outcome s takes the same value for each of the random variables.

$$X(s) = \begin{cases} 1, & s \in \{\text{HH}, \text{HT}, \text{TH}\} \\ 0, & s = \text{TT}, \end{cases}$$

DOI: 10.1201/9781003324997-13

$$Y(s) = \begin{cases} 1, & s \in \{\text{HT}, \text{TH}, \text{TT}\} \\ 0, & s = \text{HH}. \end{cases}$$

A visual representation of the sample space and mapping to the random variables is shown in Fig. 13.1.

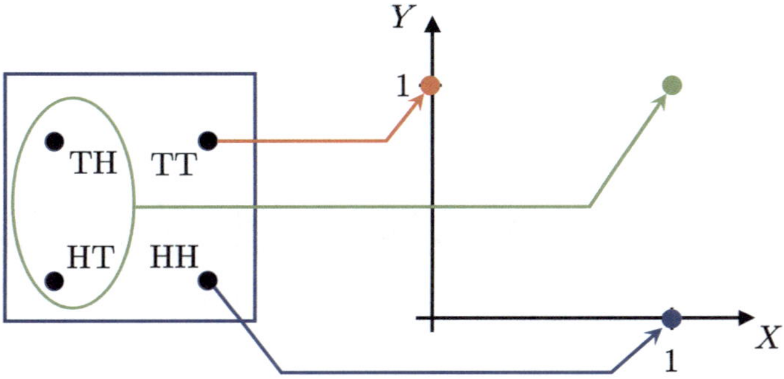

FIGURE 13.1
Visual depiction of the mapping from the sample space {HH, HT, TH, TT} to the values of the random variables X and Y.

One reason to define random variables in this way is that the random variables can be dependent, since the values of the random variables come from a common source of randomness. In Section 6.6, we defined independent and dependent events. However, we have not defined independence or dependence for two random variables. Nonetheless, take a moment to reason about whether these random variables should be independent before continuing to read.

One argument for why these random variables should not be considered independent is as follows. If an observer doesn't know anything about Y, then the PMF of X is

$$P_X(x) = \begin{cases} 3/4, & x = 1 \\ 1/4 & x = 0. \end{cases}$$

This is because there are four total outcomes from flipping the coin twice. Three have a head, and one has no heads, leading to the corresponding probabilities.

If we do not know anything about Y, then we can see that $P(X = 1) = 3/4$. Suppose instead that we know that $Y = 0$. If $Y = 0$, then the outcome must have been $s = \text{HH}$. From the definition of $X(s)$, we see that if we know $Y = 0$, then $X(\text{HH}) = 1$. Thus, we would say that the conditional probability that $X = 1$ given that $Y = 0$ is 1. Mathematically, we would write $P(X = 1 \mid Y = 0) = 1$. Since knowing the value of Y changes the probabilities of X, we would say that X and Y are *not* independent.

Referring again to Fig. 13.1, we can see that each outcome in S does not separately determine X and Y but rather determines both values simultaneously. We can group the two random variables so that each $s \in S$ determines the tuple (X, Y) or the vector $[X, Y]^T$. The tuple notation is usually only used for the case of pairs of random variables, whereas the vector form can be extended to include any number of random variables.

We say that (X, Y) are jointly distributed random variables and are defined as a function from the sample space to $\mathbb{R}^2$, the real plane. From inspection of Fig. 13.1, the function is

$$(X,Y) = \begin{cases} (1,0), & s = \text{HH} \\ (0,1), & s = \text{TT} \\ (1,1), & s \in \{\text{HT}, \text{TH}\}. \end{cases}$$

For jointly distributed discrete random variables, we can define a *joint probability mass function*:

Definition

joint probability mass function (pair of random variables)

For a pair of random variables (X, Y), the *joint probability mass function* (PMF) gives the probability that (X, Y) takes on each value $(x, y) \in \mathbb{R}^2$,

$$\begin{aligned} P_{X,Y}(x,y) &= P\left[\{s \,|X(s) = x, Y(s) = y\}\right] \\ &= P\left[X = x, Y = y\right]. \end{aligned}$$

In practice, we often write $P_{XY}(x, y)$ (where the comma between X and Y in the subscript is dropped).

Example 13.1 (continued)

For our example, the joint PMF is nonzero at three points:

$$\begin{aligned} P_{XY}\left[(1,0)\right] &= P\big[\{s \,|(X,Y) = (1,0)\,\}\big] = P\left(\text{HH}\right) = 1/4 \\ P_{XY}\left[(0,1)\right] &= P\big[\{s \,|(X,Y) = (0,1)\,\}\big] = P\left(\text{HH}\right) = 1/4 \\ P_{XY}\left[(1,1)\right] &= P\big[\{s \,|(X,Y) = (1,1)\,\}\big] = P\big(\{\text{HT}, \text{TH}\}\big) = 1/2. \end{aligned}$$

Then the overall PMF is

$$P_{XY}(x,y) = \begin{cases} 1/4, & x = 1,\ y = 0 \\ 1/4, & x = 0,\ y = 1 \\ 1/2, & x = 1,\ y = 1 \\ 0, & \text{otherwise.} \end{cases}$$

Since this PMF assigns a real value to pairs (x, y), a three-dimensional plot is required to visualize it. A stem plot[1] of the nonzero values of this PMF is shown in Fig. 13.2.

[1] I am going to include three-dimension plots in this section to illustrate functions for jointly distributed random variables, but I am omitting the code from the book. The code will be included on the website for the book.

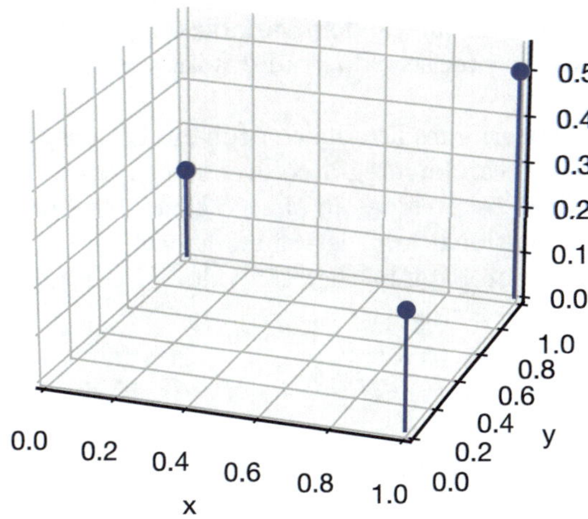

FIGURE 13.2
PMF for pair of random variables created by flipping a fair coin twice.

We can also define a joint cumulative distribution function for pairs of random variables:

> Definition
>
> **joint cumulative distribution function (pair of random variables)**
>
> For a pair of random variables (X, Y), the joint *cumulative distribution function* (CDF) is
>
> $$\begin{aligned} F_{XY}(x, y) &= P\left[\{s \,|X(s) \leq x, Y(s) \leq y\}\right] \\ &= P\left[X \leq x, Y \leq y\right]. \end{aligned}$$

The joint CDF applies to both discrete and continuous random variables. In this book, we will only use the joint CDF to define a joint probability density function for continuous random variables:

13.1.1 Jointly Distributed Continuous Random Variables

Jointly distributed continuous random variables are usually specified in terms of a joint probability density function:

> Definition
>
> **joint probability density function (pair of random variables)**
>
> For a pair of random variables (X, Y), the joint *probability density function* (pdf) is
>
> $$f_{XY}(x, y) = \frac{\partial^2}{\partial x \, \partial y} F_{XY}(x, y).$$

If you are not familiar with the mathematical notation, it refers to the partial derivatives of the CDF. These can be thought of as derivatives taken in two separate directions, each of which corresponds to one of the random variables. For our purposes, it is not necessary to understand the definition precisely. Instead, I want you to have some intuition about the meaning of the joint density.

Basically, $f_{XY}(x, y)$ determines how much probability density is at each point (x, y). As in Section 8.5, the joint pdf can be integrated over some region to determine the probability that the random variables take values in that region. The differences now are that the regions must be two-dimensional and determine a probability for the joint values of the random variables. So, the probability that the value of the random variables lies in some region in the real plane, $R \in \mathbb{R}^2$, is

$$P\left[(X, Y) \in R\right] = \iint_R f_{XY}(x, y)\ dx\ dy.$$

The properties of the joint pdf are extensions of those for the pdf for a single random variable:

1. The pdf is non-negative: $f_{XY}(x, y) \geq 0$ for all x and y.

2. The joint pdf integrates to 1 (since the total probability across the real plane must be 1):

$$\int_{-\infty}^{\infty}\int_{-\infty}^{\infty} f_{XY}(x, y)\ dx\ dy = 1.$$

3. X and Y are statistically independent if and only if the joint pdf factors as a product of the pdfs of the individual random variables:

$$f_{XY}(x, y) = f_X(x) f_Y(y).$$

The pdfs $f_X(x)$ and $f_Y(y)$ are called the *marginal pdfs* of X and Y:

Definition

marginal probability density function (pair of random variables)

For a pair of random variables (X, Y) with joint pdf $f_{XY}(x, y)$, the *marginal probability density functions (marginal pdfs)* of X and Y are the individual pdfs $f_X(x)$ and $f_Y(y)$. They can be calculated from the joint pdf as

$$f_X(x) = \int_{-\infty}^{\infty} f_{XY}(x, y) dy, \text{ and}$$

$$f_Y(y) = \int_{-\infty}^{\infty} f_{XY}(x, y) dx.$$

We are only going to introduce one type of jointly distributed continuous random variables: jointly Normal random variables.

13.1.2 Jointly Normal Random Variables

Jointly Normal random variables are the most commonly encountered types of jointly distributed continuous random variables. Let's start with the simplest case:

Zero-mean, Unit-variance, Independent Normal Random Variables

Recall that if X is Normal with zero mean and unit variance, its pdf is

$$f_X(x) = \frac{1}{\sqrt{2\pi}} e^{-x^2/2}.$$

Then applying the same form for Y and multiplying the pdfs (since they are independent) gives the joint pdf:

$$f_{XY}(x,y) = \frac{1}{2\pi} \exp\left[-\frac{x^2+y^2}{2}\right].$$

A three-dimensional visualization of this curve shows a bell shape, as shown in Fig. 13.3.

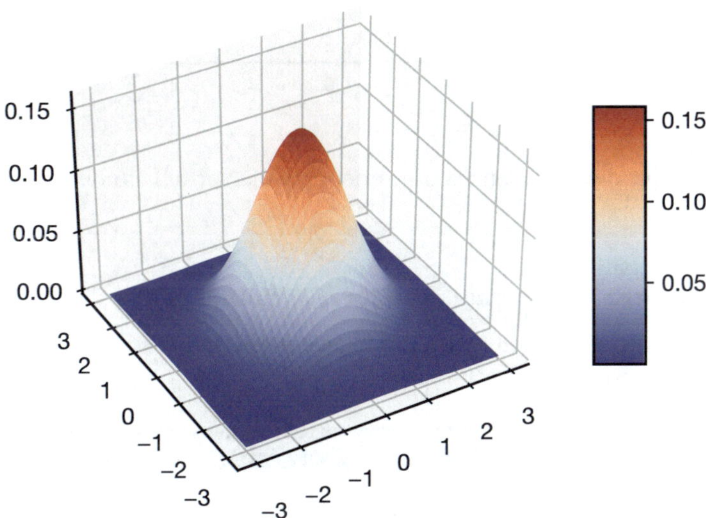

FIGURE 13.3
Surface plot of pdf of jointly Normal random variables.

Another useful view of this pdf is if we trace around the pdf and show the (x, y) pairs that all achieve the same density. This is called a *contour of equal probability density*:

> Definition
>
> **contour of equal probability density (pair of random variables)**
>
> Given some value a in the range of $f_{XY}(x,y)$, the corresponding *contour of equal probability density* $\mathcal{C}_a$ is defined as the set of (x, y) values that achieve that density. I.e., if the joint density is $f_{XY}(x,y)$, then
>
> $$\mathcal{C}_a = \{(x,y) \mid f(x,y) = a\}.$$

In a 3D plot of the joint density, a given value of the joint density corresponds to a particular height above the (x, y) plane. Thus, $\mathcal{C}_a$ can be found by taking a slice through the joint pdf at $z = a$ and drawing the resulting slice of the pdf. For independent Normal random variables with zero mean and unit variance, these contours of equal probability density are circles of different radii, as shown in Fig. 13.4.

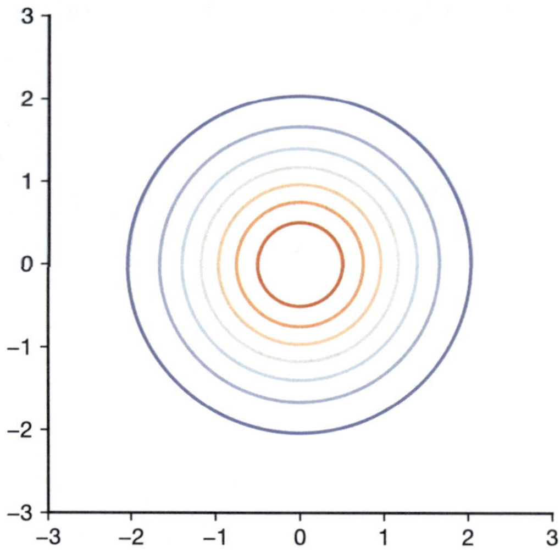

FIGURE 13.4
Contours of equal probability density for independent Normal random variables with equal variance.

General pdf for Jointly Normal Random Variables

If we have more than two random variables, then we can simplify our notation and our work by collecting them into a *random vector*:

> Definition
>
> **random vector**
>
> A random vector $\mathbf{X} = [X_0, X_1, \ldots, X_{n-1}]^T$ is an ordered collection of random variables. Formally, a random vector is defined on a probability space $(S, \mathcal{F}, P)$ and is a function $\mathbf{X}(s)$ that maps from the sample space to $\mathbf{R}^n$.

Then a general pdf for jointly Normal random variables is

$$f_{\mathbf{X}}(\mathbf{x}) = \frac{1}{\sqrt{(2\pi)^n \det \mathbf{K}}} \exp\left(-\frac{1}{2}(\mathbf{x}-\boldsymbol{\mu})^T \mathbf{K}^{-1} (\mathbf{x}-\boldsymbol{\mu})\right),$$

where $\mathbf{X}$ is a vector of jointly Normal random values and $\mathbf{x}$ is a vector of values at which to evaluate the pdf. Here, $\boldsymbol{\mu}$ is the mean vector, and $\mathbf{K}$ is the covariance matrix:

> Definition
>
> **mean vector**
>
> For a random vector $\mathbf{X} = [X_0, X_1, \ldots, X_{n-1}]^T$, the *mean vector* $\boldsymbol{\mu}$ is the n-dimensional vector whose ith entry is the mean of X_i; i.e. $\boldsymbol{\mu}_i = E[X_i]$.

Recall that in Section 12.1, we found that the mean for a numerical multi-dimensional data set was a vector. Also, we found that such a data set has a covariance matrix. Here, $\mathbf{K}$ is the covariance matrix for the random variables:

Definition

covariance matrix

For a random vector $\mathbf{X} = [X_0, X_1, \ldots, X_{n-1}]^T$, the *covariance matrix* $\mathbf{K}$ is the $n \times n$ matrix whose i, jth entry is

$$\begin{aligned}\mathbf{K}_{ij} &= \mathrm{Cov}(X_i, X_j) \\ &= E\big[(X_i - \mu_i)(X_j - \mu_j)\big].\end{aligned}$$

The covariance matrix can be calculated as

$$\mathbf{K} = E\left[(\mathbf{X} - \boldsymbol{\mu})(\mathbf{X} - \boldsymbol{\mu})^T\right].$$

For more general distributions, it is often too difficult to work with or even characterize the distributions of the data. Instead, we will often characterize these distributions using their mean vectors and the covariance matrices.

For the special case of a pair of jointly distributed random variables X and Y, the covariance matrix is

$$\mathbf{K} = \begin{bmatrix} \sigma_X^2 & \rho\sigma_X\sigma_Y \\ \rho\sigma_X\sigma_Y & \sigma_Y^2 \end{bmatrix},$$

where σ_X^2 and σ_Y^2 are the variances of X and Y, respectively. The parameter ρ is the correlation coefficient, which satisfies $-1 \leq \rho \leq 1$. If $\rho = 0$, then $\mathrm{Cov}(X, Y) = 0$, and the variables are *uncorrelated.*

Similar to how covariance matrices tabulate all the pairwise covariances between random variables, we can also define a *correlation matrix*:

Definition

correlation matrix

For a random vector $\mathbf{X} = [X_0, X_1, \ldots, X_{n-1}]$, the *correlation matrix* $\mathbf{R}$ is the $n \times n$ matrix whose i, jth entry is

$$\begin{aligned}\mathbf{R}_{ij} &= \rho_{i,j} \\ &= \frac{\mathrm{Cov}\left(X_i, X_j\right)}{\sigma_i \sigma_j}.\end{aligned}$$

IMPORTANT PROPERTY OF JOINTLY NORMAL RANDOM VARIABLES

Uncorrelated jointly Normal random variables are independent. In general random variables can be uncorrelated without being independent; it is a special property of jointly Normal random variables that uncorrelated jointly Normal random variables must be independent.

For jointly Normal random variables that are uncorrelated and have equal variances, the contours of equal probability density will be circles. The radii of those circles will vary with the variance. If these random variables also have equal means, their distributions will be identical. We say that they are *independent and identically distributed*:

Definition

independent and identically distributed (iid)

Random variables are *independent and identically distributed (iid)* if they have the same distribution (including any parameters of the distribution) and are independent.

Uncorrelated Jointly Normal Random Variables with Unequal Variances

For uncorrelated (independent) jointly Normal random variables with unequal variances, the contours of equal probability density will be ellipses whose major axes align with the x or y axes. Fig. 13.5 illustrates the joint pdf for zero-mean, uncorrelated jointly Normal random variables X and Y with $\sigma_X^2 = 3$ and $\sigma_Y^2 = 0.5$. The contours of equal probability density are ovals that are wider in the x-direction than in the y-direction (because there is more variance in the x-direction).

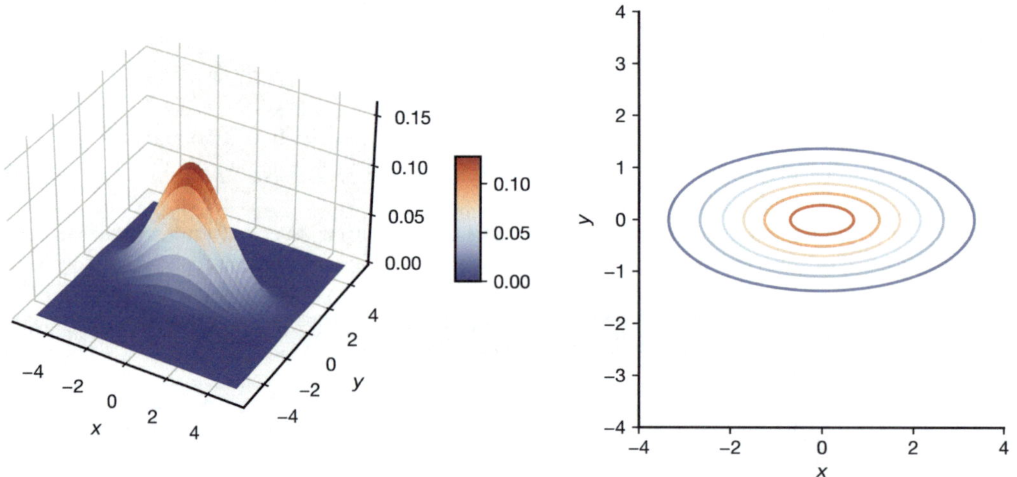

FIGURE 13.5
Surface plot and contours of equal probability density for uncorrelated Normal random variables with unequal variance.

Correlated Normal Random Variables with Unequal Variance

Now let's look at the joint pdf for correlated Normal random variables. Let X and Y be zero-mean Normal random variables with the same variances as in the previous example ($\sigma_X^2 = 3$ and $\sigma_Y^2 = 0.5$), but let the correlation coefficient be 0.82. Then the off-diagonal entries in the covariance matrix are equal to

$$0.82\sqrt{3}\sqrt{0.5} \approx 1.0.$$

The covariance matrix is

$$\mathbf{K} = \begin{bmatrix} 3 & 1 \\ 1 & 0.5 \end{bmatrix}.$$

Fig. 13.6 shows a surface plot of the joint pdf along with the counters of equal probability density for these random variables. For these correlated random variables, the major and

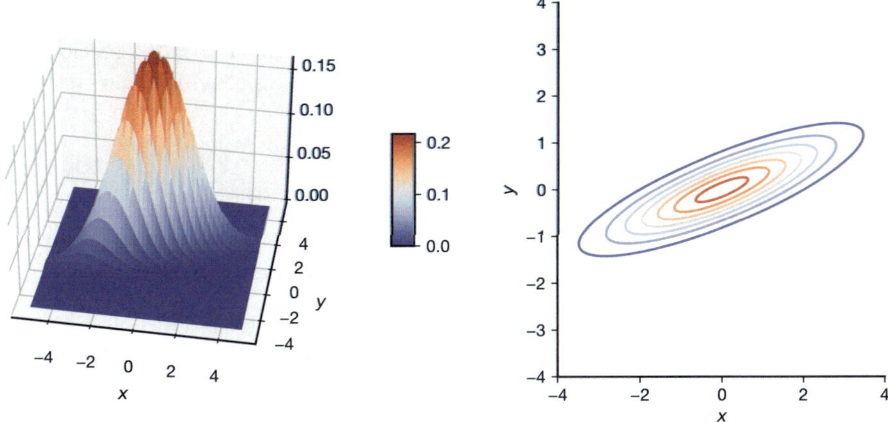

FIGURE 13.6
Joint pdf and contours of equal probability density for correlated Normal random variables with unequal variances, $\rho \approx 0.82$.

minor axes no longer line up with the x and y-axes. As in Section 12.1, a positive correlation means that the random variables generally have values in the same direction (i.e., both a positive offset or both a negative offset from the mean). A negative correlation means that the random variables generally have values in the opposite directions. Fig. 13.7 shows the joint density when we change the correlation coefficient to -0.3.

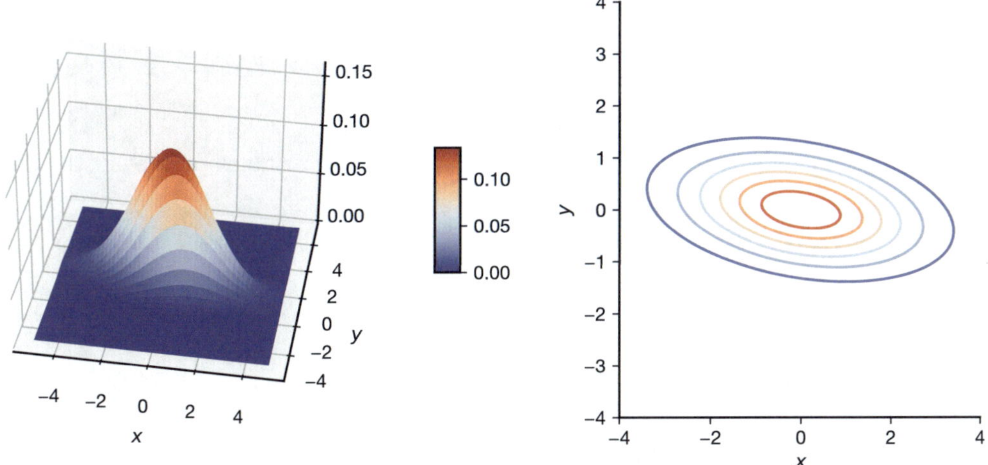

FIGURE 13.7
Joint pdf and contours of equal probability density for correlated Normal random variables with unequal variances, $rho \approx -0.3$.

I am omitting examples where the mean is nonzero because the only effect is to change the location of the center of the probability density. Thus, if we were to change the means, the plots would show the exact same shape of the distribution, but they would be shifted to be centered on the means.

Terminology review and self-assessment questions

Interactive flashcards to review the terminology introduced in this section and self-assessment questions are available at fdsp.net/13-1, which can also be accessed using this QR code:

13.2 Standardization and Linear Transformations

In this section, we will use one of the oldest and most famous data sets for classification problems, the Iris data set. This is a relatively simple data set, and we are going to simplify it more for the purposes of this section. In Section 13.4, we will formalize and generalize the techniques developed in this section and apply them to a larger data set associated with more contemporary issues.

According to the `DESCR` field of the `scikit-learn` iris dataset object:

```
This is perhaps the best known database to be found in the
pattern recognition literature.  Fisher's paper is a classic
in the field and is referenced frequently to this day....  The
data set contains 3 classes of 50 instances each, where each
class refers to a type of iris plant.
```

(To see the full description, run `print(iris{['DESCR'{]}}` after loading the Iris data set as shown below.)

The data set is from Robert Fisher's paper "The use of multiple measurements in taxonomic problems," Annual Eugenics, 7, Part II, 179-188 (1936). We can load it as follows:

```
from sklearn import datasets

iris=datasets.load_iris()
```

The data set includes a `DESCR` property that explains the variables present in the data set:

```
print(iris['DESCR'][:500])
```

```
.. _iris_dataset:

Iris plants dataset
--------------------

**Data Set Characteristics:**

    :Number of Instances: 150 (50 in each of three classes)
    :Number of Attributes: 4 numeric, predictive attributes and the class
    :Attribute Information:
        - sepal length in cm
        - sepal width in cm
        - petal length in cm
        - petal width in cm
        - class:
```

(continues on next page)

(continued from previous page)

```
        - Iris-Setosa
        - Iris-Versicolour
        - Iris-Virginica
```

As the description indicates, each data point contains four variables, which are labeled in `iris['feature_names']`. The data itself is contained in `iris['data']`:

```
print(iris['feature_names'])
print(iris['data'][:5])
```

```
['sepal length (cm)', 'sepal width (cm)',
 'petal length (cm)', 'petal width (cm)']
[[5.1 3.5 1.4 0.2]
 [4.9 3.  1.4 0.2]
 [4.7 3.2 1.3 0.2]
 [4.6 3.1 1.5 0.2]
 [5.  3.6 1.4 0.2]]
```

(An Iris flower consists of similarly colored sepals and petals, but the sepals are longer and have a bulb shape that is wider than the petals, as is indicated by the data.)

Each data point is also associated with its correct classification or classification *target*. The `iris['target']` member contains the numerical classification target, and `iris['target_names']` contains the description of each class, which in this case are three different types of Irises.

```
iris['target'], iris['target_names']
```

```
(array([0, 0, 0, 0, 0, 0, 0, 0, 0, 0, 0, 0, 0, 0, 0, 0, 0, 0, 0, 0, 0, 0,
        0, 0, 0, 0, 0, 0, 0, 0, 0, 0, 0, 0, 0, 0, 0, 0, 0, 0, 0, 0, 0, 0,
        0, 0, 0, 0, 0, 0, 1, 1, 1, 1, 1, 1, 1, 1, 1, 1, 1, 1, 1, 1, 1, 1,
        1, 1, 1, 1, 1, 1, 1, 1, 1, 1, 1, 1, 1, 1, 1, 1, 1, 1, 1, 1, 1, 1,
        1, 1, 1, 1, 1, 1, 1, 1, 1, 1, 1, 1, 2, 2, 2, 2, 2, 2, 2, 2, 2, 2,
        2, 2, 2, 2, 2, 2, 2, 2, 2, 2, 2, 2, 2, 2, 2, 2, 2, 2, 2, 2, 2, 2,
        2, 2, 2, 2, 2, 2, 2, 2, 2, 2, 2, 2, 2, 2, 2, 2, 2, 2]),
 array(['setosa', 'versicolor', 'virginica'], dtype='<U10'))
```

The first 50 elements are of class 0, the next 50 are of class 1, and the final 50 are of class 2. We will use this fact when plotting the data.

The usual goal when working with this data set is to determine a classification function that maps from the four-dimensional data to the three classes. Here, we simplify the problem to just the first two classes and the first two features:

```
class01=np.where(iris['target']<2)[0]

iris2=iris.data[class01][:,:2]
target2=iris.target[class01]
```

The reduction to two features (sepal length and sepal width) allows us to plot the data as points using a scatter plot, as shown in Fig. 13.8.

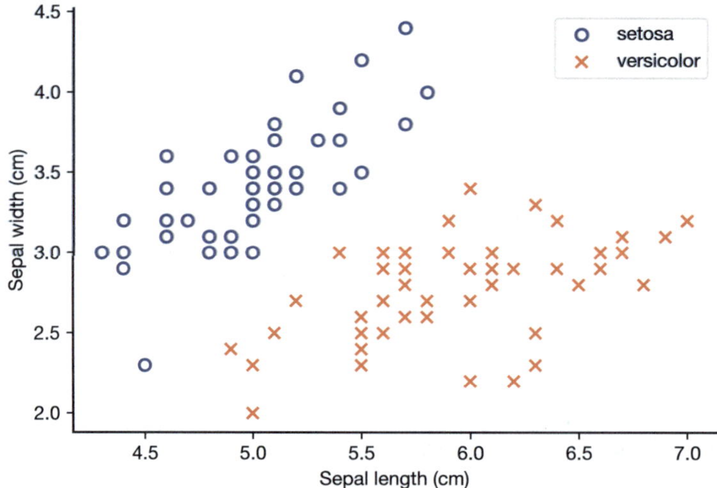

FIGURE 13.8
Plot of first two classes and first two features of Iris data set.

13.2.1 Standardizing Data

Different variables or features of a data set often differ greatly in their means and variances, and this can have significant effects on the performance of statistical and learning algorithms. Such differences can often be artificial. For instance, features representing physical measurements may be expressed using different units, and the choice of units changes the means, variances, and covariances of the data. Even when features have the same units, different features are often naturally distributed over different ranges, and increased means or variances in one feature may result in that feature being given artificial significance when used in statistical or learning algorithms. Thus, data may often be *standardized* to remove such differences. Let's begin by investigating differences in the moments of the two features considered for the Iris data set before defining standardization formally.

Standardizing Means

We begin by considering the means of the two features. The rows of `iris2` correspond to the different features, and the columns of `iris` correspond to different data points. Since the sepal length is feature 0 and the sepal width is feature 1, we can get their means separately, as shown in the following:

```
print(f'mean sepal length = {np.mean(iris2[0]):.2f}')
print(f'mean sepal width = {np.mean(iris2[1]):.2f}')
```

```
mean sepal length = 5.47
mean sepal width = 3.10
```

We can also get both means simultaneously using `np.mean()`, but we have to be careful to tell NumPy to compute the mean over the appropriate axis. With no axis specified, the mean is computed for the pooled set of features:

```
np.mean(iris2)
```

```
4.285
```

To get the desired result, we can tell `np.mean()` to average over the data points, which are in the columns (axis 1):

```
mu = np.mean(iris2, axis=1)
mu
```

```
array([5.471, 3.099])
```

When we standardize for mean, we apply a transformation to make the data have zero mean. Before working with data, consider a linear transformation of a random vector $\mathbf{X}$, which has mean $\boldsymbol{\mu}_{\mathbf{X}}$. Suppose we subtract $\boldsymbol{\mu}_{\mathbf{X}}$ from $\mathbf{X}$ to create a new random vector $\tilde{\mathbf{X}}$. Then the mean of $\tilde{\mathbf{X}}$ is

$$\begin{aligned}
E\left[\tilde{\mathbf{X}}\right] &= E\left[\mathbf{X} - \boldsymbol{\mu}_{\mathbf{X}}\right] \\
&= E\left[\mathbf{X}\right] - E\left[\boldsymbol{\mu}_{\mathbf{X}}\right] && \text{(by linearity)} \\
&= E\left[\mathbf{X}\right] - \boldsymbol{\mu}_{\mathbf{X}} && \text{(expected value of a constant vector is that constant vector)} \\
&= \boldsymbol{\mu}_{\mathbf{X}} - \boldsymbol{\mu}_{\mathbf{X}} && \text{(by the definition of } \boldsymbol{\mu}_{\mathbf{X}}\text{)} \\
&= \mathbf{0}.
\end{aligned}$$

(Note that $\mathbf{0}$ denotes an all-zeros vector of the same dimensionality as $\mathbf{X}$.)

We can see that for random vectors, subtracting the mean vector results in a random vector that has zero mean for each of its constituent random variables. The same effect holds for real data. Let's test that out using the Iris data set. We want to subtract the mean vector `mu` from every data point. This can be accomplished in NumPy via broadcasting, in which Numpy performs an operation on arrays that differ in dimensions by repeating the smaller array to make it match the dimensions of the larger array.

To use broadcasting to subtract the mean vector from all of the data points, we first have to turn the mean vector into a 2×1 column vector because each data point is a column vector. We can make a column vector from `mu` by adding an axis (which will turn `mu` into a 1×2 NumPy array) and then transposing the result into a 2×1 column vector. Here is the result:

```
mu_col = mu[np.newaxis].T
mu_col
```

```
array([[5.471],
       [3.099]])
```

Now if we subtract `mu_col` from our `iris2` data set, the `mu_col` vector will be broadcast across all of the columns of `iris`. The result of this broadcasting is that `mu_col` is subtracted from every row of `iris2`:

```
mean_standardized = iris2 - mu_col
```

The resulting data set is plotted in Fig. 13.9. Since we are standardizing across the combined set of data, I show these using a single marker style and color: We can see that the data points are now roughly centered around $(0, 0)$. Let's check the effect on the sample mean of subtracting off the mean vector:

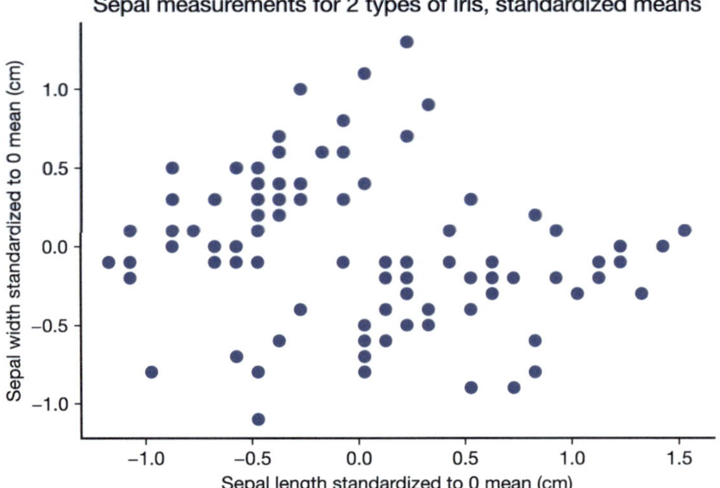

FIGURE 13.9
Iris data after subtracting off mean of each feature.

```
np.round(np.mean(mean_standardized, axis=1), 10)
```

```
array([-0., -0.])
```

After subtracting the mean vector, the new data set `standardized1` has zero mean. However, note that the data in the figure has more spread in the x-direction (corresponding to sepal length) than in the y-direction (corresponding to sepal width). In the following section, we consider how to standardize the variances of a numerical data set.

Standardizing Variances

Let's begin by looking at the variances of `iris2`. The easiest way to do this is to look at the diagonal elements of the covariance matrix:

```
K=np.cov(iris2)
K
```

```
array([[ 0.41177677, -0.06326162],
       [-0.06326162,  0.22919091]])
```

We can see that the variance of the sepal length feature is approximately twice as large as that of the sepal width feature. This is not too surprising because the sepal lengths are generally larger than the sepal widths, and this often translates into larger variances.

Before trying to standardize the variances, we should first consider the effect of subtracting the mean vector on the variances. As before, I will show the result for the random vectors; the result for covariances of the data is similar. Let's consider the covariance matrix for $\tilde{\mathbf{X}} = \mathbf{X} - \boldsymbol{\mu}_{\boldsymbol{X}}$. Let $\tilde{\boldsymbol{\mu}}_{\boldsymbol{X}}$ denote the mean of $\tilde{\mathbf{X}}$; from our previous work, we know that

$\tilde{\boldsymbol{\mu}}_{\mathbf{X}} = \mathbf{0}$. Then

$$
\begin{aligned}
\operatorname{Cov}\left(\tilde{\mathbf{X}}\right) &= E\left[\left(\tilde{\mathbf{X}} - \tilde{\boldsymbol{\mu}}_{\mathbf{X}}\right)\left(\tilde{\mathbf{X}} - \tilde{\boldsymbol{\mu}}_{\mathbf{X}}\right)^T\right] \\
&= E\left[\tilde{\mathbf{X}}\tilde{\mathbf{X}}^T\right] && \text{(since } \tilde{\boldsymbol{\mu}}_{\mathbf{X}} = \mathbf{0}\text{)} \\
&= E\left[(\mathbf{X} - \boldsymbol{\mu}_{\mathbf{X}})(\mathbf{X} - \boldsymbol{\mu}_{\mathbf{X}})^T\right] && \text{(since } \tilde{\mathbf{X}} = \mathbf{X} - \boldsymbol{\mu}_{\mathbf{X}}\text{)} \\
&= \operatorname{Cov}(\mathbf{X}).
\end{aligned}
$$

Thus, standardizing the mean does not affect the covariance matrix of a random vector. Let's check this for our standardized data set:

```
np.cov(mean_standardized)
```

```
array([[ 0.41177677, -0.06326162],
       [-0.06326162,  0.22919091]])
```

The covariances are unchanged from the original data set. In fact, since the mean of the random vector is subtracted off in calculating the covariance, the covariance is unchanged if any constant vector is added to or subtracted from all of the data points. Thus, we cannot standardize the variances by adding or subtracting some factors. However, from Section 9.3, we know that we can change the variance of a random variable by multiplying it by a constant. In particular, suppose X is a random variable with variance σ_X^2. If $\tilde{X} = cX$, then $\operatorname{Var}(\tilde{X}) = c^2\sigma_X^2$.

To standardize the variance of $\tilde{X}$, we want to choose c to make $\operatorname{Var}(\tilde{X}) = 1$. The value that achieves this is $c = \sigma_X^{-1}$, the inverse of the standard deviation. Let's test this with the Iris data. The variance of the first feature is `K[0,0]`, so we can get the standardized variance form of feature 0 as:

```
c0 =  1/np.sqrt(K[0,0])
feature0_std = c0 * iris2[0]
```

Then the variance after scaling is:

```
np.cov(feature0_std)
```

```
array(1.)
```

Similarly, the variance of feature 1 is `K[1,1]`, so we can get the standardized variance form of feature 1 as:

```
c1 = 1/np.sqrt(K[1,1])
feature1_std = c1 * iris2[1]
```

The variance after standardization is:

```
np.cov(feature1_std)
```

```
array(1.)
```

As usual, we can perform both sets of standardization simultaneously. We start by taking the covariance matrix `K` and extracting the variances into a vector:

```
variances = np.diag(K)
variances
```

```
array([0.41177677, 0.22919091])
```

Then the values of `c0` and `c1` that we need can be calculated by raising this vector to the power $-1/2$:

```
c = variances ** (-1/2)
c
```

```
array([1.55836462, 2.08882139])
```

We can apply these two factors individually to the corresponding rows of the `iris2` matrix by putting them into the diagonal of a square matrix and then performing matrix multiplication. The function `np.diag()` produces the necessary diagonal matrix when passed a vector:

```
c_mat = np.diag(c)
c_mat
```

```
array([[1.55836462, 0.        ],
       [0.        , 2.08882139]])
```

Then we can multiply each of the data features by the appropriate value of `c0` or `c1` using matrix multiplication:

```
var_standardized = c_mat @ iris2
```

The resulting data is plotted in Fig. 13.10.

Let's check the resulting covariance matrix to make sure we achieved our goal of unit variances:

```
np.cov(var_standardized)
```

```
array([[ 1.        , -0.20592576],
       [-0.20592576,  1.        ]])
```

This approach resulted in both features having variance 1, as desired. Note that a side effect of standardizing the variances is that the covariance terms $k_{0,1}$ and $k_{1,0}$ also changed.

Finally, let's put both types of standardization together by applying our variance standardization to the mean-standardized data:

```
fully_standardized = c_mat @ mean_standardized
```

The combined transformation is called *standardization*:

> Definition
>
> **standardization**
>
> The process by which numerical data is transformed such that each feature has mean zero and variance one.

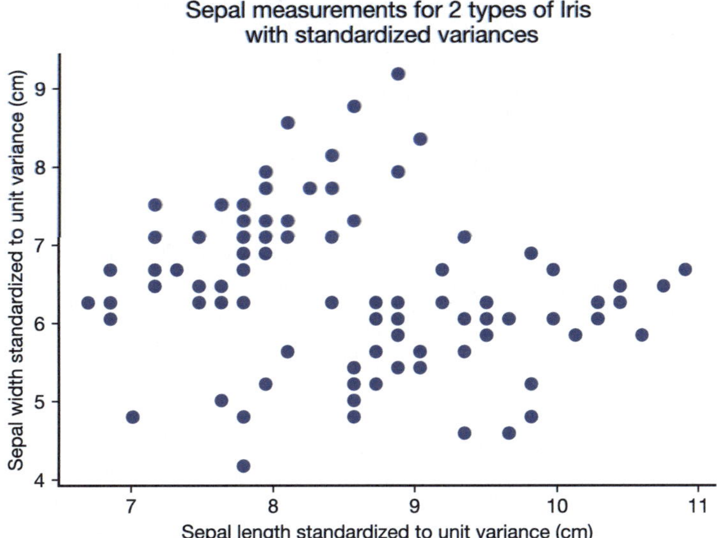

FIGURE 13.10
Iris data after scaling to normalize variances.

Since standardization of means and variances is common in data science workflows, `scikit-learn` provides the `StandardScaler` class to perform this operation. `StandardScaler` is part of the `preprocessing` submodule of `scikit-learn`. The `scikit-learn` module is called `sklearn` in Python, so we can import `StandardScaler` as

```
from sklearn.preprocessing import StandardScaler
```

The next step is to make a StandardScaler object. We can then standardize the means and variances of our data set by passing the data to the `fit_transform()` method of the `scaler` object. This method both calculates the necessary transform and applies it to the data. However, this method expects the columns to correspond to features and the rows to correspond to data points. Thus, we transpose our data at both the input and output to keep it in the form expected by the NumPy methods:

```
scaler = StandardScaler()
standardized = scaler.fit_transform(iris2.T).T
```

Let's check the mean vector and covariance matrix:

```
np.round(np.mean(standardized, axis=1), 10)
```

```
array([-0., -0.])
```

```
np.cov(standardized)
```

```
array([[ 1.01010101, -0.20800581],
       [-0.20800581,  1.01010101]])
```

The variances are not quite 1. This is because StandardScaler uses the biased variance estimator. If we compute the biased covariance matrix, we will get variance 1:

```
np.cov(standardized, ddof=0)
```

```
array([[ 1.        , -0.20592576],
       [-0.20592576,  1.        ]])
```

The difference between these two approaches to normalization should not affect most statistical or learning algorithms applied to the standardized data.

Note that after standardization, the covariance matrix is equal to the correlation matrix:

```
np.corrcoef(standardized)
```

```
array([[ 1.        , -0.20592576],
       [-0.20592576,  1.        ]])
```

The next natural question might be: **Is there a way to standardize the covariances?** The answer is yes, but we will analyze the effects of general linear transforms first.

13.2.2 General Linear Transforms

In this section, we consider the effects of arbitrary linear transforms on the mean and covariance of random variables or data vectors. To see why this might be of interest and to motivate our work in Section 13.3, consider the effect of a rotation on covariance:

Example 13.2: Rotation Effects on Covariance

Let's look at the effect of applying a rotation to the two-dimensional Iris data. The function `makerot()` function takes a parameter θ and returns a set of axes that are rotated θ degrees counterclockwise from the standard x- and y-axes:

```
def makerot(theta):
  '''return basis vectors for rotation of theta degrees

  Parameters
  ----------
  theta : number
    Rotation in degrees

  Returns
  -------
  numpy.ndarray
      Basis vector for first dimension (new x coordinate)

  numpy.ndarray
      Basis vector for second dimension (new y coordinate)
  '''
```

(continues on next page)

(continued from previous page)

```
    theta=np.radians(theta)
    return np.array([np.cos(theta), np.sin(theta)]), \
           np.array([-np.sin(theta), np.cos(theta)])
```

Consider a 40° clockwise rotation. We stack the output vectors into the columns of a matrix:

```
basis40 = makerot(-40)
rot40 = np.vstack((basis40[0],
                   basis40[1])).T
rot40
```

```
array([[ 0.76604444,  0.64278761],
       [-0.64278761,  0.76604444]])
```

Let's see what happens if we project the Iris data onto these basis vectors. This is equivalent to rotating the Iris data by 40° counterclockwise:

```
std_rotated = rot40.T @ standardized
```

The data after standardized and rotation are shown in Fig. 13.11.

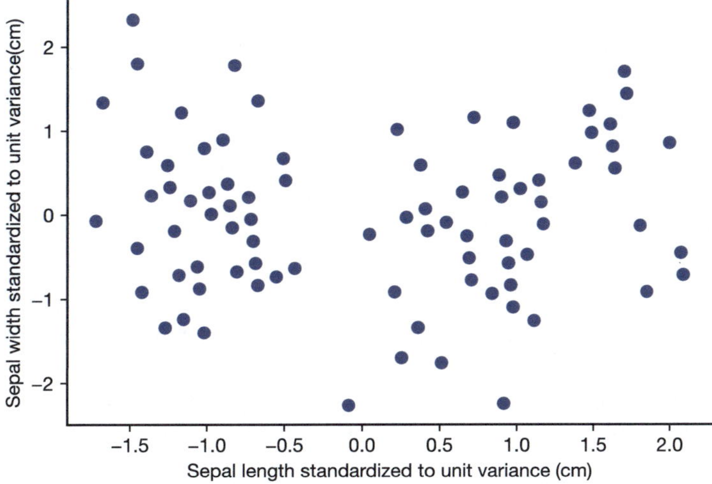

FIGURE 13.11
Standardized Iris data after 40° counterclockwise rotation.

The data looks very similar to that without the standardization, but compare the covariance before and after the rotation for the standardized data. Without the rotation:

```
K_std = np.cov(standardized, ddof=0)
K_std
```

```
array([[ 1.        , -0.20592576],
       [-0.20592576,  1.        ]])
```

After rotation:

```
K_stdrot = np.cov(std_rotated, ddof=0)
K_stdrot
```

```
array([[ 1.20279728, -0.03575863],
       [-0.03575863,  0.79720272]])
```

The covariance is reduced in magnitude from approximately 0.21 to 0.036, indicating that the data is almost uncorrelated.

Exercise

Modify the code below to change the value of X by one or two degrees to either side to try to reduce the correlation further. Find a value that gives a correlation value of approximately zero.

```
# Try to adjust the value of the angle variable to achieve near zero correlation
angle = 40

basisX = makerot(-angle)
rotX = np.vstack((basisX[0],
                  basisX[1])).T

std_rotatedX = rotX.T @ standardized

np.cov(std_rotatedX, bias=True)
```

```
array([[ 1.20279728, -0.03575863],
       [-0.03575863,  0.79720272]])
```

By experiment, we can find a value that achieves near zero covariance for this standardized data set, making the transformed data uncorrelated. However, this approach is not always practical, especially when we consider higher-dimensional data. In the following section, we develop techniques to find the optimum transform mathematically.

$$
\begin{aligned}
E[\mathbf{Y}] &= E[\mathbf{AX} + \mathbf{c}] \\
&= \mathbf{A}E[\mathbf{X}] + \mathbf{c} \\
&= \mathbf{A}\boldsymbol{\mu}_X + \mathbf{c}.
\end{aligned}
$$

Let $\boldsymbol{\mu_Y} = E[\mathbf{Y}]$. Then

$$
\begin{aligned}
\mathbf{Y} - \boldsymbol{\mu_Y} &= (\mathbf{AX} + \mathbf{c}) - (\mathbf{A}\boldsymbol{\mu_X} + \mathbf{c}) \\
&= \mathbf{A}(\mathbf{X} - \boldsymbol{\mu_X}).
\end{aligned}
$$

Using this result, we can write the covariance matrix for $\mathbf{Y}$ as

$$\begin{aligned} E\left[(\mathbf{Y}-\boldsymbol{\mu}_{\mathbf{Y}})(\mathbf{Y}-\boldsymbol{\mu}_{\mathbf{Y}})^T\right] &= E\left\{[\mathbf{A}(\mathbf{X}-\boldsymbol{\mu}_{\mathbf{X}})][\mathbf{A}(\mathbf{X}-\boldsymbol{\mu}_{\mathbf{X}})]^T\right\} \\ &= E\left[\mathbf{A}(\mathbf{X}-\boldsymbol{\mu}_{\mathbf{X}})\ (\mathbf{X}-\boldsymbol{\mu}_{\mathbf{X}})^T\mathbf{A}^T\right] \quad \text{(transpose property of matrix product)} \\ &= \mathbf{A}E\left[(\mathbf{X}-\boldsymbol{\mu}_{\mathbf{X}})\ (\mathbf{X}-\boldsymbol{\mu}_{\mathbf{X}})^T\right]\mathbf{A}^T \quad \text{(linearity of expected value)} \\ &= \mathbf{A}\mathbf{K}_{\mathbf{X}}\mathbf{A}^T. \end{aligned}$$

Terminology review and self-assessment questions

Interactive flashcards to review the terminology introduced in this section and self-assessment questions are available at fdsp.net/13-2, which can also be accessed using this QR code:

13.3 Decorrelating Random Vectors and Multi-Dimensional Data

In Section 13.2, we saw that a rotation matrix could be used to change the correlation among a pair of random variables in a random vector. This is equivalent to projecting the random vectors onto a basis that is rotated by a specified angle from the original axes. There are several reasons that we might want to project random vectors onto another basis:

1. If we can make the correlation between random variables equal to zero, then the dependence among those variables may be reduced. For instance, if the data is jointly Normal, then the random variables become independent.
2. By projecting onto another basis, the random variables' variances also change. For instance, we could project a pair of random variables onto a basis that maximizes the variance in one output random variable and minimizes the variance in the other output random variable.

We will show that both of these goals are achieved through the same approach. The major reason that this is used is for *dimensionality reduction*:

> Definition
>
> **dimensionality reduction**
>
> The process of going from a high-dimensional data set to a lower-dimensional representation of that data set, usually with the goal of preserving as much important information as possible from the original data set.

To perform dimensionality reduction, we usually want to transform our data into pieces of information that are in some sense independent of each other. We will make the data uncorrelated because we do not generally have enough information to achieve true independence. We want to preserve a set of features that have maximum variance – those where there are significant variations among the data. Features that have low variance can be well approximated by their mean.

Some applications of dimensionality reduction include:

1. Visualizing high-dimensional data

For data up to 3 dimensions, we can visualize it using scatter plots or histograms, for instance. But how can we visualize a 4-dimensional data set? Or a 100-dimensional data set? One approach is to perform dimensionality reduction and preserve only the most important 2 or 3 dimensions of the transformed data.

2. Reducing computation for high-dimensional data

Many machine-learning (ML) algorithms are computationally demanding (requiring long processing times or powerful compute engines), and the computational demands may scale exponentially in the number of features. Dimensionality reduction can lower the computational requirements by utilizing only the most important features (i.e., those containing the most information) as input to some ML algorithm.

3. Compression

In addition to computational complexity, high-dimensional data may require a lot of storage space and/or network traffic to move it between different devices for collection, storage, and processing. This issue is particularly important in the Internet of Things, where data may be generated and collected at many small devices that are very constrained on their resources. Dimensionality reduction can be used to achieve lossy compression, in which the data is reduced in size at the expense of introducing some error because part of the data is discarded.

13.3.1 Decorrelating Pairs of Standardized Random Variables

We can perform decorrelation for pairs of standardized random variables using techniques that we have already developed in this book. To give some insight into how we could achieve this, consider the plots of equal probability density in Fig. 13.12 for standardized Normal random vectors with different correlations.

The contours of equal probability density consist of ellipses for which the major axis (the line running through the center of the ellipses that connects to the most distant edges of the ellipse) takes on only two values. If $\rho > 0$, then the major axis is at 45°. If $\rho < 0$, then the major axis is at $-45°$. (Recall that if $\rho = 0$, then the contours of equal probability density are circles, not ellipses, and the data is already uncorrelated.)

In Section 13.1, we showed that for uncorrelated Normal random variables with unequal variance, the contours of equal probability density are ellipses aligned with the x- or y-axes. This suggests that we could make the correlated random variables above become uncorrelated by applying a $\pm 45°$ rotation. (This is the same number you should have found if you completed the exercise at the end of Section 13.2.) Note that if the goal is just decorrelation, then a 45° clockwise or counter-clockwise rotation should achieve the desired result. The only reason to choose one over the other is if we wanted the major axis of the transformed data to lie on a particular axis.

Let's test this observation using mathematics. We start by finding a 45° rotation matrix using the `makerot()` function:

```
basis45 = makerot(45)
R = np.vstack((basis45[0],
               basis45[1])).T
R
```

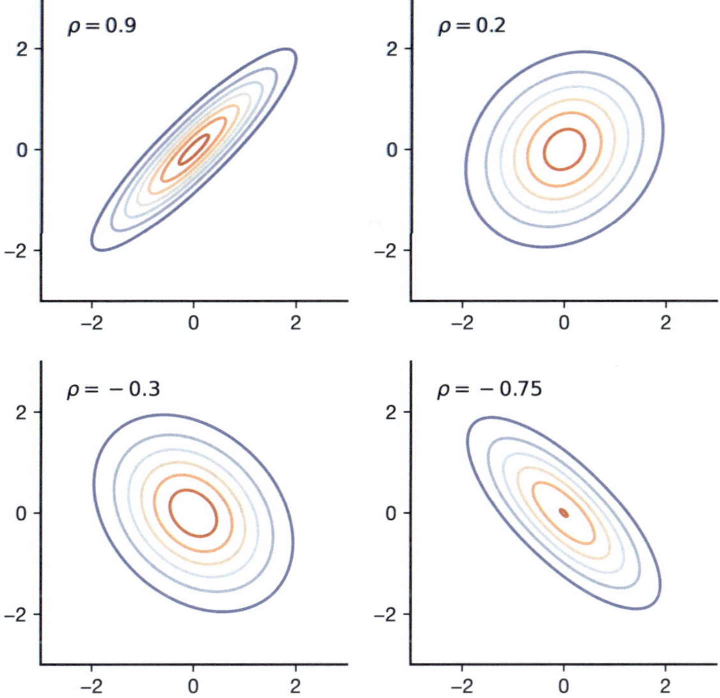

FIGURE 13.12
Contours of equal probability density for Normal random variables with different correlation coefficients, ρ.

```
array([[ 0.70710678, -0.70710678],
       [ 0.70710678,  0.70710678]])
```

Let's test this for a couple of example covariance matrices that correspond to the distributions illustrated in the previous plot. If $\rho = 0.9$, the covariance matrix is:

```
K1 = np.array([
  [1, 0.9],
  [0.9, 1]
])
```

If the random vector is $\mathbf{X}$, then we want to determine the covariance matrix for $\mathbf{Y} = \mathbf{RX}$. From Section 13.2, the covariance matrix is $\mathbf{K}_Y = \mathbf{RXR}^T$. Let's check it numerically:

```
np.round(R @ K1 @ R.T, 10)
```

```
array([[0.1, 0. ],
       [0. , 1.9]])
```

The transformed random variables are uncorrelated because the off-diagonal elements of the covariance matrix are zero. Note, however, that although the original random variables both have variance equal to 1, the transformed random variables have very different variances (1.9 and 0.1).

Let's check one more example, when $\rho = -0.3$. Then the covariance matrix is:

```
K2= np.array([
  [1, -0.3],
  [-0.3, 1]
])
```

The covariance matrix of the transform random variables is:

```
np.round(R @ K2 @ R.T, 10)
```

```
array([[ 1.3,  0. ],
       [-0. ,  0.7]])
```

Again, the random variables are now uncorrelated, and the variances are different than in the previous example. This 45° rotation will always decorrelate a pair of standardized random variables. For a given ρ, the resulting covariance matrix will be

$$\begin{bmatrix} 1-\rho & 0 \\ 0 & 1+\rho \end{bmatrix}.$$

13.3.2 Decorrelation for General Random Vectors

The technique introduced for pairs of standardized random variables only works for that special case. We would like to find a similar technique that can work for random vectors of any dimension with almost any covariance matrix. If we have a random vector $\mathbf{X}$ with covariance matrix $\mathbf{K}_X$, then we want to find a matrix $\mathbf{A}$ such that the covariance matrix of $\mathbf{Y} = \mathbf{AX}$ is a diagonal matrix.

In this book, we will only consider covariance matrices that satisfy an additional requirement: they are *nonsingular*.[2] Then using eigendecomposition, the covariance matrix can be factored as

$$\mathbf{K}_X = \mathbf{U}\mathbf{\Lambda}\mathbf{U}^{-1},$$

where $\mathbf{U}$ is the modal matrix of eigenvectors and $\mathbf{\Lambda}$ is the diagonal matrix of eigenvalues. For symmetric, nonsingular matrices, the modal matrix has a special form: it is an orthogonal matrix. One nice property of an orthogonal matrix $\mathbf{U}$ is that its inverse is just the transpose of the orthogonal matrix; i.e., $\mathbf{U}^{-1} = \mathbf{U}^T$. Thus for covariance matrices, we can write the eigendecomposition as

$$\mathbf{K}_X = \mathbf{U}\mathbf{\Lambda}\mathbf{U}^T.$$

Finally, let's left multiply both sides by $\mathbf{U}^T$ and right-multiply both sides by $\mathbf{U}$ and simplify

$$\begin{aligned}
\mathbf{U}^T\mathbf{K}_X\mathbf{U} &= \mathbf{U}^T\mathbf{U}\mathbf{\Lambda}\mathbf{U}^T\mathbf{U} \\
\mathbf{U}^T\mathbf{K}_X\mathbf{U} &= \mathbf{I}\mathbf{\Lambda}\mathbf{I} \\
\mathbf{U}^T\mathbf{K}_X\mathbf{U} &= \mathbf{\Lambda}.
\end{aligned}$$

[2]If the covariance matrix is singular, then some of the random variables can be written as a linear combination of the other random variables. Then at least one of the random variables can be dropped because it is redundant.

Thus, if we could find a linear transformation of the random vector $\mathbf{X}$ that has covariance matrix $\mathbf{U}^T\mathbf{K}_X\mathbf{U}$, then the resulting covariance matrix would be the diagonal matrix $\mathbf{\Lambda}$, and the random variables would be uncorrelated. From our previous work, if we let $\mathbf{Y} = \mathbf{U}^T\mathbf{X}$, then the covariance of $\mathbf{Y}$ has the desired form

$$\begin{aligned}\mathbf{K}_Y &= \mathbf{U}^T\mathbf{K}_X\left(\mathbf{U}^T\right)^T \\ &= \mathbf{U}^T\mathbf{K}_X\mathbf{U} \\ &= \mathbf{\Lambda}.\end{aligned}$$

This is called the (discrete) *Karhunen-Loève Transform*:

Definition

Karhunen-Loève Transform (KLT)

Suppose $\mathbf{X}$ is a random vector with nonsingular covariance matrix $\mathbf{K}_X$. Find the eigendecomposition $\mathbf{K}_X = \mathbf{U}\mathbf{\Lambda}\mathbf{U}^T$. Then the discrete KLT is $\mathbf{Y} = \mathbf{U}^T\mathbf{X}$, which gives a random vector of uncorrelated random variables with diagonal covariance matrix $\mathbf{\Lambda}$.

For an n-dimensional orthogonal matrix $\mathbf{U}$, its columns form a basis for $\mathbb{R}^n$, the set of real vectors of dimension n. Then the operation $\mathbf{U}^T\mathbf{X}$ is equivalent to projecting the random vector $\mathbf{X}$ onto an alternative basis given by the eigenvectors of $\mathbf{K}_X$.

Our derivation of the KLT can also tell us something about the nature of covariance matrices. In terms of the presentation above, the result will tell us about the nature of $\mathbf{K}_X$. Since $\mathbf{\Lambda}$ is the covariance matrix of $\mathbf{Y} = \mathbf{U}^T\mathbf{X}$, its diagonal entries must be non-negative, since they are the variances of the Y_i. Thus, $\mathbf{\Lambda}$ has non-negative values on its diagonal, and these diagonal elements are the eigenvalues of $\mathbf{K}_X$.

Note:

Recall that the determinant of $\mathbf{K}_X$ is equal to the product of its eigenvalues. The product of non-negative values is also non-negative. Therefore:

Any valid covariance matrix must have non-negative determinant.

If the determinant of a covariance matrix is zero, then some of the variables at the output of the KLT will have zero variance and essentially be deterministic. So, some of the original random variables must be linearly dependent.

Any linearly independent random variables will have a covariance matrix with positive determinant.

13.3.3 Examples

Let's conduct some experiments to see how this works.

Example 13.3: Correlated, Standardized Normal Random Variables

Let $\mathbf{X}_1$ be a pair of standardized Normal random variables with $\rho = -0.3$. The covariance matrix is:

```
K1 = np.array([
  [1, -0.3],
  [-0.3, 1]])
```

Let's find the eigendecomposition for `K1`. For convenience, let's import `numpy.linalg` as `la`. Since all covariance matrices are real symmetric matrices, we will use `la.eigh()` to perform the eigendecomposition:

```
import numpy.linalg as la
lam1, U1 = la.eigh(K1)
lam1, U1
```

```
(array([0.7, 1.3]),
 array([[-0.70710678, -0.70710678],
        [-0.70710678,  0.70710678]]))
```

We see that the eigenvalues are the values $1-\rho$ and $1+\rho$ as found previously, and the matrix `U1` is a 45° rotation. If we use the modal matrix to decorrelate the random variables, the new covariance matrix is:

```
K1y = np.round(U1.T @ K1 @ U1, 10)
K1y
```

```
array([[ 0.7, -0. ],
       [-0. ,  1.3]])
```

The covariance matrix for $\mathbf{Y}$ is equal to `lam1`, as expected. The contours of equal probability density before and after decorrelation are shown in Fig. 13.13.

Now let's test the decorrelating transform on data drawn from the distribution for $\mathbf{X}_1$. We can create this distribution using SciPy's `stats.multivariate_normal()` function. It returns an object that provides a variety of methods that can calculate the pdf, calculate the CDF, draw random values from the distribution, etc. The code below draws 100,000 pairs of random values from the distribution. Note that the `rvs()` method of the `multivariate_normal` object returns a matrix with the rows corresponding to different data points and the columns corresponding to different features. I have transposed the output to put it in our standard form. Then I have applied the decorrelating tranform to the *values* drawn from this distribution. The results are shown in Fig. 13.14.

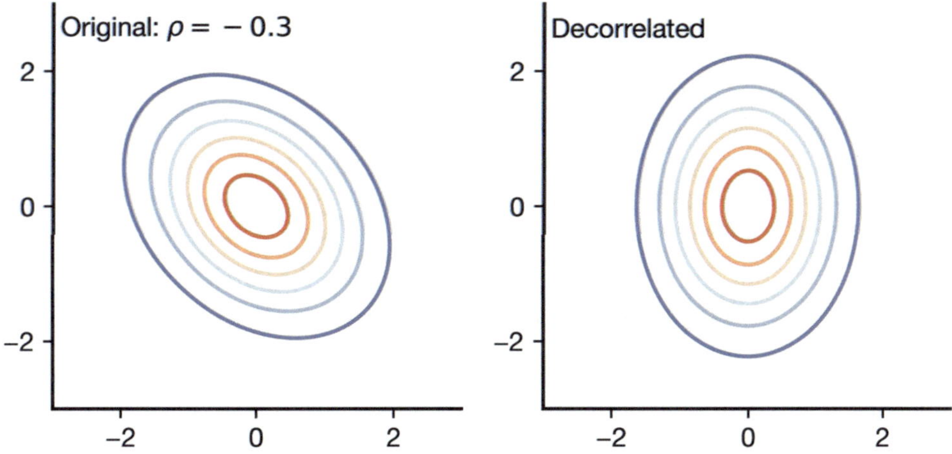

FIGURE 13.13
Contours of equal probability density for jointly Normal random variables before and after correlation.

```
from scipy import stats
import matplotlib.pyplot as plt

N1 = stats.multivariate_normal(mean=[0,0], cov=K1)
vals1 = N1.rvs(size=10_000).T
decor1 = U1.T @ vals1
```

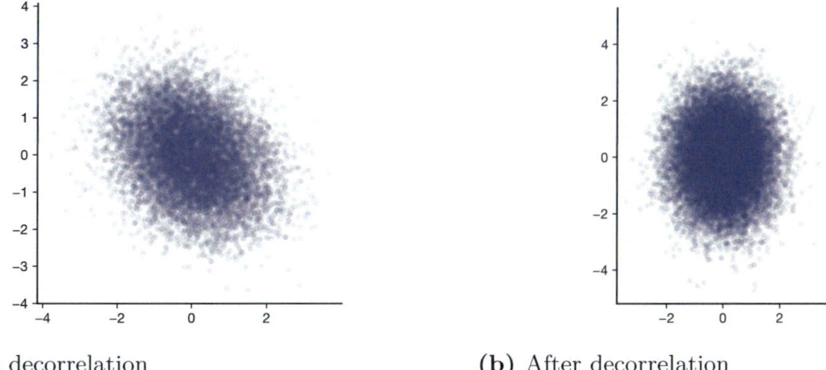

(a) Before decorrelation **(b)** After decorrelation

FIGURE 13.14
Values from correlated Normal random variables, before and after decorrelating transform.

The distribution of the values after decorrelation matches the shape of the contours of equal probability and is aligned with the axes, indicating that the data has been decorrelated. We can also check this by computing the sample covariance matrix for the decorrelated data:

```
np.cov(decor1)
```

```
array([[ 7.01559458e-01, -1.03554985e-03],
       [-1.03554985e-03,  1.30310990e+00]])
```

The values in the sample covariance matrix are close to the theoretical values for the decorrelated random vector.

Example 13.4: Decorrelating Two-Dimensional Iris Data

Let's decorrelate the Iris data using the modal matrix of the sample covariance matrix. This is known as *principal components analysis*:

Definition

principal components analysis (PCA)

Given a multi-dimensional data set, *PCA* is the process by which the data is decorrelated using the modal matrix of the **sample covariance matrix**.

Starting from the variable `standardized` containing the standardized data, the first step is to calculate the sample covariance matrix:

```
K_iris2 = np.cov(standardized, ddof=0)
K_iris2
```

```
array([[ 1.        , -0.20592576],
       [-0.20592576,  1.        ]])
```

The next step in PCA is to find its eigendecomposition. The function `la.eigh()` works on Hermitian matrices (i.e., matrices that have complex-conjugate symmetry around the main diagonal). It returns a vector of eigenvalues and corresponding unit-norm eigenvectors. The eigenvectors are stacked in the columns of a modal matrix.

```
lam_iris2, U_iris2 = la.eigh(K_iris2)
lam_iris2, U_iris2
```

```
(array([0.79407424, 1.20592576]),
 array([[-0.70710678, -0.70710678],
        [-0.70710678,  0.70710678]]))
```

Let's apply PCA and check the sample covariance for the output data:

```
decor_iris2 = U_iris2.T @ standardized
np.round(np.cov(decor_iris2, ddof=0), 10)
```

```
array([[0.79407424, 0.        ],
       [0.        , 1.20592576]])
```

We can see that the data is successfully decorrelated and that feature 1 has more variance than feature 0 in the decorrelated data. In Fig. 13.15, the left plot shows the standardized data, and the right plot shows the output of PCA applied to the standardized data. I have indicated the different classes in the right plot using different markers and colors. After standardization and decorrelation, the two classes can be separated using only PCA feature 1, except for a single data point.

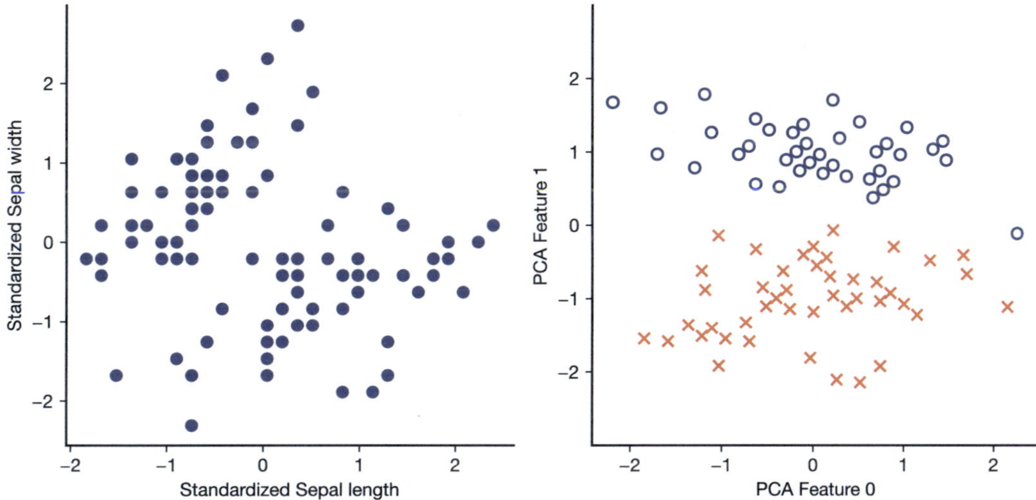

FIGURE 13.15
Standardized two-dimensional Iris data before and after applying PCA.

Terminology review and self-assessment questions

Interactive flashcards to review the terminology introduced in this section and self-assessment questions are available at fdsp.net/13-3, which can also be accessed using this QR code:

13.4 Principal Components Analysis

In this section, I show how to apply PCA for dimensionality reduction, with applications to data visualization and classification. Examples are used to provide insight into how basis functions extract important information from data. I introduce important properties of KLT/PCA, and some new plots are introduced to help determine how many features should be preserved when performing dimensionality reduction.

13.4.1 Properties of KLT/PCA

In Section 13.3, I introduced KLT and PCA as ways to decorrelate random variables and data, respectively. However, KLT and PCA also have a second important effect that is essential to their use in dimensionality reduction. It will be easiest to explain this second effect in terms of applying KLT to a random vector, but the same effect will occur when PCA is applied to data.

Let $\mathbf{X}$ be a n-dimensional random vector with nonsingular covariance matrix $\mathbf{K}_X$. Let $\mathbf{U}$ and $\mathbf{\Lambda}$ be the modal matrix and eigenvalue matrix of $\mathbf{K}_X$. Let the eigenvalues be in sorted order from largest to smallest, so $\lambda_0 \geq \lambda_1 \geq \ldots \geq \lambda_{n-1}$.

Now suppose we want to find a single feature based on $\mathbf{X}$ that contains the most information in some sense. In particular, consider finding a basis vector $\mathbf{v}$ onto which $\mathbf{X}$ should be projected to maximize the variance of the resulting feature. Since the output variance will depend on $\|\mathbf{v}\|$, we constrain $\mathbf{v}$ to have unit energy; i.e., $\|\mathbf{v}\|^2 = 1$.

Then the maximum variance that can be achieved is λ_0 (the maximum eigenvalue of $\mathbf{K}_X$), and a basis vector that achieves that is $\mathbf{u}_0$, the eigenvector corresponding to that maximum eigenvalue. If we then repeat the optimization but constrain the next basis function to be orthogonal to $\mathbf{u}_0$, then the resulting maximum variance is λ_1, and a basis vector that achieves that variance is $\mathbf{u}_1$. This process can be repeated until the final basis vector will be $\mathbf{u}_{n-1}$, and it will achieve the minimum possible variance, which is λ_{n-1}.

It is hard to visualize this process for general distributions and numbers of dimensions, but it is easy to see it for two-dimensional Normal random vectors. Fig. 13.16 shows the contours of equal probability for jointly Normal random variables with

$$\mathbf{K}_X = \begin{bmatrix} 0.5 & 0.5 \\ 0.5 & 1 \end{bmatrix}.$$

These random variables have correlation coefficient $\rho = \sqrt{2}/2 \approx 0.707$. The plot on the left is for the original distribution, and the plot on the right is after decorrelation via the KLT. After performing the KLT, the widest components of the contours of equal probability density are now aligned with the x-axis, meaning that feature 0 has the largest possible variance. Similarly, performing KLT has resulted in the narrowest components of the contours of equal probability density being aligned with the y-axis. The result is that feature 1 at the output of the KLT has the minimum possible variance.

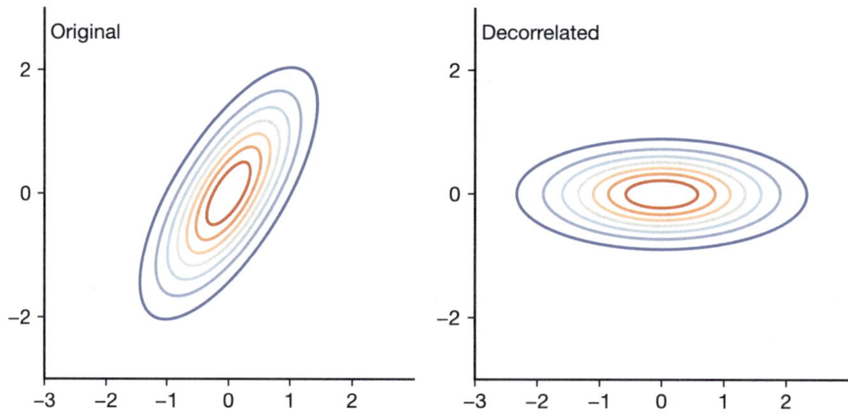

FIGURE 13.16
Visualization of effect of decorrelation on variances of a pair of Normal random variables.

Example 13.5: Feature Extraction for Handwriting Recognition

In this section, we will use PCA to perform dimensionality reduction on a common database of handwritten digits. The reason for choosing this example is that it will allow us to visualize the PCA basis functions in a meaningful way.

We will use the UCI MNIST database, which is commonly used in the study of machine learning and is included in the `scikit-learn` library. UCI stands for the University of California at Irvine, which hosts the UCI Machine Learning Repository: https://archive.ics.uci.edu/. MNIST stands for "Modified National Institute of Standards and Technology". The full MNIST database consists of 70,000 images of handwritten digits from 0 to 9. Each image is normalized to fit in a 28×28 pixel bounding box and anti-aliased. The UCI MNIST database provided by `scikit-learn` consists of 1797 data points, each of which is a processed version of the original MNIST image. The data in UCI MNIST are images of dimension 8×8.

Because I want to show you visually how the bases create features that differentiate among the digits, I will further constrain our exploration to the 537 entries that correspond to digits 0, 1, and 2. We can load the data for these classes as follows:

```
from sklearn import datasets

digits = datasets.load_digits(n_class=3)
```

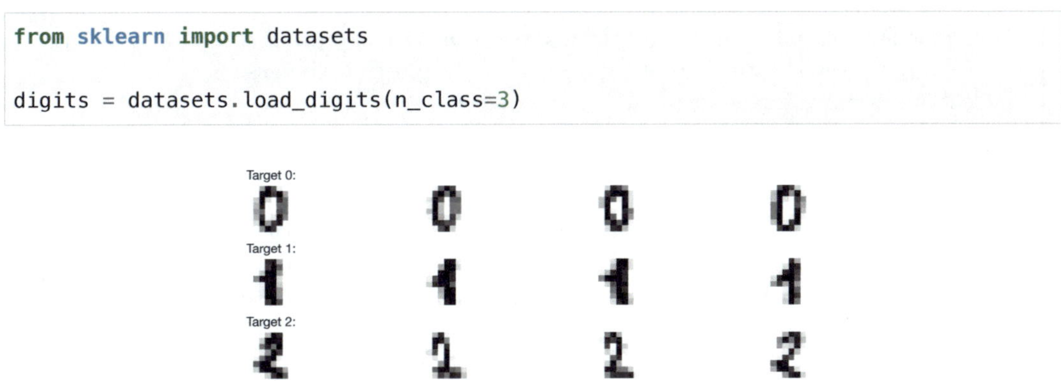

FIGURE 13.17
Example MNIST images for digits 0, 1, and 2.

Examples from the database for each of these digits are shown in Fig. 13.17. These images have overall dimension $8 \times 8 = 64$, which is much lower than the original MNIST data. However, it is still much larger than most of the datasets we have encountered so far, and we may be able to reduce the dimensionality while preserving the most-important information.

To perform PCA, we need the data in the form of vectors instead of two-dimensional images (or corresponding matrix representations). The `digits` object has a `digits.data` attribute that contains "flattened" versions of the images, in which the 8 different rows of data representing an image are concatenated to form a single length-64 vector. As is usual for `scikit-learn` data sets, each row of `digits.data` corresponds to a data point (i.e., one MNIST handwriting sample), so the resulting shape of the `digits.data` array is:

```
digits.data.shape
```

```
(537, 64)
```

We will store the transposed data for use with NumPy's methods:

```
mnist =  digits.data.T
```

At this point, we would standardize many data sets. However, for the UCI MNIST data set, every variable represents the same thing: a value from 0 to 16 that represents a downsampled version of the original MNIST data. Because each variable represents the same type of information with the same range, no normalization is required.

Now we can calculate the sample covariance matrix for the length-64 data. Because it is hard to extract meaning from looking at a 64×64 covariance matrix, we use a *heatmap* to visualize it:

Definition

heatmap

A plot that maps each value in a matrix to a corresponding color. The different matrix values are then shown together by plotting the associated colors as rectangles in an image grid.

A 64×64 heatmap of the covariance matrix for the MNIST data is shown in Fig. 13.18 (code to generate this figure is available online at fdsp.net/13-4). The red colors off the diagonal indicate that the different features are correlated. The blocking effect comes from the fact that the data represents 8×8 images that have been "flattened" into length-64 vectors. Therefore, each line of 8 pixels that starts on a boundary $8k$ represents one line in the original image. The highest correlations are among those lines of the image that are close to the center.

Let's find the eigendecomposition of `Kmnist`. In addition, we will sort the eigenvalue-eigenvector pairs in decreasing order of the eigenvalues. The function `np.argsort()` returns the indexes required to sort the values in increasing order. We will use the index `[::-1]` to reverse that order. Finally, fancy indexing is used to sort the eigenvalues and eigenvectors to create new sorted variables `lam_mnist` and `U_mnist`:

```
lam1, U1 = np.linalg.eigh(Kmnist)
lam1_order = np.argsort(lam1)[::-1]

lam_mnist = lam1[lam1_order]
U_mnist = U1[:, lam1_order]
```

The eigenvalues decrease quickly as we proceed through the sorted array. It is sufficient to look at every 4th eigenvalue to see the general trend:

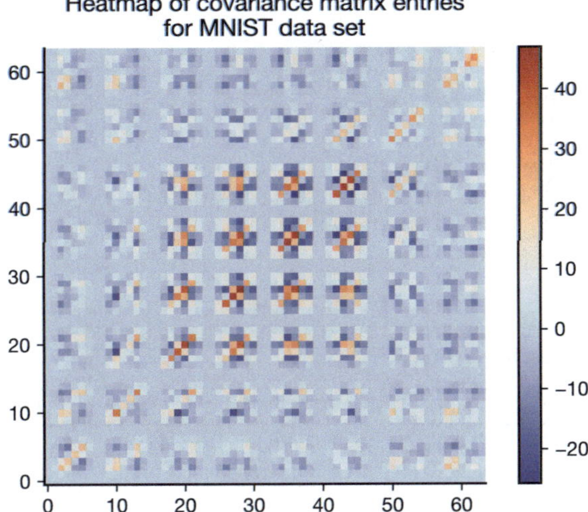

FIGURE 13.18
Heatmap for covariance matrix of MNIST data.

```
print(lam_mnist[::4])
```

```
[ 3.65431793e+02  4.86666527e+01  2.87619368e+01  1.33069906e+01
  8.23338671e+00  5.63225490e+00  3.94370157e+00  3.08597746e+00
  2.22739566e+00  1.69608693e+00  1.23473337e+00  6.71702876e-01
  4.01215007e-01  6.48154990e-03  5.18348060e-15 -2.16419372e-16]
```

A useful tool in dimensionality reduction is to visualize the relation among the eigenvalues. A typical way to do that is a *scree plot*:

Definition

scree plot

In PCA, a *scree plot* is a line plot that illustrates the eigenvalues of the covariance matrix of a multidimensional data set.

Example 13.5 (continued)

Since we have already sorted the data, we can generate a scree plot using one line: `plt.plot(lam_mnist)`. A version with axis labels is shown in Fig. 13.19. Note that the eigenvalues decrease quickly with the component number. This is the case for most data sets, and it means that most of the variation in the data can be captured using a much smaller number of features than is present in the full data set.

Since the features at the output of PCA are decorrelated, the total variance of the data set is equal to the sum of the eigenvalues:

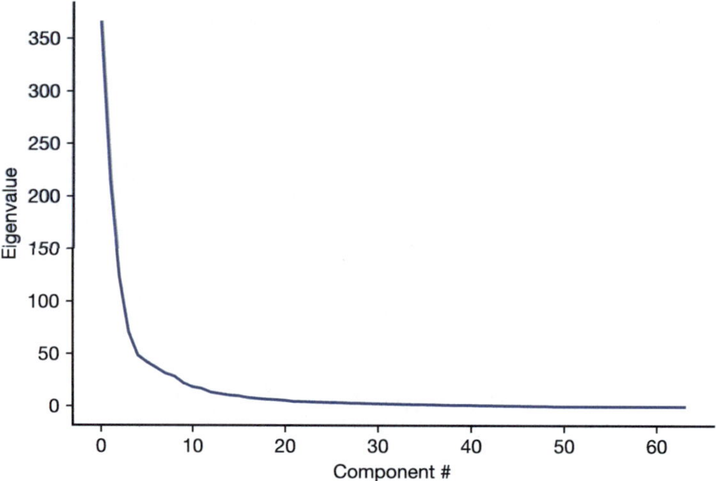

FIGURE 13.19
Scree plot for MNIST data.

```
total_variance_mnist = lam_mnist.sum()
total_variance_mnist
```

```
1160.5143139053337
```

To perform dimensionality reduction, we will utilize some $K < 64$ features with the highest eigenvalues. We can assess the impact of this dimensionality reduction in terms of the *explained variance*:

Definition

explained variance (dimensionality reduction)

In dimensionality reduction, the *explained variance* is the ratio of the total variance after dimensionality reduction to the total variance of the original data set.

Using PCA, the explained variance with K features is equal to the sum of the K largest eigenvalues. We can find the total variance after dimensionality reduction for every K by applying `np.cumsum()` to find the cumulative sums of the eigenvalues:

```
total_variances_mnist_pca = lam_mnist.cumsum()
```

Example 13.5 (continued)

A plot of the explained variance can be generated by dividing the total variances after dimensionality reduction by the total variance of the input data set:

```
plt.plot(total_variances_mnist_pca/total_variance_mnist)
plt.xlabel('No. of preserved features')
plt.ylabel('Explained variance')
plt.title('Effect of dimensionality reduction on the MNIST data set');
```

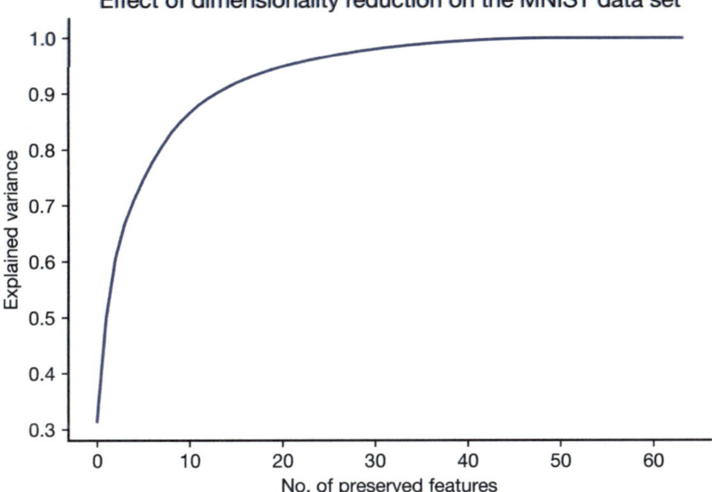

If we use only two features, the explained variance is:

```
print(f'{100 * total_variances_mnist_pca[1] / total_variance_mnist:.1f}%')
```

```
50.0%
```

Thus, 50% of the total variation among the data is captured using just two features. Using two features also has the advantage that we can visualize the data using a scatter plot.

We can apply PCA and dimensionality reduction by left-multiplying the data by the first two eigenvectors:

```
digits_pca = U_mnist.T[:2] @ mnist
```

Let's find the covariance matrix of the output data to confirm that the data is decorrelated and the variances match the first two eigenvalues:

```
np.round(np.cov(digits_pca), 10)
```

```
array([[365.43179314,  -0.        ],
       [ -0.        , 215.34653886]])
```

The transformed features are shown using a scatter plot on the 2-D data from PCA in Fig. 13.20. I have used different markers and colors to indicate the ground truth for the value of the digit for each data point. (Code to generate this graph is at fdsp.net/13-4.) Two features are sufficient to create almost completely distinct clusters among these three data classes. They cannot be separated using just lines, but more advanced machine learning algorithms can be used to easily create a classifier using just these two features.

So how can PCA extract the most important differences among the data into just two features? We can get some insight into this by plotting the basis functions as if they were images. I reshaped the length-64 basis vectors as 8×8 arrays that match the shape of the original digit data and plotted a heatmap of each of these basis arrays in Fig. 13.21. Code to generate this plot is online at fdsp.net/13-4.

For basis function 0, shown in the left plot of Fig. 13.21, pixels to the left and right of the middle make the feature more positive, while pixels in the middle make the feature more negative. A digit 0 has more pixels to the left and right and should produce a positive value, while digits 1 and 2 generally have more pixels in the middle and will produce a negative value. Compared to the data in the scatter plot, we see that these observations match the feature values for the data.

For basis function 1, shown in the right plot of Fig. 13.21, pixels just to the right of center (except at the very bottom of the pixel array) contribute to a positive feature value, while pixels in the upper and lower left and bottom right contribute to a negative feature value. Pixel values that make this feature positive are most likely to be seen in a 1 digit, whereas pixel values that make this feature negative are most likely seen in a 2 digit.

This is even more clear if we create linear combinations of these two basis functions. In Fig. 13.22, I show heatmaps of the sum and difference of these basis functions. The code to generate these figures is available online at fdsp.net/15-5. If you look closely at the plot on the left, you can probably see a rough "2" shape in blue and a rough "1" shape in red. Similarly, if you look closely at the figure on the right, you can see part of a "0" shape in blue and part of a "1" shape in red.

By inspecting the heatmaps of the basis functions as if they were images, we can see how these basis functions extract some of the unique properties of the different digits considered in this example.

Example 13.6: Feature Extraction for Breast Cancer Classification

Another PCA example using the 30-dimensional Breast Cancer Wisconsin data set is available on this book's website at fdsp.net/13-4. This example also includes information on model validation using the Train-Test-Split approach.

Terminology review and self-assessment questions

Interactive flashcards to review the terminology introduced in this section and self-assessment questions are available at fdsp.net/13-4, which can also be accessed using this QR code:

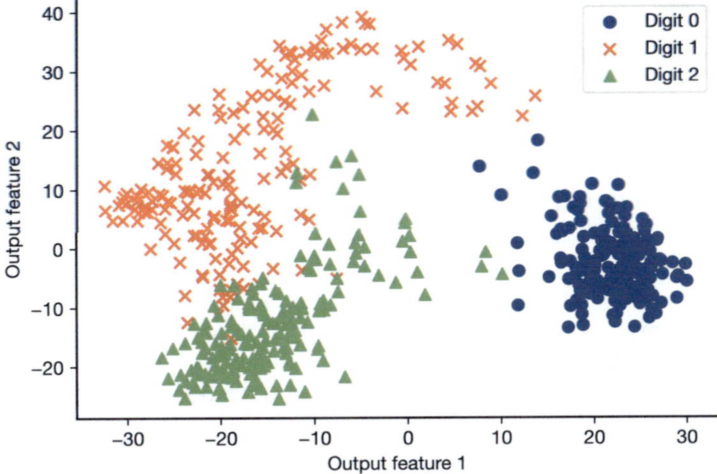

FIGURE 13.20
MNIST data reduced to two dimensions via PCA.

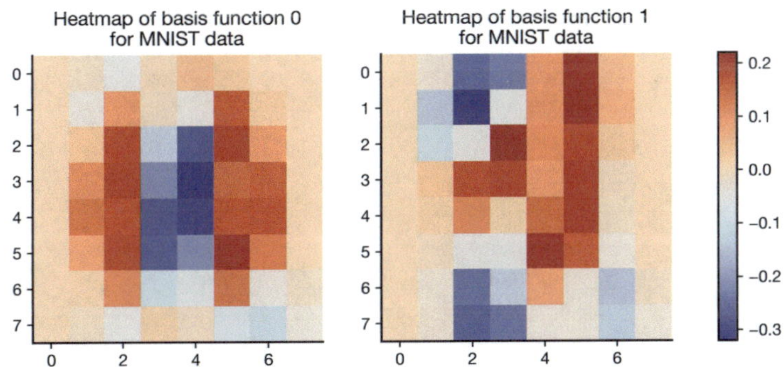

FIGURE 13.21
First two PCA basis functions reshaped to map image data and plotted as heatmaps.

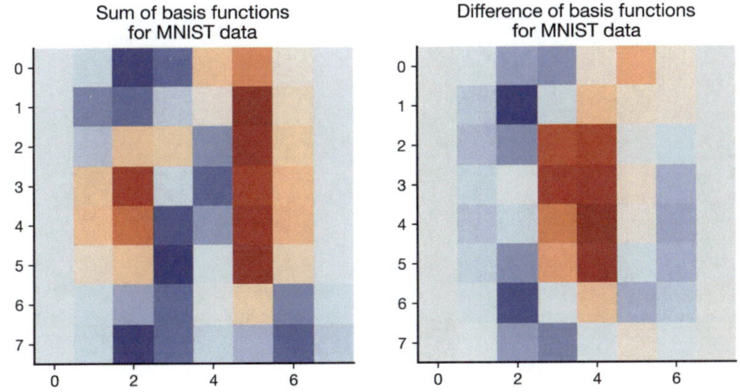

FIGURE 13.22
Heatmaps of sums and differences of basis functions for MNIST data.

13.5 Chapter Summary

In this chapter, I introduced techniques for modeling and working with dependent data. I introduced the basic concepts of jointly distributed random variables. Because of the complexity of most joint distributions, we often work with the moments of these random variables instead of the full distributions. I introduced the most important of these: the mean vector and the covariance and correlation matrices. Next, we examined the effects of linear transforms on the mean and covariance matrices of random variables and random data. Standardization was introduced as a technique to make all numeric variables in a data set have zero mean and unit variance; this is often used to prevent differences in units and innate scales of features from creating imbalances in how the data is processed and interpreted. Linear transforms were shown to be able to decorrelate two-dimensional data. I then showed that we can use eigendecomposition to find a linear transform to decorrelate a set of random variables or data points. The decorrelation process for random variables is called the discrete KLT, and the decorrelation process for data is called PCA (especially if it is also used for dimensionality reduction). Finally, we applied PCA for dimensionality reduction in two example applications (one of which is on the book's website).

Access a list of key take-aways for this chapter, along with interactive flashcards and quizzes at fdsp.net/13-5, which can also be accessed using this QR code:

Index

a posteriori probability
discrete stochastic system, 190
discrete-input, continuous-output stochastic system, 378
a priori probability
discrete stochastic system, 191

Anaconda Python distribution, 7
area under curve (AUC), 374
average
data, 55
axiomatic probability, 86
axioms of probability, 89
corollaries, 93

base rate fallacy, 193
Bayes' Rule
discrete stochastic system, 191
discrete-input, continuous-output stochastic system, 379
Bayesian hypothesis testing, 206
Bernoulli random variable, 252
bias-variance tradeoff, 333
binary communication system
optimal decision rules, 203
binary hypothesis test, 64
binomial coefficient
SciPy function, 111
Binomial random variable, 255
blinding, 114
bootstrap distribution, 142
bootstrapping, 132
Borel σ-algebra, 220
Borel field, 220
Borel sets of $\mathbb{R}$, 220

Cartesian product, 101
categorical data, 387
CDF, *see* cumulative distribution function (CDF)
chain rules, 178
Chi-squared random variable, 296
codebook, 123
coefficient of determination (simple linear regression), 430
Cohen's d, 361
combinatorics, 99
sampling with replacement and ordering, 107
sampling without replacement and with ordering, 111
sampling without replacement and without ordering, 111
compound experiment, 78
conditional independence (events), 176
conditional probability, 161
confidence interval (CI), 144
contingency table, 390
continuous random variable, 272
contour of equal probability density (pair of random variables), 451
convenience sampling, 117
correlation coefficient
data vectors, 417
random variables, 417
correlation matrix, 453
covariance
data vectors, 416
random variables, 415
covariance matrix, 453
Covid
data, 34
credible interval, 213
cross-sectional study, 117
cumulative distribution function (CDF), 237
inverse, 283

data, 2
data cleaning, 124
data points, 2
data science, 2
dataframe, *see* Pandas->dataframe
decision problem, 198
decision rule, 198
MAP, 203

minimum probability of error, 203
ML, 199
degrees of freedom
Chi-squared random variable, 296
contingency table, 393
Student's t random variable, 297
dimensionality reduction, 467
discrete random variable, 229
Discrete Uniform random variable, 248
disjoint, 47
distribution (of a random variable), 253

effect size, 362
empirical pdf, 299
estimator, 327
estimator bias, 328
estimator error, 328
event, 78
event class, 87
examples
chi-squared contingency test on exercise and student's sex, 396
chi-squared contingency test on marijuana use and student's sex, 389
communication system with continuous outputs, 375
communication system with discrete outputs, 197
correlation between height and weight, 418
decorrelating Iris data, 474
generalized likelihood decision rules, 369
goodness of fit test on MLB players' birthdays, 401
linear regression between GDP and COVID rates, 432
linear regression between year and temperature in Miami-Dade County, 431
magician's coin, 152
modeling heights of US Adults, 292
Monty Hall Problem, 185
NHST on effect of graduate education on family wealth, 138
NHST on effect of permitless carry on firearms mortality, 348
NHST on effect of states' socioeconomic factors on early COVID rates, 68
nonlinear regression: exponential growth of COVID cases, 438
number of Bootstrap samples, 107
PCA on MNIST handwritten digits, 477
polynomial regression, 437
Prostate-Specific Antigen Test, 365
rotation effects on covariance of Iris data, 464
shark attacks, 262
spurious correlations, 434
standardizing Iris data, 458
expected value
continuous random variable, 307
discrete random variable, 307
experimental study, 114
explained variance (dimensionality reduction), 480
explained variance (simple linear regression), 430
explanatory variable, 422
Exponential random variable, 281

factorial, 110
SciPy function, 110
fair experiment, 84
features, 3

Geometric random variable, 258

heatmap, 478
hidden state, 195
histogram, 22
hypothesis, *see* statistical hypothesis
hypothesis testing, 206

iid, *see* independent and identically distributed (iid)
independent and identically distributed (iid), 454
interval estimate, 334
inverse CDF, 283
itertools, 101

joint cumulative distribution function (pair of random variables), 449
joint probability density function (pair of random variables), 449
joint probability mass function (pair of random variables), 448
Jupyter, 6
magics, 15

Markdown, 13
JupyterLab, 11
interface modes, 11

Karhunen-Loève Transform (KLT), 471
kernel density estimation, 300

Law of the Unconscious Statistician (LOTUS), 317
Law of Total Probability, 182
left tail, 129
legend, 41
likelihood
discrete-input stochastic system, 190
discrete-input, continuous-output stochastic system, 302
likelihood ratio, 368
linear regression, 424
longitudinal cohort study, 117
longitudinal study, 117
LOTUS rule, *see* Law of the Unconscious Statistician (LOTUS)

MAP decision rule, 203
marginal probability density function (pair of random variables), 450
Matplotlib, 20
maximimum likelihoood decision rule, 199
mean
continuous random variable, 307
data, 55
discrete random variable, 307
mean vector, 452
median
data, 53
of a random variable, 315
minimum probability of error decision rule, 203
ML decision rule, 199
mode(s)
data, 51
of a random distribution, 128
of a random variable, 314
model-based methods, 65
model-free methods, 65
moment
nth, 316
nth central, 319
Monte Carlo permutation testing, 133

natural experiment, 115

NHST, *see* null hypothesis significance test (NHST)
nominal data, 388
Normal random variable, 285
null hypothesis, 64
null hypothesis significance test (NHST), 64

observational study, 114
one-way frequency table, 402
optimal decision rules, 203
ordinal data, 387
outcome, 77
outlier, 46

pairwise statistically independent, 171
Pandas, 33
dataframe, 34
partition, 47
of the sample space, 181
pdf, *see* probability density function (pdf)
permutation, 110
permutation testing, 133
Monte Carlo, 133
piecewise function, 241
plot
histogram, 22
scatter, 20
stem, 81
PMF, *see* probability mass function (PMF)
point conditioning, 376
point estimate, 327
Poisson random variable, 260
population, 116
population study, 116
post hoc analysis, 118
principal components analysis (PCA), 474
prior
informative, 208
uninformative, 207
probability, 18
axioms of, 89
probability density function (pdf), 274
empirical, 299
probability mass function (PMF), 230
probability measure, 89
probability space, 87
prospective study, 118
Python, 15
plotting, 20

qualitative data, 3

quantitative data, 3

random experiment, 17
random sampling, 116
random variable
  Bernoulli, 252
  Binomial, 255
  Chi-squared, 296
  continuous, 271
  discrete, 229
  Discrete Uniform, 248
  Exponential, 281
  formal definition, 220
  Geometric, 258
  informal definition, 217
  Normal, 285
  Poisson, 260
  Student's t, 297
  Uniform, 279
random vector, 452
randomized control trial (RCT), 114
range of a random variable, 220
receiver operating characteristic (ROC), 371
relative frequency, 26
repeated experiment, 79
resampling, 65
research question, 4
response variable, 422
retrospective study, 118
right tail, 129

sample, 116
sample mean
  data, 55
sample space, 77
  compound experiment, 106
  enumerating, 101
sampling distribution
  estimator, 356
  test statistic, 120
scatter plot, 20
scree plot, 479
selection bias, 119
sensitivity, 189
SF, *see* survival function (SF)
simple linear regression, 424
simulation
  computer, 18
  to estimate a probability, 31
  to estimate conditional probabilities, 164
staircase function, 243
standard deviation
  random variable, 321
standard error of the mean (exact), 338
standardization, 462
statistical hypothesis, 63
statistical power, 149
statistical regularity, 80
statistical study, 113
statistically independent
  any number of events, 171
  pairwise, 171
  two events, 168
stochastic system, 187
Student's t random variable, 298
summary statistic, 48
survival function (SF), 246

tail probability, 129
test statistic, 63
time-series data, 431
topcoding, 122
total variance (simple linear regression), 430
trial (compound experiment), 79
two-sided tail, 130
Type I error, 148
Type II error, 148

unbiased estimator, 328
uncorrelated, 415
Uniform random variable, 280
unimodal distribution, 128
uninformative prior, 207
union, probability of, 95

variables, 3
variance
  data vector, 333
  random variable, 320

word problems, translating to mathematics, 188

Made in United States
Troutdale, OR
05/10/2024